Potato Production Systems

Jeffrey C. Stark • Mike Thornton • Phillip Nolte
Editors

Potato Production Systems

Springer

Editors
Jeffrey C. Stark
Idaho Falls Research and Extension Center
University of Idaho
Idaho Falls, ID, USA

Mike Thornton
Parma Research and Extension Center
University of Idaho
Parma, ID, USA

Phillip Nolte
Idaho Falls Research and Extension Center
University of Idaho
Idaho Falls, ID, USA

ISBN 978-3-030-39156-0 ISBN 978-3-030-39157-7 (eBook)
https://doi.org/10.1007/978-3-030-39157-7

This Springer imprint is published by the registered company Springer Nature Switzerland AG
The registered company address is: Gewerbestrasse 11, 6330 Cham, Switzerland

Preface

There have been significant technological advancements in potato production in recent years that have remarkably improved potato yields in North America and throughout the world. However, substantial increases in transportation, fuel, fertilizer, pesticide, and processing costs, as well as changing consumer preferences, have created an even greater need for information that can be used to improve potato production efficiency and sustainability. The successful development and implementation of sustainable potato production systems requires the integration of a wide array of cultural and pest management practices that are well adapted to local environments. This is a process that is both challenging and rewarding, given the importance of the potato crop in the world's food systems.

With this goal in mind, we have endeavored to bring together the latest information on the science and practice of potato production with emphasis on North American production systems. Obviously, management recommendations vary across production regions and environments, and it is not practical to include all of them in a single book. However, we have tried to present representative management approaches that have broad application across the major production regions in North America, which, in turn, can be adapted to local production environments.

The second edition of *Potato Production Systems* represents the combined efforts of over 39 potato scientists from the USA and Canada. We have endeavored to make this new edition as comprehensive as possible, covering all aspects of potato production from field preparation, varietal selection, seed production, and planting, through pest management, fertilization, irrigation, harvesting, and storage. There are chapters focusing specifically on disease, nematode, weed, and insect management, as well as chapters on marketing and economics.

The editors express their appreciation to all of those who contributed information, insight, and images to this effort. This book would not have been possible without their participation.

We also thank Barbara Gronstrom-Smith for the outstanding job she did reviewing and editing this book and helping us manage the many details associated with its development. Her advice, guidance, and support have made an immeasurable contribution to the success of this effort.

For those seeking more information on potato production systems, we have included suggestions for further reading for most chapters at the end of the book.

Idaho Falls, ID, USA Jeffrey C. Stark
Parma, ID, USA Mike Thornton
Idaho Falls, ID, USA Phillip Nolte

Contents

About the Editors

Jeffrey C. Stark is a professor in the Department of Plant Sciences at the University of Idaho. During his 38-year career, he has conducted research on irrigation and nutrient management in potato cropping systems and has authored over 100 research publications on these topics. He has served as director of the Potato Variety Development Program at the University of Idaho since 2005, coordinating the development, release, and commercialization of 19 new potato varieties and developing management guidelines for these varieties. Together with Dr. Stephen Love, he edited the first edition of *Potato Production Systems*, which was published in 2003. He has also served as division chair of the Plant Science and Horticulture Divisions at the University of Idaho from 1999 to 2013.

Mike Thornton is a professor of plant science working on potatoes for the University of Idaho at the Parma Research and Extension Center. His research program focuses on sustainable production of new varieties, management of in-season pest problems, and reduction of losses during storage. He has worked closely with key influencers in the potato industry (growers, commodity commissions, and processors) to document and address the most important issues they face. Mike has over 35 years' experience in the potato industry in North America and has worked both in academia and industry. This allows him to see problems from several perspectives and develop effective research and extension programs.

Phillip Nolte is an extension seed potato specialist, University of Idaho (emeritus). His programs not only focused on seed potatoes but also included seed-related problems in commercial production and general potato disease diagnosis and management. His areas of study included investigations on the potato mosaic virus complex (PVY and PVA), management of potato late blight, fungicide resistance studies in Fusarium dry rot, and the effect of chemical application on wound healing (suberization) in cut seed. He served as the technical editor for *Potato Grower* magazine and as a contributing editor for *American Vegetable Grower* magazine. He was also president of The Potato Association of America 2009–2010. He began working on potatoes in 1979.

Chapter 1
A Short History of Potato Production Systems

Stephen L. Love, Kurt Manrique-Klinge, Jeffrey C. Stark, and Edgar Quispe-Mamani

Contents

S. L. Love (✉)
Aberdeen R&E Center, University of Idaho, Aberdeen, ID, USA
e-mail: slove@uidaho.edu

K. Manrique-Klinge
Inter-American Institute for Cooperation on Agriculture, Port of Spain, Trinidad, West Indies
e-mail: kurt.manrique@iica.int

J. C. Stark
Idaho Falls R&E Center, University of Idaho, Idaho Falls, ID, USA
e-mail: jstark@uidaho.edu

E. Quispe-Mamani
Quispe Family Farm (Former), Sierra Vista, AZ, USA

J. C. Stark et al. (eds.), *Potato Production Systems*,
https://doi.org/10.1007/978-3-030-39157-7_1

Introduction

In the early 1500s when Spanish explorers stepped onto the continent of South America, they discovered a civilization consisting of 10 million people spread along 2000 miles of the Andes. They also discovered that this relatively advanced culture subsisted largely on a previously unknown crop that came to be known as the potato. Far beyond being just something to eat, explorers recorded that potatoes infused every aspect of Andean culture. This crop was incorporated into many local creation myths, served a central role in religious rites, became a common subject of artwork, and was central in daily and seasonal rituals. Over an undefined period of history, potato production in the Andes evolved in sophistication and productivity. Thousands of years before this crop became a staple in Europe, the Andean culture had developed a complex, efficient potato production system. All aspects of what we commonly assume are modern agricultural principles were addressed within this system: crop rotation, soil fertility, soil preparation, irrigation, cultivar improvement, seed management, pest control, judicious harvest protocols, long-term storage, production of processed products, and complex marketing schemes. Evidence suggests that efficacious potato production gave rise to the Incan Empire and vastly improved quality of life for its citizens. As modern producers consider options for improving potato production practices, comparison of ancient Andean potato production methodologies may be both enlightening and constructive.

History

Spanish conquerors traveling into the Andean highlands in the early 1500s recorded observations about the production and consumption of a previously unknown crop, the potato (Fig. 1.1). Chronicler Pedro Cieza de León wrote, "Of the natural resources that the Indians use for sustenance, apart from corn, there are two principle foodstuffs; one of these is called the potato, which is somewhat like a truffle, yet after cooking it becomes soft on the inside like a roasted chestnut; like a truffle, it does not have a shell or bone, because it grows beneath the earth." At the time, Spaniards were not particularly impressed with this lowly underground crop. Little did they know that potatoes would become vastly more important to world economics than the tons of gold they transported out of South America. Writing about the potato in Scotland during the 1700s, Thomas Garnett claimed that, "…this useful root, for which we are indebted to America,… is more valuable than all the gold of Mexico, all the diamonds of Golconda, or all the tea of China."

When the first Spanish explorers stepped onto the South American continent, the Inca Empire comprised 10 million people and was spread along 2000 miles of the Andes Mountains. This Empire was a complex culture that incorporated extensive trade and incredible advancements in agriculture, music and the arts, mathematics, and medicine. Potatoes were central to the food security of the region, and many historians express the opinion that this vast and sophisticated Inca Empire was

Fig. 1.1 High-elevation Andean agriculture in Ayacucho, Peru, in the region where potato production had its beginnings. This modern scene is little changed from the terraced agricultural systems discovered by the Spanish conquerors. (Photo credit: Kurt Manrique-Klinge)

raised on the back of this humble crop. Adoption of potato agriculture ensured a constant supply of nutrient-rich food that could be produced on relatively small amounts of land, while giving people time and energy to pursue other interests. It was the primary source of nutrition for noblemen and peasants alike. Freeze-dried potatoes, known as chuño, provided sustenance during times of famine and were the food of choice for mobile Incan armies.

Far beyond being just something to eat, potatoes made their way into every aspect of Andean culture. They were incorporated into many local creation myths, served a central role in religious rites, became the subject of artwork (Fig. 1.2), and were integrated into many of the activities of everyday life. According to Incan lore, the universe was divided into three worlds: the upper world (Hanan Pacha) inhabited by the primary gods such as the sun, moon, stars, lightning, and the rainbow; the underworld (Uku Pacha) inhabited by death, spirits, diseases, and minor Gods, such as Huatiacuri (personification of the potato); and the world that makes up the zone of life (Kay Pacha) where men, animals, and plants exist. These three worlds coexist and their natural and spiritual entities interact. The potato belongs to Uku Pacha because it grows and lives in darkness beneath earth. Huatiacuri was a lowly God in that he lived below the Earth's surface, wore ragged clothes, and was covered with dirt and purple flowers. Yet, he possessed the hidden power to protect the entire Incan universe. The Andean culture celebrated ceremonies to mitigate the Gods, including Huatiacuri, where an early potato planting was completed in late August to awaken the Uku Pacha. At harvest, if double or coalescent tubers were found, they were kept and revered as a sign of fertility and a profitable future. Such was the reverence Andean cultures had for the potato.

Fig. 1.2 An Andean pottery artifact, approximately 2000 years old, modeled after a potato tuber. (Photo credit: Stephen Love)

Fig. 1.3 Homes amidst agricultural terraces on the Island of Taquile in Lake Titicaca. This region is the proposed center of origin and earliest known site of potato production. (Photo credit: Stephen Love)

Based on evidence from modern genetic studies, it is now accepted that the center of origin and site of domestication of potatoes is the high Altiplano region (11,000–13,000 ft. elevation) of Peru near Lake Titicaca (Fig. 1.3). Recent archeological evidence suggests a history of potato production and use in this area that dates back at least 7000, and possibly as many as 13,000, years. Rich genetic resources, in the form of wild relatives, are common in the region. Native cultures took advantage of these resources and began a process of adapting these naturally occurring species into something of greater societal value. As the value of potatoes increased, the amount of effort dedicated to their improvement also grew. The rich local gene pool, derived primarily from eight local species of the genus *Solanum*, evolved into over 4000 unique, locally produced potato cultivars. The incredible diversity infused into this crop can still be observed as a kaleidoscope of colors and multitude of shapes found in any modern Peruvian marketplace. Conservationist Andean farmers in Peru hold a collection of potato cultivars estimated at more than 2800 native landraces. The International Potato Center (CIP) preserves the world potato collection with almost 5000 cultivars.

Indigenous people in the Andes continue to maintain a close relationship with potato genetic diversity. Stephen Brush, professor at the University of California at Davis, wrote that in a single valley in the Peruvian Andes, peasant communities may grow between 70 and 100 distinct potato cultivars, and a typical Andean household may keep up to 50 cultivars from several potato species for home consumption and eventual exchange (Brush et al. 1990) (Fig. 1.4). This diversity contributes to potato

Fig. 1.4 Potato harvest in the uplands of Peru, with an example in the foreground of the diversity of potato cultivars grown on this farm. (Photo credit: Kurt Manrique-Klinge)

Fig. 1.5 Artwork of the native Peruvian chronicler, Guaman Poma, depicting the practice of planting potatoes as recorded in the year 1615. (Public domain image from Wikimedia Commons)

utilization by providing adaptation to a wide range of production conditions and conversion to a wide range of culinary uses, although most of these native potato cultivars are conditioned to grow in the high Andes and don't adapt easily to areas of lower elevation.

Sixteenth century drawings created by Guaman Poma, a native Peruvian chronicler, depict scenes of potato production systems and technology developed by the Incas (Figs. 1.5 and 1.6). In many regions, Andean potato production has changed very little in the intervening years. From a modern point of view, both historical and current methods superficially appear primitive. On closer inspection, it becomes clear that the culture that contributed the potato to the human family also developed many of the productions systems used in modern agriculture. Ancient Andean

Fig. 1.6 Artwork of the native Peruvian chronicler, Guaman Poma, depicting the practice of harvesting potatoes as recorded in the year 1615. (Public domain image from Wikimedia Commons)

potato growers understood their environment, the demands of the crop, and the available resources. They also learned to use these factors to their advantage. A closer look will reveal how sophisticated Andean potato production systems became prior to the destruction that accompanied the Spanish intervention.

The one thing early Andean potato producers lacked, and often still lack, is a source of inexpensive energy. As a result, hand labor substituted for the mechanical aids that have become synonymous with modern potato production. If we look past this one element of agricultural advancement, the sophistication of Andean production systems becomes much more evident. For example, below are some of the production principles developed by primitive Andean potato growers.

Production Principles

Crop Rotation

Although lacking an understanding of the biological imperatives for crop rotation practices, the necessity of proper farm management was fully accepted. A typical crop rotation for highland-grown potatoes was 1 year of potatoes; 1 year of another endemic root crop, such as isaño, oca, olluco, maca, or mashua; 1 year of the grain crop quinoa, maize, or kiwicha; followed by 2–9 years of fallow, depending on personal need or market demand.

Soil Fertility

Without modern fertilizer products, many primitive cultures worldwide relied on the use of legume-based green manures to manage soil fertility and replace nutrients lost with crop removal. Lacking even this resource, Andean growers developed unique procedures that included pasturing animals on fallow ground to provide manure, adding additional manure taken from animal enclosures, burning weeds and plant refuse on the fields to provide phosphorus and potassium, and adding nutrient-rich bog soils to their fields. In coastal areas, fertility practices included the burial of fish carcasses near plants and application of *guano*, which is the dried remains of birds' semisolid urine. It makes an excellent fertilizer, a mechanism for giving plants nitrogen.

Soil Preparation

One of the greatest agricultural achievements of Andean farmers was the development of extensive terracing systems that not only allowed production on steep land, but also alleviated erosion and permitted water capture. Additionally, proper seedbed preparation was understood and practiced. Rather than using a modern moldboard plow or ripper, ancient Peruvians used a foot plow (*chakitaqlla*), and rather than a harrow, they used a hand-held clodbuster (*waqtana*). Regardless, the result was a mellow, aerated planting medium (Fig. 1.7).

Irrigation

Prior to the Spanish invasion, Andean farmers in sloped regions built the amazing *qocha* (reservoir) irrigation systems consisting of channels, aqueducts, and cisterns to deliver and store rainwater for use in their terraced fields. Most of these delivery

Fig. 1.7 Peruvian farmer on the Island of Taquile plowing a field with a traditional foot plow (*chakitaqlla*) in preparation for planting potatoes. (Photo credit: Stephen Love)

systems were destroyed by the Spaniards, and only a small portion of them have been restored to full function. We arc just now beginning to understand the incredible ingenuity behind these systems. To meet a different set of conditions in the flat, lowland areas around Lake Titicaca, the *waru-waru* irrigation system was developed. Long beds, 10 or more ft. wide, were raised above the natural ground level by excavating adjacent soil. Due to a high water table, the result was a series of mounded planting beds interspersed with water channels up to 4-ft deep. The channels served not only to provide water during dry periods through soil wicking, but also helped drain excess water after heavy storms, served as a source of nutrient-rich silt, and eliminated or reduced frost damage by serving as a heat reservoir. Additionally, the channels were often used to farm fish.

Hilling

Ancient Andean farmers developed methods for row planting and hilling for potatoes very similar to what is used in modern production (Fig. 1.8). The primary objective was to improve soil drainage and keep the tubers from becoming water logged. Further, they also recognized soil temperature advantages that result from furrow orientation. Farmers living in drier regions pushed up small hills, hoping to conserve water. Those living in areas with greater rainfall or high water tables

Fig. 1.8 Potatoes near Huancayo, Peru, planted in a linear hill arrangement to provide appropriate drainage. (Photo credit: Jeffrey Stark)

pushed up very high hills, sometimes creating furrows up to 2-ft deep, to drain excess water from the fields.

Cultivar Improvement and Selection

Farming in a period prior to the advent of modern breeding did not prevent Andean farmers from taking advantage of genetic diversity. Each farm served as an on-going experimental cultivar evaluation site. At the end of every season, each grower, often in consultation with local residents, reviewed the performance of each cultivar and determined which ones to grow the following year. In antiquity, cultivar choice was closely related to culinary use, and Peruvian farmers developed classes of potatoes for very specific purposes. They created *chaucha* for early fresh table use; *hatum* for main crop fresh table use; *siri*, a bitter potato used for making chuño; *moraya* for making tunta; and cultivars intended solely for boiling or baking. Beyond local selection of landraces, there is evidence Incan "scientists" (known as *amautas*) devised evaluation procedures to improve potatoes. John Earls, in a study published in 1998, concluded that concentric terraces at an archeological site in Moray served as an experimental center for improving crop production on the Inca state terraces (Fig. 1.9). One line of thought is that the site was used to acclimatize crops to new eco-climatic conditions and create new cultivars and sub-cultivars of adapted crops (Earls 1998).

Fig. 1.9 Archeological site near Moray, Peru, showing the concentric ring terraces thought to be an Incan agronomic research site. (Photo credit: Kurt Manrique-Klinge)

Seed Management

Although lacking complex seed certification procedures, Andean potato farmers understood the need for high-quality seed. At the end of each growing season, before consuming or selling the crop, they selected the best tubers from the previous crop to serve as seed for the subsequent year. Seed tubers were stored carefully and strictly reserved for production the following year.

Pest Control

It cannot be argued that advancements in insect and disease management are mostly modern. However, ancient Andean producers recognized pest issues and developed more or less effective management strategies. The most important approach was to mix and grow multiple cultivars with a range of resistance responses to ensure that infestation by a destructive insect or disease pest would not destroy an entire crop. The unique practice of encouraging frogs to live and proliferate in fields was used to control the destructive tuber worm by their consumption of adult moths. In storage, the Incas utilized the deterrent effect of certain Andean herbs; e.g., muña

(*Minthostachis mollis*) and aya manchana (*Lantana camara*) that have active, fragrant essential oils to deter damaging insects.

Harvest

Rather than using a six-row, self-propelled mechanical harvester, Andean farmers historically (and many still do) used manual labor and the native version of the mattock (raucana) to harvest potatoes. Regardless of tools, they understood principles and developed practices to optimize maturity and reduce handling injury.

Storage

Ground storage (simply leaving the potatoes in the ground during the dry season) was possible and often practiced in the climatic conditions of the high Andes where cool conditions prevailed but ground frost was rare. However, more sophisticated storage facilities were commonly built and utilized. Ancient potato cellars found in Huanuco Pampa provide an example of the astounding storage technology of the Inca civilization. Control of storage temperature was accomplished by manipulating three factors in the storage environment: ventilation, insulation, and the selection of adequate warehouse locations. Morris and Thompson (1985) described these ancient buildings as being located in cool locations at the top of cliffs and mountains. They were built with thick walls and thatched roofs to insulate against heat during warm days and excessive cold at night. Ventilation was provided either by construction of windows placed on opposite sides of the storage or by construction of crevices in the stone floors connected to the outside by ducts or vents (Fig. 1.10).

Processing

Andean potato farmers developed a range of useful, and more importantly, storable, potato products (Fig. 1.11). The most widespread and important of the processed products was—and still is—chuño. Ancient growers took advantage of natural climatic conditions of the high mountainous regions to produce a natural lyophylization process. Chuño was made by placing dark-skinned, bitter, small potatoes in a single layer on the ground and allowing them to alternately freeze at night and thaw during the day. As the potatoes began to weep, people walked on them to press out the moisture. Within a few weeks, the potatoes were completely freeze-dried (the first dehydrated potato products). Another freeze-dried product, *tunta*, was made using a more complex procedure. Potatoes tubers were spread out on the ground to freeze but covered during the day to keep them in the dark. After 2 days of alternate freezing and drying, the tubers were immersed in frequently changed water for 6–8 weeks and ultimately dried to create a product that retained its bright white

Fig. 1.10 Ruins of an ancient Peruvian potato storage building located near the top of a steep slope where climatic conditions are cool. (Photo credit: Kurt Manrique-Klinge)

Fig. 1.11 Examples of freeze-dried potato processing products known as chuño (left) and tunta (right). (Photo credit: Stephen Love)

color. Anciently, *tunta*, also referred to as *moraya*, was considered to be a superior product and was made specifically for nobility. Another processed product, called tocosh (CIP 2003), was made by soaking freshly harvested tubers until they fermented, then drying the softened tubers in the sun. Tocosh is still produced and utilized nowadays as a natural antibiotic because penicillin is produced during the fermentation process. Simply drying pre-baked tubers was also a common practice. Processing supplied food through the dry season, during times of famine and war, and in cases of crop failures.

Exchange

The majority of potatoes grown by Andean farmers was used for subsistence or for paying homage to nobility. The Incas didn't know the concept of money; however, they had a good understanding of food availability for social welfare. Therefore, barter was a common practice among regions and communities to exchange a diversity of food products. Food distribution in the Inca Empire was a mandatory cooperative system based on community work (*ayni, minka*) and on work dedicated to the Inca state terraces and fields (*mita*). This system, still practiced by native communities, ensured that every subject in the empire (including elderly and ill people, children, widows, and the disabled) received sufficient food. Thus, famine never occurred at that time.

As people began to understand the value of potatoes, they spread throughout the world. Initially, they were adopted into the cultures of nearby regions of South America, then into Central America and Mexico. We used to think this was the limit of potato dissemination, until European explorers took them across the Atlantic. However, more recent evidence suggests a much wider ancient distribution. People of the Navajo Nation in the southwestern U.S. still occasionally cultivate and eat tubers from a naturalized species. This is obviously not a part of European potato succession. One tribe of Native Americans in Alaska retains two very old cultivars accompanied by traditions that one came from South America and the other from Hawaii. The Maori of New Zealand claim to have grown potatoes long before the arrival of the first European explorers.

The potato began its journey into the old world as Spanish explorers returned home with the spoils of war. There is still some argument as to when potatoes made their way into Europe, but we have record of a shipment of potatoes from the Canary Islands into most likely Belgium around 1567, suggesting production on the islands began several years earlier. We have written records of potatoes being present in Spain by 1570 and England by 1580.

For many years, the European populace failed to see the value in potatoes, and they were usually collected and transported as botanical oddities rather than foodstuffs. They did not become economically and nutritionally important for almost two additional centuries. People were suspicious of any plant from the generally poisonous nightshade family and were put off by something that came out of the dirt. However, potatoes had their occasional supporters among the European elite. In 1586, Diego Dávila Briceño, a Spanish official living in Huarochiri, Peru, wrote that, "...if in our Spain, these [potatoes] were to be grown as they are here, they would be a great solution in the years of famine." Over time, this opinion was held by nobility and common folk alike, but only after time broke down the barriers of neglect. After adoption by nobility in France, followed by a rapid shift in public image, the potato became a staple of European diets. This had an immediate impact on many cultures, as potato crops provided greater food security and better nutrition. Since then, the potato crop has saved the lives of millions around the world during times of human tragedies and natural disasters. After WWII refugees and

Fig. 1.12 Post-WWII refugees planting potatoes as a subsistence crop. (Public domain image from Wikimedia Commons)

displaced populations in Europe depended on the potato to survive (Fig. 1.12). This enabled populations to increase, and people found more time for activities beyond basic survival.

The influence of potatoes went far beyond simple nutrition; it changed cultures and shifted world politics, as evidenced by the potato famine of Ireland. The introduction of late blight, a devastating disease of potatoes, combined with an unstable political situation, led to mass starvation and upheaval. One million Irish residents died of starvation, and another 2 million migrated to other countries during this devastating period in 1852; the majority to the U.S. On the positive side, the potato is often credited with advancing the industrial revolution. It is currently grown in at least 148 countries, more than any other crop, except corn. It is the fourth most important food crop worldwide. It is a critical nutrient and energy source in nearly every temperate country, has been adapted to the highlands of the tropics, and most recently has been found to be a suitable crop for the dry season in tropical lowlands.

As potatoes made their way around the world, the knowledge concerning production systems followed. Soil preparation, fertility management, seed management, and hilling practices developed by the ancient Andean farmers were, for the most part, duplicated. But some important practical management differences emerged. Rather than using genetic diversity to manage pests and environmental issues, single cultivars were grown. Individual fields became larger. Crop rotations were shortened or eliminated. Many of these practices have been retained in the modern era. The consequence is a need for more intensive practices to manage pests

Fig. 1.13 Early generation tractor and single row potato harvester. (Public domain image from Wikimedia Commons)

and problems. Widespread devastation caused by late blight and subsequently the Colorado potato beetle led to the evolution of the modern crop protection industry. The industrial revolution led to the design and manufacture of equipment that mechanized many of the drudgery aspects of potato production. The discovery of fossil fuels and the combustion engine provided the power to operate these mechanical wonders (Figs. 1.13 and 1.14). Farms grew in size and complexity. Agriculture, in general, and potato production in particular, quickly changed.

And the end is not in sight. As time passes, potato growers will be required to adapt to new situations and challenges, such as climate change. Imminent concerns include the loss of fertilizers and protection products as a result of ever-increasing concern over safety and environmental issues. As temperature increases, water supply will become more and more limited. Improved, but very different, potato cultivars will become major components of production. Market specifications will become more stringent. The need for fiscal sustainability will dictate that farming operations become even larger and more intensely managed. Problems we have not yet encountered will become significant barriers to profitable production. Advancements in our understanding of basic agricultural principles will produce new tools, but only for those with the knowledge to use them.

The purpose of this book is to detail the current status of potato production science and technology. It will serve as an educational tool to help producers, consultants, educators, students, and anyone else involved in making efficacious potato production decisions. As the production environment rapidly evolves, finding balance on the cusp of potato science will be critical to its continued success. But as things change, they, in many ways, remain the same. Just like our Andean forefarmers, success will come as we gain understanding of our environment, the demands of the crop, and the available resources. We then must learn to use these factors to

Fig. 1.14 Modern four-row potato harvester. (Photo credit: William Bohl)

our advantage. The details will change, the tools will evolve, but the overarching principles will remain the same.

Acknowledgement Unless otherwise noted, photographs, graphics, and data were adapted from collections of University of Idaho Extension educators, scientists, and researchers, who wrote the chapters of the first edition of this textbook.

References

Brush SB, Taylor JE, Bellon Corrales M (1990) Biological diversity and technology adoption in Andean potato agriculture. Paper presented at American agricultural economics association annual meeting, Vancouver, BC, Canada

CIP (2003) White chuño, rescuing an ancestral flavor INCOPA newsletter, year 1, November. Available at: www.cipotato.org/papandina/publicaciones/Tunta%20flyer%20eng%20I.pdf

Earls J (1998) The character of Inca and Andean agriculture. Department of Social Sciences, Pontificia Universidad Católica del Perú. http://macareo.pucp.edu.pe/~jearls/documentosPDF/theCharacter.PDF

Morris CE, Thompson D (1985) Huanuco Pampa: an Inca city and its hinterland. Thames and Hudson, London

Chapter 2
Potato Growth and Development

Mike Thornton

Contents

Introduction

This chapter describes the somewhat unique characteristics, structures, and growth patterns of the potato plant and is designed to enhance the understanding of management strategies described in later chapters

Characteristics

The potato plant is distinct among major food crops in that it is almost always propagated by planting whole or cut pieces of the tuber (i.e., seed pieces) instead of true seeds. This form of propagation is called "vegetative." Vegetative propagation means that new growth must arise from axial buds (commonly called the "eyes"), as opposed to a fully formed embryo as occurs with seeded crops. As a result, potato crops often emerge slower than seeded crops, but subsequent development is faster due to the relatively large energy reserves contained within the seed piece in the form of carbohydrates.

M. Thornton (✉)
Parma Research and Extension Center, University of Idaho, Parma, ID, USA
e-mail: miket@uidaho.edu

J. C. Stark et al. (eds.), *Potato Production Systems*,
https://doi.org/10.1007/978-3-030-39157-7_2

Sidebar 2.1: The Potato Fruit

Most commercial potato varieties will drop their flowers after a few days. However, under proper environmental conditions, flowers can pollinate and develop into a mature fruit or seed ball (Fig. 2.1). These fruits look like small, green tomatoes, and they can develop tiny seeds that are about one-fourth the size of tomato seeds (Fig. 2.2). Under field conditions, these fruits are rarely seen on varieties such as Russet Burbank, but occasionally a few plants will form potato fruits.

Fig. 2.1 Potato fruit resembles small, green tomatoes

Fig. 2.2 True potato seed is small and similar in size to tomato seed

Each seed inside of a potato fruit is genetically different, and each seed is potentially a new potato variety. Potato breeders use these fruits to produce new varieties, and they're often frustrated when the flowers drop before they can be pollinated. To improve the retention of flowers, potato breeders spray the plants at flowering time with a plant hormone called gibberellic acid, which improves fruit set. Another frustration for the potato breeder is that some varieties, such as Russet Burbank, are male sterile, meaning that they do not produce useable pollen. As a result, the Russet Burbank variety can be used only as a female parent for genetic crosses.

Potatoes also produce a fruit that contains true seeds that can be used in propagation. See Sidebar 2.1. However, each seed is genetically unique, and tubers produced from true-seeded crops are not uniform enough to meet requirements of most markets. However, breeding of new varieties relies on this genetic variation to introduce new traits.

Below-Ground Structures

Sprouts

The first visible growth after seed pieces are planted is a swelling in the axial nodes, or eyes. These structures are called "sprouts," and consist of stem tissue and meristems where growth occurs. Under dark conditions the sprouts elongate until light is reached, then leaves form. When sprouts begin to grow while exposed to light, they generally form short stems with leaves.

Roots

Root development begins shortly after sprouting in the nodes that develop above the seed piece. Potatoes have a relatively shallow, sparse root system, with up to 70% of the root system developing in the upper 12 in. of soil. Compared to sugar beets and cereal crops, potatoes produce 25–50% less total root length, and the root system contains a smaller portion of root hairs. This has major consequences in terms of managing nutrients and soil moisture, as outlined in Chaps. 8 and 13.

Stolons

Stolons are modified stems that grow horizontally. Within the first 3 weeks after emergence, plants will generally begin producing stolons at the underground nodes above the seed piece. A common cultural practice with potatoes is the process of "hilling" or mounding soil at the base of the plants so that stolons will form underground.

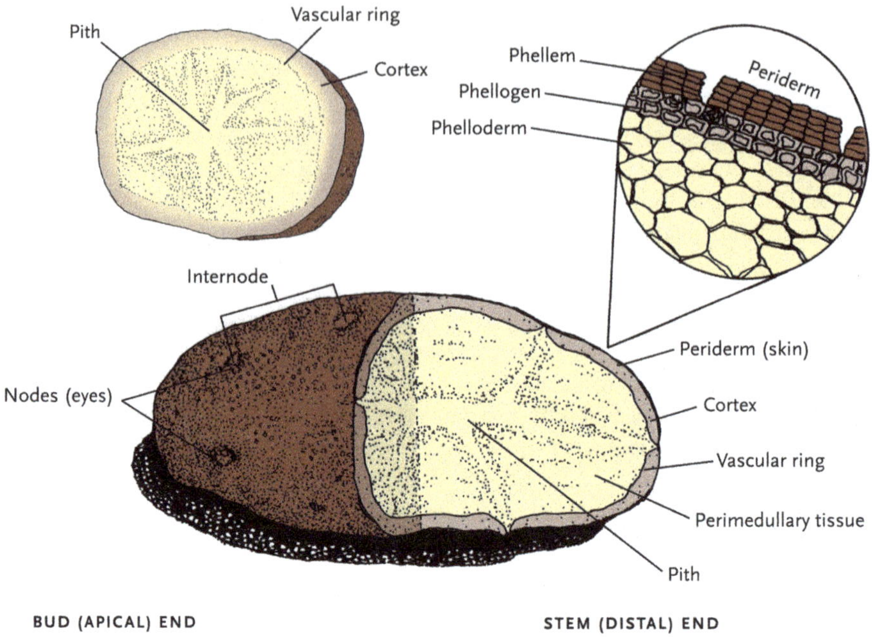

Fig. 2.3 The anatomy of a potato tuber. (Adapted from Dean 1994)

In some instances, stolons will continue to elongate, emerge through the soil surface, and develop into a leafy stem (see discussion of heat runners in Chap. 14). However, with appropriate conditions the tips of the stolons will soon "hook" and begin to swell, resulting in the initiation of new tubers.

Tubers

A potato tuber is actually a modified stem that has all of the internal and external structures that are characteristic of stem tissue. The primary function of a potato tuber is to store chemical energy produced by the leaves in the form of carbohydrates. Once the tuber starts to sprout, these energy reserves are mobilized to drive plant growth.

Internal Structures

A cross section through a tuber shows an outer region of storage tissue (the cortex), a ring of vascular tissue, a region of inner storage tissue (perimedullary), and a somewhat translucent or light-colored inner ring (the pith) (Fig. 2.3). Each of these structures has different characteristics, such as starch and nutrient content, as well as cell size.

External Structures

The outer tuber surface is covered in skin tissue (periderm) and a number of small pores called "lenticels." The skin protects the tuber from moisture loss and invasion by some pathogens, while the lenticels allow for gas exchange (e.g., oxygen and carbon dioxide) in and out of the tuber. When the soil is habitually saturated with water, lenticels can enlarge and may provide an entry point for opportunistic pathogens. Also visible on the surface are the eyes, which are actually undeveloped leaf buds. Each eye has three buds, one being primary and the others secondary (or axillary).

Tuber shape can vary from long and skinny (length to width ratio greater than 2) to round (ratio close to 1). Regardless of shape, all tubers have two ends—the stem end where the stolon attaches to the tuber, and a bud end. The stem end holds the first formed cells (i.e., oldest), contains the vascular connections to the rest of the plant, and tends to be susceptible to expression of stress-related disorders. The bud end contains the last formed (i.e., youngest) cells, has the highest concentration of eyes, and is usually the first portion of the tuber to develop sprouts after dormancy is broken.

Immediately after formation, potato tubers enter a dormant stage where they will not sprout. Dormancy is a survival mechanism meant to ensure that tubers survive adverse environmental conditions during the winter months. The length of dormancy is dependent on many factors, with variety and temperature being the most important. Dormancy is considered to be complete when tubers placed in favorable conditions begin to sprout.

During dormancy, the eye on the bud end of the tuber produces hormones that prevent sprouting of other eyes. This control of sprouting is termed "apical dominance." A characteristic of young seed that has just come out of the dormant stage is that only a single bud in the dominant eye will sprout, producing a plant with a single stem. As the tuber ages, apical dominance is lost, and multiple eyes, and even multiple buds within a given eye will sprout, leading to multi-stem plants. See Chap. 7 for further discussion on physiological age.

Above-Ground Structures

Stems and Leaves

Terms such as "haulm," "canopy," and "vine" are often used interchangeably to describe the stem and leaf tissues of a potato plant. As potato stems elongate, leaves form in a spiral pattern; the first leaves forming at the bottom of the plant and younger leaves developing from the top as the stem continues upward growth (Fig. 2.4). Potato plants develop a compound leaf at each stem node. The compound leaf consists of a terminal leaflet and a row of opposite leaflets all attached to a supporting structure called the "petiole." The petiole tissue is often collected and analyzed as a guide to nutrient management. See Chap. 8.

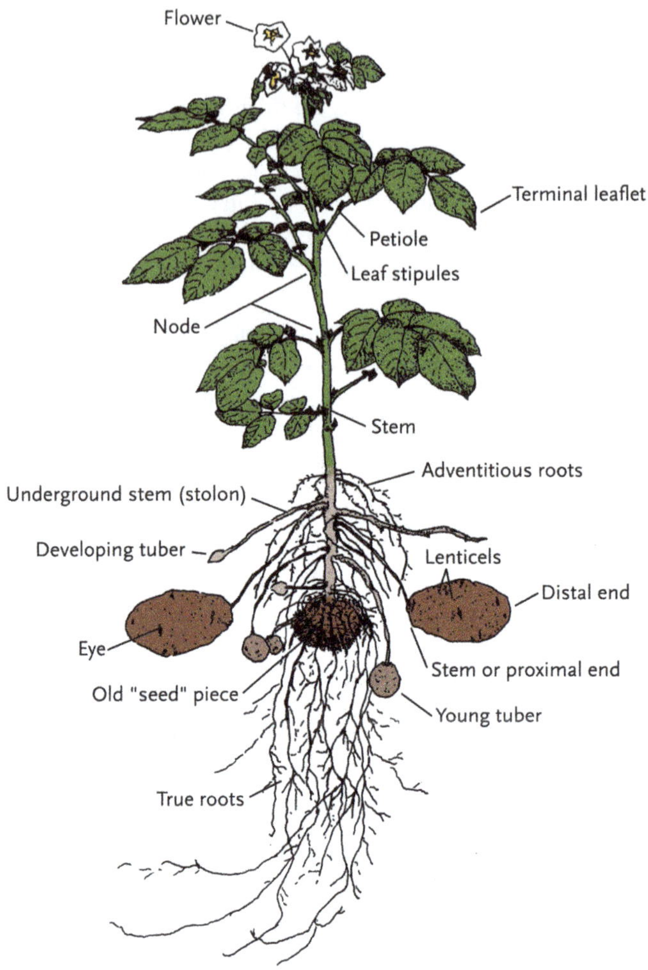

Fig. 2.4 A potato plant developing from a seed piece cut from a tuber. Tubers develop from the enlarged tips of stolons (underground stems). Tubers have eyes (dormant buds) which can develop into shoots, and lenticels pores) through which air penetrates to interior tissues. (Adapted from Thornton and Sieczka 1980)

The primary function of leaves is to capture light energy and turn it into chemical energy in the form of carbohydrates through photosynthesis. The three primary functions of stems are to provide structural support to the leaves, act as the conduit for transport of water and nutrients taken up by the roots to the vine tissues, and to translocate carbohydrates produced in leaves down to tubers for storage. As such, the vascular system is a very important structural component of stems. The potato vascular system consists of three separate bundles of xylem and phloem tissue located just inside the periderm that can be seen when the stem is cut in cross section.

Fig. 2.5 Growth stages of the potato. (Adapted from Western Regional IPM Project 1986)

Flowers

Potato plants will generally form 10–13 leaves and then produce a flower (also called an "inflorescence"). The extent and timing of flowering varies greatly among varieties. Potato flowers have both female (pistol) and male (stamen) parts, and the outer portion (corolla) may be white, pink, red, blue, or purple.

Growth Stages

Growth of a potato plant can be separated into five stages: sprout development, plant establishment, tuber initiation, tuber bulking, and maturation (Fig. 2.5). The transition from one stage to another is not always readily apparent. Timing of growth stages varies depending upon factors such as temperature, availability of moisture, variety, and geographic location. At northern latitudes, emergence of new plants can occur as early as March or as late as June, and harvest typically occurs between August and late October.

Sprout Development (Growth Stage I)

The rate of initial sprout development after seed pieces are planted is dependent on the level of seed dormancy. Dormancy is determined by how much time has passed between harvest of the seed and subsequent re-planting, as well as extent of exposure of the seed crop to warm temperatures. In general, the longer the time since harvest and the longer the exposure to temperatures above 45 °F (during growth, in storage, and during seed handling), the less dormant the seed will be and the more rapidly it will start to sprout. This process of accumulating chronological age and temperature exposure is termed "physiological aging." See Chap. 7 for more extensive discussion of physiological age.

Once the seed piece sprouts, the rate of sprout growth is dependent on soil temperature. Sprout growth rate increases linearly with temperatures from about 45–68 °F, and can actually begin to decline at very high soil temperatures due to damage to the growing point. Depth of planting is also an important factor, as planting seed pieces more than 6–8 in deep increases the amount of soil cover the sprout has to grow through before emerging.

Plant Establishment (Growth Stage II)

Plant establishment refers to the growth period from sprout emergence until initiation of new tubers occurs, and this includes development of both roots and shoots. This stage is also sometimes referred to as "vegetative growth."

Both temperature and moisture conditions play key roles in determining the rate of plant development during this growth stage. Stem development is optimum at temperatures above 68 °F,0 while leaf growth is optimum at about 77 °F. If soil temperatures are too warm during this stage (above 82 °F), then root growth is reduced, and an imbalance between root and vine development can occur. Root and vine development are usually closely tied together, and any management practices that disrupt root growth will eventually result in a reduction in vine growth, and vice versa.

Growth Stage II is also the first phase of plant development when soil nutrient levels can impact growth. Research has shown that the potato plant relies primarily on the nutrients and energy stored in the seed piece to drive plant growth up until reaching a height of 8–10 in. After that, the plant transitions to relying on nutrient uptake by the roots and energy production from the leaves. See Sidebar 2.2.

Sidebar 2.2: Role of the Seed Piece in Early Plant Growth

The large amount of stored energy contained in a potato seed piece gives it an advantage over other crops that are planted using true seed. One early study reported that the seed piece contains 21 grams of nutrients in the form of starch, sugar, and protein per 100 grams of fresh weight. Upon sprouting (Fig. 2.6), these nutrients are rapidly mobilized to provide resources that fuel plant growth. After emergence, plants gradually begin to transition from relying on nutrients and energy in the seed piece to uptake of nutrients from the soil and generation of energy in the leaves through photosynthesis.

This raises an important question—How long does the plant rely on the seed piece? To answer this question, scientists at the University of Idaho conducted a study where they carefully removed seed pieces from Russet Burbank plants at emergence, 8-in plant height, and at first bloom. To account for the effects of root disturbance, they also excavated the checks, but did not remove the seed piece. The results indicate that the earlier the seed piece is removed, the larger the decrease in yield (Fig. 2.7). Tuber size was impacted more than total yield, resulting in slightly more reduction in U.S. No. 1 yield compared to total yield.

The study clearly showed that seed pieces are contributing to plant growth beyond when plants reach a height of 8 in. If the seed piece deteriorates due to decay prior to the end of this transition, yields will be reduced.

Fig. 2.6 Sprout growth relies on nutrients stored in the seed tuber

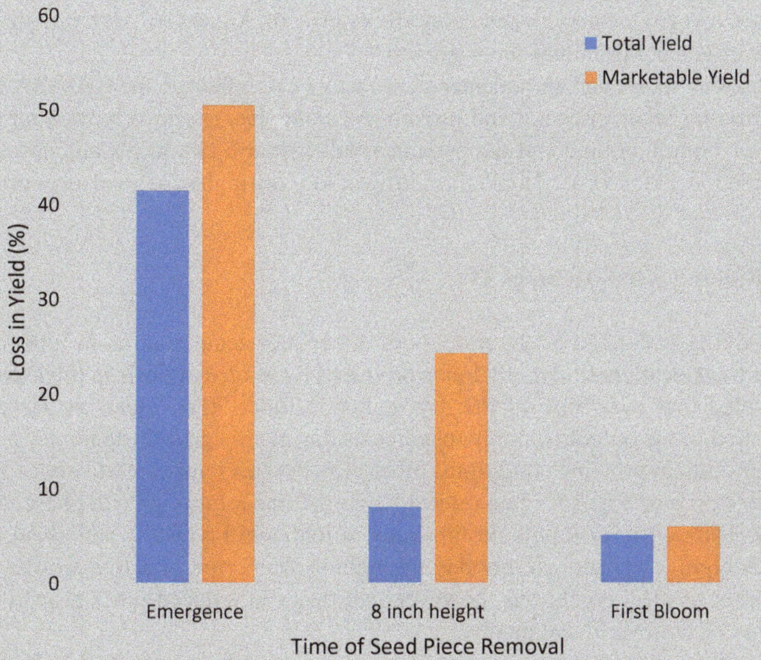

Fig. 2.7 Effect of seed piece removal timing on total and marketable yield loss of Russet Burbank potatoes. (Adapted from Bohl et al. 2001)

Tuber Initiation (Growth Stage III)

Tuber initiation is defined as the process during which the stolon stops growing and the end swells to twice the diameter of the stolon. For many potato varieties, including Russet Burbank, this occurs during early flowering, although there's no causal relationship between the two events. Tuber initiation is a key stage in potato development, as it signals a transition from the plant using all available energy for growth to partitioning some energy towards storage in the tuber.

In most instances, tuber initiation occurs over a relatively short period of time (less than 2 weeks), but this process is heavily influenced by temperature, soil moisture, day length, and levels of available nitrogen. Tuber initiation is favored by moderate temperatures (59–68 °F is optimum), moderate to high soil moisture (above 65%), high light intensity, and moderate to low N availability in the soil.

Several natural plant hormones are involved in tuber initiation, and two of these are particularly important relative to crop management practices. One is called gibberellic acid (GA). It's a growth promoter, and at high levels in the plant it promotes vine growth while delaying tuber growth. The other is abscisic acid (ABA), which is a growth inhibitor that will slow vine growth while promoting tuber growth. The ratio of GA promoter to ABA inhibitor determines the growth response. A high GA/ABA ratio favors vine growth and delays tuber growth. A low GA/ABA ratio has the opposite effect and promotes tuber growth.

Several environmental and management factors can influence the GA/ABA ratio, thus impacting tuber initiation and growth. As examples, excessively high nitrogen fertility or high nighttime soil temperatures will increase GA levels and can delay tuber initiation. Potatoes need moderate nitrogen and cool nights for good tuber growth.

Tuber Bulking (Growth Stage IV)

Tuber bulking is defined as the process of dry matter accumulation in tubers as a result of translocation of carbohydrates produced by photosynthesis in the leaves. It is estimated that over 90% of the dry matter in tubers (i.e., yield) at harvest is directly tied to the output from photosynthesis during this growth stage.

Under optimal growing conditions, tuber growth rates remain relatively constant during this period, which is often referred to as the linear tuber growth phase. As an example, Russet Burbank potatoes in southern Idaho will typically add about 6–10 hundredweight (cwt) per acre per day throughout the period of active growth. Any interruption of ideal conditions, however, can result in reduced tuber growth rates and losses of both yield and quality.

Research has shown that two major factors influence tuber yield: (1) the photosynthetic activity and duration of the leaf canopy, and (2) the length of the linear tuber growth phase. The longer a canopy is able to produce photosynthate at a relatively high rate, and the longer tubers bulk at their maximum rate, the higher the resulting yield.

Tuber bulking rate and duration can be influenced by several environmental and cultural factors. As with tuber initiation, temperature, soil moisture, and nutrient availability are the most important considerations during tuber bulking. Potato plants need warm days and cool nights for good tuber growth (Fig. 2.8). The optimum soil

Fig. 2.8 High soil temperatures can disrupt tuber growth and create abnormal growth patterns

temperature for tuber growth is about 61 °F, while the optimum air temperature for vine growth is about 77 °F. However, with full leaf canopy shading the soil, it's possible to have 77 °F air temperatures at the same time as 61 °F soil temperatures.

Developing healthy plants necessary for maximum tuber growth requires that all essential nutrients be supplied at optimal rates. Both deficit and excess fertilizer situations can reduce tuber bulking rates. Nutrient deficiencies limit canopy growth and shorten canopy duration, resulting in reduced carbohydrate production and tuber growth rates. Excessive fertilizer applications can cause nutrient imbalances that delay or slow tuber growth rates.

Maturation (Growth Stage V)

As potato vines start to senesce (Fig. 2.9), several important changes in both physical and chemical characteristics happen to the tubers. The skin or periderm thickens and becomes more fully attached to the underlying cells, which provides greater protection to tubers during harvest and handling, and blocks entry of pathogens to the tuber. The ideal temperature range is 70–75 °F, while temperatures below 45 °F or above 90 °F actually hinder skin development. Excessive nitrogen or potassium applications late in the season will also delay or prevent skin development. See Chap. 16 for more information on skin set.

During tuber maturation specific gravity (dry matter) increases, which improves quality for both processing and fresh market consumption. In addition, free sugars are converted to starch, which allows for lighter colored, better quality chips and fries. Also, with proper maturity tubers in storage have lower respiration rates, remain dormant longer, and consequently sprout later. Properly matured tubers also have greater resistance to pathogens in storage.

If, however, tubers remain too long in the soil after vine death, they can become over mature. In such cases, starch converts back to sugar, and specific gravity declines. Over-mature tubers will often have higher respiration rates in storage, will break dormancy and sprout earlier, and will be more susceptible to rot development.

Fig. 2.9 Vine death, accompanied by tuber maturation, is the culmination of senescence

Growth Habit

Potato varieties can be classified as early-, mid- and late-maturity types. Early maturing varieties are said to have a "determinate" growth habit, in that they complete their growth phases relatively early in the growing season. Mid- and late-maturing varieties have a "semi-determinate" or "indeterminate" growth habit that results in later completion of their growth phases. See Chap. 3 for a full description of maturity classes of common varieties.

Determinate potato varieties are characterized by relatively early onset of tuber initiation, bulking, and flower production. Flowers form at the 10th–13th leaf node, and new leaf production stops or greatly slows after this point. Cessation of new leaf production and root growth coincides with the start of rapid tuber bulking. As a result, determinate varieties tend to have a much more limited canopy size and rooting depth compared to late-maturing varieties. They also enter the tuber maturation stage relatively early and may be ready to vine kill by late July to mid-August in northern production regions.

In contrast, indeterminate varieties generally initiate tubers somewhat later, and vine growth and flower production continue throughout the season. As with short-season varieties, the first flowers will form at the 10th–13th leaf node. However, growth does not cease at this point, and one of the side or axillary buds just below the flower will begin elongating into a new stem, which will produce another 10–13 nodes with a second flower at the tip. With a sufficiently long season, as well as good nutrition and growing conditions, this process may repeat itself again resulting in production of a third set of flowers. The terms "first bloom" and "second bloom" are often used in reference to these successive periods of flowering. However, since all plants in a field don't progress at exactly the same pace, this is most commonly seen as "flushes" of flowers, with dense blooms at the peak of the first inflorescence, then a reduction in flower numbers, followed by another "flush" at the peak of the second inflorescence, and so forth.

Because new leaf production is closely tied to root development, indeterminate varieties tend to have more extensive, deeper rooting systems compared to early-maturing varieties. They are also characterized by relatively long tuber bulking phases, and if the growing season is long enough, can produce much higher yields compared to determinate varieties. These varieties may not enter the tuber maturation stage until very late in the season, and thus may not be suitable for growing regions with limited frost-free periods.

Impact of Management at Each Growth Stage

Understanding the growth of the potato crop at each development stage will promote (or allow) appropriate and timely management decisions. By knowing how any management activity will affect the plant, proper decisions can be made that result in maximum harvest yield and highest quality tubers.

It may be helpful to think of the potato crop as a type of starch factory. The factory takes inputs (water, nutrients, and CO_2) and uses energy (sunlight) to turn them into a product (potato tubers that are primarily starch), while releasing oxygen and water as byproducts. The keys to making this "factory" productive are to build the factory as quickly as possible, keep it running as efficiently and for as long as possible, and package the product for shipment when the factory shuts down.

Building the factory primarily takes place during sprout development and plant establishment (Stages I and II). Rapid plant development requires planting healthy seed pieces into soil conditions that favor rapid wound healing and sprout growth. Anything that results in slow sprout development or causes seed decay will reduce plant growth rate. The most important management practices include using appropriate fungicide treatments to control dry rot and Rhizoctonia stem canker, as well as planting into moist (but not saturated) soil that is between 45 and 65 °F to promote rapid wound healing. The optimal planting date varies by region, but in all areas growers should wait to plant potatoes until daytime soil temperature warms to 45 °F or higher. Once the plants emerge, it is important to provide optimum soil moisture and nutrient availability, while not damaging root or leaf production by making inappropriate herbicide applications or performing late cultivation.

The beginning of tuber initiation (*Growth Stage III*) signals a shift from building the factory to operating it to produce tubers. This is the most sensitive stage for quality problems to develop, and the focus should be on maintaining adequate, but not excessive, moisture and nutrient availability.

Tuber bulking (*Growth Stage IV*) is the longest phase for the factory, lasting upwards of 120 days in some of the warmer production regions. Any condition that limits growth of healthy foliage, disrupts tuber growth, or shifts dry matter partitioning from the tubers to the foliage will limit the number of days the factory operates, and thus decrease yield potential. Some of the key management factors that affect tuber bulking are seed physiological age, plant spacing, fertilization, irrigation, and pest management.

Seed Physiological Age

Aged seed tends to produce potato plants with numerous stems that sprout and develop rapidly and die early. High stem numbers usually result in a high number of tubers per plant, which reduces average tuber size by reducing the amount of carbohydrate available to each tuber during bulking. Early death also shortens bulking time and limits overall productivity.

By comparison, plants from physiologically young potato seed begin to bulk later than those from aged seed, which may shorten the linear tuber growth phase in areas with a short growing season. Although controlling seed age is difficult, it is possible to avoid seed lots that were stressed during production or that were sprouted or exposed to warm temperatures in storage. Any stress during seed production or storage can accelerate physiological aging.

Plant Spacing

Closer than optimal plant spacing has a similar effect on tuber growth as does aged seed in that it increases tuber density relative to canopy size, thereby limiting the photosynthetic capacity to bulk each tuber. Although total yields may not be reduced, bulking rates of individual tubers decrease, resulting in smaller tubers and lower marketable yields.

Wider than optimal spacing can lengthen the time it takes to reach full canopy, which reduces carbohydrate supply to the tubers. Optimal spacing varies among cultivars. See Chaps. 3 and 7.

Fertilization

Excessive nitrogen applications can delay the onset of tuber bulking in indeterminate varieties, such as Alturas and Russet Burbank, which can reduce the length of the linear tuber growth period in a short-season environment. High levels of nitrogen can delay tuber initiation and growth by 10 days or more. At a growth rate of 6–10 cwt per acre per day, this delay can result in potential yield losses of 60–100 cwt/acre under conditions where the growing season is short, and lost growing time cannot be recovered later in the season.

Lower nitrogen levels early in the season can have the opposite effect and can result in earlier tuber initiation and bulking. Early (determinate) potato varieties generally initiate tubers earlier in the growing season, and this earlier development impacts nitrogen management strategies. See Chap. 8.

Regardless of the variety grown, the objective of the fertilizer program should be to maintain a healthy, green canopy with 100% ground cover so that capture of sunlight is maximized within the limits of the growing season.

Irrigation

Allowing soil moisture to drop below critical levels reduces or stops canopy and tuber growth during the stress period and for several days thereafter. This effectively shortens the tuber bulking period and can also cause a variety of internal and external tuber defects. Excessive irrigation can also reduce tuber growth by restricting plant physiological activity and nutrient uptake and increasing disease susceptibility.

Pest Management

Any insect or disease that damages leaves can reduce the amount of light intercepted by the canopy and limit tuber growth. Among the most serious of these pests are Colorado potato beetle, late blight, early blight, and Verticillium wilt. See Chaps. 9 and 11. Growers need to take appropriate measures to control these pests under conditions that will limit plant growth.

Conditions during the maturation stage (*Growth Stage V*) determine to a large extent how well the tubers will store after harvest. Critical management practices include cutting off nutrient applications, moderating soil moisture conditions, and properly timing vine kill procedures to optimize skin set and resistance to skinning and bruising. See Chap. 16.

Acknowledgements Unless otherwise noted, photographs and graphics were taken from collections of University of Idaho Extension educators, scientists, and researchers who are authors of this text.

References

Bohl WH, Love SL, Thompson AL (2001) Effect of seed piece removal on yield and agronomic characteristics of Russet Burbank potatoes. Am J Potato Res 78:397–402

Dean BB (1994) Managing the potato production system. Haworth, Binghamton, NY

Thornton RE, Sieczka JB (1980) Commercial potato production in North America. Supplement to volume 57 Am Potato J

Western Regional IPM Project (1986) Integrated pest management for potatoes in the western United States, publication 3316. University of California, Oakland, CA

Chapter 3
Variety Selection and Management

Jeffrey C. Stark, Asunta (Susie) L. Thompson, Richard Novy, and Stephen L. Love

Contents

Introduction

Selecting a potato variety to grow and successfully market has become more complex in recent years, as the number and range of market types and acceptable attributes has greatly increased. Changing consumer preferences have also been a significant driving force behind this change, resulting in a much more colorful market produce section than in the past. Descriptions are provided for some of the major

J. C. Stark (✉)
Idaho Falls R&E Center, University of Idaho, Idaho Falls, ID, USA
e-mail: jstark@uidaho.edu

A. (S). L. Thompson
Department of Plant Sciences, North Dakota State University, Fargo, ND, USA
e-mail: Asunta.Thompson@ndsu.edu

R. Novy
Small Grains and Potato Germplasm Research Unit, USDA-Agricultural Research Service, Aberdeen, ID, USA
e-mail: Rich.Novy@usda.gov

S. L. Love
Aberdeen Research and Extension Center, University of Idaho, Aberdeen, ID, USA
e-mail: slove@uidaho.edu

J. C. Stark et al. (eds.), *Potato Production Systems*,
https://doi.org/10.1007/978-3-030-39157-7_3

varieties grown in North America, along with their morphological and agronomic characteristics, incentives for production, strengths and weaknesses, and key management considerations. Although variety acceptance in the marketplace is a dynamic process, many of these varieties have remained and will likely remain in commerce for years to come.

The Purpose of Potato Variety Development

To remain competitive, potato production efficiency must improve to offset increasing transportation and finishing costs. Improvement and sustainability of potato production systems depend greatly on the introduction of new varieties, because the dominant varieties, such as Russet Burbank, Russet Norkotah, Red Norland, and Yukon Gold, have significant production and quality limitations and are currently being grown at near peak efficiency.

Variety development offers the most effective approach for addressing issues of food supply, nutrition, and impact of agriculture on the environment. With current or even reduced levels of management, new varieties can offer improved yields, processing quality, and disease and pest resistance, while reducing the demand for fertilizer and water resources.

In recognition of this fact, cooperative potato breeding and cultivar improvement programs have received public support in the U.S. since the late 1920s when, through legislative action, a national potato breeding program was developed. There are many regional and state potato improvement programs currently operating in the U.S. Working together, the United States Department of Agriculture-Agricultural Research Service (USDA-ARS) and state experiment stations focus on potato cultivar improvements needed for specific production areas across the U.S.

Since the introduction of new russet varieties and processing methods in the early 1990s, the historical choice to grow only a select few varieties has begun to change. Markets have been established for many other types of potatoes, and growers can now make the choice to include multiple varieties in their operations.

Specialty potato cultivars with red, yellow, purple, and multi-colored skin and/or flesh combinations occupy a small, but increasingly important, market niche. These potatoes have strong consumer appeal due to their unique combinations of color, shape, culinary quality, and nutrient content.

Many characteristics must be considered when choosing a variety. Market acceptance and economic advantage are still the major factors growers consider when choosing a variety; however, yield, quality, pest resistance, and adaptability to local growing conditions are also key considerations.

If the market will accept a new variety, the opportunity for inclusion in a grower's operation is available, but many additional factors become important to the decision-making process. Each variety has characteristics that present distinct strengths and weaknesses. Factors to be considered include: (1) yield potential in the area of intended production, (2) conformity to market specifications that will result in price incentives, and (3) resistance to common, as well as region-specific, defects, diseases, pests, and stress-related problems. Defining these characteristics is critical so that an informed decision can be made regarding the cost-effectiveness of producing a variety.

At least three essential criteria must be met in selecting a variety to grow:

- The intended market place exists and will accept the variety. A contract should preferably be in hand until wide acceptance is achieved.
- The variety will perform as intended. It is a wise practice to conduct initial testing on limited acreage in the first few years to avoid costly surprises.
- Sufficient information is available on managing the new variety. The full potential of a new variety may not be realized unless cultivar-specific management practices are implemented.

Potato Market Classes

Potato market classes have little basis in genetics, meaning they have a historical rather than scientific basis. There is no practical reason that a good french fry processing variety cannot have red skin, a good boiling potato cannot have russet skin and long shape, or that the ideal baker cannot be round with white skin.

Table 3.1 Characteristics of some commonly grown potato varieties in North America

Variety	Maturity	Tuber characteristics	Primary usage
Alturas	Very late	Lightly russeted, oblong	Processing
Atlantic	Medium	White (buff), light netting, round	Chips
Blazer Russet	Early-mid	Russet, oblong	Fresh, french fries
Chieftain	Medium	Medium red, oblong-round	Fresh
Clearwater Russet	Medium-late	Russet-oblong	Frozen processing
Dakota Pearl	Medium	White, round	Chips, fresh
Goldrush	Medium	Russet, oblong/blocky	Fresh
Norland	Early	Medium to dark red, round	Fresh
Ranger Russet	Medium-late	Russet, long	Frozen processing
Red LaSoda	Medium	Light red (pink), oblong	Fresh
Russet Burbank	Late	Russet, long	Fresh, frozen processing
Russet Norkotah	Early	Russet, long	Fresh
Shepody	Early-medium	White, long	Frozen processing
Snowden	Medium-late	White, round	Chips
Teton Russet	Early	Russet, oblong-long	Fresh, frozen processing
Umatilla Russet	Late	Russet, long	Frozen processing
Yukon Gem	Medium	Yellow, round-oval	Fresh
Yukon Gold	Medium	Yellow skin and flesh, oval	Fresh

Note: This publication does not provide all available information on the characteristics and management requirements for the varieties. However, it does review practices that are known, unique, and critical to attain successful production of each variety

In North America, historical regional preferences for certain potato varieties, combined with local culinary practices, led to the development of performance expectations based on tuber appearance. The consumer expects red potatoes to be round, first to market, and have good boiling quality. Russet potatoes are expected to have a long shape and make good baked potatoes and french fries. Round, white potatoes are expected to either make good potato chips or to boil well, but not necessarily both.

Long russets, round whites, and reds make up most of the North American potato market, although other specialty potatoes are making significant gains in the fresh potato industry. Current key potato markets and examples of some common varieties sold in those markets (Table 3.1) include the following:

- Russets (dual purpose)—Used for french fries and tablestock. Varieties: Russet Burbank, Blazer Russet, Teton Russet.
- Russets (processing)—Used for french fry processing. Varieties: Russet Burbank, Ranger Russet, Umatilla Russet, Clearwater Russet.
- Russets (fresh market)—Used for tablestock (mostly baking). Varieties: Russet Norkotah, Goldrush.
- Long whites (processing)—Used mostly for french fries and dehydration processing. Varieties: Alturas, Shepody.
- Round whites (chipping)—Used for making potato chips. Varieties: Atlantic, Dakota Pearl, Snowden.
- Reds (fresh market)—Used for tablestock (mostly boiling). Varieties: Red Norland/Dark Red Norland, Chieftain, Red LaSoda.
- Specialty potatoes (fresh market)—Used for home preparation of various products. Varieties: Yukon Gold, Yukon Gem (yellow skin and flesh).

Widely Grown, Commercially Available Varieties

Russet Varieties for Fresh Use or French Fry Processing (Figs. 3.1–3.4)

Blazer Russet

General Information

Parentage: A7816–14 x NorKing Russet.

Developers: Northwest (Tri-State) Variety Development Program.

Plant Variety Protection: Northwest Potato Variety Development Program (administered by PVMI).

Morphological Characteristics

Plant: Small to medium-sized, semi-erect vine expressing medium maturity with white flowers.

Tubers: Oblong, with a moderate, tan russet skin and shallow eyes.

Fig. 3.1 Blazer Russet. (Photo credit: Potato Variety Management Institute (PVMI))

Fig. 3.2 Clearwater Russet. (Photo credit: PVMI)

Fig. 3.3 Russet Burbank. (Photo credit: PVMI)

Fig. 3.4 Teton Russet. (Photo credit: PVMI)

Incentives for Production

Blazer Russet is an early to mid-season variety notable for its high U.S. No. 1 yield of medium-russeted tubers and its good processing and culinary qualities. It is very suitable for processing into french fries and other frozen potato products directly from the field or from extended storage, with higher merit than Russet Burbank and Ranger Russet in processing and post-harvest evaluations. Blazer Russet also has high merit for use in the fresh market, with sensory evaluations comparable to those of Russet Burbank.

Agronomic Characteristics

Vine Maturity: Early to mid.

Yield Potential: Typically produces higher total and U.S. No. 1 yields than Russet Norkotah and Shepody for early harvest production. In full-season trials in the western U.S., total yields for Blazer Russet were slightly lower than Ranger Russet and slightly higher than Russet Burbank, but U.S. No. 1 yields were greater than Ranger Russet and Russet Burbank. Blazer Russet also produces a high percentage of U.S. No. 1 tubers, similar to Russet Norkotah.

Specific Gravity: Specific gravities are comparable to Shepody and higher than Russet Norkotah and Russet Burbank, but lower than Ranger Russet.

Culinary Quality: Blazer Russet is suitable for both fresh pack and french fry processing. Fry color was consistently acceptable in comprehensive processing evaluations following long-term storage of tubers obtained from trials conducted in Washington, Idaho, and Oregon. Fry color is light and uniform, with a low incidence of sugar ends. Post-harvest processing ratings for Blazer Russet in the Pacific Northwest have been higher than those for Ranger Russet and Russet Burbank. Blazer Russet has compared favorably to Russet Burbank in both pre- and post-storage sensory evaluations of baked potatoes.

Diseases/Pests/Physiological Disorders: Blazer Russet is resistant to sugar ends, tuber malformations, and most internal and external defects; the exception being its moderate susceptibility to hollow heart. It is resistant to common scab, tuber powdery scab, and potato virus X (PVX), and has moderate resistance to blackspot bruise and tuber late blight infections. It is moderately susceptible to powdery scab root galling, potato virus Y (PVY^{o}), early blight infection of the tuber, Erwinia soft rot, and Fusarium dry rot. Blazer Russet is susceptible to

Verticillium wilt, pink rot, potato leafroll virus (PLRV) net necrosis, corky ringspot, and foliar early and late blight.

Storability: Stores well. Natural tuber dormancy is approximately 40–50 days shorter than for Russet Burbank.

Key Management Considerations

Total seasonal nitrogen (N) requirements for Blazer Russet are about 10% less than Russet Burbank, but a higher proportion should be applied early in the growing season to facilitate earlier tuber development. Nitrogen uptake decreases substantially after August 1, so applications should not be made after that time. Studies in Idaho indicate that petiole nitrate sufficiency levels for Blazer Russet are similar to those for Russet Burbank. Adjust N timing and rate to deplete soil N and allow plant to mature naturally for an early harvest. To reduce shatter bruise, do not over-fertilize or over-irrigate late in the season. Allow plants to mature and skins to set for at least 14 days prior to harvest.

Strengths: High early-yield potential, dual-purpose with excellent processing quality and resistance to tuber powdery scab and common scab.

Weaknesses: Verticillium wilt and PLRV susceptibility and moderate hollow heart susceptibility.

Clearwater Russet

General Information

Parentage: Bannock Russet x A89152-4.

Developers: Released in 2008 by the USDA-ARS and the Agricultural Experiment Stations of Idaho, Oregon, and Washington, and represents a variety of the Pacific Northwest Potato Variety (Tri-State) Development Program.

Plant Variety Protection: Pacific Northwest Potato Variety Development Program (administered by PVMI).

Morphological Characteristics

Plant: Medium-sized, semi-erect vine with purple flowers having white tips.

Tubers: Oblong to long, medium russeted, with white flesh and shallow eyes.

Incentives for Production

With high marketable yield and a low incidence of external tuber defects, Clearwater Russet is suitable for fresh or processing use. Cold-sweetening resistant with a low incidence of sugar ends, Clearwater Russet can be stored at 45 °F for up to 250 days and provide acceptable french fries. Clearwater Russet is also notable for its higher tuber protein content.

Agronomic Characteristics

Vine Maturity: Medium-late.

Yield Potential: Total yield comparable to industry standard varieties, with U.S. No. 1 yield being comparable to Ranger Russet and substantially higher than Russet Burbank.

Specific Gravity: Averaging in the mid- to high-80s across western sites.

Culinary Quality: Good sensory ratings for both processed products and for fresh use with sensory ratings of baked potatoes being comparable to Russet Burbank.

Diseases/Pests/Physiological Disorders: Resistant to PVX and tuber late blight infection; moderately resistant to Verticillium wilt and common scab; susceptible to Fusarium dry rot.

Storability: Tuber dormancy is approximately 55–60 days shorter than for Russet Burbank with comparable tuber shrinkage to Russet Burbank. Fresh market storage can be up to 9 months at 42–45 °F in the absence of problematic storage disease development. With its cold-sweetening resistance, Clearwater Russet can be stored at 45–48 °F for optimum processing quality. To minimize dry rot formation in storage, bruising and wounding of tubers should be minimized during harvest and subsequent handling.

Key Management Considerations

With 34–36-in. row spacings, plant seed pieces at 10–11 in. for fresh market potatoes and 12–13 in. for processing potatoes. Smaller tuber size is noted for Clearwater Russet in the Columbia Basin region of the U.S., with the recommendation that 10–12-in. seed spacing be used and the crop be allowed to grow >150 days to ensure adequate tuber size. In Idaho, seasonal N requirements for Clearwater Russet are 25% less than for Russet Burbank, with 1/3–1/2 of the seasonal N recommended to be applied by row closure, and the remainder applied by early August. Tuber skinning and damage during harvest should be minimized as much as feasible to limit the potential for dry rot infection and subsequent development in storage.

Strengths: Attractive tubers and a low incidence of external tuber defects; excellent processing qualities with cold-sweetening resistance and a low incidence of sugar ends.

Weaknesses: Fusarium dry rot susceptibility with internal brown spot being noted as problematic in the southern Columbia Basin.

Russet Burbank

General Information

Parentage: Sport of Burbank Seedling identified in California in 1902. Burbank Seedling was an open pollinated seedling of Early Rose.

Developers: Bred by Luther Burbank in Massachusetts and released in 1876.

Plant Variety Protection: No.

Morphological Characteristics

Plant: Large, spreading vine with sparse white flowers.

Tubers: Long, slightly flattened, with medium russet skin and moderately shallow eyes.

Incentives for Production

Russet Burbank is the industry standard for french fry processing and the russet fresh market. Consequently, it is routinely sold into commodity markets. If managed properly, it produces a high yield of quality potatoes. The tubers store exceptionally well due to long dormancy and moderate resistance to storage rots.

Agronomic Characteristics
Vine Maturity: Late.
Yield Potential: Moderately high.
Specific Gravity: Medium.
Culinary Quality: Excellent for making fried products and baking.
Diseases/Pests/Physiological Disorders: Russet Burbank is very susceptible to net necrosis caused by PLRV, a problem that commonly causes losses in crop value. It is also susceptible to PVX, PVY, Verticillium wilt, foliar late blight, foliar early blight, and Fusarium tuber decay. Russet Burbank is resistant to tuber late blight rot and common scab. This cultivar is susceptible to most tuber quality defects, including hollow heart, blackspot bruise, secondary growth, growth cracks, and sugar ends.
Storability: One of the best attributes of Russet Burbank is its ability to maintain excellent tuber quality for periods of storage up to 11 months. Long dormancy reduces the need for applied sprout inhibitors for tubers stored for packing at relatively cold temperatures.

Key Management Considerations
For 36-in. row spacing, plant seed pieces 11–13 in. apart. Control management-related stresses, such as nutrient or water deficits. Minimize net necrosis by applying an insecticide to control colonizing green peach aphids. Use soil fumigation or incorporation of a green-manure mustard crop in fields with historical problems of early dying or nematodes. Potatoes for processing should be stored at 47–48 °F, and sugars should be monitored for early onset of senescent sweetening.
Strengths: Russet Burbank's most positive attribute is recognition in the marketplace for superior cooking qualities. It also possesses excellent storage characteristics.
Weaknesses: Russet Burbank is a high-input cultivar, making it less economically efficient to produce than many of the newer cultivars. Production requires management of physiological susceptibilities, including environmental stresses that cause tuber defects, such as malformations, hollow heart, brown center, and sugar ends. It is susceptible to net necrosis and early dying and produces a large proportion of undersized tubers.

Teton Russet

General Information
Parentage: Blazer Russet x Classic Russet.
Developers: Released in 2011 by the USDA-ARS and the Agricultural Experiment Stations of Idaho, Oregon, and Washington, and represents a variety of the Pacific Northwest Potato Variety (Tri-State) Development Program.
Plant Variety Protection: Pacific Northwest Potato Variety Development Program (administered by PVMI).

Morphological Characteristics
Plant: Semi-erect, medium-sized vine with white flowers.
Tubers: Oblong-long, with medium russeting of skin, white flesh, and shallow eyes.

Incentives for Production
An early-maturing russet variety with early-harvest marketable yields comparable to or higher than Russet Norkotah. Attractive tubers for fresh pack use and acceptable french fry quality following up to 8 months of storage at 48 °F makes Teton Russet a good dual-purpose variety. Teton Russet is resistant to common scab and Fusarium dry rot and is moderately resistant to tuber net necrosis resulting from PLRV infection. It has higher protein and vitamin C content than most standard varieties.

Agronomic Characteristics
Vine Maturity: Early.
Yield Potential: Medium to high and comparable to Russet Norkotah. Teton Russet produces a high percentage of U.S. No. 1 tubers.
Specific Gravity: In early-harvest trials conducted in the western U.S., specific gravities were in the high 70s, which were comparable to values for Ranger Russet and Russet Burbank; specific gravities tend to be higher for Teton Russet relative to Russet Norkotah.
Culinary Quality: Suitable for both fresh pack and french fry processing from the field and from storage. Fry color is light and uniform, with a low incidence of sugar ends. A formal taste and sensory panel overseen by Washington State University Food Sensory Lab over a 3-year period rated Teton Russet baked potatoes as comparable to Russet Norkotah and Russet Burbank for aroma, flavor, texture, aftertaste, and overall acceptance.
Diseases/Pests/Physiological Disorders: Low incidence of internal and external defects, but growth cracks have been observed when soil moisture levels are allowed to fluctuate and are not uniformly maintained. Resistant to Fusarium dry rot and common scab. Moderately resistant to tuber net necrosis associated with PLRV. Teton Russet is considered susceptible to other potato diseases and shatter bruise.
Storability: Tuber dormancy is approximately 35 days shorter than for Russet Burbank and is comparable to the dormancy of Ranger Russet. For the frozen processing market, it is recommended that Teton Russet tubers be stored at 48 °F to retain the lightest fry color and minimize fry mottling following long-term storage.

Key Management Considerations
Total seasonal N requirements for Teton Russet are approximately 20–30% less than Russet Burbank for the same yield goal. For southern Idaho, About 65% of fertilizer N should be applied by tuber initiation, with the remaining N applied via sprinkler irrigation prior to the last week of July. To promote skin set, N applications should be completed at least 30 days prior to harvest. To reduce shatter bruise, do not over-fertilize or over-irrigate late in the season; reduced irrigation also minimizes lenticel enlargement on tubers. Allow plants to mature and skins to set for at least 10 days prior to harvest.

Strengths: High early-yield potential, dual-purpose with good processing quality and resistance to Fusarium dry rot and common scab. Also good nutritional qualities with higher vitamin C and protein content than standard varieties.

Weaknesses: Growth cracks when wide fluctuations in soil moisture and shatter bruise susceptible.

Russet Varieties for French Fry Processing (Figs. 3.5 & 3.6)

Ranger Russet

General Information

Parentage: Butte x A6395-3.

Developers: Released in 1991 by the USDA-ARS and the Idaho, Oregon, Washington, and Colorado Agricultural Experiment Stations. Experimental designation was A7411-2.

Plant Variety Protection: No.

Morphological Characteristics

Plant: Large, spreading vine with dark lavender flowers.

Tubers: Long, slightly flattened with moderately deep eyes and medium russet skin.

Fig. 3.5 Ranger Russet. (Photo credit: PVMI)

Fig. 3.6 Umatilla Russet. (Photo credit: PVMI)

Incentives for Production

Ranger Russet is a consistent producer of high-quality potatoes that are preferred for french fry production. Yield potential is high for second-early harvest. Ranger Russet is resistant to most physiologically based tuber defect issues that detract from fry quality, including hollow heart, internal brown spot, sugar-ends, and general malformations. Ranger Russet exhibits exceptionally high vitamin C content in comparison to other cultivars.

Agronomic Characteristics

Vine Maturity: Late, although early tuber yields are high.

Yield Potential: Moderately high to high.

Specific Gravity: Moderately high.

Culinary Quality: Excellent for french fry production. Baking and boiling quality is good, although Ranger Russet is not commonly used for fresh sales.

Diseases/Pests/Physiological Disorders: Ranger Russet is resistant or highly resistant to Verticillium wilt, Fusarium tuber rot, PVY, PVX, and net necrosis caused by PLRV. It is susceptible to common scab, foliar and tuber late blight, foliar early blight, and root-knot nematode. Resistance to physiological defects, including hollow heart, secondary growth, and growth cracks is high to very high. Tubers are susceptible to blackspot bruise.

Storability: Although Ranger Russet is not commonly produced for long-term storage, tubers retain good quality for several months. Dormancy is medium length, and tubers held for more than 2–3 months will require applications of sprout inhibitors. Tubers are prone to pressure bruising under conditions of low humidity in storage.

Key Management Considerations

Avoid cutting and planting heavily sprouted seed. For 36-in. row spacing, plant seed pieces 8–10 in. apart. Reduce the incidence of blackspot bruise by maintaining green vines up to vine kill, maintaining high soil moisture through vine kill and maturation, and using the best anti-bruise practices during harvest. Many growers utilize green-dig procedures for Ranger Russet tubers going into storage. Utilize full-season late blight control where applicable. Avoid tuber chilling before harvest. Potatoes for processing should be stored at 47–48 °F, and sugars should be monitored for early onset of senescent sweetening.

Strengths: Ranger Russet consistently produces high yields of potatoes with excellent french fry processing qualities. This cultivar is resistant to internal defects, such as hollow heart, brown center, net necrosis, and sugar ends; has a high proportion of large tubers; is resistant to PVY and net necrosis; and is moderately resistant to early dying.

Weaknesses: Ranger Russet is susceptible to blackspot bruise, a problem that requires special considerations during harvesting and delivery. It is also very susceptible to late blight tuber rot and moderately susceptible to stress-induced tuber constrictions.

Umatilla Russet

General Information

Parentage: Butte x A77268-4.

Developers: Released in 1998 by the USDA-ARS and the Agricultural Experiment Stations of Idaho, Oregon, and Washington, and represents a variety of the Pacific Northwest Potato Variety (Tri-State) Development Program.

Plant Variety Protection: Pacific Northwest Potato Variety Development Program (administered by Oregon State University).

Morphological Characteristics

Plant: Medium-sized, semi-erect vine, with blue-violet flowers that tend to purple-violet on inner surface of petals.

Tubers: Long, medium-russeted skin, white flesh, with a tendency for tapering on apical ends.

Incentives for Production

Umatilla Russet produces high U.S. No. 1 (marketable) yields, with tubers having consistent specific gravities and acceptable fry colors from storage temperatures as low as 45 °F.

Agronomic Characteristics

Vine Maturity: Late.

Yield Potential: High.

Specific Gravity: Typically in the low- to mid-80s in western trial sites.

Culinary Quality: Good, especially for french fry production.

Diseases/Pests/Physiological Disorders: Resistant to PVX and common scab; moderately resistant to Verticillium wilt, tuber late blight infection, and net necrosis caused by PLRV. Less susceptible to hollow heart and growth cracks relative to Russet Burbank, but with greater susceptibility to Fusarium dry rot, blackspot, and shatter bruise.

Storability: Tuber dormancy is approximately 30 days shorter than for Russet Burbank. Umatilla Russet can be stored at 45–48 °F for optimum processing quality. To minimize dry rot formation in storage, bruising and wounding of tubers should be minimized during harvest and subsequent handling.

Key Management Considerations

Plant emergence can be slow and non-uniform, but delayed emergence does not impact final yield or market value in regions with longer growing seasons, such as in the Columbia Basin. Minimize tuber wounding during harvest to mitigate the development of Fusarium dry rot in storage. Nutrient and irrigation management are similar to guidelines developed for Russet Burbank.

Strengths: Umatilla Russet produces high yields of marketable tubers with uniform specific gravity and good fry color from as low as 45 °F storage. It is resistant to most internal and external tuber defects, PVX, and common scab, and has moderate resistance to Verticillium wilt, tuber late blight infection, and net necrosis caused by the PLRV.

Weaknesses: Umatilla Russet occasionally exhibits pointed tubers as a result of stress and is susceptible to shatter bruise, which can promote associated dry rot infection of tubers.

Russet Varieties for Fresh Use (Figs. 3.7 & 3.8)

Goldrush

General Information

Parentage: ND450-3Russ x Lemhi Russet.

Developers: The North Dakota Agricultural Experiment Station.

Plant Variety Protection: Goldrush was protected via a plant patent (administered by the North Dakota State University Research Foundation).

Morphological Characteristics

Plant: Semi-erect, medium-large sized vine with light red-purple flowers.

Tubers: Oblong and block, with medium heavy golden russet skin, and well-distributed shallow eyes.

Fig. 3.7 Goldrush. (Photo credit: PVMI)

Fig. 3.8 Russet Norkotah. (Photo credit: PVMI)

Incentives for Production

Attractive fresh market russet and processing variety with early-yield potential, mid-maturity, good culinary quality, and resistance to hollow heart.

Agronomic Characteristics

Vine Maturity: Mid-season.

Yield Potential: Medium to high. Typically produces about 90% U.S. No. 1 tubers.

Specific Gravity: Medium; similar to Russet Norkotah.

Culinary Quality: Suitable for fresh pack and possibly for some early processing from the field. In sensory panels conducted at North Dakota State University, Goldrush rated similarly to Russet Norkotah and Russet Burbank for boiling, baking, and microwave cooking for flavor and mealiness. Tuber glycoalkaloids are low.

Diseases/Pests/Physiological Disorders: Low incidence of internal and external defects; resistant to hollow heart. It is moderately resistant to blackspot bruise. It has good resistance to common scab, and moderate tolerance of Verticillium wilt and silver scurf. Goldrush is susceptible to PVY and bacterial ring rot, showing good symptoms of both. It is also susceptible to early blight, late blight, soft rot, and Fusarium dry rot.

Storability: Stores well. Natural tuber dormancy is shorter than for Russet Burbank.

Key Management Considerations

Under non-irrigated conditions, U.S. No. 1 yields are comparable to Russet Norkotah, but substantially higher than those of Russet Burbank. Under irrigation, U.S. No. 1 yields of Goldrush are comparable to Russet Burbank. It is not well suited for processing into fries, although it can be used for such purposes with early harvest and limited storage.

Strengths: High early-yield potential and high pack-out, very white flesh, and excellent culinary quality, with some potential for processing out of the field. It has resistance to hollow heart and common scab.

Weaknesses: Occasional off-type tubers have been noted (hearts, twins), and a reddish blush on the apical end of the tuber may be observed after storage.

Russet Norkotah

General Information

Parentage: ND9526-4Russ x ND9687-5Russ.

Developers: North Dakota Agricultural Experiment Station.

Plant Variety Protection: Released prior to plant variety protection (PVP). Several line selections have been granted PVP (administered by varying entities).

Morphological Characteristics

Plant: Semi-erect, medium-sized vine with white flowers.

Tubers: Long and blocky, with medium to heavy golden russet skin and shallow eyes.

Incentives for Production

Russet Norkotah is suited for the fresh market. It possesses high-yield potential, produces a high percentage of U.S. No. 1 tubers, and tubers size early.

Agronomic Characteristics

Vine Maturity: Mid-season.

Yield Potential: Medium-high. Russet Norkotah produces a high percentage (>90%) of U.S. No. 1 tubers.

Specific Gravity: Similar to Russet Burbank.

Culinary Quality: Suitable for fresh packing. Formal taste and sensory panels have occasionally noted off-flavor associated with storage conditions. Tuber glycoalkaloids are low.

Diseases/Pests/Physiological Disorders: Low incidence of external and internal defects, though it will occasionally produce protruding eyes, and it is susceptible to hollow heart. Russet Norkotah is susceptible to foliar early blight, Verticillium wilt (line selections tend to be moderately susceptible), and the early dying complex, blackleg, PLRV, PVY (often producing few or very mild symptoms), PVX, bacterial soft rot, late blight, Fusarium dry rot, Pythium leak, pink rot, and silver scurf. It is resistant to tuber net necrosis associated with PLRV and moderately resistant to common scab, tuber early blight, and Rhizoctonia (black) scurf. It is moderately resistant to blackspot bruise. Russet Norkotah may exhibit a physiological symptom related to toxic seed piece decay when the seed piece disappears during hot periods around tuber initiation or shortly thereafter.

Storability: Stores well. Natural tuber dormancy is shorter than for Russet Burbank.

Key Management Considerations

Nitrogen requirements are similar to Russet Burbank unless stressful conditions are common or early dying pressure is present, in which case apply up to 120% of that needed by Russet Burbank. Apply all N preplant or make the last application by the time flowering is completed. Avoid over-irrigation late in the season. For irrigated production, at maximum evapotranspiration (ET) an irrigation interval of about 2.5 days is recommended.

Strengths: High early-yield potential, with a high percentage of U.S. No. 1 tubers.

Weaknesses: Susceptibility to early dying and PVY.

Long White Varieties for Processing (Figs. 3.9 & 3.10)

Alturas

General Information

Parentage: A77182-1 x A75188-3.

Developers: Released in 2002 by the USDA-ARS and the Agricultural Experiment Stations of Idaho, Oregon, and Washington, and represents a variety of the Pacific Northwest Potato Variety (Tri-State) Development Program.

Plant Variety Protection: Pacific Northwest Potato Variety Development Program (administered by PVMI).

Fig. 3.9 Alturas. (Photo credit: PVMI)

Fig. 3.10 Shepody. (Photo credit: Chelsey Lowder)

Morphological Characteristics

Plant: Large, semi-erect vine, with white flowers.

Tubers: Oblong, lightly russeted, with white flesh and a high set and smaller average size.

Incentives for Production

Alturas has very high yield potential, especially in areas with an extended growing season, and high specific gravity; it is used primarily for processing and dehydrated products. Alturas is resistant to Verticillium wilt and foliar early blight, moderately resistant to net necrosis caused by PLRV, and resistant to most internal and external defects. Alturas also has relatively low production input requirements, especially N, which are approximately 60–70% of the N requirements of Russet Burbank.

Agronomic Characteristics

Vine Maturity: Late to very late.

Yield Potential: High to very high.

Specific Gravity: High, with averages in western trials slightly higher or comparable to Ranger Russet.

Culinary Quality: Alturas has good sensory attributes for fries, as well as fresh-pack use, although fresh-pack use could be limited by the light russeting of tubers.

Diseases/Pests/Physiological Disorders: Resistant to Verticillium wilt and foliar early blight, which both contribute to early die, moderately resistant to net necrosis caused by PLRV, and resistant to most internal and external tuber defects.

Storability: At storage temperatures of 45 °F with no sprout inhibitor application, tuber dormancy averages 45 days less than that of Russet Burbank. Cold-sweetening resistant and can be stored at 42 °F for dehydrated processing and 45–48 °F for frozen processing.

Key Management Considerations

Nitrogen requirement is about 60–70% of Russet Burbank. In short season areas, all N should be applied pre-plant to allow tubers to mature by harvest. In longer season areas, split N applications can be used, but all N should be applied before July 31 to avoid delaying tuber maturity. Irrigation requirements are 15–20% higher than Russet Burbank.

Strengths: High yields and specific gravity; cold-sweetening resistant with the ability to be stored at colder temperatures than Russet Burbank and still provide acceptable processed and dehydrated products.

Weaknesses: Smaller tuber size with light russeting and short dormancy; late maturity, with late season N applications delaying maturing even further; greater water usage requirements than Russet Burbank.

Shepody

General Information

Parentage: Bake King x F58050.

Developers: Released in 1980 by Agriculture Canada in Fredericton, New Brunswick.

Plant Variety Protection: No. Public variety.

Morphological Characteristics

Plant: Medium-sized, spreading vine, light violet flowers with white tips.

Tubers: Long, white-skinned, white fleshed, sometimes with light netting on skin.

Incentives for Production

High early yields of larger tubers have made it useful for early fry processing directly following field harvest or from short-term storage.

Agronomic Characteristics

Vine Maturity: Early.

Yield Potential: High.

Specific Gravity: Low to medium; average of 1.083 in replicated trials conducted over 4 years in three sites in eastern Canada.

Culinary Quality: Good flavor for both processing and fresh use, with boiling and baking quality similar to that of Kennebec.

Diseases/Pests/Physiological Disorders: Reported as having moderate resistance to Rhizoctonia and Fusarium dry rot, and plant tolerance to heat stress. It is susceptible to common scab and should not be planted on acreage where common scab can be problematic. It displays poor visual symptoms of infection by PVY, making removal of infected plants difficult in seed acreage.

Storability: Limited—typically used for processing directly from the field or from short-term storage at temperatures of 50–55 °F to reduce tuber sugar accumulation.

Key Management Considerations

Applying excessive N rates, especially late in the growing season can produce specific gravities that are too low for processing. Shepody requires approximately 80% of the N required by Russet Burbank. Tuber greening can be an issue with the larger tuber size of Shepody, so careful hilling operations to ensure season-long coverage of tubers is important. Shepody is very sensitive to metribuzin.

Strengths: High early yields of tubers suitable for fry processing, resistance to hollow heart, and plants tolerant of heat stress.

Weaknesses: Common scab and metribuzin susceptible; poor visual symptoms of PVY infection in leaves can make removal of PVY-infected plants difficult for seed growers. Misshapen tubers and too large of size—careful monitoring of tubers during the growing season is required to mitigate.

Round White Varieties for Processing into Chips (Figs. 3.11–3.13)

Atlantic

General Information

Parentage: Wauseon x B5141–6 (Lenape).

Developers: Released in 1978 by the USDA-ARS-Beltsville.

Fig. 3.11 Atlantic. (Photo credit: PVMI)

Fig. 3.12 Dakota Pearl. (Photo credit: PVMI)

Fig. 3.13 Snowden. (Photo credit: PVMI)

Plant Variety Protection: No. Public variety.

Morphological Characteristics

Plant: Medium maturing with a medium-large, upright vine, pale lavender flowers.

Tubers: Round to oval, buff-colored with light netting and high specific gravity. May produce a large percentage of oversized tubers.

Incentives for Production

Suitable for processing into potato chips directly from the field and from short-term storage.

Agronomic Characteristics

Vine Maturity: Mid-season.

Yield Potential: Moderately high to high.

Specific Gravity: High.

Culinary Quality: Excellent chip quality when processed directly from the field or from short-term storage.

Diseases/Pests/Physiological Disorders: Susceptible to internal necrosis when grown in sandy soils under hot, dry conditions, as well as hollow heart.

Susceptible to late blight with medium susceptibility to early blight, black leg, and PLRV, as well as most potato viruses. Atlantic has medium susceptibility to soft rot, as well as common scab, dry rot, and pink rot. It is resistant to tuber net necrosis.

Storability: Tuber sugar content in Atlantic tubers readily increases when storage temperatures drop below 50 °F, so it is preferable to process directly from the field or within 3–4 months of harvest.

Key Management Considerations

With a 36-in. row spacing, plant seed pieces 9–11 in. apart. If intended for late harvest, use N application rates and timings that are appropriate for Russet Burbank. If intended for early harvest, use 20% less N and complete seasonal applications 4–5 weeks before the intended harvest date. Use weed control strategies that do not include metribuzin, or if metribuzin is applied, use minimum rates and incorporate with a minimal amount of water (less than 0.5 in.). When feasible, avoid storage by marketing directly from the field.

Strengths: Good yield potential and high specific gravity across environments. Atlantic is tolerant to early dying and is resistant to pinkeye, PVX, race A of the golden nematode, and tuber net necrosis.

Weaknesses: Tubers have short dormancy. Atlantic is susceptible to shatter bruise and associated storage rots, as well as hollow heart/brown center, internal heat necrosis (particularly in sandy soils in warm dry seasons), and is moderately susceptible to common scab and metribuzin herbicide injury.

Dakota Pearl

General Information

Parentage: ND1118–1 x ND944–6.

Developers: The North Dakota Agricultural Experiment Station.

Plant Variety Protection: North Dakota State University and the North Dakota Agricultural Experiment Station (administered by the North Dakota State University Development Foundation).

Morphological Characteristics

Plant: Semi-erect, medium-sized vine with white flowers.

Tubers: Round, bright white skin, and shallow eyes.

Incentives for Production

Good yield of bright white, smooth, uniform, round tubers. Produces light chips from the field and from 42 °F storage.

Agronomic Characteristics

Vine Maturity: Mid-season.

Yield Potential: Medium to high.

Specific Gravity: Medium; lower than for industry standards Atlantic and Snowden.

Culinary Quality: Suitable for both chip processing and tablestock. Chip color is light and uniform, with a low incidence of defects. Dakota Pearl is considered a cold chipping industry standard due to its ability to reliably chip from colder storage temperatures than other cultivars. It may also double as a tablestock cultivar due to attractive tuber appearance and acceptable baked, boiled, and microwaved taste panel ratings. Tuber glycoalkaloids are low.

Diseases/Pests/Physiological Disorders: Low incidence of internal and external defects; hollow heart has occasionally been noted in the northern plains and heat necrosis in production areas such as North Carolina. Dakota Pearl has no notable disease resistances or susceptibilities. It is susceptible to late blight (foliar and tuber), PVY, and bacterial ring rot (expressing typical symptoms for both foliage and tubers).

Storability: Stores well. It develops low levels of glucose. Natural tuber dormancy is shorter than for Snowden. Most chip processors will utilize by the March–April window.

Key Management Considerations

Planting physiologically young seed is important to avoid high tuber sets. It responds best when planted at soil temperatures above 45 °F. Plant spacing should be 12 in. for commercial production and 11 in. for seed. Proper hilling is required to avoid tuber greening. Dakota Pearl requires a rather high rate of N and supplemental foliar feeding during the growing season. It also will perform best with a consistent supply of water.

Strengths: Good yields of bright white skinned, uniformly sized tubers; low sugar accumulation in storage.

Weaknesses: Susceptible to heat necrosis and hollow heart occasionally noted.

Snowden

General Information

Parentage: Lenape (B5141–6) x Wischip.

Developers: University of Wisconsin.

Plant Variety Protection: No.

Morphological Characteristics

Plant: Large, vigorous, and semi-erect vine with white flowers.

Tubers: Round, slightly flattened tubers with a slight net. Eye depth is deeper than for Dakota Pearl, including on both bud and stem ends. Eye distribution is uniform.

Incentives for Production

High yield, produces light chips from the field and from 45 °F storage; long-term storage industry standard.

Agronomic Characteristics

Vine Maturity: Full season.

Yield Potential: High, similar to Atlantic.

Specific Gravity: High, similar, but slightly lower, than for Atlantic.

Culinary Quality: Excellent chip processing quality from long-term storage. Total tuber glycoalkaloids are acceptable—perhaps on the high side—and some note a slight aftertaste.

Diseases/Pests/Physiological Disorders: Snowden is less susceptible to soft rot, Fusarium dry rot, and bruising than Atlantic. It is susceptible to common scab, early blight, late blight, pink rot, and *Phytophthora nicotianae*. Snowden is resistant to Pythium leak.

Storability: Stores well. Dormancy is medium.

Key Management Considerations

Susceptible to metribuzin. Due to heavy set of tubers and long stolons, a 12-in. within-row spacing is recommended. Irrigation about every 2.5 days is recommended at maximum ET. Excessive late season irrigation may increase late-season hollow heart and has also been linked to internal brown spot. Proper mid-season irrigation minimizes common scab on tubers.

Strengths: Snowden is high yielding, with excellent chip processing quality; particularly from long-term storage and colder (45 °F) storage than many industry standard cultivars.

Weaknesses: Sets high so tubers are often undersized, some question about glycoalkaloid levels (on the high end of acceptable).

Specialty Varieties for the Fresh Market (Red Flesh) (Figs. 3.14–3.16)

Chieftain

General Information

Parentage: La 1354 x Ia 1027-18.

Developers: Released in 1966 by the Iowa Agriculture and Home Economics Experiment Station and the USDA-ARS.

Plant Variety Protection: No. Public variety.

Morphological Characteristics

Plant: Medium sized, spreading vine and light violet flowers.

Tubers: Oblong to round with medium-red skin and white flesh.

Incentives for Production

A higher yielding variety, Chieftain does well in differing production environments and is a good variety for fresh-pack.

Fig. 3.14 Chieftain (Photo credit: Chelsey Lowder)

Fig. 3.15 Norland (Photo: Chelsey Lowder)

Fig. 3.16 Red LaSoda (Photo credit: Chelsey Lowder)

Agronomic Characteristics

Vine Maturity: Medium.

Yield Potential: High.

Specific Gravity: Low—similar to other red-skinned varieties.

Culinary Quality: Good quality for fresh use with less after-cooking browning reported relative to Norland following boiling.

Diseases/Pests/Physiological Disorders: Resistant to stem end browning and net necrosis from infection of tuber by PLRV; moderate resistance to common scab.
Storability: Noted as having medium tuber dormancy.

Key Management Considerations

Chieftain is susceptible to metribuzin, so application should be applied prior to plant emergence. With a tendency to skin, it is recommended that tubers not be harvested directly from the field without prior vine kill to promote skin set.
Strengths: Chieftain is adapted to a wide range of environments. It is high yielding, has moderate resistance to common scab, and resistance to stem end browning and net necrosis from infection of tuber by PLRV.
Weaknesses: Tubers tend to skin during harvest and handling, with some susceptibility to growth cracks. Eyes can tend to be deep, depending on environment.

Norland

General Information

Parentage: Redkote x ND626.
Developers: North Dakota State University and the North Dakota Agricultural Experiment Station.
Plant Variety Protection: No.

Morphological Characteristics

Plant: Medium and spreading vine with a determinate growth habit and red-purple flowers.
Tubers: Oblong, slightly flattened, and smooth with red skin. Several darker-red skinned strains of Norland have been selected since Norland was released in 1957. Eye depth is shallow to medium and uniformly distributed.

Incentives for Production

Widely adapted, early maturing, several strains with darker-red skin color, high proportion of marketable tubers.

Agronomic Characteristics

Vine Maturity: Early.
Yield Potential: Medium with a high proportion of marketable tubers.
Specific Gravity: Low to medium.
Culinary Quality: Fresh market standard, suitable for boiling, mashing, potato salads, and soups particularly. Total tuber glycoalkaloids are low.
Diseases/Pests/Physiological Disorders: Norland has a low incidence of internal and external defects. It is moderately resistant to pink rot caused by *Phytophthora erythroseptica* and has a moderate response to common scab (less susceptible than Red LaSoda) and net necrosis associated with PLRV. Norland is susceptible to early dying including Verticillium wilt, PVY, PVX, PLRV, black leg, silver scurf, early blight, late blight, Fusarium dry rot, *Phytophthora nicotianae*, and Pythium leak.

Storability: Stores well, though dormancy is short, and it can be susceptible to pressure bruise if not properly hydrated. Skin color also fades in storage or may become brownish in color by silver scurf and/or black dot.

Key Management Considerations

Susceptible to metribuzin. Sensitive to moisture stress resulting from drought and excess moisture. If producing Norland under irrigation, it is recommended every 2.5 days at maximum ET. Avoid excessive irrigation as plants are senescing.

Strengths: Norland is widely adapted, produces early yields of uniform tubers, and has versatile cooking quality.

Weaknesses: Low specific gravity and low to medium yields if compared to late season cultivars, such as Red LaSoda and Red Pontiac. Susceptibility to silver scurf and black dot often results in brown tubers in storage. Red skin color fades in storage.

Red LaSoda

General Information

Parentage: Red LaSoda is a selected red-skinned mutant of LaSoda. Parentage of LaSoda is Triumph x Katahdin.

Developers: Released in 1953 by the USDA and Louisiana State University.

Plant Variety Protection: No.

Morphological Characteristics

Plant: Upright to spreading, medium-sized vine with light pink to lavender flowers.

Tubers: Round to oval, medium-light red skin, with deep eyes.

Incentives for Production

Red LaSoda is considered the standard for red potato varieties in many regions of the U.S. and is a recognized entity in the marketplace. It is a widely adapted cultivar that produces early high yields of attractive tubers. It is also adapted to winter production conditions. Vines hold up well in the heat, and production is good under stressful conditions. Good flavor and waxy texture make it a choice cultivar for boiling and salads.

Agronomic Characteristics

Vine Maturity: Medium-late, although tubers size early.

Yield Potential: High.

Specific Gravity: Moderately low, but higher than most modern red cultivars.

Culinary Quality: Good for boiling, making salads, and canning. Red LaSoda can be used for baking.

Diseases/Pests/Physiological Disorders: Red LaSoda is susceptible to PLRV, PVX, and PVY. Response to PVY is extreme, causing plant collapse and ultimately death. It is also susceptible to late blight, Fusarium dry rot, and common scab. Resistance to Verticillium wilt and early blight is moderate. Red LaSoda is less prone to swollen lenticels in wet soils than most red cultivars. It is susceptible to hollow heart, growth cracks, and general tuber malformation. Deep eyes become a quality issue in large tubers.

Storability: It is not typically produced as a storage potato. Tuber dormancy is medium to short, and tubers retain quality in storage if sprouting is controlled.

Key Management Considerations

For 36-in. row spacing, plant seed pieces 5–8 in. apart as a means to control size. Apply N fertilizers primarily preplant or early in the season to encourage a late-season N deficit that will improve tuber skin set. Avoid high levels of soil moisture late in the season to minimize swollen lenticels and enhance tuber color. Applications of low rates of 2,4-D herbicide are sometimes used to improve color. Kill vines when tuber size profile is optimal. Allow a 2–3 week maturation period after vine-kill for tubers to set skins.

Strengths: Red LaSoda is adapted to a wide range of production conditions. It produces high yields of tubers with good culinary quality when baked, canned, or used in salads. Variety recognition in the marketplace creates good early-season sales appeal.

Weaknesses: Tubers of Red LaSoda are susceptible to several defect problems, including deep eyes, general malformations, and growth cracks. Red color of the tubers tends to fade with time.

Specialty Varieties for the Fresh Market (Yellow Flesh) (Figs. 3.17 & 3.18)

Yukon Gem

General Information

Parentage: Brodick x Yukon Gold.

Developers: Northwest Potato Variety Development Program, including the USDA-ARS and the Agricultural Experiment Stations of Idaho, Washington, and Oregon.

Plant Variety Protection: Northwest Potato Variety Development Program.

Morphological Characteristics

Plant: Medium-sized, erect vine with medium red-purple flowers. Flowers tend to abort, and pollen production is limited.

Tubers: Round to oval; light yellow skin with a pink splash around the eyes. Eyes are intermediate in depth and number and are evenly distributed. Tuber set is low to medium, setting approximately two tubers more per plant than Yukon Gold. Tubers are medium in size, slightly smaller than Yukon Gold, on average.

Incentives for Production

Higher yield potential than Yukon Gold, with improved resistance to several diseases and physiological disorders. Additionally, it chips acceptably and has similar culinary quality when compared to Yukon Gold. The tuber size profile is slightly

Fig. 3.17 Yellow Gem (Photo credit: PVMI)

Fig. 3.18 Yukon Gold (Photo credit: PVMI)

smaller, and Yukon Gem produces more tubers per plant than Yukon Gold. Yukon Gem has potential for the organic market.

Agronomic Characteristics

Vine Maturity: Mid-season.

Yield Potential: Higher yield potential than Yukon Gold, with a smaller size profile.

Specific Gravity: Moderately low, averaging 1.075, compared to 1.085 for Yukon Gold.

Culinary Quality: Excellent baked, boiled, and microwaved. Tubers exhibit little or no after-cooking darkening. Yukon Gem produces acceptable potato chips; however, the low specific gravity may limit its use for processing. Total tuber glycoalkaloids are low.

Diseases/Pests/Physiological Disorders: Yukon Gem is notable for its PVY^{O} and tuber early and tuber late blight resistances. Additionally, it is resistant to PLRV net necrosis. It is moderately resistant to Fusarium dry rot and moderately susceptible to moderately resistant to foliar late blight. It is moderately susceptible to common scab and Pectobacterium soft rot. Yukon Gem is rated as susceptible to Verticillium wilt, foliar early blight, PVX, and PLRV (foliar infection). It has resistance to internal and external defects, including growth cracks, secondary growth, and hollow heart.

Storability: Stores well. Dormancy is medium.

Key Management Considerations

Optimal seed piece spacing for 36-in. wide rows is 9–11 in., while an 11–13 in. seed piece spacing should be used if an increased proportion of large tubers is desired. Adequate soil needs to be applied to the surface of the hill at final hilling to minimize the potential for tuber greening. Total seasonal N requirements for Yukon Gem are about 85–90% of Russet Burbank per cwt of yield produced. Typically, 50% of the seasonal N requirement should be applied by row closure, with subsequent in-season applications based on petiole nitrate concentrations. Vines should be killed 2–3 weeks before harvest to allow for proper skin maturation.

Strengths: Yukon Gem is a high yielding, round to oval yellow-fleshed cultivar with attractive appearance, good culinary quality, and chip potential. Yukon Gem has PVY^{o} resistance and is resistant to early and late tuber blight.

Weaknesses: Low specific gravity.

Yukon Gold

General Information

Parentage: W5279-4 x Norgleam.

Developers: Agriculture Canada and the University of Guelph.

Plant Variety Protection: No.

Morphological Characteristics

Plant: Medium-early maturing, erect, and medium-large to large-sized vine, with some tendency to spread at senescence. Yukon Gold has a determinate growth habit and light red-purple flowers.

Tubers: Slightly oval and flattened, smooth tubers with yellow skin, and shallow pink eyes. Eyes are few and not uniformly distributed.

Incentives for Production

Yukon Gold is widely adapted, with medium-early maturity, attractive appearance, yellow flesh color, and excellent culinary quality.

Agronomic Characteristics

Vine Maturity: Medium early.

Yield Potential: Medium with a high proportion of marketable tubers.

Specific Gravity: Medium (about 1.085 across most northern tier states).

Culinary Quality: Fresh market standard, suitable for baking, boiling, mashing, potato salads, soups and stews; can be used for specialty fries out of the field. Total tuber glycoalkaloids are low.

Diseases/Pests/Physiological Disorders: Yukon Gold is moderately resistant to PLRV. It is susceptible to PVY, common scab, early blight, late blight, silver scurf, black scurf, Fusarium dry rot, and soft rot, and reportedly is tolerant of PVX. It is resistant to net necrosis associated with PLRV infections. Yukon Gold is susceptible to air pollution (ozone), hollow heart, and internal heat necrosis.

Storability: Stores well, with medium to long dormancy. Care should be taken at harvest to minimize bruising, as Fusarium dry rot may be problematic. Storage should be monitored for soft rot development, as Yukon Gold has fairly large-sized lenticels and may be predisposed to swelling if excessive late season irrigation or rainfall at harvest time.

Key Management Considerations

Tubers have few eyes that are not uniformly distributed; thus, using whole (single-drop) seed is often recommended (as well as warming to aid in dormancy break) for better stands. Additionally, due to low eye numbers, using whole seed will increase stem numbers resulting in more tubers per plant and minimizing oversizing, which often results in increased occurrence of hollow heart. Good hill conformation is also important to minimize greening, as tubers are often set high in the hill. Irrigation approximately every 2–3 days is recommended at maximum ET. Excessive late season irrigation may increase soft rot in storage due to enlarged lenticels, or tubers may appear to have freckles, minimizing their attractiveness for fresh marketing. Producers should monitor tuber size to minimize oversized tubers and associated hollow heart.

Strengths: Yukon Gold has high early yields of attractive round to oval, yellow-fleshed tubers. Culinary quality is excellent.

Weaknesses: Sets high in the hill so tuber greening may be an issue. Due to low tuber numbers per stem and rapid tuber bulking, oversized tubers may result if not monitored. Hollow heart has been associated predominantly with oversized tubers.

Chapter 4
Seed Potato Production and Certification

Alan Westra, Phillip Nolte, Jonathan L. Whitworth, and Jenny Durrin

Contents

A. Westra (✉)
Idaho Crop Improvement Association, Inc., Idaho Falls, ID, USA
e-mail: awestra@idahocrop.com

P. Nolte
Department of Entomology, Plant Pathology and Nematology, University of Idaho, Moscow, ID, USA

J. L. Whitworth
Agricultural Research Service, USDA, Aberdeen, ID, USA
e-mail: jonathan.whitworth@usda.gov

J. Durrin
CALS, University of Idaho, Moscow, ID, USA
e-mail: jsdurrin@uidaho.edu

J. C. Stark et al. (eds.), *Potato Production Systems*,
https://doi.org/10.1007/978-3-030-39157-7_4

Introduction

Successful commercial potato production is highly dependent upon a consistent supply of quality seed. While the production and management of seed potatoes is very similar to commercial crop production, a major difference is that seed potato producers focus on the production of a crop that meets specific quality (purity and phytosanitary) standards. Seed production, therefore, can be considered a specialized sector of the potato industry. The vast majority of seed potatoes are produced within certification programs that define those quality standards, and most major potato production areas have laws requiring that commercial potato producers plant certified seed potatoes. This chapter will discuss seed potato production within the context of seed certification programs.

Seed Certification

Essentially, seed certification is a quality control program. Seed certification differs from private quality control programs in that it is an independent, third-party certification conducted by an official certification agency. Official certification agencies are designated by statute or regulation, and these same laws grant protection to the terminology and indicia of official certification. Internationally, seed certification is conducted at the federal level. In the U.S., authority to conduct official seed certification resides at the state level and is conducted by state Departments of Agriculture, land-grant universities, and/or grower associations.

Regardless of the specific agency conducting certification, seed potato certification is a process that consists of the following basic elements:

1. Approved Planting Stocks Only documented planting stocks of known origin and that meet the required purity and phytosanitary standards are permitted for the production of certified seed.

2. Limited-Generation System Seed potato production is performed under a scheme in which the number of generations of greenhouse and field increase is limited. Limited-generation systems typically include a classification system based on disease levels and require a flush out of seed when disease tolerances or the maximum number of field generations has been exceeded (Table 4.1).

3. Inspections Seed potatoes are subject to inspection at all stages of production. The inspection regime will depend upon the stage of production and is performed on a visual basis according to prescribed methods. Visual inspections may be supplemented and/or confirmed by laboratory testing.

4. Post-Harvest Testing The harvested crop is subject to post-harvest testing for viruses and other factors. Post-harvest testing may consist of an off-season grow-out, laboratory testing, or some combination thereof.

5. Grade Inspection Seed potatoes are inspected at the shipping point to ensure conformity with defined seed potato grades. These grades are based on the U.S. No. 1 grade for seed potatoes.

Table 4.1 Seed potato generation sequence

Generation	Usual source material	Production facility
Pre-nuclear	In vitro plantlets	Laboratory
Nuclear	In vitro plantlets, micro-tubers, stem cuttings	Greenhouse
FG-1 (field generation 1)	Greenhouse mini-tubers	Early-generation seed producer
FG-2	Field-grown tubers	Early-generation seed producer
FG-3	Field-grown tubers	Certified seed producer
FG-4	Field-grown tubers	Certified seed producer
FG-5	Field-grown tubers	Certified seed producer
FG-6	Field-grown tubers	Certified seed producer
	Commercial potato production	

Seed Potato Production

Seed potato production consists of a series of sequential increases of approved planting stocks intended to provide the commercial potato industry with sufficient quantities of seed meeting appropriate disease tolerances and purity standards. This is accomplished through the combined efforts of public and private sources and involves laboratory, followed by greenhouse, and then field production. The terms pre-nuclear and nuclear are used to refer to laboratory and greenhouse stocks in some systems. This process may be vertically integrated and involve all stages of production from variety development to commercial production. More commonly, however, individual operations specialize in specific stages of production; e.g., laboratory and/or greenhouse production. Another common area of specialization is in the production of early-generation seed. In this case, a seed operation will perform one to two field increases before selling seed to another seed operation for further increase or to a commercial potato grower.

Introductory Materials

A basic requirement for the production of certified seed potatoes is that the planting stocks originate from pathogen-tested in vitro materials (Fig. 4.1a, b).

Thus, the first stage of production involves the introduction of parent material into tissue culture. This stage of production is performed in laboratories capable of maintaining aseptic conditions and requires the use of specialized equipment. The parent material, usually tubers, is surface-sterilized and used as a source of meristems or nodal cuttings for introduction into tissue culture. Once successfully established as in vitro material, the resultant plantlets are tested for pathogens specified in seed certification rules, typically potato viruses A, M, S, X, Y, potato latent virus, potato leafroll virus (PLRV), potato mop top virus (PMTV), potato spindle tuber viroid (PSTV), tobacco rattle virus (TRV), *Pectobacterium* spp., and *Clavibacter sepedonicus*. Tissue culture plantlets testing positive for any of these pathogens either are discarded or subjected to a combination of chemo- and thermotherapy for pathogen elimination. This cycle of testing and therapy is continued until the plantlets test free of the pathogens in question. Finally, greenhouse and/or field growouts are conducted to assess trueness to type and ensure that the in vitro plantlets retain the characteristics of the original parent material.

Tissue culture materials are subject to inspection and testing by the certification agency. The above-described pathogen testing is required every time any parent material is introduced into tissue culture. There is a zero-tolerance for the specified pathogens in tissue culture plantlets intended for certified seed potato production. Only when the tissue culture material tests negative for the pathogens prescribed in seed certification regulations and is approved by the certification agency, can it used for certified seed production in the greenhouse or field.

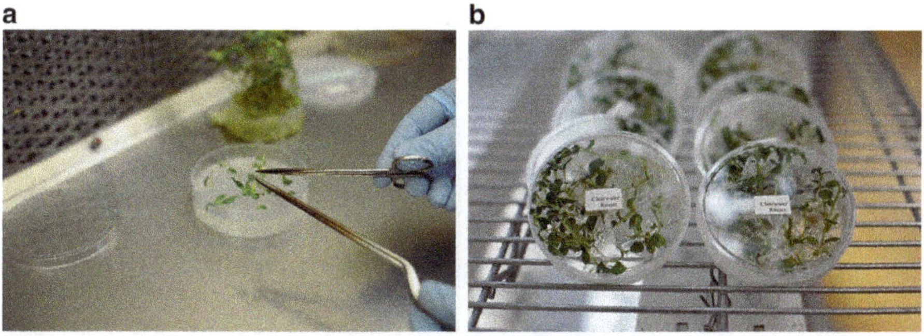

Fig. 4.1 (**a**) Dividing potato seedling into single-leaf cuttings for tissue culture production. (**b**) Tissue culture plantlets, Clearwater Russet. (Photo credit: Jenny Durrin, University of Idaho)

While the vast majority of certified seed originates from pathogen tested, in vitro planting stocks, there are instances where it is desirable to produce certified seed of material for which these planting stocks do not exist. Specific examples of this can include breeding lines, obsolete varieties, and selections made from existing varieties. Available planting stocks may be limited to true potato seed or tubers. Seed certification rules do make allowances for these special circumstances by requiring pretesting of this type of planting stock and designating its progeny in a way that differentiates it from other certified seed; e.g., the use of an "Experimental" class.

Nuclear Material

The next stage in certified seed potato production is the production of nuclear material. Most typically, this involves the production of mini-tubers in a greenhouse utilizing pre-nuclear in vitro plantlets as planting stocks. Less frequently, true seed, plant cuttings, or in vitro micro-tubers serve as planting stocks.

The amount of pre-nuclear material available is usually limited, and one or more laboratory increases of the so-called "mother" plantlets is required prior to greenhouse production. These increases are subject to inspection and testing by the certification agency. Seed certification rules typically require pathogen testing of the basal portion of a percentage (e.g., 1%) of these mother plantlets to ensure that the planting stocks are clean. These testing requirements are often less extensive than at the introductory level and usually include the following pathogens: potato viruses A, X, Y, PLRV, *Pectobacterium* spp., and *Clavibacter sepedonicus*. As with the original pre-nuclear material, there is a zero tolerance for these pathogens in the in vitro stocks used for the production of nuclear material.

After multiplication and sufficient growth, tissue culture plantlets are transplanted into the greenhouse (Fig. 4.2).

Logistically, growers are able to produce a greenhouse crop of mini-tubers every 3–4 months. However, most greenhouses typically produce one or two crops (i.e., a

Fig. 4.2 Mini-tuber production occurs in the controlled environment of a greenhouse. (Photo credit: Jonathan Whitworth)

spring and/or a fall crop) per year because of the cost constraints and lower yields associated with winter production. The specifics of the management of the growing crop are dependent upon the system of culture used. However, because the mini-tubers will serve as the basis for future generations of field-grown seed, good sanitation and pest exclusion are critical to the successful production high-quality mini-tubers:

- Greenhouses should be insect-proofed. All intake and exhaust openings should be covered with an aphid-proof mesh hardware cloth. Doors should remain closed at all times; a double-door entry system will minimize insect entry (Fig. 4.3a, b).
- Greenhouses must be thoroughly cleaned and disinfected before planting. Organic residue; e.g., plant debris and soil, should be removed prior to washing and treatment with disinfectant.
- Pots or bedding areas should be cleaned and sanitized prior to use. A suggested method of disinfection is to wash the pots in a concentrated solution of laundry detergent (one-half cup in five gallons of water), rinse in clean flowing water, and finally dip briefly into a 2% solution of sodium hypochlorite (Clorox™ or similar product) or a prescribed solution of chlorine dioxide (Oxidate™) or quaternary amine (Sanitol™). If an open-bed system is used, growers should wash and disinfect the bed. This can be done by thoroughly steam cleaning the beds until no organic residue remains, then spraying to complete wetness with a disinfectant solution, such as those described for cleaning pots.
- Precautions must be in place to assure that the mini-tubers are produced under sanitary growing conditions:
  - Entry should be restricted to authorized personnel only.
  - Personnel entering or working in greenhouses should never do so after spending time in cellars or fields.
  - Footbaths (a shallow pan or tray containing a disinfectant solution about 1-in. deep) should be placed near the entry of each greenhouse. An effective and inexpensive alternative is to place disposable boots at each entry point for use by anyone entering the facility.
  - Tools should be sanitized between planting units and between greenhouses.

a b

Fig. 4.3 Optimum growing conditions for nuclear seed production require screened openings (**a**) and controlled entry (**b**). (Photo credit: Jonathan Whitworth)

Production Systems

Greenhouse production most commonly involves the use of soilless potting mixes; e.g., Peat-Lite™. These mixes typically contain peat and perlite or vermiculite to aid in drainage and usually a small charge of starter fertilizer. Pots, soil bags, or raised beds may all be used in these systems. The soilless mix can be prepared by the grower and pasteurized by heating for 24 h at 240 °F. Alternatively, premixed, commercial potting mixes may be purchased. These mixes should also pasteurized before use. Some growers are moving away from organic soilless media, such as peat, because it is a potential inoculum source for pathogens. Sterilization can be a labor-intensive process, and as an alternative, pure perlite or pumice could also be used (Fig. 4.4).

To avoid the potential introduction of plant pests, soil or compost should never be used in the potting mix. Further, the potting mix should never be reused for subsequent crops.

The choice of whether to use pots, bags, or beds will depend upon a number of factors, including the design of the greenhouse. It must be noted that the choice of container does influence both yield and tuber size. In evaluations conducted at the University of Idaho's Tetonia Research and Extension Center, 6-in. diameter pots gave the optimal number of tubers while also limiting tuber size. Similar results were achieved by using 8–10-in pots with 2–3 plantlets per pot. If this system is used, additional soil should be added to the pots as the plantlets grow to allow for more subsurface stolon development and an increase in tuber numbers. If larger tubers and higher yield per unit area are desired, potato producers should consider growing transplants in raised beds. Experimentation with the potting system under a grower's own conditions is essential to allow the best use of resources to get the desired yields. For example, use of square pots reduces wasted space and water,

Fig. 4.4 Mini-tuber production using perlite as a growth media. (Photo credit: Jenny Durrin, University of Idaho)

especially if watering is done by an automatic overhead system. If size is not as important as the number of tubers needed overall, then different configurations of pot shape and size can be used to determine the best combination.

Increasingly, mini-tubers are produced using aero- or Nutrient Film (NFT) Hydroponic production systems. These systems can have higher initial setup costs when compared to conventional systems using soilless potting mixes, are technically more challenging, and require personnel with specialized training. These systems have, however, demonstrated to produce higher yields on a per-square-foot basis, because precise control over the timing of nutrient applications and the amounts administered can be achieved.

Also, because NFT systems allow for the continuous harvest of mini-tubers, tuber size can be more precisely controlled by the grower (Fig. 4.5).

Transplanting

Maintaining sanitary conditions during the transplanting operation will mitigate the risk of potential contamination of the planting stocks. Disposable gloves should be used during transplanting and should be changed frequently; minimally between varieties or lots, preferably between individual plantlet containers. Weak plants and containers that are contaminated with fungal or microbial growth should be discarded. Transplant shock can be avoided by hardening the plantlets prior to transplanting (Fig. 4.6).

If this is not feasible, transplanting late in the afternoon or on a cloudy day can help to reduce transplant shock. It is necessary to moisten pots or open beds before

Fig. 4.5 (**a**) Nutrient Film Technique (NFT) mini-tuber production. (**b**) Harvesting NFT mini-tubers. (Photo credit: Jenny Durrin, University of Idaho)

Fig. 4.6 Hardening of plantlets in the intended growth media or an alternative, such as rock wool under a dome, can decrease the chance of transplant shock. (Photo credit: Jenny Durrin)

transplanting. When transplanting is complete, the watering system should be adjusted to ensure adequate moisture for the plants while at the same time avoiding overwatering. Additionally, clear plastic cups with a ventilation hole can be placed over each transplant to provide a high humidity environment as the plant acclimates.

Fertilization

Fertilizers that contain both macro and micronutrients are necessary to maintain good growth of the transplants. Production systems utilizing soilless potting mixes typically use a combination of a slow-release fertilizer; e.g., Osmocote™, applied at soil mixing and a readily soluble fertilizer applied through the irrigation system. A standard formulation is 26–16–8 (NPK), plus micronutrients and calcium. The need for fertilization is minimal until stolon initiation, after which the fertilizer requirements increase because of demands on the plant associated with tuber development. Fertilizer applications should be scheduled based on petiole nitrate-nitrogen (N) levels. Normal levels of petiole nitrate-N in the greenhouse are lower than what is normal in the field. Optimal levels are near 12,000 ppm until stolon formation. During early tuber development, the petiole nitrate-N levels can be raised to about 18,000 ppm but should then be allowed to drop slowly until tubers are about 1/2 oz. in size. At this point, no additional fertilizer should be applied. This fertilization schedule aids in vine maturation, skin set, and tuber storage. If early applications of nitrogen are limited, plant height can be restricted to about 14 in. Careful attention to plant size can also help reduce large canopies that favor high humidity and greater chance for fungal disease establishment. See Chaps. 9 and 11. Seed certification standards for nuclear material are extremely stringent. The maintenance of an insect-free environment in the greenhouse, especially aphids, is critical in preventing the introduction and spread of viruses, such as potato virus Y (PVY). Plants should be inspected regularly for insects, and the placement of "sticky traps" can be useful in monitoring greenhouses. A preventative insect control program, including the application of a systemic insecticide during soil mixing or just after planting is recommended. Foliar diseases can also be an issue in greenhouse production. Disease scouting and the regular application of protectant fungicides are recommended. If foliar diseases are of concern, a drip irrigation system should be considered. These systems work well, especially during cooler months, because they reduce foliar wetting and the potential for foliar diseases.

Harvest and Storage of Pre-Nuclear Tubers

To prepare for harvest, greenhouse-grown plants are often artificially killed. Vine kill may be performed chemically or mechanically, and watering schedules may be manipulated to accelerate vine maturity while controlling tuber growth. Whatever method is used, the timing of the vine killing procedure is based on tuber size and is done after certification inspections are completed. If vines are killed before harvest, growers should remove the tops and leave the crop in the pots until the tubers are mature. Instruments used to remove vines should be dipped into a disinfecting solution between each plant or test unit. The harvesting process can be simplified by dumping pots onto an expanded metal screen, sifting through the potting soil, and separating the tubers.

Tubers can be stored in any kind of open mesh bags at 39 °F and a relative humidity of 95% until the following spring. If mini-tubers are green dug, they should be cured for 2 weeks at 55–60 °F before cooling to the final storage temperature. Seed from late fall or winter greenhouse crops and those from later harvests of aeroponically produced mini-tubers may express dormancy beyond planting time. Pre-warming of this seed to 55–60 °F for several weeks before planting and/or treatment with gibberellic acid may be required for dormancy break and proper stands in later field plantings.

Certification of Greenhouse Crops

Greenhouse crops are subject to inspection and testing during production and after harvest. Normally, two inspections are performed while the crop is actively growing in the greenhouse. The first inspection is usually performed when the plants are at least 12-in tall; the second shortly before vine kill. Factors that are considered during these inspections include identification and isolation of individual seed lots; overall condition of the crop; and the presence of insects, weeds, and disease. It is common to collect leaf samples during the second inspection; typically 2% of the plants in the crop are tested for potato virus X, Y, A and PLRV (Fig. 4.7).

After harvest, inspectors gather a sample of tubers, equivalent to 1% of the crop, that are tested for *C. sepedonicus* (bacterial ring rot) and *Pectobacterium* spp. (bacterial soft rot). Finally, mini-tubers to be sold are subjected to inspection at the shipping point to ensure conformity with seed potato grades. Mini-tubers meeting the requirements prescribed in the certification standards are eligible for further production of certified seed potatoes.

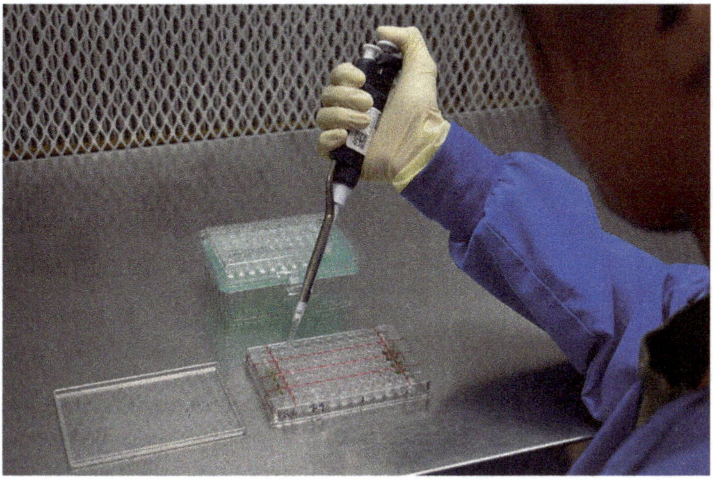

Fig. 4.7 Laboratory testing for potato virus X, Y, A, and PLRV. (Photo credit: Phillip Nolte)

Field Production of Certified Seed Potatoes

As noted above, field production of certified seed potatoes involves many of the same practices employed in the production of commercial potatoes. Certified seed potato production practices differ in that they must take into account the necessity of meeting seed certification requirements. In general, successful production of seed potatoes requires increased attention to detail and higher inputs than does the production of commercial potatoes. Good seed potato production practices generally require a sacrifice of total yield to produce a crop of the required quality and tuber size. Key areas in which differences between seed and commercial potato production occur include:

1. Identity Preservation The basic unit of certification is a seed lot. A seed lot is usually comprised of a single generation of each variety of seed potatoes produced. Seed lots may range in size from a few plants to 100 acres or more. Each seed lot is given a unique identifier; i.e., a certification number, which is used to track the results of inspection and testing and to establish a pedigree. Certification standards require proper identification and strict physical separation of seed lots at all stages of production to prevent admixture and mitigate the spread of disease. Degradation or loss of identity at any stage of production can result in a significant financial loss due to downgrading or rejection of a seed lot from certification.

2. Planting Stocks Only planting stocks that meet the minimum requirements of certification rules are eligible for the production of certified seed potatoes. Typically, planting stocks are limited as to maximum tolerances for disease and admixture, maximum generation, and grade requirements. Eligibility of planting stock is documented by official tags or certificates issued by a certification agency. This documentation must be provided to the certification agency when application for certification is made.

3. Sanitation Sanitation is of paramount importance to the production of quality seed potatoes. All surfaces with which seed potatoes come in contact must be cleaned and disinfected using recommended practices and products to mitigate the potential for disease spread. Ideally, all handling equipment is disinfected between each unit, lot, or variety of seed. At the beginning and end of each operation, all harvesters, truck beds, storages, cutters, planters, and handling equipment should be thoroughly cleaned and decontaminated. The practice of frequently cleaning equipment will help control the spread of bacterial and fungal diseases and may also help control spread of some virus diseases.

4. Field Restrictions and Rotation Requirements In order to prevent disease spread and admixture, certification rules place restrictions on the choice of fields used for planting certified seed. Minimally, certified seed production is not permitted in fields that were planted to potatoes in the previous season. In cases where specific diseases are found on farms, the rotation requirement may be increased to as long as 3 years (e.g., bacterial ring rot), or may be prohibited entirely (e.g., Columbia root knot nematode, corky ring spot disease).

There are isolation requirements placed upon fields used for certified seed production. When more than one seed lot is planted in a field, the seed lots must be physically separated by a blank row or a crop other than potatoes. Also, restrictions are placed on the proximity of seed and commercial potato fields. Usually a minimum distance between seed and commercial fields is specified; an extreme example occurs in Idaho, which has designated seed production areas where commercial potato production is prohibited.

5. Storage Requirements Certification standards place restrictions on the storage of seed potatoes. It is common for certification rules to include sanitation inspections and preapproval of storages prior to harvest. These rules also prohibit the use of storages in which sprout inhibitors have been used, the storage of certified seed with commercial potatoes, or the storage of seed lots that are infested with specified diseases; e.g., bacterial ring rot, Columbia root knot nematode.

6. Inspection and Testing of the Seed Crop In addition to routine scouting of the crop performed by the grower or crop consultants, seed potatoes are subject to inspection and testing by the certification agency. A typical inspection regime will include a minimum of two field inspections and inspection of the harvested crop, post-harvest storage inspections, post-harvest testing, and a grade inspection at the shipping point.

7. Field Inspections are performed according to prescribed procedures, including minimum plant counts, and are intended to assess conformity with seed certification field standards. Although the focus of each field inspection may differ, they will each include examination for varietal identity and purity, virus diseases (leafroll and mosaic), bacterial diseases (bacterial ring rot, blackleg), and other miscellaneous factors. Field inspections may be supplemented by laboratory testing; examples of supplemental testing include confirmatory testing of suspect plants and routine screening of early-generation seed lots for potato virus X (PVX) (Fig. 4.8).

a b

Fig. 4.8 (**a**) Field inspection of certified seed potato field. (**b**) Plants with suspected disease are flagged. (Photo credit: Phillip Nolte)

8. Storage Inspections are performed at harvest or shortly thereafter. These inspections focus on proper storage conditions, including required lot identification and physical separation. Other factors that may be considered during these inspections include admixture and diseases, such as bacterial ring rot.

9. Post-harvest testing is required prior to final certification and may include a post-harvest grow out (field or greenhouse), laboratory testing, or some combination thereof. Currently, most certification programs conduct post-harvest grow outs of grower-submitted samples in FL or HI. Post-harvest testing is used to estimate virus disease (mosaic and leafroll); varietal mixture; and other factors, such as herbicide damage. The estimates obtained in the post-harvest test will determine eligibility of a seed lot for both additional seed production (recertification) and certification for commercial seed sales.

10. Shipping Point Inspection A grade inspection at the shipping point, usually performed by the Federal-State Inspection Service, is the final step in the certification process. Seed potato grades are based on the USDA-Agricultural Marketing Service standards and focus on quality issues, such as size and specified defects. The final grade is indicated by tag color, with blue tags representing the highest grade (Fig. 4.9).

Fig. 4.9 A tag attached to a container of seed potatoes assures the buyer that the contents meet quality standards. Colors represent different grades. (Photo credit: Phillip Nolte)

Cultural Practices for the Field Production of Seed Potatoes

Cultural practices specific to seed potatoes are implemented to ensure seed quality, including minimizing disease, appropriate size profile, and meeting grade requirements. The following are practices that are routinely used to meet these requirements.

Isolation

As noted above, there are minimum physical isolation requirements mandated by seed certification rules. However, there are also several other recommended forms of isolation that can be employed by seed growers to mitigate disease risk.

- *Temporal isolation* of seed crops can be achieved by following the standard recommendation to "plant early and kill early." In this case, the goal is to limit the total potential exposure to insects that vector disease, especially aphids, which are vectors of PVY and PLRV. This form of isolation necessarily involves a tradeoff between limiting potential disease exposure and obtaining proper seed size and maximum yields.
- *Physical isolation* beyond that required by certification rules can be useful in controlling the spread of viral diseases, including PLRV, PVY, and PVA. Isolation of seed fields from other potatoes using a distance of at least 1/4 mile will mitigate risk of infection with PVY and PVA. Isolation for control of PLRV may require distances of one mile or more. The use of green border crops, such as spring-planted winter wheat, has been shown to be effective at slowing spread of viruses into seed fields. Mesh row covers have also been effective when conditions permit their use (Fig. 4.10).
- *Intergenerational isolation*, both in the field and in storage, is often overlooked as a disease control strategy. When possible, plantings should be grouped by generation. Early-generation seed should be isolated from all other potato fields by the greatest distance possible (two or more miles is best). This is especially critical for controlling the spread of PVY and PVA in varieties that are susceptible to those viruses. Similarly, field operations, such as roguing and spraying, should begin with early-generation fields and proceed in order of increasing generation. Proper sanitation practices should be employed when moving from one generation to another.

Where feasible, potato growers should store each generation of seed or at least early-generation seed in a completely separate facility. Storage of several generations of seed in any building with a common air system may result in failure to break the disease cycle and may also result in contamination of early-generation seed with fungal and bacterial pathogens.

Fig. 4.10 Insect-proof mesh row cover can prevent spread of viruses into valuable early-generation seed potatoes. (Photo credit: Jill Randall)

Selection of Planting Stocks

True seed, tissue culture plantlets, mini-tubers, and field-grown tubers can all serve as planting stocks for field production of certified seed potatoes. Planting stocks are selected based on the intended purpose or market of the resultant crop; e.g., early-generation seed, recertification, or commercial seed production. Regardless of the type of planting stock used, care must be taken to select the highest quality stock that is available. Identifying a source of quality seed for multiplication is paramount for success, as any disease problems present in early generations of seed likely will be magnified in later generations. Planting seed with disease problems can result in failure to meet certification with that particular lot and can also jeopardize an entire seed operation. Certification records, including a North American Certified Seed Potato Health Certification (Fig. 4.11), should be obtained to ensure that the generation and disease status of the planting stocks are suitable for the intended purpose. If possible, the buyer should also inspect the field and storage facility of the source of seed to be purchased.

NORTH AMERICAN CERTIFIED SEED POTATO HEALTH CERTIFICATE - CROP YEAR 2019

	Grower	Importer
Name	Seed Potato Producer	
City, State/Prov.	Anytown, ID	

Variety Russet Burbank **Acres** 85 **Quantity Shipped**

Size

Lot Certification

Certification #	19-101	**Lot origination from tissue culture**	No	Yes X
Seed Class/Gen.	FY4	Year micropropagated for planting		2015
Certifying State/Prov.	ID	by	Tissue Culture Lab	

Production environment pedigree: Fill 1 column per production year, use different initials in Greenhouse and Field boxes for different farms (e.g. JSF for John Smith Farms); indicate a tuber-united lot with a '+" after farm initials; describe other footnotes in notes below.

2012	*2013*	*2014*	*2015*	*2016*	*2017*	*2018*	*2019*	Year of Production
			Pre-nuclear					Greenhouse (insect excluding) & sterile soil
				FY1	FY2	FY3	FY4	Field (note special measures below)
			PN-15-10	16-150	17-189	18-200	19-101	Certification No.
			ID	ID	ID	ID	ID	Certifying State

Number of years produced in field soil: 4

Post harvest readings

Location: N/A

Summer Field Readings

1st	2nd	3rd	Final		FINAL
0.0000%	0.0000%	N/A	N/A	%LEAF ROLL	0.75%
0.0000%	0.0000%	N/A	N/A	%MOSAIC	0.00%
			N/A	%VARIETAL MIXTURE	N/A
		Less Than	N/A	%BLACKLEG	
		Less Than	N/A	%VERT + %FUSARIUM + %EARLY BLIGHT	

Sample No. 19-101

Plant Count 400

Winter ELISA Test Results

%PVY N/A %PVX N/A

BRR Test Results NEGATIVE

Other Diseases	Not known to occur in growers area	No. of years since last found on this grower's farm, or NONE ON RECORD if free > 5 years	Not found this year during normal certification field inspections
Bacterial Ring Rot		NONE ON RECORD	
Late Blight			X

Notes: An official (ICIA) sample of 400 tubers tested negative for Clavibacter michiganensis subsp sepedonicus using CelA primers described by Gudmestad et al (Plant Disease 93:649-659).

The above information is accurate to the best of our knowledge:

Program official / title — Date: 2/3/2020

Agency: Idaho Crop Improvement Association — Telephone: 208-522-9198

FAX: 208-529-4358

Approved for use by the Certification Section of the Potato Association of America

Fig. 4.11 North American Certified Seed Potato Health Certification https://cvp.cce.cornell.edu/submission.php?id=253

Planting Stocks for Early-Generation Production

Mini-tubers are the most common type of propagative material for the production of first field generation seed potatoes. Mini-tubers may be planted by hand or with mechanical planters. In either case, it is critical to use strict sanitary practices, and disinfection of equipment between seed lots is essential. Mini-tubers tend to have longer dormancy than field-grown tubers and will tend to emerge slowly; therefore, mini-tubers are planted as early as possible to take advantage of the full growing season. Dormancy issues can be mitigated by prewarming and/or treatment with gibberellic acid. If dormancy has been properly broken, tubers planted 4 in deep will emerge in 3–5 weeks, depending on the size of the mini-tuber. Mini-tuber sizes from 0.25–1.0 oz. produce acceptable results. When mini-tuber size is larger, they can be still be planted as whole mini-tubers. Seed cutting should be avoided to mitigate the potential for disease spread.

True seed (TPS) and tissue culture plantlets are much less frequently used as planting stocks for field production. Both may be planted directly in the field. However, TPS is difficult to plant because of its small size, and a so-called "field plantlet introduction" using tissue culture plantlets entails greater risk due to the potential for high plant mortality. In both cases, better results are obtained when the materials are started in the greenhouse; e.g., in 2 × 2-in cell packs. When the plants have reached 6–7 in, they should be hardened and can then be hand or mechanically transplanted in the field. During transplanting, care should be taken to minimize root damage, and appropriate sanitation measures should be taken. A suggested transplanting method involves 6–7 in tall plants that are planted about 5 in deep, leaving just the top leaves out. Irrigation should be applied immediately after transplanting, followed with light, frequent irrigations until the plants become established. No herbicides or systemic insecticides should be applied during planting because transplants may be susceptible to chemical injury. Row covers can be used to provide added protection from both wind and freezing temperatures for transplants (Fig. 4.10).

After the transplants are established, potato producers may cultivate the field and create hills. At this time, applications of fertilizer and insecticides can be made using side dressing or chemigation. Weed control should be done by hand, with cultivation, or with light amounts of herbicides that are safe for post-emergent use. Once the plants are completely established and 8–10 in tall, further mechanical operations should be avoided to minimize disease spread. The crop should be managed as any other seed potato crop in terms of irrigation timing and frequency.

Later Generation Increases

Tubers harvested from first-year production serve as the basis for subsequent seed increases. Although seed potatoes may be increased for five or more field generations, increases typically cease after the fourth field year, at which time the seed is sold for commercial potato production. Regardless of the generation produced, proper seed health and grade, including size profile, are important in the selection of these planting stocks. Except as noted below, the handling and planting of this seed is similar to that of commercial potatoes.

Units of Production

Uniting; i.e., establishing discrete, identifiable groups of plants is important in seed production. Beyond the minimum requirements specified in seed certification rules, it may be advantageous to further subdivide seed lots to minimize risk of disease spread. Although uniting may result in more expense during certification, in cases where disease is present, a lack of uniting may result in the loss of an entire crop. The size of the unit largely depends upon the size of the seed lot and the degree of risk the grower is willing to assume. Minimally, units from the previous generation should be maintained; i.e., planting stocks should never be mixed.

For field year one production, growers should maintain units from nuclear production. Depending on the size of the seed lot, it may be advantageous to divide the lot further into so-called "family" units of 1–4 rows each. A blank row left between units will facilitate roguing, certification inspections, and leaf sampling. The family units should be harvested and stored separately, and a post-harvest test sample submitted for each family unit. If disease is detected in any plant(s) in the unit at any time, the entire unit should be removed.

In later generations planted with field-grown tubers, the size of the seed lot will determine the degree of uniting that is practical. Smaller, earlier generation lots may benefit from the planting of family units and, when cut seed is planted, tuber uniting. As lots become larger in size, family uniting becomes impractical. In the case of very large lots that are planted in more than one field, benefit may be realized from treating each field as a separate certification unit. This is especially the case when fields are separated by large distances and have different exposure to potential risk; e.g., proximity to commercial potato fields.

Planting and Nutrition

In general, the practices employed in planting and fertilizing a seed potato crop mirror those of commercial plantings. In addition to the recommendation for early planting, the following important differences apply to seed potato production:

- *Seed cutting* is a very effective method for spreading bacterial, fungal, and some viral diseases. Therefore, a standard recommendation is for seed growers to avoid, when possible, cutting seed. When seed cutting is necessary, proper sanitation must be practiced. Minimally, seed cutters should be cleaned between seed lots.

- *Seeding rates* may be higher for some varieties than in commercial plantings. Higher plant populations are used to maintain yields while achieving the smaller tuber size profile desired for a seed crop. Higher plant populations are usually achieved by decreasing plant spacing on standard-width rows. Higher plant populations can also be achieved by planting seed potatoes in beds with narrow rows; e.g., 30 in. Seeding rates in these systems can be up to twice the normal rates typical of commercial potato production.
- *Fertilization rates*, particularly nitrogen, may be lower than those recommended for commercial production. Seed potatoes generally are not grown to full season and, therefore, will not require the full nutrient rates of a commercial crop. Additionally, reduction of nitrogen aids in vine killing and in a more fully mature plant and tuber crop. Tuber diseases can enter through wounds when immature tubers are subject to skinning during handling at harvest and storage.

Disease and Insect Control

Control of diseases and insects, particularly aphids, is necessary to prevent the introduction and spread of disease and, thus, to meet certification requirements. Routine scouting of the crop, plus the use of a preventative insecticide and fungicide program is a cornerstone of a good management program. See Chaps. 11 and 9. This will minimize early infestations and mitigate risk as the season progresses. It should be noted that the actionable threshold for some potato pests may be lower in seed potatoes than in a commercial planting.

- *Roguing*, a term used to describe the physical removal of undesirable plants from a field, can also be an important component of a comprehensive disease control strategy. When admixtures or off-types are present, roguing is also necessary for maintaining the varietal purity of seed. Roguing should be performed "early and often." In the case of virus diseases, roguing usually begins when the plants are 8–12 in tall, or as soon as symptoms are observed. Rouging for varietal mixtures may require additional plant growth, up to and including flowering, to allow proper identification of admixtures. Fields are commonly rogued twice; additional roguing may be done in early-generation seed and seed lots with more issues. When roguing, all vines and tubers should be removed from the field.

  While roguing can be an effective tool in disease management, it is not a substitute for purchasing planting stock with low disease levels. It may be difficult or impossible to rescue a seed lot due to cost and time constraints. It should also be borne in mind that, if the initial disease levels are high enough, the rate of current season spread of virus may outstrip the ability of the roguing crews to remove diseased plants. Finally, roguing for disease after row closure is of limited value, as many diseased plants will be hidden, and current season spread is likely to have already begun

Vine Killing

Early vine killing is recommended for seed potato production. Vine killing should be performed as soon as possible after target yield and size profile have been achieved in order to avoid late season aphid flights and the resultant spread of viruses vectored by them. Early vine kill also ensures plenty of opportunity for tubers to mature before harvest. Proper skin set will help to prevent the spread of diseases, such as soft rot, that may impact the quality of the seed crop.

- Early vine kill on seed crops may be difficult to achieve because of vigor and lack of natural senescence. Since most vine-killing chemicals act slowly and vines are vigorous, killing vines usually requires repeated chemical applications or some type of mechanical vine treatment before the chemical is applied. Growers need to be cautious in using mechanical treatment, however, because the use of machines increases the potential for virus movement into the tubers. A vine-killing product that kills vines rapidly, such as sulfuric acid, should be selected. If satisfactory vine killing can be achieved with the application of the vine-killing agent alone, mechanical treatments should not be used.
- Despite the difficulty, potato growers must take steps to ensure that the vines are killed as quickly and completely as possible without damaging the tubers. Early-generation seed fields should be killed earlier than surrounding fields so they do not become an "oasis" for late-season aphid vectors. Any green within killed fields should be eliminated as soon as possible for the same reason.

Harvesting and Storage

Other than the restrictions placed on storage by certification requirements, seed potato storage recommendations are similar to those of commercial potatoes. Storage temperatures are generally lower; however, proper conditioning, airflow, and humidity all are important to prevent disease spread and maintain the grade of the crop. Seed potatoes should be harvested before any danger of frost injury occurs. Tubers that are damaged by frost or exhibiting other signs of breakdown should be removed to prevent the potential spread of disease in storage.

During harvest, the units established at planting may or may not be maintained, depending on logistics, availability of storage, and other factors. Whenever possible, the original planting units should be maintained. Bulking of units, especially during harvesting of early-generation fields, is not always the appropriate course of action. Often, the best strategy is to harvest in units that are as small as possible and then, if necessary, units can be combined after the outcome of post-harvest testing is known. The process of uniting ensures that virus or other problems are restricted to a small, identifiable portion of the total seed lot.

During harvest of the first field generation seed, all tubers from a family unit should be kept together in sacks or bins. Each of these units should be given an identification code that is maintained throughout storage. Units should be stored off the cellar floor in mesh bags or slatted boxes to ensure good air flow. Immediately after harvest, the storage temperature should be held at 50–55 °F and 95% relative humidity for 2 weeks, which will promote wound healing. After 2 weeks, the storage temperature should be lowered as quickly as possible (without causing condensation) to about 38 °F and maintained there until spring. Fluctuations in storage temperature should be avoided.

Chapter 5
Field Selection, Crop Rotations, and Soil Management

Jeffrey C. Stark and Mike Thornton

Contents

Introduction

In developing a sustainable, efficient potato cropping system, several key decisions need to be made before planting, including: (1) selecting a suitable field for production, (2) developing an appropriate tillage and residue management program to optimize soil health, and (3) choosing a compatible sequence of crops for the rotation

J. C. Stark (✉)
Idaho Falls R&E Center, University of Idaho, Idaho Falls, ID, USA
e-mail: jstark@uidaho.edu

M. Thornton
Parma Research and Extension Center, University of Idaho, Parma, ID, USA
e-mail: miket@uidaho.edu

J. C. Stark et al. (eds.), *Potato Production Systems*,
https://doi.org/10.1007/978-3-030-39157-7_5

that favors the production of a healthy, productive potato crop. Since each field has its own unique set of characteristics, growers need to adapt management plans on a field-by-field basis to maximize production efficiency. The key factors that need to be considered in developing sustainable field management plans that promote soil and plant health include: soil physical and chemical characteristics, topography, crop rotation history, previous pest problems, tillage management practices, and production costs.

Field Selection

The suitability of a field for potato production depends on a wide range of characteristics, including soil physical and chemical properties, topography, cropping history, and previous pest problems, all of which influence soil health. Ideally, potato fields should have low to moderate slopes and soils that are deep, well drained, and friable with good water-holding capacity and low soluble salt and sodium concentrations.

Soil Health

Some of the key factors influencing soil health include the accumulation of adequate levels of soil organic matter for promoting good soil tilth, good soil structure and minimal soil compaction to provide adequate aeration and drainage for root growth, and optimal soil chemical characteristics that enhance nutrient availability and uptake.

Soil Texture and Organic Matter

Some of the key soil physical properties influencing potato growth and development include texture, which is related to the proportions of sand, silt, and clay; structure, which is influenced by the arrangement of individual soil particles into definable aggregates; and bulk density, which is soil mass per unit volume (g/cm^3).

Potatoes grow effectively in soils with textures ranging from sands to clay loams, although yield potentials can vary appreciably across textural classes. Soils with relatively fine textures, such as clay loams and silty clay loams, generally have much lower water infiltration rates than coarse-textured soils and, therefore, are more prone to runoff and unequal soil water distribution. This can be a substantial problem in fields with slopes greater than 5%.

Runoff problems can result in excessively dry soils on upper slopes and ridges and waterlogged soils in low areas, unless addressed through specialized tillage operations, such as basin tillage. Excess water can erode soil and leach soluble nutrients out of the root zone. Waterlogged conditions can also reduce soil aeration, which contributes to reduced root physiological activity and increased disease susceptibility and tuber disorders.

The relatively low infiltration rates characteristic of fine-textured soils make them more suitable for set-move sprinkler irrigation systems, such as wheel lines and hand lines, or solid-set systems. Surface irrigation is also more effective on fine-textured soils.

Sandy soils, on the other hand, typically have high infiltration rates and low water-holding capacities, which are more suitable for center-pivot and linear-move irrigation systems. In addition, sandy soils are generally more susceptible to nutrient leaching than fine-textured soils because of their lower water-holding capacity.

Soil water-holding capacity, permeability, and tilth tend to improve as soil organic matter content increases. However, many soils in potato-producing areas have an organic matter content less than 1%. As a result, growers must make consistent efforts to increase soil organic matter levels by returning crop residues to the soil and minimizing residue burning and unnecessary tillage operations.

Soil Compaction

Compaction can result from any process that breaks down soil structure and increases bulk density, which reduces the amount of pore space available to hold water and air. Compacted soil layers typically result from excessive wheel traffic or heavy axle loads, particularly when equipment is operated on wet soils. Natural hardpans also occur and can have similar effects on soil water movement and crop growth. These zones of high soil bulk density can decrease soil water-holding capacity, infiltration, and drainage, which can result in excessively wet and dry areas in the field and increased runoff and erosion.

Soil compaction can substantially reduce potato production efficiency. Compacted soils and hardpans interfere with potato root and tuber growth because of increased mechanical resistance of the soil. Compaction also increases the amount of energy required for plant emergence and root extension and reduces the plant's ability to absorb water and nutrients.

Another effect of compacted soil is that potato plants are typically less vigorous and more susceptible to stress-induced tuber defects than those grown in non-compacted soils. Limitations on root growth, coupled with uneven soil water distribution in the field, can also hasten the onset of early dying and increase the incidence of several other potato diseases, including Rhizoctonia, pink eye, and pink rot.

Generally, clay soils compact more readily than sandy soils, although hardpans in sands can also severely limit potato root penetration. Compaction of clay and loam soils can lead to clod problems at harvest, which can increase labor requirements, interfere with harvester operations, and increase tuber bruise damage. In addition, wet soils are more susceptible to compaction than dry soils; so if possible, fields should be allowed to dry to moderate soil moisture levels before conducting tillage operations.

Compaction can occur at different depths in the soil profile. Compacted layers in the upper part of the root zone can be loosened by tillage operations, such as ripping or chisel plowing, as well as by root penetration by deep-rooted crops, such as alfalfa. Loosening deeper compacted layers is more difficult, but the effects on potato growth and yield are usually less severe.

Growers can use the following management practices to help minimize or reduce soil compaction in potato rotations:

- Avoid field operations, such as tillage, planting, and harvesting, when soils are too wet. Also, consider performing primary tillage operations directly after harvest in areas where soil moisture is typically drier in the fall than in the spring.
- Chisel or rip when soils are relatively dry at a depth that will adequately disrupt the compacted layers. Vary the depth of tillage operations from year to year to avoid developing tillage pans.
- Reduce axle loads by using lighter field equipment or equipment with axle loads distributed over several axles rather than one or two.
- Minimize repeated passes across the same wheel tracks, and avoid unnecessary field operations.
- Use deep-rooted rotation crops, such as alfalfa, small grains, and corn that help break up hardpans and improve soil structure. These crops also return relatively large amounts of crop residue to the soil.
- Increase the number of years between potatoes and other row crops, such as sugar beets, onions, and beans in the rotation. Row crops return little crop residue to the soil, and the heavy equipment traffic associated with their production can increase soil compaction.
- Incorporate green manure crops, cover crops, manure, composts, and other crop residues to help increase soil organic matter and improve soil structure and drainage.

Topography

The development of center-pivot and linear-move irrigation systems, together with basin tillage, has greatly expanded the range of topographic conditions under which potatoes are grown. Selection of an appropriate field site should take into account the availability of field equipment necessary to manage the field and the potential for soil erosion and nutrient and pesticide leaching and runoff. Strongly sloping ground increases the potential for soil erosion and uneven water distribution. It also presents severe limitations to planting, cultivating, and harvesting operations (Fig. 5.1).

Slopes greater than 5% substantially increase runoff and erosion potential unless growers use some form of basin tillage, such as dammer diking (Fig. 5.2), to intercept and hold water in place. Even with basin tillage, however, runoff can be significant in wheel tracks. Slopes greater than 10% create problems with slippage and alignment of planting, cultivation, and harvesting equipment.

Fig. 5.1 Steep slopes can present challenges for planting, cultivating, and harvesting operations. (Photo credit: Potato Grower Magazine/Harris Publishing)

Fig. 5.2 Dammer diking equipment form small catchment basins in furrows that help reduce runoff and improve soil water uniformity on sloping ground. (Photo credit: Potato Grower Magazine/Harris Publishing)

Misalignment and unequal spacing of rows can cause plant damage during cultivation and harvesting operations and can also create a hazard for equipment operators. The lower areas of the field also tend to accumulate excess moisture, resulting in a greater risk of diseases, while the ridges tend to be too dry, resulting in reduced yield and quality.

Soil Chemical Characteristics

The primary soil chemical characteristics that affect potato production are pH, cation exchange capacity (CEC), soil salinity, and sodicity.

Soil pH

Potatoes grow well under a fairly wide range of soil pH levels, although soils with near neutral pH (6.5–7.5) generally provide maximum nutrient availability. However, potatoes frequently are grown on high pH soils (greater than 7.5).

At high soil pH levels, the availability of phosphorus, iron, zinc, and manganese typically are reduced, which creates a need for growers to modify fertilizer management practices that improve nutrient uptake efficiency. Under acidic soil conditions (pH less than 6.0) nutrient availability is also reduced, and soils frequently need lime applications to raise soil pH to the optimal range for nutrient uptake.

Cation Exchange Capacity

Cation exchange capacity (CEC) is an important soil property that strongly influences the availability of positively charged ions (cations), such as potassium, ammonium, calcium, magnesium, iron, zinc, and manganese. Since clays contain the vast majority of cation exchange sites in soils, CECs are proportional to clay content. Although CEC is usually not a limitation to potato production, soils with low CECs, such as sands, usually have a greater need for in-season nutrient applications than those with high CECs.

Soil Salinity

Saline soils, which have relatively high soluble salt concentrations, can reduce potato growth and yield by decreasing the plant's ability to absorb water from the soil. Salinity can also interfere with plant physiological processes involved in dry matter production and distribution within the plant. This condition may cause reduced vine and tuber growth. Saline soils can be reclaimed if infiltration and drainage characteristics are adequate and sufficient amounts of low-salt water are available to leach salts out of the root zone.

For long-term irrigation management considerations, growers should adjust water application amounts to maintain adequate leaching of salts out of the root zone. The amount of leaching required, known as the leaching requirement (LR), is the amount of drainage water (DW) that must be produced, expressed as a fraction of the amount of water applied to the soil (IW).

The LR depends primarily on the ratio of the salinity (electrical conductivity [EC]) of the irrigation water (EC_{iw}) to the acceptable level of salinity in the soil solution as measured in the DW (EC_{dw}). The LR can be determined using the following relationship: LR = ECiw/ECdw.

Recommended long-term LRs for potatoes grown on saline soils are usually about 6–8%. For potatoes, the EC of a saturation extract of the soil should not exceed 1.7 dS/m to avoid yield loss, while the EC of the irrigation water should not exceed 1.1 dS/m. The amount of water that will be needed to leach salts out of the soil will depend on the depth, texture, and permeability of the soil and the proportion of the initial salinity that growers must remove.

Sodicity

Sodic soils have high sodium concentrations adsorbed on clay particles that disrupt soil structure, which results in a substantial reduction in soil permeability. Low soil permeability associated with sodicity interferes with soil water availability and root growth. High soil sodium concentrations can also become toxic to plants by disrupting physiological processes and damaging tissue. The combined effects of sodicity on soil physical conditions and plant physiological processes can cause severe stunting and even death of potato plants.

The sodium adsorption ratio (SAR) provides a good indication of the sodicity hazard in irrigation waters. The SAR, which is a standard test conducted by commercial laboratories, represents the ratio of the sodium concentration in the irrigation water to the concentrations of calcium and magnesium. If the SAR is below 6, there should be no significant reduction in soil permeability. When the SAR increases to between 6 and 9, there is an increasing hazard to soil permeability. Severe permeability problems are likely when the SAR exceeds 9.

To reclaim sodic soils, sodium has to be displaced from the cation exchange sites by calcium. Growers can accomplish this by adding a soluble source of calcium, such as gypsum, or adding acidifying agents, such as sulfuric acid or elemental sulfur, to dissolve existing $CaCO_3$ (lime) in the soil. Reclamation of sodic soils typically requires large amounts of amendments, often in the range of several tons per acre or more.

Once the sodium has been displaced, it needs to be leached out of the root zone by application of adequate amounts of low-sodium water. As with salt removal, the amount of water required to leach sodium out of the soil profile will depend on the concentration of sodium in the soil, as well as the depth, texture, and permeability of the soil.

Field History

In addition to soil characteristics and topography, growers should consider the history of the field when selecting appropriate sites for growing potatoes. In particular, growers need to know the cropping history of the field, previous pest populations, and pesticide use practices.

Growers are advised to not have potatoes follow potatoes too closely in the rotation. Longer rotation intervals help avoid buildup of soilborne potato diseases, such as Verticillium wilt and Rhizoctonia. Growers should also avoid planting potatoes after crops such as sugar beets and onions, which are intensively managed with heavy equipment, because of the negative effects of soil compaction on potato yield and quality. When following alfalfa, canola (Fig. 5.3), or other forage legumes, growers should take care to manage nitrogen fertilization because of the large amounts of nitrogen mineralized from plant residues.

In evaluating the pest history of a field, growers need to gather accurate information on the presence of nematodes, wireworms, and problem weeds. If field records or other information on nematode populations are not available, growers would be wise to take field samples using appropriate procedures and submit them to public or commercial laboratories for a quantitative analysis of pest levels in the soil.

The potential for herbicide carryover should always be carefully considered when determining whether or not to plant potatoes in a particular field. Many

Fig. 5.3 Rotation crops, such as canola, help break pest cycles and improve soil physical conditions. (Photo credit: Potato Grower Magazine/Harris Publishing)

herbicides used to control weeds in common rotation crops can cause injury to potatoes if applied within 12 months prior to potato planting.

To facilitate field management in the future, detailed information on pesticide applications, including materials applied, plant-back restrictions, application rates, and timings should be maintained along with pest levels, field cropping histories, and important climatic information, such as soil temperature and irrigation and precipitation amounts and timings.

Crop Rotations

When considering crop rotation options, short-term economic returns must be balanced against the potential long-term benefits to the entire agricultural enterprise. Crop management systems should promote and maintain sustainable, economically efficient production throughout the rotation. When planning crop rotations, potato growers should also consider the potential for improving soil conditions and minimizing pest levels in addition to evaluating production costs and potential economic returns.

A primary benefit of a well-designed crop rotation is improved soil health. The addition of large amounts of crop residue to the soil over time can improve soil tilth, nutrient availability, and aeration, as well as increase soil water-holding capacity, infiltration, and drainage. These improved soil conditions can increase potato yield and quality while reducing water runoff and erosion.

The potential effects of previous crops on weed, insect, disease, and nematode populations are other key considerations in planning a rotation. For example, the inclusion of non-host crops in a rotation can reduce populations of pests that attack potatoes, particularly when the number of years between potato crops is increased.

A well-known example of this response is Verticillium wilt, which is greatly affected by rotation length. Rotation lengths of at least 4–5 years are typically required to adequately reduce soil inoculum levels of this disease and build up populations of beneficial soil organisms. Increasing the length of the rotation also helps reduce the severity of Rhizoctonia, black dot, common scab, and silver scurf. In addition, the use of non-host crops and longer rotations are primary management tools for reducing nematode populations in soil.

Opportunities to control weeds are also affected by the cropping sequence. For example, nightshade is more easily controlled in a wheat-sugar beet-barley-potato cropping sequence than in a wheat-potato-wheat-potato sequence. Most perennial weeds should be controlled by using a combination of herbicides and tillage during the years prior to potato production. In some cases, it may be advisable to delay potato production for a year or two to implement appropriate weed control strategies.

Any rotation that introduces new classes of herbicides and unique weed control practices has the potential to reduce weed problems in potatoes. However, care should always be taken to follow herbicide label restrictions on plant-back intervals to avoid herbicide carryover damage.

Cover Crops and Green Manures

The terms cover crop and green manure are often use interchangeably and refer to crops planted specifically with the intended purpose of managing soil quality, reducing erosion, improving fertility, and/or reducing pests. The inclusion of various cover crops and green manures in the potato cropping rotation has been shown to have beneficial effects on pest suppression as well as tuber yield and quality. These crops may be grown to maturity and harvested, or flailed and incorporated into soil to maximize the addition of organic matter. In some production systems, cover crops are grown over an entire season; in others they are planted in the spring prior to a cash crop or after harvest of cereals or other crops in late summer. The management of green manures and cover crops are covered in detail in Chap. 6.

Crop Sequences in Potato Cropping Systems

Crop sequences that include potato vary widely from region to region and farm to farm depending on local marketing options, climate, and other factors. Rotations can be as short as 2–3 years (ex., wheat-potato and wheat-wheat-potato), to well over 5 years in regions with high pest pressure. Almost all rotations include high-residue crops, such as small grains and corn, but may also include crops such as beans, peas, sugar beets, or alfalfa.

A 2-year wheat-potato rotation is far too short to provide adequate suppression of common soilborne diseases and nematodes. Consequently, growers typically treat these fields with fumigants, such as metam sodium, metam potassium, 1,3-dichloropropene, or chloropicrin in the fall before planting potatoes in an effort to reduce the effects of early dying and other diseases.

Short rotations can have significant impacts on potato crop quality and yield. Research conducted in eastern Idaho showed that average yields for 3-year wheat-wheat-potato rotations were usually about 3000–4000 lb./ac higher than those for 2-year wheat-potato rotations. A 4-year rotation of potatoes, sugar beets, and 2 years of grain usually performs better with a year of grain between the sugar beets and potatoes because of reduced soil compaction and disease problems.

When alfalfa is included in the potato rotation for 2–4 years, it often is followed directly by potatoes. However, a grain crop is frequently grown between alfalfa and potatoes to moderate the effects of high nitrogen release from alfalfa residue on the subsequent potato crop and to reduce plant residue management problems.

About 50% of the nitrogen mineralized from an alfalfa crop is released during the first year after incorporation. An additional 25% of the nitrogen is released during the second year.

Preplant Tillage Management

The selection of preplant tillage operations will depend, to a large extent, on soil type and structure, soil erosion susceptibility, and energy costs, as well as on the residue management requirements of the previous crop. The tillage system should be designed to effectively incorporate crop residues, break up compacted soil layers, improve water infiltration, and prepare the soil for planting, while minimizing the potential for soil erosion. A properly designed tillage system can also help improve the control of annual and perennial weeds and volunteer potatoes. See Chap. 12.

Fall Tillage

For years, potato growers have used moldboard plowing as a primary tillage method before potatoes (Fig. 5.4), particularly on relatively flat ground where water and wind erosion are not significant problems. Plow depth is an important consideration. It varies somewhat with soil type and the amount and type of crop residue, but usually is about 8–12 in. Growers should avoid plowing wet fields in order not to develop compacted layers at the bottom of the plow shear.

Fall plowing is usually preferred over spring plowing because potato producers can till ground over a longer period, and it also allows more time for crop residues to decompose. Fall plowing, however, can increase susceptibility to wind erosion because of the lack of crop residue on the soil surface.

Fig. 5.4 Moldboard plowing is often used as a primary tillage method before potatoes. (Photo credit: Potato Grower Magazine/Harris Publishing)

Growers who plow potato ground in the fall often leave fields with a rough surface over winter. Fields that have only been plowed in the fall are usually disked and roller-packed or harrowed in the spring after application of broadcast fertilizer. Some growers prefer to cultivate in the fall, especially after broadcasting fertilizer or prior to fall bedding. Producers typically use disks and/or roller packers, as well as bedding equipment for these secondary tillage operations.

Potato growers commonly use chisel plows, with chisel points or sweeps, in the fall to break up hardpans and compacted soils and improve water infiltration during the fall and winter months. For grain fields, growers typically irrigate, chop the crop residue with a straw beater, and then chisel plow the ground in the late summer after harvest. The grain stubble is then either disked or left standing over winter to improve soil moisture distribution and reduce erosion.

Although growers use several types of chisel implements under a range of field conditions, they often use sweeps in fields where hardpans are a problem. Growers prefer using chisel points on sloping ground that is prone to erosion.

Growers can disrupt compacted soil layers by sub-soiling below the plow layer with deep rippers or chisel plows when soils are relatively dry. Sub-soiling when soils are too wet will usually not adequately fracture the compacted layer and may, in some cases, make the compaction problem worse. For chiseling to be effective, tillage depth must exceed compaction depth so that hardpans are adequately disrupted. Ideally, growers need to set shanks at an operating depth that will lift and shatter the compacted soil layer without exceeding the sheer stress value of the soil. Growers should set shank spacing so that the entire surface layer of soil is disrupted by the tillage operation. An integrated approach to reducing compaction should also include increasing soil organic matter levels through the use of soil-building crop rotations and appropriate residue management practices.

After a grain crop, some growers will disk the field twice in the fall at opposite angles. They will then establish beds and fertilize during the marking operation. Fall bedding (Fig. 5.5) has increased in popularity in recent years. Growers usually irrigate potato fields, fertilize, and then chisel or use a moldboard plow before forming beds in the fall. Fall bedding allows more time for soil preparation work to be done when conditions are usually good and a grower has more time. Fall soil preparation also provides added time in the spring for growers to focus on seed preparation, pest management, and planting operations.

Some fall field preparation, such as minimum or conservation tillage, may need to be performed on sandy soils to help reduce wind erosion. This may involve leaving previous crop residues partially incorporated on the surface of the soil to help hold the surface soil in place. This approach can be effective in reducing soil erosion, but can also make it more difficult to incorporate preplant, broadcast-applied fertilizer and herbicides.

Spring Tillage

During spring tillage operations, growers apply broadcast fertilizer over the field, then disk and harrow, and mark the rows with shanks mounted on a tool bar or other implement.

Fig. 5.5 Beds can be fertilized and formed in the fall to allow more time for other operations in the spring. (Photo credit: Potato Grower Magazine/Harris Publishing)

Pre-plant bedding or "marking out" is typically performed in the spring prior to planting, particularly to facilitate accurate rowing spacing with large, multi-row planters. Global Positioning System (GPS) guided tractor steering systems have greatly improved row position accuracy and essentially eliminated errors during the row marking operation. This, in turn, helps prevent plant root and tuber damage during subsequent hilling and harvesting operations.

When planting, it is important to avoid leaving excessive crop residues on the surface that can be incorporated into the seed row. Large amounts of undecomposed crop residues, particularly from green manure and animal manures, can favor the development of tuber diseases, such as common scab, Pythium seed piece decay, and bacterial soft rot.

Dammer diking, or reservoir tillage, is a tillage operation growers perform after planting at final cultivation that forms small catchment basins in the furrows (Fig. 5.2). These basins are designed to increase water infiltration and reduce water runoff, which generally improves soil water uniformity across sloping ground. The basins also help to reduce soil erosion and surface movement of fertilizers and pesticides. When used properly, basin tillage significantly improves water use efficiency on sloping ground.

Fumigation

For many producers, fumigation is an important step in the ground preparation process prior to planting potatoes. Estimates of the proportion of potato acreage that is fumigated run from around 50% in Idaho, to over 80% in the Columbia Basin. Growers in other production regions in the Midwest and Eastern U.S. also fumigate prior to potatoes, with the proportion of treated acres depending on factors such as cropping history, length of rotation, yield potential, and variety choices.

Economics is a major factor in the decision on whether or not to fumigate, as it can be one of the single most expensive crop production inputs. That cost has to be balanced against the economic benefits gained from increased tuber yield and quality. For example, fumigation is estimated to increase yields by an average of 8–10% in some production regions. Slightly stronger yield responses are expected in long-season production areas due to increased pest pressure. Even bigger impacts on tuber quality attributes, such as size and the incidence of external and internal defects, have been reported in some studies. In some situations, fumigation is one of the few effective tools to reduce losses, or outright rejection, of the crop due to damage from pests, such as root knot nematode and wireworm.

The decision on whether or not to fumigate should be made on a field-by-field and year-to-year basis. The key information needed to make these decisions are field histories of pest damage and results of soil sampling to determine populations of damaging nematode species and soilborne pathogens, such as Verticillium. Since many of the pests that damage potatoes are not uniformly distributed throughout a given field, it is important that the sampling process is extensive and takes into account differences in soil type and crop rotation. See Chap. 10 for more specific recommendations on soil sampling procedures.

Fumigation is generally done in the fall prior to potatoes, and requires extensive ground preparation for optimum results. Because these products move laterally and upward through the soil from the point of application, anything that impedes that movement, or allows rapid loss of the product from the soil, reduces efficacy. It is important that clods and compaction layers are broken up and crop residues are finely chopped and incorporated. Failure to do this may allow disease organisms to survive in clods and large pieces of plant debris where the fumigant cannot penetrate. Plant debris that is not well incorporated also allows the fumigant to escape from the soil before pests are exposed to an adequate concentration for the duration needed to achieve control. More specific recommendations on soil preparation, soil moisture, and application procedures of fumigants are given in Chap. 10.

It is important to understand that fumigation is not a cure-all, and it cannot take the place of a good crop rotation plan. Fumigants reduce populations of both soilborne pests and beneficial microorganisms. As a result, populations of pests can rebound to damaging levels after only one planting of a susceptible crop, resulting in the need to fumigate again the next time potatoes are grown in that field. The suppression of beneficial soil microorganisms resulting from fumigation can alter the response to fertilizers and result in higher levels of herbicide residues. These effects occur because fumigation suppresses the populations of soil microorganisms responsible for nutrient cycling and herbicide breakdown.

Chapter 6
Organic Potato Production

Amber Moore, Dan M. Sullivan, Nora Olsen, Pamela J. S. Hutchinson, Phillip Wharton, and Erik J. Wenninger

Contents

A. Moore (✉) · D. M. Sullivan
Department of Crop and Soil Science, Oregon State University, Corvallis, OR, USA
e-mail: amber.moore@oregonstate.edu; Dan.Sullivan@oregonstate.edu

N. Olsen
Department of Plant Sciences, University of Idaho, Kimberly Research and Extension Center, Kimberly, ID, USA
e-mail: norao@uidaho.edu

P. J. S. Hutchinson
Department of Plant Sciences, University of Idaho, Aberdeen Research and Extension Center, Aberdeen, ID, USA
e-mail: phutch@uidaho.edu

P. Wharton · E. J. Wenninger
Department of Entomology, Plant Pathology and Nematology, University of Idaho, Aberdeen Research and Extension Center, Aberdeen, ID, USA
e-mail: pwharton@uidaho.edu; erikw@uidaho.edu

J. C. Stark et al. (eds.), *Potato Production Systems*,
https://doi.org/10.1007/978-3-030-39157-7_6

Introduction

Organic crop production is founded in management practices that promote biodiversity, soil biological activity, minimal off-farm inputs, and restoration of ecological harmony. Organic potato production is on the rise in the U.S., with certified organic potato production acreage more than doubling from 8000 to 17,000 ac between 2008 and 2016, and organic potato sales increasing fivefold from $30 to $150 million over this same time period. The goal of this chapter is to provide crop management information to organic growers who are considering potato production and to conventional potato growers who are interested in select organic management techniques. Topics to be discussed include regulations, the transition from conventional to organic management, and cover crops as they apply to organic potato production. We will also describe how variety selection, nutrient management, pest management, vine kill, tuber harvest, and storage practices can be adapted to comply with organic production methods while optimizing tuber yield and quality. Recommendations provided in this chapter complement standard crop management practices, which are described in detail throughout the other chapters of this production guide.

Photo credit: Mike Heath, organic potato grower, Buhl, ID

Background

As described by the U.S. National Organic Program (NOP), organic crop production promotes: (1) ecological production systems that promote and enhance biodiversity, biological cycles, and soil biological activity; (2) minimal off-farm inputs; and (3)

Table 6.1 Changes in U.S organic potato production from 2008 to 2016

Parameter	2008	2014	2016
Harvested area of organic potatoes in the U.S.	7989 ac	12,082 ac	17,244 ac
Quantity of harvested organic potatoes in the U.S.	2.3 million cwt/ac	3.6 million cwt/ac	4.7 million cwt/ac
Value of organic potato sales in the U.S.	30.0 million	61.8 million	150.6 million
Number of farms growing organic potatoes in the U.S.	1331	953	681

Source: USDA NASS (2010, 2015, 2017) Census of Agriculture 2008, 2014, and 2016 organic surveys

management practices that restore, maintain, and enhance ecological harmony. In contrast to conventional methods, organic production does not permit the use of man-made chemicals and processes, or other methods or amendments that are believed to compromise soil health and production sustainability.

According to the United States Department of Agriculture, National Agricultural Statistics Service (USDA NASS), organic potato acreage in the U.S. increased by 115% from 2008 to 2016 (Table 6.1). Organic potato sales increased by 402% during the same time period (Table 6.1). The growth in organic potato acreage and organic potato sales illustrates a rising interest in organic potato production by growers and a rising demand for organic potatoes by consumers. Despite the increase in sales, the number of farms producing organic potatoes decreased by 48% between 2008 and 2016, illustrating that few growers have introduced organic potato production into their farming systems in recent years. Providing information on the management of potatoes using organic methods may help to reverse this trend.

In this chapter, we will discuss regulations, the transition from conventional to organic management, and cover crops as they apply to organic potato production. We will also describe how variety selection, nutrient management, pest management, vine kill, tuber harvest, and storage practices can be adapted to comply with organic production methods while optimizing tuber yield and quality. Although this discussion specifically addresses organic potato production systems, conventional growers interested in improving their sustainability and long-term productivity may consider some of these practices as well. Recommendations in this chapter adhere to the 2015 USDA NOP standards. As organic regulations are altered and updated on a continual basis, growers are advised to refer to the most recent NOP standards before adopting a specific practice or product into their programs. Find these most up-to-date standards online at: http://www.ams.usda.gov/rules-regulations/organic.

Organic Regulation

Federal governments have developed programs for the certification of cropping systems and other facilities for crops grown using organic methods. In the U.S., the USDA NOP establishes and adapts organic regulations and practice standards on a regular basis. Fields, production plans, and other aspects of the supply chain must be re-certified on an annual basis. Accredited certifying agents assist with this process through annual inspections and organic certification application reviews.

Transitioning to Organic

Crop fields must be managed with organic management practices for a 3-year transition period before being eligible for organic certification. Nutrient deficiencies and pest pressure tend to be more severe during the transition period than they are in fields that have been certified organic for several years. The inclusion of beneficial cover crops and/or animal manure or compost application practices commonly associated with organic production promotes accumulation of soil organic matter (SOM), plant nutrients, and development of diverse microbial populations. These, in turn, can reduce crop nutrient deficiencies while improving plant growth.

In the Pacific Northwest, alfalfa hay is commonly grown during the transition period from conventional to organic production. The reasons for this are: (1) alfalfa fixes N, which is a benefit to transitioning fields that are no longer able to receive conventional nitrogen (N) fertilizer applications; (2) alfalfa crops are effective weed suppressants, which is a benefit when conventional herbicides are not an option; and (3) a grower can achieve comparable organic alfalfa yields to conventionally produced alfalfa, which prevents significant profit losses during the transition period. Other options for transition crops include grass, hay, and small grains, which also have lower N requirements and compete better in weedy conditions than row crops.

Varieties

Selecting potato varieties that are best suited for optimal yield and quality under organic systems is key to success with organic potatoes. A working knowledge of the production attributes of varieties will guide organic growers in choosing the most appropriate variety for the intended market. Table 6.2 provides specific varietal attributes to consider for organic potato production.

Seed and Planting

Seed and planting management relies upon the same fundamental strategies as described in Chap. 7. The NOP (as of 2015) requires an organic potato producer to source and plant organic certified seed. Exemptions to the requirement include

Table 6.2 Specific varietal attributes to consider for organic potato production

Interest or concern	Target varietal attributes
Disease and insect pressure	Short-season varieties to avoid late blight disease pressure Resistance to diseases or insects that are common in the growing region
Fresh market potatoes	Color, skin finish, size, shape, and other attributes determined by the market
Late-season release of N	Early maturing varieties
Limited N	Short-season varieties Varieties with reduced fertility requirements
Long-term storage	Long-dormancy varieties, tolerance to cooler storage temperatures, and low weight loss in storage conditions
Processing potatoes	High specific gravity Low reducing sugar concentrations
Weed pressure	Early and vigorous emergence Indeterminate varieties

situations in which the quantity, specific variety, and desired quality (e.g., low virus) cannot be sourced from an organic seed producer. Accredited certifying agents can provide more information on the documentation needed for an exemption if organic seed is not commercially available. Only certified seed that meets the standards for the generation, grade, and disease level should be planted. Seed can be cut and planted without the addition of an organically approved seed treatment, but treatments, such as fir bark, may prevent cut seed pieces from sticking together and causing inconsistent planter performance.

Nutrient Management

Phosphorus, potassium, sulfur, iron, and zinc requirements for potato plants under organic production can usually be met through additions of composts and other byproduct materials that are both affordable and easily accessible for most growers. Accessing N sources for organic production, on the other hand, is much more challenging. We will focus heavily on N management in this section, with emphases on SOM, and soil testing, petiole testing, composts, manures, specialty fertilizers, green manure crops, and nutrient credit calculators can be an aid to anticipating N release from composts, manures, and byproducts. Similar to conventional methods, excessive amounts of any nutrient can cause environmental and/or agronomic issues; therefore, we recommend following nutrient guidelines closely from Chap. 8 to prevent excessive nutrient accumulations in the soil.

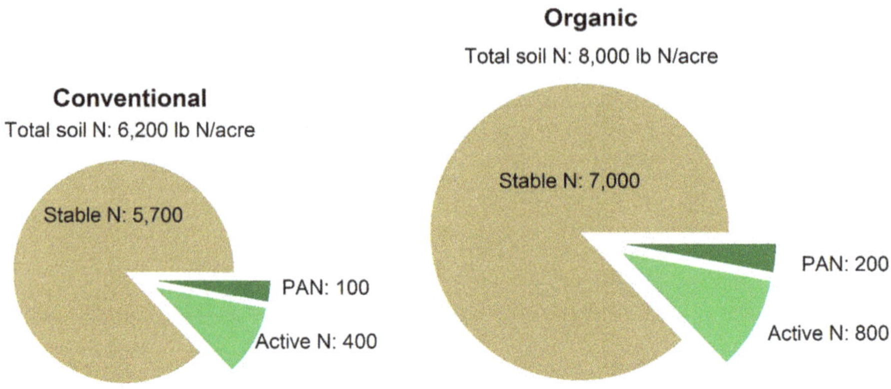

Fig. 6.1 Increasing SOM increases total soil N and biologically active N (0–12 in depth). This example is based on Willamette Valley (OR) silt loam soils, where SOM typically increases from 3 to 4% under organic management. PAN mineralized from SOM typically doubles under long-term organic management. (Source: N mineralization monitoring, conventional and organic grower fields 2002–2013 [D. Sullivan, Oregon State University (UI); unpublished])

Soil Organic Matter (SOM)

Organic production systems often have greater amounts of SOM than conventional systems due to a reliance on alfalfa residues, green manures, animal composts/manures, and other organic nutrient sources that are rich in organic matters as nutrient sources. The fraction of total SOM that is "biologically active" is also often increased. Soils with higher organic matter usually have improved soil tilth and increased water storage capacity, which can pay dividends in increased tuber size and grade. Decomposition of the organic matter (and thus organic N) may significantly increase plant-available N (ammonium and nitrate) (PAN) levels in the soil (Fig. 6.1).

In-Season Soil Testing

In-season soil testing for nitrate-N and ammonium-N is a valuable tool for organic production systems. Organically managed soils often contain significant amounts of stable organic N compounds. As temperatures warm, microbes convert these compounds to PAN. Refer to Chap. 8, for in-season soil test recommendations. When midseason soil test nitrate-N is less than established soil nitrate thresholds listed in Chap. 8, growers may want to consider applications of N-rich specialty products to provide additional PAN. When mid-season N additions are not feasible, in-season soil test data are a useful assessment of the present management system in supplying PAN to the potato plant.

Soil P and soluble salts electrical conductivity (EC) should be closely monitored, as repeated applications of composts and other organic materials can cause accumulation of P and salts to environmentally and/or agronomically detrimental levels. Soil test P thresholds are set by individual states. Potatoes are salt-sensitive plants that begin to show yield loss when EC exceeds 1.7 dS/m.

Monitoring N Supply via Petiole Testing

Testing the petiole of the fourth leaf from the top of the potato plant for nitrate concentrations is commonly used to monitor the N status of the plant. See Chap. 8. However, it is not appropriate for organic growers to aim for the same petiole nitrate standards as their conventional counterparts. Petiole nitrate goals were established by monitoring nitrate levels for plants that were fertilized with conventional N fertilizers that contain 90–100% available N. For example, petiole nitrate levels ranged between 800 and 18,000 ppm during tuber bulking in a certified organic potato research study in Kimberly, ID, which is a much lower range than the 15,000–20,000 ppm nitrate-N concentration recommended for conventional potatoes (Moore et al. 2011; Moore and Olsen 2010). However, these plants still yielded between 350 and 380 cwt/ac, which is an acceptable yield for organic potatoes in this region.

Organic potato growers may still want to monitor petiole N throughout the season to determine N status of the crop. We suggest the organic potato growers rely heavily on pre-plant and in-season soil tests when determining both pre-plant and in-season organic fertilization rates.

Petiole testing for P, K, and other macro and micronutrients may be helpful for recording nutrient status and responses to specific management practices. Petiole nutrient testing for P, K, and micronutrients may be helpful for identifying nutrient deficiencies and determining what in-season organic nutrient sources are available to address the deficiency. However, petiole testing for these nutrients may not be beneficial for many organic growers because: (1) manures and composts applied pre-plant generally supply the majority of the P, K, and other nutrients needed by potatoes, and (2) effective organic sources of these nutrients for in-season application are extremely limited.

Animal Manure and Compost

Animal manure and animal manure composts may be applied to organic potato production systems in the U.S. (as of 2015). Composting is the active management of manure and/or other organic materials to aid in the decomposition of organic materials and destruction of pathogens and weed seeds by microorganisms. Finished compost is considered lower risk for human pathogen or weed seed transmission than manure, because manure is sanitized by heat during composting. For certified organic potato production, management practices specified in the National Organic Standards must be followed, including:

- A waiting period of 120 days between the application of fresh manure and the harvest of high-risk crops, such as potatoes.
- For compost production, temperatures above 131 °F must be maintained for a minimum of 3–15 days, depending on the type of composting system used. See the NOP or consult with your organic advisor for specifics.

Composting transforms manure to a product that is relatively uniform and easier to store and apply than manure. Manure is often available at the cost of transport, with little or no additional charges. Composters charge a fee for the compost and an application charge for custom spreading. Manure volume is typically reduced by more than half by composting due to the loss of water, carbon, and nitrogen during the composting process. Most plant nutrients (except N) are retained during composting. Compost typically has higher P, K, and other nutrient concentrations than does manure.

Manure generally has greater immediately PAN than compost, because N is lost as ammonia or converted to more stable organic N compounds during composting. Manures that are highest in PAN usually contain considerable ammonium-N, which can be lost as ammonia gas if not incorporated immediately by tillage. After the application year, manure or compost provide similar amounts of PAN via mineralization (when applied at same total N application rate).

Several countries consider the application of fresh manure an unacceptable organic crop production practice; therefore, growers marketing internationally may be subject to additional regulations. Refer to PNW publication, "Fertilizing with Manure and Other Organic Amendments," for more information (PNW 533).

Specialty Product Fertilizers Specialty product fertilizers that are rich in N can be of use to organic growers who are: (1) growing potatoes or other crops with a high demand for N, and (2) have relatively low reserves of organic N in the soil. PAN release from specialty products containing more than 4% total N is very rapid.

Examples of specialty product fertilizers that typically contain more than 4% N include sodium nitrate, pelleted or granulated fish, seed meal, feather meal, and poultry manure products. It should be noted that sodium nitrate applications are limited to less than 20% of the total N input applied during a growing season (as of 2015), due to sodium plant toxicity and sustainability concerns. Also, some international markets may not consider sodium nitrate as an organic-approved fertilizer.

Specialty products are often expensive; therefore, long-term management should focus on reducing or eliminating their use. Nutrient composition information regarding a wide variety of organic specialty products is listed in University of Idaho publication, "Organic Potato Production in Idaho: Nutrient Management and Variety Selection," bull 885 (2013).

Green Manure Crops

Green manure crops are commonly used in organic production systems to provide N and other nutrients to row crops like potatoes. The term "green manure" refers to growing plants that are tilled into the soil as a means of supplying plants to the succeeding crop. Green manure contributes significant amounts of organic matter and N to the soil when the plants are able to accumulate large amounts of biomass (Figs. 6.2 and 6.4). Green manure crops can also provide the benefit of reducing soil

Fig. 6.2 (continued) when most potatoes (yellow cultivars including Nicola, Carola, Yukon Gold, and Island Sunshine) were 4–8 oz size. Source: OSPUD project (Western SARE Grant SW05–091). https://agsci.oregonstate.edu/ospud-participatory-organic-potato-project/participatory-approach-ospud-project. Data collected by John McQueen and Lane Selman of Oregon State University [2007, unpublished]

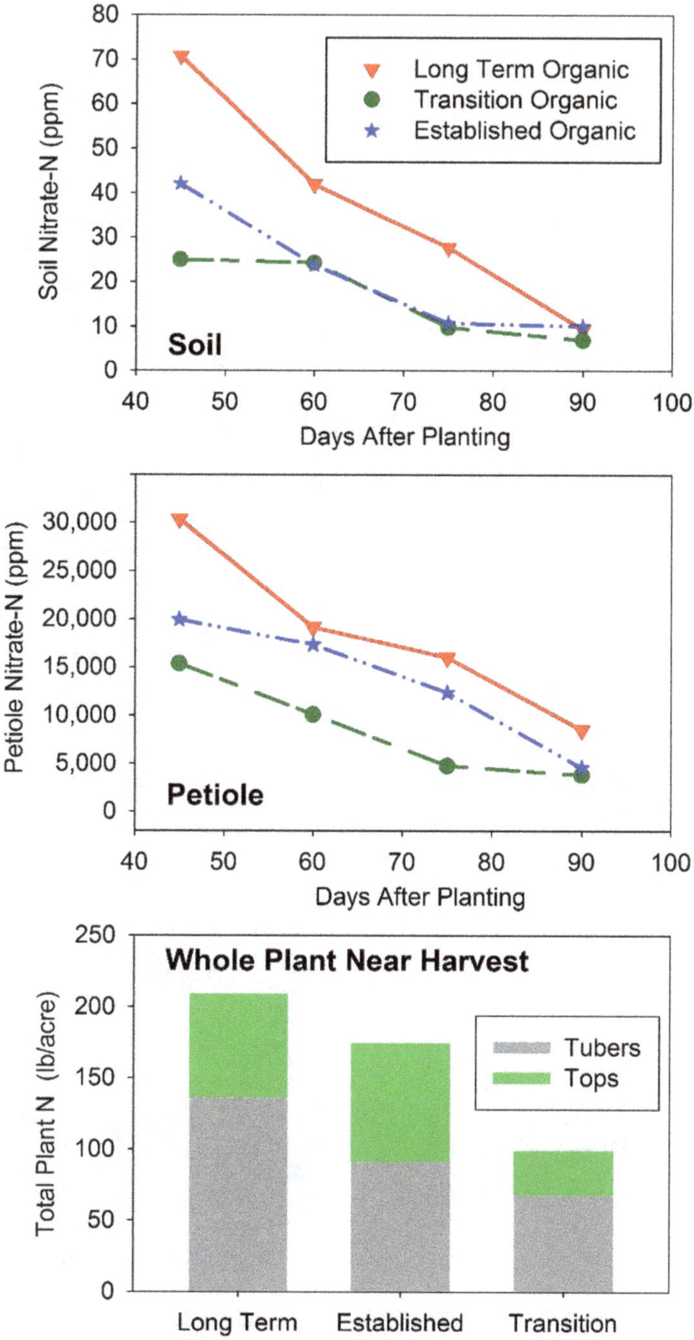

Fig. 6.2 Measurements of soil nitrate (top) and petiole nitrate (middle) can help to understand the supply of N provided to the potato crop via mineralization (bottom). This example demonstrates a buildup of mineralizable N in soil resulting from continuous organic management for less than 3 years (Transition), approximately 3–10 year (Established), and 15+ year (Long Term). Data source: Field locations in western OR on medium-textured soils with 3–4% organic matter (0–12 in depth). Whole plant N uptake determined in grower fields at 105 days after planting (Aug 22–27),

erosion losses, especially over the winter and spring months. In areas like Southern Idaho, most row and grain crops are planted in early spring, before there is enough time for significant biomass accumulation (Fig. 6.3). In these cases, soil erosion control may become the primary benefit of the green manure crops, with organic matter and N a minor benefit in comparison (Fig. 6.4).

The benefit of the May cover crop kill is erosion control, with the option to plant potatoes and other early season crops in cool climate regions like Southern Idaho. The benefits of the July cover crop kill is erosion control, mineralizable N source, organic matter source, and other soil building properties, but with limited options for succeeding crops in cool climates. (Photo credits: Amber Moore (left) and Megan Satterwhite (right), University of Idaho)

Alfalfa is commonly grown in rotation with potatoes in organic production systems. As alfalfa is a legume, N-fixing rhizobia bacteria in the root nodules convert dinitrogen (N_2) gas from the atmosphere into ammonium that can be used by the plant (Fig. 6.5). It has been estimated that an alfalfa stand, in combination with the SOM, can contribute between 155 and 240 lb N/ac (Westermann and Crothers 1993). Westermann and Stark (1993) illustrated tuber yield increases following cover crops, with the magnitude of the yield increase dependent on the type and the management of the preceding cover crop (Table 6.3). Yield increases in this study were the most evident for green manure leguminous cover crops that were incorporated into the soil (Table 6.3). Many organic growers in the Pacific Northwest have been successful with maintaining a minimum of 50% of their crop fields in alfalfa for a minimum of 3 years in order to build nitrogen reserves, as well as improve soil structure and organic matter content.

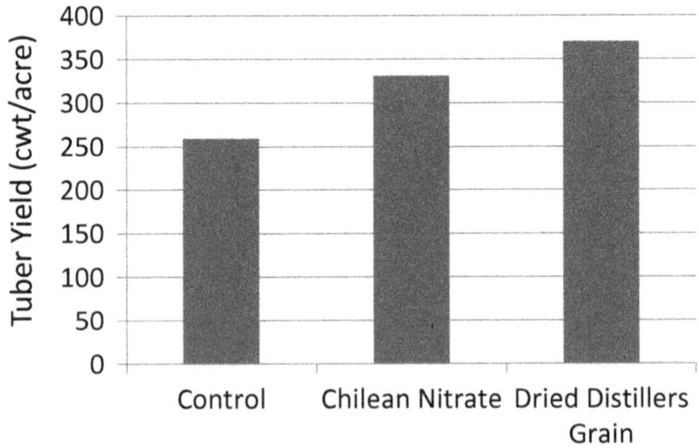

Fig. 6.3 Organic Russet Burbank potato tuber yield response to additions of either Chilean nitrate (48 lb N/ac, legal limit by NOP standards as of 2014) or dried distillers grain (1.1 ton/ac, or 95 lb N/ac) in Kimberly, ID. Canola and soybean meals have similar fertilizer nutrient compositions to dried distillers grains; distiller grains are no longer approved as a certified organic production practice by the NOP (as of 2016)

Fig. 6.4 Left photo: May cover crop kill, Austrian winter pea; Right photo: July cover crop kill, Austrian winter pea

Fig. 6.5 Hairy vetch root nodules, pink coloring and large nodule size indicate active N fixation. (Photo credit: Amber Moore, University of Idaho)

Calculating Nutrient Credits With Online Calculators

When relying on cover crops as an N source, growers often search for simple tools to help them to predict how much N they can expect to be released from the plowed-under plant material. Tools currently available to growers include the Oregon State University Organic Fertilizer and Cover Crop Calculator (for Western Oregon growing conditions) available at: https://extension.oregonstate.edu/organic-fertilizer-cover-crop-calculators, and the University of Idaho Cover Crop Calculator (for

Table 6.3 Tuber yields for conventionally grown potatoes in Aberdeen, ID, as affected by management of various leguminous cover crops (Adapted from Westermann and Stark 1993)

Preceding crop	Treatment	Tuber yield (cwt/ac)
Wheat	Tilled under	402
Alfalfa	Harvested for hay	470
Alfalfa	Tilled under	511
Austrian winter pea	Harvested for seed	455
Austrian winter pea	Tilled under	491

Table 6.4 Predicted PAN release at 10 weeks after cover crop termination, based on Oregon State University and University of Idaho cover crop calculator studies

Example cover crop residue properties		% predicted PAN		Estimated lb N/dry ton residue	
Total N (% N, dry wt.)	Approximate C: N	Western Oregon	Southern Idaho	Western Oregon	Southern Idaho
1.0	40	<0	12	<0	2
2.0	20	20	21	8	8
3.0	13	40	27	24	16
4.0	10	55	33	44	26

Southern Idaho growing conditions) available at: www.extension.uidaho.edu/nutrient/CC_Calculator/Cover_Crop_Main_page.htm. These calculators were created to allow growers to quickly and easily predict N availability from green manure crops based on tissue N content, total biomass, and dry matter content. An example of predicted PAN based on total N concentration or C:N ratio for using the OSU and UI cover crop calculators is illustrated in Table 6.4.

Weed Management

Weeds compete with potatoes for light, water, and nutrients (Fig. 6.6). Weeds also host other pests that are detrimental to potatoes, can cause tuber damage, and even interfere with potato harvest operations. As in conventional potato production, growers should use an integrated approach for weed control in organic potatoes. Combinations of biological, cultural, and mechanical tactics can provide successful weed control.

Cultural Weed Management Practices

Organic potato growers should plan on a multi-year approach for controlling weeds and plan ahead for potato production in a given field. Tactics include:

Fig 6.6 Late-season weed pressure can be a major issue for potato growers. This photo shows kochia, common lambsquarter, hairy nightshade, and annual sowthistle weed species becoming visible during potato plant senescence. (Photo credit: Amber Moore, University of Idaho)

- Avoiding fields with histories of excess weed pressure.
- Preventing weeds from producing viable seeds.
- Using a diverse crop rotation.
- Including cover crops in the rotation that can compete with weedy species.
- Selecting competitive cultivars.
- Planting at optimal dates and row spacings.
- Optimizing soil fertility and irrigation for a healthy, competitive crop.
- Cleaning weed seeds and vegetation from machinery routinely.

It is recommended that organic growers avoid planting potatoes in fields with a known history of weed problems. For example, a field with a history of quackgrass pressure should not be selected for organic potato production, as the quackgrass stolons can cause damage by growing into the tubers. Weeds should be prevented from producing viable seeds.

A diverse crop rotation can disrupt the life cycles of weeds and prevent species from dominating. Growers should consider including both summer annual and winter annual crops in the rotation and, if possible, alternate early- and late-planted crops and short- and long-season crops to break up the life cycles of weeds. This type of rotation can prevent weeds from germinating and establishing during the same periods year after year.

Growers should select potato cultivars that are competitive against weeds at the time of crop emergence. The critical period for controlling weeds in potatoes is usually between potato emergence and when the vegetative canopy closes over the rows. Therefore, growers should choose varieties that not only produce a full canopy quickly, but also maintain the canopy as long as possible. Wisconsin researchers reported that 96% or greater shading by potatoes is necessary for general weed control. A canopy that opens/falls flat can allow for late-season weed emergence and viable seed production. Indeterminate potato varieties (e.g., Russet Burbank, Ranger Russet, Alturas, Century Russet, Russet Nugget, and German Butterball) usually produce larger canopies that close sooner and compete with weeds better than determinate varieties, such as Russet Norkotah, Red Lasoda, Red Norland, Sangre, and Yukon Gold. For instance, University of Idaho has shown that Russet Norkotah, a short-season, determinate growth cultivar, provided less shade and was less competitive with hairy nightshade season-long than Russet Burbank, a full-season cultivar with a large canopy. The presence of one hairy nightshade plant per meter-row in a Russet Norkotah production field reduced U.S. No. 1 and total tuber yields 21 and 16%, respectively, compared with weed-free yields. Russet Burbank yields were not reduced significantly until competing season-long with two hairy nightshade plants per m row. Some growers can use a stale seedbed for which the ground is prepared in advance of potato planting and irrigated to promote an early flush of weeds. Those weeds can be controlled before the potato crop planting or emergence by shallow cultivation or flaming.

Adjusting within row spacing can encourage quick canopy development and row closure. Bed planting can be used to provide earlier canopy cover that can shade and delay weed germination and slow weed growth. In bed plantings, potatoes are uniformly spaced in an offset, equidistant grid, and canopy closure is 15–20 days earlier than traditionally planted hills.

Green manure and cover crops that have been reported to inhibit weed seed germination through the release of allelopathic compounds when they are incorporated into the soil. Green manure crops with weed suppressive properties include rye (*Secale cereale*), oats (*Avena sativa*), rapeseed, mustards (*Brassica juncea*, *Sinapis alba*, *B. napus*), barley, sorghum-sudangrass hybrids, and buckwheat, as all of these crop species can grow quickly to out-compete many weeds species. In a University of Idaho study, planting and incorporating a 75% white mustard and 25% oriental mustard mixture in the fall significantly reduced hairy nightshade biomass in potatoes growing the following season, in comparison to biomass in potatoes following no green manure, oilseed radish, or a 50/50 oilseed radish/oriental mustard mixture (Hutchinson unpublished).

Equipment sanitation to prevent movement of weed seed and vegetation between fields is also important. Organic growers in irrigated regions should also consider screening irrigation water to remove weed seeds.

Fig. 6.7 Cultivating organic potato production field with a Lilliston rolling cultivator prior to germination. (Photo credit: Amber Moore, University of Idaho)

Mechanical Weed Management Practices

Mechanical weed management tactics include cultivation, hoeing-hand weeding, flaming, mulching, and mowing. Properly timed cultivations can control early germinating annual weeds, as long as weeds are small (two to three true leaf stage) at the time of cultivation (Fig. 6.7). Cultivation also may reduce tuber exposure to sunlight, which reduces tuber greening. Potato seed pieces are planted deeper than most annual crops and can withstand cultivations to remove weeds before potato plants emerge. However, timely cultivation may be difficult on large acreages or in wet weather. Multiple cultivations may cause soil compaction, reducing aeration and potato growth, and producing clods that bruise potatoes at harvest. Late cultivation may also directly damage potato foliage and roots, resulting in lower tuber yields. Sometimes only a flash of light is needed for weed seed germination. Cultivation at night can prevent germination of some small-seeded annual weeds.

Studies in Idaho have shown that when weed populations are low and a competitive potato variety is grown, cultivating when weeds are small (0.5-in tall) and potatoes are 4–6 in tall can provide economical weed control. However, when heavy weed populations exist, multiple cultivations are needed, and weed control may still be inadequate to prevent yield loss. Also, multiple cultivations have been shown to reduce tuber yields in conventional production fields. The direct effects of cultivation alone on tuber yield and quality were studied in weed-free experiments.

U.S. No. 1 yields were 12–17% less in plots that were hilled and cultivated twice (potatoes were 12–14 in. tall at the last cultivation) than in weed-free plots that were hilled or hilled and cultivated once when potatoes were 4–6 in. tall (Eberlein et al. 1997).

Fields should be cultivated under drier soil conditions when possible. Cultivated weeds die more readily in dry soil than in moist soil; at least one day of lying uprooted in dry soil will kill most weeds in the 2–3 true leaf stage or smaller. Less soil compaction occurs on drier soil, which minimizes yield reduction caused by compaction and reduces formation of clods that cause bruising.

When cultivating, adjust the cultivator to ensure it is throwing 1–2 in of soil over the entire hill in order to cover and kill weeds which have emerged there. Less than 1–2 in will not be enough soil for uniform coverage of the hill and weeds. Too much soil will build the hill too high for a late cultivation to do the same.

Hand weeding has been found to be an effective method for preventing weed competition with the crop and for preventing weeds from forming seed that add to the weed seed bank in the soil. Hand weeding is labor intensive and a costly investment for growers, but it does eventually deplete the weed seed bank and can lead to lower weeding costs in the future. If needed, this is the primary weed control of weed escapes after canopy closure. Hand weeding is crucial for keeping the weed seed bank low, because it removes weeds before they produce seeds.

Flaming can be used for weed control. However, this method is expensive and can cause air quality problems. Early-season weeds can be destroyed by flaming immediately, prior to, or shortly after potato emergence. Minor damage to emerging potato plants will not negatively impact tuber production. Unlike cultivation, flaming does not bring new weed seeds to the surface, where they can germinate. However, the operation of planting potatoes disturbs the soil, so the advantage conferred by flaming is lost during planting. In addition, flaming is not effective against grass weeds.

At season-end, potatoes beginning to senesce can be mowed or chopped with a vine beater to prevent weeds from producing seed. Mowing or vine beating can also reduce equipment problems during harvest due to weed vegetative biomass and reduce harvest losses.

Organic Herbicides

Use of organic herbicides should be explained in the farm's Organic System Plan (an organic farm plan required for organic certification). Growers may have to explain that the use of the other tactics for weed control in organic potatoes, such as cultural and mechanical practices, are insufficient and that the use of an organically approved herbicide is, therefore, required.

The Organic Materials Review Institute (OMRI) Products List is a directory of products suitable for weed suppression in certified organic production. These products are all foliar, contact, and nonselective. In other words, organic herbicides can only control actively growing, emerged, green vegetation. A repeat application would be needed to control weeds emerging after the product is applied. Organic herbicide effectiveness depends on the severity of weed pressure, weed species composition, and weed size at the time of application. The degree of control is reduced when the plants are inactive, mature, or biennial/perennial types. Multiple applications can also improve control, in some cases. Organic herbicides are generally ineffective for suppression of grass weeds. When using any of these products in a crop, they should be directed to the base of the crop using drop nozzles and not applied over the top of the crop canopy in order to minimize plant tissue damage. Shielded sprayers also may provide some level of protection to emerged crops.

In order for a compound to be used as an herbicide in an organic crop, the active ingredients must either have a tolerance for the intended use or must be exempt from the tolerance requirement. In many cases, these herbicides are not labeled for in-crop use and, therefore, can only be used around field margins and for other uses. Growers should check with their certifying organization before using any organic herbicides.

Acetic acid and vinegar are commonly discussed as potential herbicides in organic crop production. The US Environmental Protection Agency (EPA) defines vinegar as being 8% or less acetic acid. Household vinegar contains about 5% acetic acid; when applied at post-emergence to small weed seedlings, the vinegar can injure and sometimes kill them. Anything with greater than 8% acetic acid is referred to as acetic acid. Acetic acid is not allowed because it is synthetic; however, there may be sources of nonsynthetic acetic acid, which can be used as herbicides if the requirements of the NOP Rule: 205.206(e) are met. Even though vinegar is a nonsynthetic product, it is not currently registered as a pesticide by the EPA. Research has shown that when using vinegar or acetic acid as an herbicide, control is greatly improved when acetic acid concentration is 20% or greater. The corrosive effects of acetic acid can be serious. Emerged potatoes can recover from early, directed applications of acetic acid.

Products containing corn gluten have also been marketed as organic herbicides. Corn gluten-based herbicides inhibit root development of germinating susceptible seedlings and are applied as a preemergence application. As with mustard vegetation and mustard roots, mustard seed meal contains glucosinolates that break down to isothiocyanates and ionic thiocyanate that can inhibit weed seed germination and establishment. Mustard seed meal is not currently registered for use as an organic herbicide, but can be applied as an organic fertilizer source.

There are also organic herbicides that contain cinnamon and/or and clove oil. These products are exempt from tolerance thresholds, and are allowed for use as herbicides in organic crops.

Biological Weed Management Practices

Biological weed management practices involve the use of insects and pathogens for weed suppression. The primary example of this for organic potatoes is the use of beetle banks. Weed seed predation by beetles may provide some weed suppression and can be encouraged by planting permanent strips of vegetative cover (aka "beetle banks") as habitat for the weed-seed eating beetles. Research conducted by Oregon State University can be seen at www.beetlebank.org. Seed predators being studied are carabid beetles, including *Harpalus* spp., *Pterostichus melanarius*, and others. Seed predation is not always achievable, since many weed seeds are buried too deeply to be available to surface-scavenging beetles. A possible drawback to the use of beetle banks in organic systems is that they can become a reservoir for perennial weeds and weed seeds.

Disease Management

Potato growers face an array of pre- and post-harvest diseases that can be a serious constraint to potato production even under conventional agricultural conditions. According to the American Phytopathological Society publication, *Compendium of Potato Diseases*, 2nd ed., there are 35 economically important bacterial, fungal, and oomycete potato pathogens worldwide (Stevenson et al. 2001). Crop losses due to late blight (caused by *Phytophthora infestans*) alone have been estimated as high as 211 million in the U.S., with control costs totaling 77 million for fungicides. Organic potato growers don't have the array of synthetic fungicides on which they can rely to produce high-quality potatoes to meet market demand. As such, disease management in organically grown potatoes relies heavily on sanitation practices to reduce initial inoculum levels before planting, and cultural practices to prevent or reduce the rate of disease development during the growing season and going into storage. Potatoes have been grown for at least 8000 years, and over the millennia sustainable agricultural practices have been developed for managing pests and diseases. These practices include changing plant and crop architecture, burning, adjusting crop density, depth or time of planting, planting at increased elevation, fallowing, flooding, mulching, multiple cropping, planting without tillage, using organic amendments, planting in raised beds, rotation, sanitation, and manipulating shade (Thurston 2004).

As in all crops, the occurrence of disease in the potato crop is a function of the interaction between host, pathogen, and the environment. These three components interacting together are commonly known as the disease triangle. Shifts in production techniques, such as shortening of rotation intervals between crops, may favor the buildup of soilborne inoculum of potato pathogens, such as *Verticillium dahliae*, leading to situations that require the development of new management options. On the other hand, organic growers can manipulate the disease triangle using production practices, such as good sanitation measures, which result in environmental conditions that are not favorable to pathogen reproduction.

Successful organic potato production begins with good field selection. Fields with a history or moderate to high levels of plant parasitic nematodes or diseases caused by soilborne pathogens, such as *V. dahlia*, should be avoided for organic potato production. Soils can be tested for nematodes prior to planting to determine their level in the field. Fields with heavy soils, or low areas where soil moisture is highest and foliage stays wetter longer due to high moisture, dew, or relative humidity, are more favorable for disease development. Practices that encourage good soil structure and drainage, such as the addition of organic matter, and carefully managed irrigation practices can help reduce the length of periods favorable for disease development.

Once a field has been selected, sanitation practices are the first line of defense against disease development. The main purpose of sanitation practices is to reduce or eliminate sources of pathogen inoculum, which can initiate disease outbreaks. Field sanitation practices may include biofumigants, soil solarization, crop rotation, and eliminating cross-contamination.

Biofumigation is a production practice that is similar to the use of green manure crops. However, in this case the crop is used for management of soilborne diseases. Mustard green manures are an example of a green manure crop that have traditionally been used for their soil quality benefits. However, mustard and other *Brassica* crops have been shown to suppress nematodes and soilborne fungal pathogens. Research has shown that green manures preceding potatoes can suppress Rhizoctonia, common scab, powdery scab, Verticillium wilt, and *Pythium* populations.

Soil solarization and leaving a long period of time between potato crops (i.e., rotating with other crops) can also reduce some soilborne pathogens. However, the crops grown in rotation have to be non-hosts, and weed-hosts must be controlled. For example, in Idaho, potatoes are commonly grown in rotation with sugar beet. Both crops can get diseases caused by *Rhizoctonia solani*, Rhizoctonia stem canker, and black scurf in potatoes and seedling damping off and crown rot in sugar beets. Research at the University of Idaho has shown that *Rhizoctonia* isolates from sugar beet can cause disease on potato. As such, following potatoes with sugar beets would not reduce the levels of *Rhizoctonia* in the soil and may actually increase inoculum levels.

It is very important to avoiding cross-contaminating fields with infested farming equipment, such as potato planters and harvesters, which may carry soil and plant debris from one field to another. Steam or pressure washing to sanitize all equipment before it enters the field is essential to minimize the transfer of soil and plant debris, which can harbor pathogens from other fields. Certain soaps, hypochlorite, peroxide, and other materials approved for organic production are available and should be used to thoroughly disinfect tuber handling equipment, such as planters, harvesters, truck beds, storage facilities, and conveyor belts. Check the label of any product before use to ensure it is certified for organic use.

Seed Certification, Sanitation Measures, and Disease-Resistant Cultivars

Cultural practices form the foundation of any disease management program in organic potato production. After good field selection and sanitation, the second most important factor in preventing disease development is to plant only clean seed. In most cases, this will mean planting only certified seed. In organic potato production, this means certified for organic potato production, as well as certified for varietal purity and disease standards by regulating authorities, normally a state seed certification authority. Such seed is grown under regulated conditions and historical aspects of the seed, such as generation, source, year, grow-out tests, and field observations (e.g., any diseases the crop encountered) are recorded. Once a good source of certified seed has been located, it is important to carry out cultural practices to ensure that the seed does not become diseased before planting. These include, thoroughly cleaning and disinfecting seed storage facilities, not storing seed near potential sources of inoculum (e.g., cull piles), keeping seed lots as separate as possible, and checking for signs of damage during transit (odors and liquefaction). After careful unloading, it is important to store seed at 45 °F and 85–90% relative humidity and keep it well ventilated. Prior to cutting seed, slowly raise the temperature to 50–55 °F. During cutting, seed cutters should be cleaned and disinfected regularly, especially between seed lots. After cutting, seed should be piled no more than 6-ft high, stored at 50–55 °F, and ventilated to promote wound healing. It is important to allow cut seed time to suberize before planting to avoid infection through the cut surface.

Once planted, scouting fields weekly after emergence is key to early disease detection and identification. The earlier that disease is detected, the quicker control measures can be implemented, and the better likelihood of disease control. This is particularly important in organic potato production, since the available products are usually not as effective as synthetic pesticides. Once pathogen populations build up past certain threshold levels, it may not be possible to control them with any organic measures, including organic fungicides. Available fungicides registered for organic production, as of 2016, are protectants (Table 6.5), meaning that they must be present on the plant surface before disease inoculum arrives to prevent infection rather than being curative. In addition, most organic fungicides are biological control products and, as such, must be handled carefully to keep the microbes alive. Biological control products that contain living microbes have a much narrower window of climatic conditions for optimal efficacy, which usually corresponds to the optimal growth conditions for the biological control organism. For example, if you plant potatoes in cold, wet soil (below 55 °F), those tubers are much more susceptible to diseases like Fusarium dry rot and Rhizoctonia stem canker. Applying an in-furrow fungicide will provide good disease control at temperatures well below 55 °F. However, if the optimal growth temperature of a biological control organism is 60 °F and it doesn't do well in wet soil, then it may not provide effective disease control at 55 °F.

Moving towards harvest, potatoes need at least 7–21 days between vine kill and harvest to promote tuber maturation and good skin set. Tubers with good skin set where the tuber skin has fully formed are protected from most common storage diseases, can resist skinning and bruising during harvest and transport, and will be able to store longer with less shrinkage. Late in the season, practices that reduce exposure to damage during harvest, storage, and post-storage are important in the control of diseases, such as bacterial soft rots. Pre-harvest factors, such as washing digging equipment; timing of crop desiccation (dependent on canopy and tuber maturity); storage preparation including inspection, repair and cleaning of insulation, ducts, fans and humidifiers, doors, sensors, and control panels are all very important in the prevention of diseases going into storage. Modifications to harvest

Table 6.5 Biofungicides currently registered for use in organic potato production in the Pacific Northwest[a]

Product type[b] name (active ingredient)	Fusarium dry rot	Silver scurf	Rhizoctonia diseases	Pink rot	Pythium leak	Common scab	Powdery scab	Verticillium wilt	Bacterial soft rot	Late blight	Black dot	White mold	Grey mold	Early blight	Viruses
Biocontrol bacteria															
Double Nickel 55® (*Bacillus amyloliquefaciens* str. D747)	IF, ?		IF, ?	IF, ?	IF, ?			IF, ?	F, ?	F, ?**		IF, F, ?	F, ?	F, ?**	
Double Nickel® LC (*Bacillus amyloliquefaciens* str. D747)	IF, ?		IF, ?	IF, ?	IF, ?			IF, ?	F, ?	F, ?**		IF, F, ?	F, ?	F, ?**	
Optiva® (*Bacillus subtilis* str. QST 713)										F, ?**		F, ?	F, ?	F, ?**	
Serenade® ASO (*Bacillus subtilis*)										F,?**		F, ?	F, ?	F, ?**	
Serenade® MAX (*Bacillus subtilis*)										F,?**		F, ?	F, ?	F, ?**	
Serenade® Optimum (*Bacillus subtilis*)									IF, ?			F, ?	F, ?		
Serenade® Soil (*Bacillus subtilis*)	IF, ?		IF, +	IF, +	IF, +			IF, ?	IF, +	IF, +					
Taegro® Biofungicide (*Bacillus subtilis* var. *amyloliquefaciens* str.n FZB24)	ST, IF, ?		ST, IF, ?												

Table 6.5 (continued)

Product type[b] name (active ingredient)	Fusarium dry rot	Silver scurf	Rhizoctonia diseases	Pink rot	Pythium leak	Common scab	Powdery scab	Verticillium wilt	Bacterial soft rot	Late blight	Black dot	White mold	Grey mold	Early blight
Biocontrol fungi														
ActinoGrow® (*Streptomyces lydicus*)	ST, IF, ?		ST, IF, ?	ST, IF, ?	ST, IF, ?			ST, IF, ?	F, ?	ST, ?		ST, IF, F,?	F, ?	F, ?
Actinovate® AG (*Streptomyces lydicus*)	ST, IF, ?		ST, IF, +	ST, IF, ?	ST, IF, ?			ST, IF, +	F,?			F, ?	F, ?	F,
Actinovate® STP (*Streptomyces lydicus*)	ST, ?		ST, ?	ST, ?	ST, ?			ST, ?						
BIO-TAM® (*Trichoderma asperellum, Trichoderma gamsii*)	IF, ?		IF, ?	IF, ?	IF, ?			IF, ?				IF, ?		
Contans® WG (*Coniothyrium minitans*)												IF, F, +		
MycoStop® Biofungicide (*Streptomyces griseoviridis*)	IF, ?													IF, ?
Prestop® Biofungicide (*Gliocladium catenulatum* Str. J1446)	SD, ?		SD, ?	SD, ?	SD, ?			SD, ?					F, ?	

RootShield® Granules (*Trichoderma harzianum* Rifai strain T-22)	IF, ?		IF, ?		IF, ?									
RootShield® WP (*Trichoderma harzianum* st T-22)	ST, IF, +		ST, 3		ST, IF, ?									
RootShield® PLUS+ WP (*Trichoderma harzianum* str. T-22, *Trichoderma virens* str. G-41)	ST, IF, +		ST, IF, 3	ST, IF, ?	ST, IF, ?					ST, +				

[a]This list is not extensive and does not include copper and hydrogen peroxide products that are certified for use in organic potato production

[b]Listing a pathogen on a fungicide label does not assure the fungicide's effectiveness. The registration status of a fungicide can and does change. At the time of writing, all fungicides listed were registered for use in organic production of potatoes. Application timing: *ST* seed treatment, *IF* in-furrow at planting treatment, *SD* soil drench, *F* foliar treatment. Efficacy: ? = efficacy not tested in University of Idaho field trials, and or no research available on efficacy. + = product tested in University of Idaho field trials and showed some efficacy in at least one trial. ** = Provides disease suppression only

equipment are important factors in managing soft rot bacteria and other fungal pathogens that enter through damaged tuber periderm, such as late blight, pink rot, Pythium and Fusarium dry rot. Knowles and Plissey (2007) identified the harvester as being the major source of mechanical damage to tubers and made recommendations that could reduce damage. These included adjustments to the digging blade, reducing drop heights, and use of padding on hard surfaces to decrease bruising. The potential for damage continues from harvest to loading of tubers onto bulk trucks and to the storage-bin piler, and similar recommendations based on the use of improved padding and flow speeds were made. Preferably, crops should be harvested when tuber pulp temperature is in the range 45–65 °F to eliminate temperature gradients that can promote the development of condensation that, in turn, enhances conditions that are conducive for the development of soft rot.

Other important factors that reduce the risk of soft rot developing during the early storage period include:

- Limiting the pile size to a height of 16–18 ft.
- Quickly cooling the tubers to the final storage temperature.
- Running fans to dry the tubers as much as possible, without reducing humidity too much to dehydrate the tubers.
- Not harvesting low spots or other areas that have elevated levels of tuber decay.
- Sorting out rot during storage filling, aiming to keep infection levels below 3%.
- Piling high-risk lots in areas of the storage that can be removed quickly if rot begins to develop.
- Daily monitoring for high-risk areas with elevated temperature and/or moisture.

Once the desired storage temperature is reached, continuous airflow should be maintained to dry the tubers only if a severe rot potential still exists. Excessive airflow rates, especially at low relative humidity, will dehydrate tubers and interfere with the wound healing process and, thus, disease suppression. There is a fine line between too much moisture and too little in storage. Relative humidity should be as high as possible without causing condensation on the tuber surface and storage structure, as free water will promote pathogen sporulation and development.

Insect Management

Successful management of insect pests of potatoes, whether with an organic or conventional approach, requires understanding of the basic principles of Integrated Pest Management (IPM) and Insecticide Resistance Management (IRM), as well as knowledge of the basic biology, behavior, and ecology of insect pests and their natural enemies. These topics are covered in detail in Chap. 11. Here, we expand upon these principles from the standpoint of organic management of insect pests. Therefore, the reader is encouraged to read Chap. 11 before studying this discussion of organic management of insect pests.

Conventional growers may at times be able to reduce their adherence to IPM recommendations for insect pests and compensate for this, to some extent, by relying on the relatively vast arsenal of conventional insecticides that are available. Organic potato producers, however, generally must adhere vigilantly to a solid IPM program that includes both cultural and biological control in order to ensure successful management of insect pests.

Different approaches to management of insect pests may be categorized as Cultural, Biological, Physical, or Chemical control. Management options within each of these categories are described below with respect to insect pests of potatoes in general or particular species.

Cultural Control

Cultural approaches to insect pest management involve changing the way the crop is grown in order to make it less suitable for the pest and/or to enhance the ability of the crop to withstand attack. Thus, cultural control tactics affect insect pests indirectly via the crop or crop environment. Colorado potato beetle (CPB) is one of the most studied insect pests in the world; therefore, far more is known about management options for this species than for other insect pests of potatoes (Fig. 6.8).

Fig. 6.8 CPB pressure can be intense for organic potato production. Photo shown here illustrates the difference in defoliation between untreated plots (left) and plots treated with an organic insecticide (right). (Photo credit: Erik J. Wenninger, University of Idaho)

Colorado Potato Beetle (CPB)

Crop Rotation Though crop rotation is challenging to use as the primary means of management of CPB, some organic producers have been successful with such an approach. Each year these growers plant potatoes in a field that is several miles from fields that were planted to potatoes during the previous year, which exploits the limited dispersal tendencies of this species.

Early or Late Planting Both early and late planting can reduce the effect of the second generation of beetles on the crop. Early planting can avoid the second generation if the crop has matured and is being harvested when the second generation emerges. Assuming alternative food sources are not available locally earlier during the season, late-planted potatoes can shift the beetle's life cycle to later in the season; the short-day photoperiod then stimulates reproductive diapause, which greatly reduces the effect of the second larval generation on the crop. Implementation of such tactics may be complicated by economic and agronomic logistical issues, including duration of the growing season in a given area and whether the cultivar being grown is a long- or short-season variety.

Trap Crops Trap crops may be used to attract beetles away from a potato crop so that they can be treated within a smaller area. A trap crop may consist of a border crop surrounding the main field, comprised of a more attractive potato variety, possibly with an added kairomone attractant. The trap crop may intercept colonizing and dispersing adults, but would have to be sprayed or otherwise treated to prevent subsequent movement of beetles into the main crop. This tactic can substantially reduce the area that requires treatment; however, the logistics of design and implementation may be prohibitive to most growers.

Straw Mulches Beetle damage can be reduced in potatoes if a straw mulch is applied to a field. Straw mulch appears to inhibit the ability of beetles to find potato plants, increase dispersal away from the field by flight, and increase predation on beetle eggs and larvae.

Manure Amendments to Soil Use of manure with reduced amounts of synthetic fertilizer has been shown to reduce beetle damage, reproduction, and survival, as well as increase potato yields relative to use of full rates of synthetic fertilizer. Specific recommendations on manure application rates are not available, and effects on beetles were robust, but small. Therefore, use of manure amendments would not work as a standalone practice for beetle management.

Wireworms

Crop Rotation Crop rotation to reduce wireworm damage is critical in conventional production, but is of particular importance in organic systems because there are no organic insecticide options available currently for wireworm management. This tactic is discussed in more detail in Chap 11.

Biological Control

Numerous arthropod natural enemies may attack the various insect and mite pests of potatoes. The primary means by which most conventional producers may promote these natural enemies is by simply limiting the use of broad-spectrum insecticides. Fortunately for the organic producer, organic insecticides are generally less detrimental to beneficial arthropod populations; however, many organic insecticides are still toxic. Their relatively low effect on beneficial arthropods may be a result of shorter residual activity and/or lower acute toxicity. Insecticides still should be applied only when justified by scouting of pest populations.

In addition to limiting and targeting insecticide use, additional steps may be taken to encourage populations of beneficial arthropods. Such conservation biological control can be achieved by providing stable habitat and alternative food resources for beneficial arthropods within and adjacent to fields, as follows.

Flowering Plants Many beneficial insects use flowering plants during at least part of their life cycle. For example, lady beetle larvae are voracious predators, but adults feed on pollen in addition to other insects. Parasitoid wasps that lay their eggs in or on pest insects (and whose larvae consume their host) generally feed on nectar in the adult stage. Maintenance of a diverse community of flowering plants adjacent to or within agricultural fields can provide natural enemies with these alternative food resources, thereby increasing the abundance of natural enemies that disperse into fields and provide biological pest control. There are numerous plant species that will provide appropriate floral resources for beneficial insects, but in general, plants with small flowers will be most useful, especially for tiny parasitic wasps. Another important consideration is that the plant community should include a mix of species that flower at different times during the year so that some floral resources are available at all times during the season.

Beetle Banks More commonly used in Europe, beetle banks are comprised of narrow strips of perennial plants (often bunchgrasses) that cut across fields, providing stable habitat in which predatory beetles and other epigeal predators may persist during habitat-disrupting farming operations like tilling, planting, and harvesting. Following a farming operation that disrupts habitat within a field, predators will recolonize a field that contains one or more beetle banks more quickly than if they were colonizing only from the outside edges of fields. A drawback of beetle banks includes the logistics of maintaining a strip of vegetation that often is incompatible with the herbicide regime being used across the rest of the field; however, this is less of a concern in organic systems. Giving up space that would otherwise be planted to a crop also may be considered as a drawback; however, beetle banks can be effective while taking only a very small portion of the field. One 6-ft wide strip for every 75 ac of field area may be sufficient to conserve beneficial arthropods across the field.

Producers who are interested in using part of their farms to create habitat for beneficial arthropods can work with the Natural Resources Conservation Service (NRCS) for logistical and financial assistance. Local NRCS offices can provide more information on available programs, including the best plant communities to use for a given area.

Physical Control

Some physical control methods (i.e., physical actions used directly against a pest) have proven effective against CPB in certain growing areas. These include the use of plastic-lined trench traps, propane flamers, and vacuums. Though labor intensive, these approaches might be useful for some growers—especially for smaller fields or where insecticide resistance limits options.

Plastic-Lined Trench Traps Plastic-lined trenches dug along field edges can intercept beetles that are walking into the field. This can reduce populations of both the spring and summer generations of adults by roughly 50%.

Manipulation of Overwintering Habitat Temperature An experiment showed that applying straw mulch to the overwintering habitat of beetles, then removing that mulch in January, along with the layer of snow that was covering it, significantly reduced survival of overwintering beetles, apparently via the rapid drop in soil temperatures. Successful implementation of this strategy would require knowledge of the local overwintering habitat.

Heat Treatment Application of heat using propane-fueled flames has been shown to cause 30–100% beetle mortality. However, potato plants greater than 10 cm in height also exhibit serious damage from such heat treatments.

Vacuum A vacuuming approach consists of a tractor-mounted machine that supplies bursts of air to dislodge beetles from foliage and a vacuum to suck them up for disposal. Depending on beetle life stage, about 25–50% of individuals may be successfully removed with this method.

Heat Treatment and Vacuum A combination of the heat treatment and vacuum approach can be more effective than either alone and can achieve efficacy similar to the use of insecticides. Many of the beetles dislodged by the blower/vacuum machine fall to the soil surface. Immediately scorching these beetles on the ground with a propane flamer reduces the number of beetles that are able to return to a plant after the vacuum passes.

Chemical Control

Any mention in this chapter of specific chemistries or classes of chemistries is used for informational purposes only. No endorsement of any named products is intended, nor is criticism implied of similar products not mentioned.

There are a number of insecticides approved for organic potato production. Local Cooperative Extension Service offices and extension entomologists should be contacted for information on registered insecticides specific to your area. Because organic insecticides, in general, may be less acutely toxic to insects and/or have shorter residual activity relative to many conventional insecticides, timely application (i.e., before populations develop to very high levels) is important to improve efficacy.

As stated above, far fewer insecticides are approved for organic production than for conventional production. This means that not only must organic producers rely more heavily on other non-chemical components of IPM, but they also must be exceptionally conscientious stewards of the products available. IRM may be even more critical in organic systems because few products are available and few new products are likely to be developed. Unfortunately, CPB have been shown to exhibit cross resistance between neonicotinoid insecticides and at least one insecticide that has organic formulations (spinosad). In other words, a population of CPB that has developed resistance to the conventional neonicotinoid insecticides also will exhibit resistance to spinosad, even with no previous exposure to the latter chemical.

See Chap. 11, for more information on use of insecticides in potatoes.

Vine Kill and Harvest

To facilitate skin set in order for potatoes to endure the rigors of harvest and handling, minimize bruising and skinning damage, and reduce potential disease development and weight loss in storage, potato vines are either vine killed or allowed to naturally senesce. Depending upon variety and maturity of the plant, vines should typically be dead for approximately 10–21 days before harvest to set the skin. Under an organic system, vines can either be mechanically removed either by shredding, chopping, or flailing or by desiccation with burning or flaming the vines. All these methods are easier to accomplish with vines that do not have vigorous foliage and in which natural senescence has already begun. Mechanical vine removal also helps facilitate harvest by removing both potato foliage and weeds to minimize debris that may interfere with the harvester chains and movement of the crop, as well as limit debris entering the potato storage.

Harvesting organic potatoes follows the same management strategies of conventional potatoes, with particular attention given to soil moisture, proper pulp temperatures, and harvester operation. Some organic potato growing operations may have smaller harvesters or lifters, but the principles remain the same regarding equipment maintenance, cushioning of the crop handling system, and minimizing tuber drops to less than 6 in.

Additional information on skin set, vine kill, and harvest can be found in Chaps. 15 and 16, respectively.

Storage

Organic potato storage management relies upon the same fundamental strategies as described in Chap. 17. The foundation of management revolves around ventilation, humidity, and temperature. Minimizing disease, weight loss, and sprouting are important in maintaining quality in storage. Knowing the end use of the potatoes and the criteria necessary to meet that quality will determine storage management and temperature recommendations. Potatoes destined to be processed into fried products need to be stored in a manner to minimize reducing sugar accumulation that contributes to darker fry color. Depending upon the variety, processing potatoes typically need to be stored above 45 °F to lessen the accumulation of sugars. It is ideal to store potatoes at the coolest temperature possible (>38 °F) and still maintain quality for the intended market. Fresh and seed potatoes can be stored at these cooler temperatures to prolong dormancy length, suppress sprouting, and minimize disease development.

The length of dormancy differs by cultivar and storage temperature. Understanding the cultivar differences in dormancy length is important in choosing the appropriate storage temperature for the cultivar and knowing when to apply a sprout control product. It can also help in making decisions on which cultivar to grow and market throughout storage.

One method to delay sprout development is with cold temperature storage (38–42 °F). These cooler temperatures can lengthen the marketing window by retarding sprout development, but potatoes may begin to rapidly sprout once placed in a warmer environment. Fluctuations in tuber temperature may also promote sprouting; for that reason, storage temperatures should remain as consistent as possible.

A second method to minimize sprout development is with the application of an organically approved sprout suppressant. This strategy can also be used in conjunction with cooler storage temperatures. Oils of some herbs and spices (essential oils) have been shown to reduce sprouting in potatoes and could be applied to certified organic crops. These compounds are volatile plant derivatives, such as spearmint oil, peppermint oil, and clove oil. Volatile oils physically damage developing sprouts with a high concentration of the product in the surrounding headspace in the potato storage. Visible damage (blackened) to the emerging sprout is evident after application. Since new sprouts continue to develop, repeat applications are required at 2–3 weeks intervals or on a continuous basis while in storage. Sprout control products are typically applied with a thermal applicator at high temperatures to create an aerosol or thermal fog that is circulated in the storage ventilation system. Timing of application is critical with these sprout suppressants and are most effective when applied at "peeping" or before sprouts are 0.25 in. long. Delay of application may result is sprout suppression failure. Application methods will need to be fine tuned for individual growers, seasons, and cultivars. Some cultivars that sprout rapidly and vigorously may not respond well to these alternative sprout control methods. Cultivars need be assessed on an individual basis for proper rate, timing, and frequency of application. A spray application of an essential oil (e.g., clove) can also be made when potatoes are being packed. This will provide some sprout suppression as potatoes are transported to customers. Depending upon the time of year and the dormancy of the variety packed for retail, potatoes may begin to sprout within 2 weeks of a spray application. Transportation conditions and travel time to customers need to accommodate the potentially narrow retail window.

See Chap. 17, for more information on disease management in storage. Cooler storage temperatures are beneficial in minimizing particular disease development in storage. Selection of varieties tolerant or less susceptible to storage diseases should be integrated into storage management decisions. There are organically approved post-harvest disease products that can be applied directly to potatoes as conveyed into storage or applied through the ventilation system. Approved products for disease suppression include biological fungicides and general disinfectants that may contain chlorine materials, hydrogen peroxide, and peracetic acid. The biological fungicides may be specific to a pathogen, such as silver scurf or Fusarium dry rot, whereas the disinfectants are used as a general biocide. Check with your state organic certifying agency prior to use of these materials.

References

Eberlein CV, Patterson PE, Guttieri MJ et al (1997) Efficacy and economics of cultivation for weed control in potato (*Solanum tuberosum*). Weed Technol 11:257–264

Hutchinson P (unpublished) Weeds specialist. University of Idaho, Aberdeen Research and Extension Office

Knowles NR, Plissey ES (2007) Maintaing tuber health during harvest, storage, and post-storage handling. In: Johnson DA (ed) Potato health management. American Phytopathological Society Press, St. Paul, pp 79–99

Stevenson W, Loria R, Fran G, Weingartner D (2001) Compendium of potato diseases. American Phytopathological Society, St. Paul

USDA NASS (2010) 2007 census of agriculture, organic survey (2008), vol 3, special studies, part 2. https://www.nass.usda.gov/Publications/AgCensus/2007/Online_Highlights/Organics/ORGANICS.pdf. Accessed 20 Apr 2019

USDA NASS (2015) 2012 Census of agriculture, organic survey (2014), vol 3, special studies, part 4. https://www.nass.usda.gov/Publications/AgCensus/2012/Online_Resources/Organics/ORGANICS.pdf. Accessed 20 Apr 2019

USDA NASS (2017) Certified organic survey 2016 summary. https://www.nass.usda.gov/Publications/Todays_Reports/reports/census17.pdf. Accessed 20 Apr 2019

Westermann D, Crothers S (1993) Nitrogen fertilization of wheat no-till planted in alfalfa stubble. J Prod Agric 6:404–408

Westermann DT, Stark JC (1993) Nitrogen balance in cropping sequences. In: Proceedings of the University of Idaho Winter Commodity Schools, Moscow, ID, pp 87–90

Chapter 7
Seed and Planting Management

Phillip Nolte, Nora Olsen, William H. Bohl, and Stephen L. Love

Contents

P. Nolte (✉)
Department of Entomology, Plant Pathology and Nematology, University of Idaho, Moscow, ID, USA

N. Olsen
Kimberly R&E Center, University of Idaho, Kimberly, ID, USA
e-mail: norao@uidaho.edu

W. H. Bohl
Bingham County Extension Office, University of Idaho, Blackfoot, ID, USA

S. L. Love
Aberdeen R & E Center, University of Idaho, Aberdeen, ID, USA
e-mail: slove@uidaho.edu

J. C. Stark et al. (eds.), *Potato Production Systems*,
https://doi.org/10.1007/978-3-030-39157-7_7

Introduction

Proper seed management can significantly impact subsequent crop growth. Selecting high-quality seed is an essential first step for growing a quality potato crop. Planting certified seed will minimize seed-related problems, but growers also need to cut seed pieces to the correct size and plant them accurately and efficiently. This chapter will discuss seed selection, seed piece preparation, and effective planting protocols.

Guidelines for Selecting Seed

The purchase of certified seed is the first step in ensuring suitable vigor and yield potential in a crop. For a seed grower purchasing seed for recertification, it also provides assurance that the seed is within tolerance for economically important diseases. However, just because seed is certified does not guarantee it is equal to all other certified seed, nor that it is free from disease. Certification provides evidence that the seed lot has been inspected, tested, and is within tolerance to the grade standard. This section provides some ideas on how to obtain high-quality potato seed (Fig. 7.1).

Selection of High-Quality Seed

Much information can be obtained about the quality of a prospective seed lot by visiting the seed grower's farm. Visits should be made during the growing season to visually check seed fields. After harvest, equipment, storages, and the seed lot itself can be inspected. Growers should do the following:

Fig. 7.1 A quality crop of potatoes starts out with high-quality seed. (Photo credit: Nora Olsen)

Inspection of Equipment and Storage Facility

Seed handling equipment should be in good repair and clean. Clean equipment is especially important to minimize the spread of diseases, such as bacterial ring rot. Likewise, the storage facility should be in good working order to protect the seed potatoes from temperature fluctuations and light. The area around the storage facility should be free of cull piles and other potato debris, which are sources of disease.

Many seed producers have temperature-recording devices to keep accurate records of storage conditions. These records can be reviewed to determine that storage temperature has been managed properly. Adequate ventilation and humidity are also important. Large fluctuations in storage conditions may lead to increased physiological aging (discussed later in this chapter) and decreased seed performance.

Inspection for Sprouting or Mechanical Damage

Sprouted seed potatoes may suffer performance problems. Broken sprouts often produce excessive and weaker stems, which, in turn, produce more tubers per plant with reduced tuber size at harvest. Bruised or damaged seed is an indication of rough handling during harvest and transport and is associated with physiological aging and increased levels of disease.

Inspection for Diseases

The presence of several important diseases can be detected by visual inspection or with a simple test. It is important to plant and maintain healthy seed because the seed piece contributes to the establishment of the plant. A study with Russet Burbank showed the importance of maintaining seed health and that seed remaining attached to the plant contributes to U.S. No. 1 yield beyond the time plants reached about 8 in tall. Although in this study the seed pieces were physically removed, the study provides evidence that the longer seed pieces remained intact, the higher the yield may be at harvest. This study was with only one cultivar, but it offers insight that other cultivars may possibly react the same way. Pictures and descriptions of disease symptoms discussed here are presented in Chap. 9.

Late Blight Tolerances for late blight in seed are typically included in the same category with other tuber rots. This is a result of the difficulty of detection and identification rather than a reflection of its potentially destructive impact on the subsequent crop. Also, secondary tuber-rotting organisms often invade late blight-infected tubers and mask the late blight symptoms. Consequently, low levels of late blight decay may not be noticeable nor will these levels prevent the seed from being certified.

A careful visual inspection of any seed produced in an area where reports of late blight have occurred during the previous growing season must be completed.

Fusarium Dry Rot Symptoms of severe infestations of *Fusarium* dry rot can be visually detected, but seed will not be certified if it contains more than 2% serious damage by dry- or moist-type *Fusarium* dry rot. Seed lots without severe visual symptoms may still have the potential to develop this disease. Dry rot potential can be determined by conducting a simple "bag test." See the Sidebar 7.1.

Sidebar 7.1: Testing for *Fusarium* Dry Rot Potential in Seed

Randomly select 40–60 tubers from the seed lot in question. Using a sterilized knife, cut 20–30 of the tubers into seed pieces the same size as those produced by your seed cutter. Place the cut seed pieces in a large paper bag (like the bags from a grocery store), fold the top over, and shake vigorously for 1 min.

Place the paper bag inside a large plastic trash bag and fold the top over, but do not seal it because some air must be able to enter the bag. Keep the bag at approximately 70 °F for 3–4 weeks.

Cut the other 20–30 tubers in the same manner, but after placing them in another paper bag, add the seed piece treatment intended for use on the seed. Once again, shake the bag vigorously for 1 min, place the paper bag in the plastic trash bag, and store as described above.

At the end of the 3-week incubation period, examine the seed for *Fusarium* dry rot decay. If the untreated seed pieces have *Fusarium* decay, check to see if the seed piece treatment prevented the growth of *Fusarium* on the treated seed pieces.

Remember, a seed piece treatment will not stop the growth of a preexisting *Fusarium* infection. It will only prevent the development of new infections on healthy seed pieces.

Use proper protective equipment when working with the seed piece treatment.

Soft Rot Some soft rot will be found in most seed lots, but the level of infestation should not exceed 1%. More than this amount could be an indication of potential problems for seed piece decay. Seed tubers coated with dried "slime" that resulted from earlier rot problems in storage will be more prone to soft rot infection after cutting. A test performed by a qualified laboratory can determine soft rot potential.

Rhizoctonia Canker and Black Scurf A simple visual inspection at the seed grower's storage can detect potential problems with Rhizoctonia. On seed potatoes, *Rhizoctonia* sclerotia (black scurf) should not cover more than 20% of the tuber surface.

Sclerotia are seldom responsible for more than cosmetic damage to the infected tuber but are the source of inoculum for the more damaging canker form of Rhizoctonia, which has the potential to cause losses in the field. Rhizoctonia cankers may girdle underground sprouts, thereby stunting or killing stems. The result is a poor stand, lower than expected stem numbers, or both. Developing stolons can also be infected, which can lead to a lower number of tubers per plant.

Bacterial Ring Rot Seed certification programs implement a zero tolerance for this pathogen. Sampling strategies that maximize the probability of detecting the organism in conjunction with sensitive DNA-based testing, such as polymerase chain reaction assays, are commonly used to determine eligibility of a seed lot for certification. Serological testing, such as enzyme-linked immunosorbent assays or immunofluorescent staining assays, have also been used, but validated DNA-based tests are generally considered to be more sensitive and specific.

Silver Scurf Silver scurf does not usually cause yield losses, but its presence may result in cosmetic defects and elevated weight loss, leading to reduced quality in fresh-packed potatoes. Transmission of the disease can occur through seed; therefore, examination for silver scurf symptoms is recommended to decrease the likelihood of the disease becoming a problem in a commercial potato crop.

Scab In general, seed displaying excessive levels of either common scab or powdery scab should be avoided.

Virus Certification standards for allowable virus content vary by state. Tolerance levels for potato leafroll virus (PLRV) are generally below 0.25%. Tolerances for potato virus Y (PVY) are higher but also vary by state. Seed containing in access of 10% PVY should be avoided.

Examination of Certification Records

Before making a final decision on a seed lot purchase, growers should examine seed certification records. Growers can obtain seed certification records from the seed certification agency in their state. Seed buyers should examine the summer field-inspection reports, the storage-inspection reports (available in January), and, if available, results of a winter grow-out test conducted in an area with warm winter weather favorable for potato growth (available in early March). This information is obtained from the North American Certified Seed Potato Health Certification. See Chap. 4, Fig. 4.11). These certificates are available by request from the agency in the state or province where the seed was certified.

All seed potatoes should have passed a shipping-point inspection and must be sealed and properly tagged by a federal-state inspector. All transport trucks must be sealed with a metal seal by a shipping-point inspector. Growers must verify the certification number before accepting a shipment to be certain the potatoes being delivered are the same ones purchased.

Seed Tuber and Seed Piece Size

Seed performance is affected by seed piece size, and the size of seed pieces is highly dependent on the size of the uncut tubers. Because of this, growers should consider the proportion of large tubers when selecting a seed lot.

Seed Tuber Size

Tubers used for cutting into seed pieces should be 3.5–10 oz. This size range is used for cutting because of limitations in the number and nature of cuts that can be made by a mechanical seed cutter. A 3-oz tuber cut exactly in half would yield two seed pieces in the acceptable size range (1.5–2.5 oz); however, it is not likely that a seed cutter would cut these tubers exactly in half, so it is recommended that growers set cutting equipment to leave tubers that are 1.5–3 oz. in size uncut, and plant them as single-drop seed potatoes.

Utilizing tubers larger than 10 oz. increases the likelihood of producing "blind" seed pieces—those with no lateral buds or "eyes." This happens because the number of eyes on a seed tuber increases only slightly as tuber size increases.

The number of eyes per seed piece influences the number of stems per plant. Every eye on a seed piece or whole tuber has the potential of producing at least one stem, although there are physiological factors that can prevent all eyes from producing a stem. Seed pieces cut from large tubers—those more than 10 oz—may not contain enough eyes to produce the desired number of stems per plant. Large tubers also tend to produce seed pieces that may be too large for some planters to accurately plant, which causes skips during planting (Fig. 7.2).

Average Seed Piece Size and Distribution

Two important aspects of cutting seed are to: (1) obtain the appropriate average seed piece size and (2) achieve the proper seed piece size distribution. All cutting operation managers should know both the average seed piece size and the seed piece size distribution of each seed lot being cut. The seed cutter needs to be adjusted, as needed, to optimize average seed piece size and distribution (Fig. 7.3).

Fig. 7.2 Seed pieces need to be uniform in size to avoid planter skips. (Photo credit: Phillip Nolte)

Fig. 7.3 Careful attention to seed cutting and planting practices will result in a uniform stand of healthy, vigorous plants. (Photo credit: Nora Olsen)

Average Seed Piece Size

Emergence, seedling vigor, subsequent plant growth, and final yield are all related to seed piece size. Research has shown that larger seed pieces result in higher total yield than smaller seed pieces. However, the benefit of larger-sized seed pieces diminishes as the seed piece size increases above approximately 2.5 oz. The optimum seed piece size depends on factors such as cultivar, seed availability, cost of seed, in-row spacing, and market incentives.

For most cultivars, planting seed pieces averaging 1.5–2.5 oz. in size will provide optimum agronomic and economic returns. Growers are advised to eliminate tubers and seed pieces smaller than 1.5 oz. during sorting and cutting. Seed pieces less than 1.5 oz. are less productive, resulting in lower yields than larger seed pieces, because of a lower amount of reserves available for sprout growth. Researchers at Washington State University calculated that if only 10% of the total weight of seed pieces was less than 1 oz., it would result in approximately 20% of the planted area having seed pieces with limited yield potential. Several research studies have conclusively shown that seed pieces smaller than 1.5 oz. will produce significantly less yield than larger-sized seed pieces. Generally, growers should avoid planting seed pieces weighing more than 3 oz. because of increased seed costs and reduced planter accuracy.

Planting seed pieces averaging 1.5–2.5 oz. is acceptable for most cultivars, but acceptability of this size range for individual cultivars depends on the number of eyes per seed tuber as discussed above. Cultivars that produce an adequate number of eyes per tuber should produce seed pieces with at least one eye per tuber when cut into seed pieces weighing 1.5–2.5 oz. Cultivars with a low number of eyes per tuber will need to be cut into seed pieces weighing closer to 3 oz. to avoid having blind seed pieces. The seed cutter operator needs to carefully check each cultivar being cut to ensure there are no blind seed pieces.

When determining the amount of seed to purchase, growers need to consider average seed piece size, row width, and seed piece spacing to be certain enough seed is purchased. Table 7.1 shows the amount of seed needed on a per-acre basis when planted in rows spaced 36 in apart. For seed pieces being planted in rows spaced closer or further apart than 36 in, the quantity of seed purchased will need to be increased or decreased proportionally. Alternatively, for rows spaced closer or further apart than 36 in., the spacing between the seed pieces within the row can be increased or decreased such that an equal number of seed pieces per acre is planted, resulting in the same quantity of required seed per acre. The effect of seed piece size on yield and grade is presented in Table 7.2. The effect of seed piece size on stem numbers per acre is presented in Table 7.3.

Table 7.1 Amount of seed needed to plant 1 ac of potatoes at within-row spacings from 8 to 13 in and average seed piece sizes from 1.5 to 3.5 oz. in rows spaced 36-in apart[a]

	Within-row spacing in 36-in rows					
	8	9	10	11	12	13
Seed piece size (oz)	(cwt per acre)					
1.5	20.4	18.2	16.3	14.8	13.6	12.6
2.0	27.2	24.2	21.8	19.8	18.2	16.8
2.5	34.0	30.3	27.2	24.7	22.7	20.9
3.0	40.8	36.3	32.7	29.7	27.2	25.1
3.5	47.6	42.4	38.1	34.6	31.8	29.3

[a]Amount should increase by 10% to account for cutting losses

Table 7.2 Effect of different seed sizes on yield and size profile of Russet Burbank potatoes

	% U.S. No. 1		% U.S. No. 2			
Treatment	>10 oz.	4–10 oz.	>10 oz.	4–10 oz	% undersize	cwt/acre
2 oz	39	36	7	11	7	462
3 oz	33	38	7	13	9	479

Adapted from Kleinkopf and Thornton (1989)

Table 7.3 Effect of different seed piece sizes on number of stems per plant and stems per acre at 3-in row spacings in 36-in wide rows

Treatment	Spacing (in)	Seed (cwt/ac)	Stems/plants	Stems/ac
2 oz	9	24	3.1	60,016
3 oz	12	27	3.9	56,628
4 oz	12	36	4.2	60,980

Adapted from Kleinkopf and Barta (1991)

Seed quantities presented in Table 7.1 are for post-cut seed, so the actual amount of seed that growers will purchase should be increased by about 10% to account for cutting waste and occasions when seed pieces too small for planting are eliminated. Note that even a small increase in average seed piece size results in a fairly large increase in the amount of seed needed. For example, with an average seed piece size of 2.0 oz., it takes 18.2 cwt to plant 1 ac at in-row spacing of 12 in. in rows spaced 36 in, but it takes 22.7 cwt to plant 1 ac with an average seed piece size of 2.5 oz., an increase of 4.5 cwt per acre (Table 7.1). This increase in seed quantity can significantly increase seed costs for a grower, so it is essential growers pay close attention to the average cut seed piece size.

Determining Average Seed Piece Size

Average seed piece size should be frequently checked during the cutting operation. To do this, collect and weigh a sample of approximately 12–15 lbs. (192–240 oz) of cut seed pieces, count the number collected, and then divide the weight of the seed pieces in oz. by the number of seed pieces to determine the average size.

Seed Piece Size Distribution

Not only is it important to have the correct average seed piece size, but it is also important to have a minimum of at least 72% of the seed pieces within the desired size range, which for most cultivars is 1.5–2.5 oz. It is possible to cut tubers into seed pieces that have an acceptable average seed piece size but do not have an acceptable size distribution. (Fig. 7.4a, b).

A seed lot could, in theory, contain equal numbers of only 1 and 3-oz seed pieces, which is an average seed piece size of 2 oz. However, the size distribution of this cut seed would be unacceptable because seed pieces would be either smaller than 1.5 oz. or larger than 2.5 oz.

Seed Cutting and Seed Piece Treatments

The correct average seed piece size and distribution, as discussed earlier, cannot be obtained unless the seed cutter is properly adjusted and maintained. In addition to properly cutting the seed, a seed piece treatment should be used that minimizes

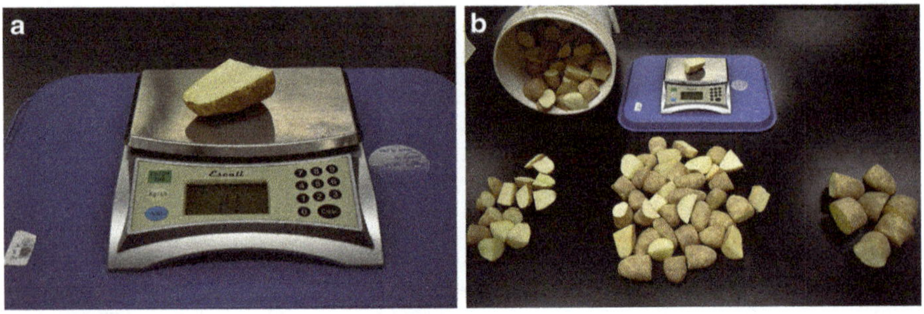

Fig. 7.4 Determination of seed piece size distribution requires weighing (**a**) and categorizing (**b**) individual seed pieces. (Photo credit: Phillip Nolte)

disease spread. Seed piece treatments are also available that will protect the crop against some insects, such as Colorado potato beetle (CPB), and fungal pathogens, such as *Fusarium* dry rot.

Seed Cutter Maintenance and Adjustments

The seed cutter must be maintained, properly adjusted, and continually monitored during cutting to produce quality seed pieces for planting.

Cutter Knives

Sharp cutter knives will produce smooth cuts on all seed pieces. A dull knife leaves an uneven cut surface, much like "fish scales," that will not heal as rapidly as a smoothly cut surface. If fungi, such as *Fusarium* dry rot, are present, there is a greater possibility for the pathogen to infect a seed piece. Also, disease pathogens under these ragged surfaces are less likely to be controlled by a seed piece treatment.

Cutter Adjustments

Initial alignment of the sponge drum, cross knife, and cutting discs should be made before cutting any seed tubers. The seed cutter owner's manual will provide proper settings and procedures. Cutting discs should be set according to the owner's manual and then adjustments made to cut the desired seed piece size. These adjustments will likely need to be changed when changing seed lots because of different tuber sizes. A seed cutter operator should continually monitor the average seed piece size and profile and make adjustments, as necessary.

Fig. 7.5 The frequent cleaning and disinfecting of seed cutting and handling equipment is an excellent way to avoid disease problems in the subsequent crop. (Photo credit: Nora Olsen)

Cleaning and Sanitation

Potato diseases can readily spread during the cutting operation. Of particular concern are bacterial ring rot, *Fusarium* dry rot, and soft rot. The frequency at which a seed cutter needs to be cleaned and disinfected is not easily determined. However, the more often it is cleaned, the less likely a disease pathogen will be spread via the cutter.

A good practice is to thoroughly clean and disinfect the cutter at least once a day and certainly when changing seed lots (Fig. 7.5). Also, if a tuber is found that has bacterial ring rot, it is critical to immediately stop the cutting operation to clean and disinfect the cutter and associated equipment.

The cutter and equipment can be thoroughly cleaned with a power washer that contains an industrial detergent and warm to hot water. This will remove dirt, dried plant sap, and bacterial slime that may be on the cutter and equipment. After completely cleaning, an approved disinfectant should be applied.

All seed cutter surfaces to be disinfected must remain moistened with the disinfectant solution for a minimum of 10 min. The disinfectant solution must contact the bacteria or fungi to kill them. If the solution cannot contact the disease organism, then the disinfecting process is of little or no value.

Several disinfectants are labeled for use on cutting equipment, but because labeled products change, obtain a list of the latest available labeled products from a reputable dealer or from your state Department of Agriculture. Some products can

Fig. 7.6 Application of a seed piece treatment can help reduce or eliminate many seed-related problems in the field. (Photo credit: Nora Olsen)

be used only on non-porous surfaces, so growers must consider this when selecting a disinfecting product. Also, before selecting a disinfectant, growers must consider its activity when diluted in hard water or effects of organic matter, corrosiveness to metal, and worker safety.

Seed Piece Treatments

Because several days are required for cut seed pieces to heal, researchers often recommend seed piece treatment fungicides to protect cut seed until the wound barriers can be established. Several products are on the market, each having advantages and disadvantages. Some seed piece treatments may also contain an insecticide for control of certain insect pests.

A seed piece treatment should be selected that is effective against the problem that is of most concern. It may be necessary to use different seed piece treatment products for different seed lots or for seed planted in different fields.

Regardless of which seed piece treatment is selected, some general guidelines need to be followed to make the most efficient use of the product. Complete coverage of the seed piece is likely the most essential factor. Adequate coverage ensures the most protection (Fig. 7.6).

Regardless of coverage, it is important to realize that a seed piece treatment will not stop an infection that has already occurred from spreading. Treatment will only protect healthy seed pieces from becoming infected.

Cutting and Healing Seed Before Planting (Pre-Cutting Seed)

Potato planting season can be difficult to coordinate and plan when growers want to get the crop planted within a certain time frame. Wet weather can delay planting, which will leave even less time for planting within the desired time frame. Part of the planting operation involves cutting seed tubers into seed pieces, and delays in the cutting operation can delay planting. For that and other reasons, some producers may prefer to "pre-cut" the seed. Pre-cutting has some advantages, but growers also need to recognize its challenges.

Advantages

The big advantage to pre-cutting is that it allows growers time to focus on the cutting operation without being concerned about planting. Growers can take more time to ensure tubers are being cut into properly sized seed pieces and size profile. If soft rot has been a problem in the field, then pre-cutting offers an advantage because the seed pieces can heal (see discussion of wound healing under *Challenges*, below, and in the Sidebar 7.2), which will stop soft rot from invading the seed pieces. Soft rot may be more of a problem when seed pieces are trying to heal a cut surface while in the soil in a field.

Sidebar 7.2: Wound Healing Process
Understanding the wound-healing process after cutting seed tubers into seed pieces will help growers have a better understanding of how planting conditions can affect emergence and plant population.

An early event in the wound-healing process is "suberization," which consists of several steps. Within the outer two or three layers of intact cells just beneath the cut surface, a complex fat-based compound called "suberin" is deposited. The chemical structure of suberin—common bottle cork is 70% suberin—has not yet been fully determined, but it is the origin of the term suberization. A suberin layer protects a seed piece from bacterial pathogens, such as soft rot, and the cut surface loses less moisture than before the healing process began.

The wound-healing process is finished after a new wound cork layer, called the "phellem" layer is developed. Cells beneath the suberin layer divide several times, which results in 4–6 layers of flattened, brick-shaped cells. After the phellem layer is complete, these cells also become suberized in the same manner as the first suberin layer.

The previously formed suberin layer dies and collapses because the cells are cut off from the moisture supply within the seed piece. The seed piece has now developed a new wound barrier similar to the original tuber skin or periderm that appears like the original skin, and better yet, protects the seed piece like the original skin.

Suberization is completed in 2–4 days. The remainder of the process may be completed in a week under ideal conditions, but usually takes longer. Ideal conditions are a temperature of 50–55 °F, plenty of oxygen, and high relative humidity.

Planting in cool, wet soils will delay the wound-healing process, which allows more time for seed-decaying organisms to invade the freshly cut seed piece.

Using an effective seed piece treatment is recommended for both cut-and-plant and pre-cut seed.

Challenges

Just as a seed piece can decay in a field, it can also decay in a pile after cutting. The biggest challenge to pre-cutting is to provide the necessary conditions for wound healing. Seed should never be piled higher than approximately 6 ft. It is also important to supply adequate amounts of moist air, because seed pieces will not heal properly in lower relative humidity, and wound healing requires oxygen. The temperature of the pile needs to be carefully regulated, keeping it at 50–55 °F. Application of a seed piece treatment to combat *Fusarium* seed piece decay is also highly recommended.

Seed pieces can be safely handled and planted 3–4 days after cutting if the proper storage conditions have been met. About 5–14 days after cutting, the healing surfaces of the seed pieces become extremely vulnerable to damage when seed is handled. For this reason, seed that cannot be planted within 3–4 days after cutting may need to be held until 14–18 days after cutting to ensure that the entire wound healing process is complete.

Seed Age

Chronological and physiological are two terms that may be used to describe seed tuber "age." Seed age has a direct impact on crop response.

Chronological Age

Chronological seed age is the duration of time from seed tuber harvest the previous fall to planting the following spring. A chronologically older seed tuber—14 months, for example—will exhibit different performance characteristics compared with a younger seed tuber of only 7 months. This is because the chronologically older seed tuber will also be physiologically older.

Physiological Age

While chronological age is easy to determine, physiological age is not. Physiological age of a tuber involves a complex interaction of environmental and cultural conditions that occurs during the seed growing season and storage, in conjunction with chronological age. Physiological age can be broadly defined as the physiological status of the tuber as it affects productivity.

In general, the most influential factor affecting physiological aging is the accumulation of heat units or exposure to warmer temperatures. Typically, the greater the heat unit accumulation, such as warmer growing temperatures surrounding the seed crop or storage conditions, the physiologically older the seed will be. Fluctuating seed storage temperature can also cause earlier dormancy break and advanced physiological age. Unfortunately, exact relationships between occurrence, type, and duration of temperature exposure and physiological age, and thus impact on crop performance, are difficult to assess and predict.

Although physiological age has been extensively researched, no indicator or test exists to accurately predict it. Researchers have analyzed specific biochemicals in seed tubers, but no direct correlation between these biochemicals and a particular age or performance capacity has been established.

However, four general visual sprout development patterns are associated with distinct stages of physiological age. The youngest and first of these stages is the single sprout, where there is typically one sprout per eye. The second stage is the development of multiple sprouts per eye and loss of apical dominance. Branching of individual sprouts marks the beginning of the third stage. The fourth and oldest stage occurs when the formation of sprouts stops, and small tubers are formed at the eyes. These visual stages will vary with cultivar.

Performance of Physiologically Aged Seed

Some performance characteristics of physiologically older seed include earlier emergence, multiple stems, an increased number of tubers per plant, and earlier senescence. Several characteristics of physiologically younger and older seed are listed in Table 7.4.

Extremely young seed exhibits slower emergence and early season plant development but higher overall yield potential. Planting older seed will result in a reduction in plant stand, season-long vigor, and yield potential.

Table 7.4 Characteristics of physiologically younger vs. older seed

Younger seed	Older seed
Slower emergence	Faster emergence
Fewer stems (sprouts/eye)	Multiple stems (sprouts/eye)
Fewer tubers per plant	More tubers per plant
Later tuber initiation	Earlier tuber initiation (at lower leaf area index)
More foliar production	Less foliar production
Later plant senescence	Earlier plant senescence
Larger-sized tubers	Smaller-sized tubers

Adapted from Iritani et al. (1972)

Factors Influencing Physiological Age of Seed

Growing Conditions of the Seed Crop

Many environmental and cultural conditions during the growing season may influence seed physiological age; unfortunately, solid relationships have yet to be established. Some research has indicated sub-optimal fertility and irrigation, early plant senescence caused by disease, insects, and frost or any other plant stress may all contribute to physiological aging of seed. Weather conditions during the growing season, especially soil and air temperatures, have the potential to age seed tubers. In general, the warmer the growing season, the greater the potential there is for aged seed.

Handling of Seed Tubers

Generally, as seed tubers are handled the physiological age of the tuber increases. Rough handling that promotes bruising, wounding, or stress on the tuber can impact seed age.

Temperature During Seed Storage

Typically, seed potatoes are stored at approximately 38 °F to minimize sprouting, decrease transpiration (water losses), and minimize physiological aging (Fig. 7.7).

Warmer storage temperatures will increase physiological age. Depending on the year and cultivar, small differences in storage temperatures may or may not significantly impact seed performance. Warming the seed at the end of the storage season will promote advanced sprouting and physiological aging. Warmer temperatures

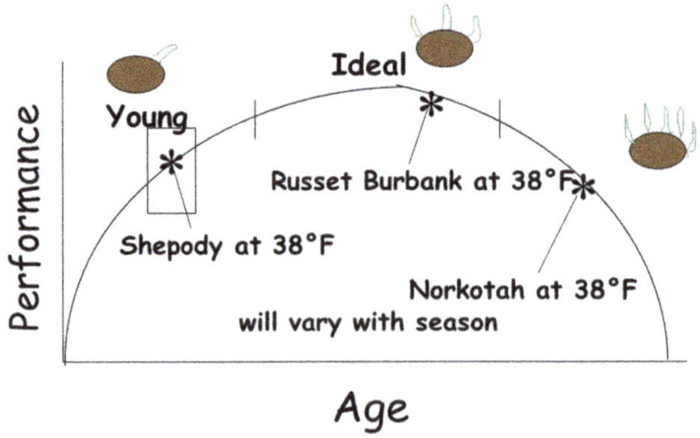

Fig. 7.7 Potential seed age and performance as influenced by seed storage temperature. (Image credit: Nora Olsen)

during early seed storage can also impact physiological age and performance. The best practice to avoid both aging and handling damage is to warm seed to at least 45 °F just before handling, cutting, and planting.

Cutting and Transporting

When seed potatoes are cut, the internal apical (bud end of the tuber) dominance within the seed piece diminishes, thereby allowing sprouting of eyes on other areas of the tuber. When transporting seed, handlers should minimize extremes and fluctuations in temperatures and limit the amount of time in trucks. Temperature variations and excessive time in transport can contribute to advancing physiological age of the seed, and lack of ventilation can increase the potential for pathogen infection.

Minimizing Physiological Age of Seed

Because a predictable method for measuring the physiological age of seed has yet to be developed, it is difficult to know how any particular lot of seed will perform. Thus, using physiological age of seed in an attempt to control crop performance is unpredictable. Therefore, the best strategy for assuring predictable seed performance is to adopt practices that keep seed physiologically young. This includes purchasing seed from growers that utilize optimal seed production practices and store seed at consistently low temperatures, minimizing bruising and tuber damage during transport and seed-piece cutting, and providing holding conditions prior to planting that minimize sprouting.

Planting

Optimizing yield requires using clean, correctly adjusted equipment; planting fields under proper conditions; and precisely planting the seed pieces (Fig. 7.8).

Establishing a Uniform Plant Stand

Several factors must be considered when establishing a uniform plant stand.

Time of Planting

Planting date should be a function of soil temperature rather than calendar date. Soil temperature should be above 45 °F to minimize seed piece decay and encourage rapid emergence. For example, Russet Burbank seed pieces planted with a soil

Fig. 7.8 Optimizing yield requires using clean and correctly adjusted equipment, planting fields under proper conditions, and precisely planting the seed pieces. (Photo credit: Nora Olsen)

temperature of 45 °F may take nearly 6 weeks to emerge, but will emerge in about 4 weeks from the same depth when planted in soil with a temperature of 50 °F.

Planting in cool soil will not likely get the crop out of the ground sooner than waiting for warmer soil temperatures. However, the initial planting date should be scheduled to get all potatoes planted in a timely manner.

Soil should be moist before the crop is planted. If necessary, producers could irrigate fields before planting. Be aware that water applied too soon after planting seed pieces will likely enhance seed piece decay.

Seed Piece Spacing

The ideal seed piece spacing depends on cultivar and end use. For example, potato processing companies that purchase potatoes to be made into french fries usually stipulate a lower size limit and desire larger potatoes. Conversely, the market for red cultivars favors smaller-sized tubers. Growers need to know the end use and the most desirable size category of the harvested crop.

The seed piece spacing that provides the desired harvested tuber size may be different for the same cultivar produced in multiple growing regions in the U.S. Grower experience becomes critical in determining optimum spacing in situations of variable growing conditions, changing market specifications, and differing cultivar requirements. Historical records of past planting practices may be very useful in planning optimum spacing protocols, and keeping such records is very important.

There is also a relationship between seed piece size and in-row seed piece spacing. Research has been reported indicating that, at least for some cultivars, equal quantities of seed planted per acre will produce equal total and U.S. No. 1 yields.

That is, smaller seed pieces planted at narrower in-row spacing will produce the same yields as larger seed pieces planted at wider in-row spacing.

Planting Depth

A common planting depth for many cultivars is 6 in. In general, that is 3 in below the soil line with an additional 3 in formed in a hill above the seed piece. It may be tempting to plant seed pieces deeper than this to reduce the amount of field-green tubers. Tubers exposed to the sun will turn green, which is considered a grading defect. However, research has shown that planting deeper may not reduce the amount of field-greening and may cause a yield reduction in some cultivars. Also, planting deeper than recommended may delay the time it takes for plants to emerge.

Growers will generally form a hill after planting, which will place the seed piece deeper in the soil when measured from the top of the hill to the top of the seed piece. Hilling practices may be marginally effective for controlling field-greening of tubers.

Planting Seed Pieces Accurately

Seed piece size distribution and planting speed are the two main factors that contribute to planting accuracy. Other minor factors influence how accurately a planter places seed pieces, but paying particular attention to these main factors will help optimize seed piece placement. See Sidebar 7.3.

Sidebar 7.3: Importance of Seed Piece Spacing

The seed piece spacing interval that will produce tubers in the desired size range at harvest depends on the cultivar grown and end use. For example, a cultivar that will be used for fresh pack may need to be planted at narrower spacing than if it is intended to be processed into frozen fries. Consequently, a grower needs to know the preferred seed piece spacing interval for each cultivar.

More importantly, however, is making certain the planter is placing the seed pieces at the intended interval. To accurately determine seed piece spacing, it is essential to uncover the seed pieces in at least 25 ft of row behind each planter unit and measure the seed piece intervals.

Generally, as the seed piece spacing within the row is increased, the average size of the harvested tubers will increase. However, extending the seed piece spacing beyond that which is ideal for a particular cultivar will likely not produce a higher percentage of large-sized tubers. Factors affecting harvested tuber size include cultivar, length of growing season, and production area.

Seed Size Distribution

Potato planters will generally place seed pieces more evenly when the seed pieces are of uniform size.

Planting Speed

No single planting speed is "best" for all planters, but generally as planting speed increases, planting accuracy decreases. Planting speeds may vary from about 2–5 mph. However, adjusting planting speed will likely solve many planter performance problems. Planters that are properly adjusted and operated correctly should place 75% or more of the seed pieces within the desired spacing.

Seed pieces placed in the row within 3 in of the desired spacing are considered accurately planted. Planting too fast causes more skips because seed pieces may roll after they are dispensed or seed pieces are not picked up by the cup or pick.

Other Planter Adjustments

A number of minor planter adjustments can be made to achieve higher planting accuracy. When using cup planters, the seed level in the bowl should be even with the conveyor delivering the seed. Chain and idler spring tension can also influence seed piece placement. Air cup and belt-type planters may also require fine tuning to produce the proper results.

References

Iritani WM, Thornton R, Weller L et al (1972) Relationships of seed size, spacing, stem numbers to yield of Russet Burbank potatoes. Am J Potato Res 40:463–469

Kleinkopf GE, Barta JL (1991) Seed quality for commercial and seed growers. In: Proceedings of the Idaho Potato School, 23:255–257. Pocatello, ID, 16–17 January 1991

Kleinkopf GE, Thornton RE (1989) Potato seed tuber quality factors: physiological age and physical characteristics. In: Proceedings of the 21st Annual Idaho Potato School, Pocatello, ID, 18–19 January 1989

Chapter 8
Nutrient Management

Bryan G. Hopkins, Jeffrey C. Stark, and Keith A. Kelling

Contents

B. G. Hopkins (✉)
Department of Plant and Wildlife Sciences, Brigham Young University, Provo, UT, USA
e-mail: hopkins@byu.edu

J. C. Stark
Idaho Falls R&E Center, University of Idaho, Idaho Falls, ID, USA
e-mail: jstark@uidaho.edu

K. A. Kelling
Department of Soil Science, University of Wisconsin-Madison, Madison, WI, USA
e-mail: kkelling@wisc.edu

J. C. Stark et al. (eds.), *Potato Production Systems*,
https://doi.org/10.1007/978-3-030-39157-7_8

Introduction

Crops providing food, fuel, and fiber for a growing world population require nutrients. Plants growing in the wild get these naturally from minerals breaking down into soil, decomposition of dead organisms, and air and water deposition. This process of nutrient cycling is a key component of agricultural production systems as well, but is inadequate to efficiently provide for the rapidly increasing world population. Modern crops produce biomass and harvestable yields at a substantially higher level than their wild relatives. Harvesting removes nutrients from soil as the crops are transported from the field. Without proper nutrient management, the soils eventually become infertile. Therefore, fertilization is essential for maintaining an adequate food supply. Efficient potato nutrient management systems ensure essential plant nutrients are available at the right rates, timing, and placement to provide for optimal growth. Nutrient deficiencies potentially reduce yield and tuber quality, but excessive applications can also reduce yield and quality, as well as cause unnecessary expense and increase the risk of air and water pollution. To optimize these considerations, potato growers must understand nutrient: (1) needs and uptake patterns, (2) availability factors, (3) assessment practices, and (4) management practices.

Potato Nutrient Requirements and Uptake Patterns

All plants require at least 17 essential nutritional elements to complete their life cycles (Table 8.1). Additionally, there are several non-essential, but beneficial elements, including: cobalt, selenium, silicon, sodium, and vanadium. Many of the essential nutrients, and all of the beneficial elements, are not needed to be added by farmers because of their abundance in the environment and/or miniscule amounts

Table 8.1 Essential plant nutrients and available diagnostic tests

Element	Form taken up by plants[a]	Major source	Reported concentration ranges for crop plants (dry weight)	Amount in 450 cwt of potato, lb./ac lb./ac (dry wt)	Diagnostic test available	
					Soil	Plant
Non-mineral nutrients						
Carbon (C)	Carbon dioxide (CO_2)	Atmosphere	~45%	4500	N/A	N/A
Oxygen (O)	Oxygen (O_2)	Atmosphere, water	~43%	4300	N/A	N/A
Hydrogen (H)	Water (H_2O)	Water	~6%	600	N/A	N/A
Primary nutrients						
Nitrogen (N)	Nitrate (NO_3^-), ammonium (NH_4^+)	Organic matter, atmosphere	0.5–6.4%	200–240	1	1
Phosphorous (P)	Phosphate ($H_2PO_4^-$, $HPO_4^=$)	Soil minerals, organic matter	0.1–1.3%	25–35	1	1
Potassium (K)	Potassium (K^+)	Soil minerals	0.3–14%	280–320	1	1
Secondary nutrients						
Calcium (Ca)	Calcium (Ca^{++})	Soil minerals, irrigation water	0.03–6%	50	1	2
Magnesium (Mg)	Magnesium (Mg^{++})	Soil minerals, irrigation water	0.02–5%	40	2	2
Sulfur (S)	Sulfate ($SO_4^=$)	Organic matter, irrigation & precipitation water	0.08–1.4%	18–24	3	2
Micronutrients						
Boron (B)	Borate ($H_2BO_3^-$)	Organic matter	1–200 ppm	0.18	1	2
Manganese (Mn)	Manganese (Mn^{++})	Soil minerals	1–2000 ppm	0.9	1	2
Zinc (Zn)	Zinc (Zn^{++})	Soil minerals, organic matter	5–400 ppm	0.11	1	2
Copper (Cu)	Copper (Cu^{++})	Soil minerals, organic matter	1–60 ppm	0.09	2	2
Iron (Fe)	Iron (Fe^{++}, Fe^{+++})	Soil minerals, organic matter	2–1820 ppm	1.8	3	2
Molybdenum (Mo)	Molybdate ($MoO_4^=$)	Soil minerals	0.01–20 ppm	0.03	4	2
Chlorine (Cl)	Chloride (Cl^-)	Precipitation	0.05–3%	40	2	2
Nickel (Ni)	Nickel (Ni^{++})	Soil minerals	0.1–10 ppm	0.03	4	4

[a]Other forms of these nutrients are present in the soil. This list includes only forms used by plants

[b]For diagnostic tests, 1 = yes, with good correlations and interpretation data; 2 = yes, but with minimal correlation/interpretation data available, 3 = yes, but poor correlations with yield response, 4 = no (Adapted from Bryson et al. 2014, Schulte et al. 2009, and Westermann 2005)

needed by plants relative to what is found in soil. However, some of the nutrients are frequently limiting and need to be added as fertilizers for sustainable potato production.

Nutrients are the building blocks of plant structures and are used in chemical reactions. For example, nitrogen is a structural component of amino acids found in proteins and DNA, and phosphorus is a constituent of energy transfer reactions. A deficiency of any nutrient hinders growth and may kill the plant or prevent it from completing its life cycle.

Plants primarily obtain non-mineral nutrients from air and water and mineral nutrients from soil. Table 8.1 provides a list of these nutrients along with the forms taken up by plants, as well as primary source, typical concentrations in plants, amounts in a typical potato crop, and those nutrients for which diagnostic tests for soil and/or plant tissue are available.

Plants are comprised mostly of water and succulent, young tissues can be over 90% water. Carbon, hydrogen, and oxygen, which are the non-mineral elements in plants, usually comprise over 90% of the dry matter that remains after all the water is removed. Hydrogen is ubiquitous in the environment, with more atoms in living terrestrial ecosystems than any other element. Carbon and oxygen are plentiful in the atmosphere, with ~400 ppm carbon dioxide and 21% O_2. As such, plant shoots (aboveground parts) have ample supplies of the non-mineral nutrients. Plant roots also obtain adequate amounts of carbon and hydrogen from the soil and from sugars generated in leaves, which are transported downward through the phloem tissue. However, there is not enough oxygen transported downward through the phloem conductive tissues to the roots to meet plant needs. Fortunately, there is generally enough gas exchange in well-drained soils between soil and the atmosphere for roots to have ample oxygen (and for the soil to rid itself of potentially toxic accumulations of other gases). However, a lack of oxygen in poorly drained soils frequently limits yields. This is remedied by proper irrigation, drainage, and in some cases, tillage to increase soil pore space.

Therefore, farmers do not need to fertilize for the non-mineral nutrients; rather insure a healthy balance of air and water in the soil. Adding non-mineral nutrients as fertilizer materials is sometimes promoted, but this is a misguided and unfounded practice.

Alternatively, despite making up a small percentage of the plant, it is vital that mineral nutrients are added as fertilizer materials. This is an essential key to successful potato production. Potato requires optimal nutrient levels throughout the growing season to ensure rapid, steady tuber growth and normal tuber development. Nutrient uptake rates for potato follow an s-shaped pattern, with slow initial uptake due to minimal growth, followed by high uptake rates during the rapid vegetative growth period and through early stages of tuber development before leveling off during tuber maturation late in the growing season. Figure 8.1 shows the nutrient uptake patterns for nitrogen, phosphorus, and potassium. The other nutrients are not shown, but they tend to follow the pattern for phosphorus. Additional information on potato nutrient use at various growth stages is provided in Chap. 2.

Potato is unique, as it is planted as a seed piece (whole or portion of a tuber from which sprouts grow) rather than a true seed for commercial production. In contrast to most crop plants that emerge within a few days of planting, potato sprouts take 3–4 weeks to emerge and then several additional weeks before beginning rapid growth (~5–7 weeks after planting). Also, in contrast to most other crops, the potato seed piece has a relatively large amount of nutrient storage, which is available for early plant growth as compared to typical seeds. This factor is important in nutrient management because of the significant delay between planting and the time when potato needs a large supply of nutrients.

Once potato plants begin rapid vegetative growth and uptake of nutrients, they amass about 40–50% of seasonal nitrogen and potassium requirements and 30–40% of phosphorus and sulfur requirements by the time tuber bulking begins at about 9-10 weeks after planting (Fig. 8.1).

Factors Influencing Mineral Nutrient Availability

Soil chemical, physical, and biological properties impact nutrient availability in complex and interacting ways. Additionally, pest and pathogen pressure and cropping systems, including variety grown, will variably impact nutrient availability.

Soil Composition Plants take up a majority of their nutrients from minerals dissolved in the soil solution. Fertile soil has sufficient reserves to maintain a high enough concentration of nutrients in the soil solution to permit optimum plant growth throughout the growing season. A major factor influencing a soil's available

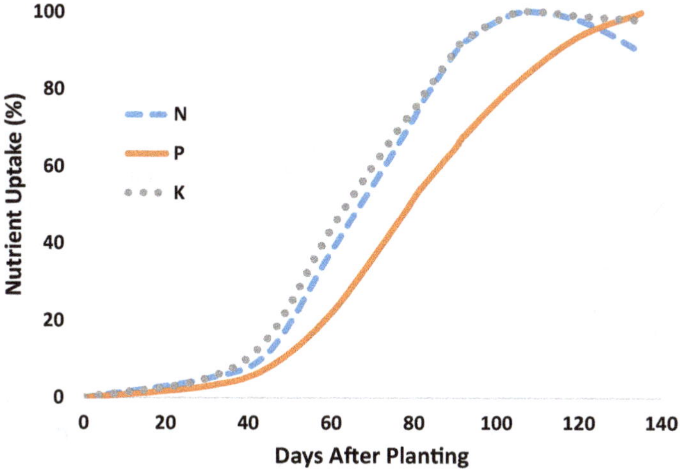

Fig. 8.1 Total nitrogen (N), phosphorus (P), and potassium (K) uptake by Russet Burbank potato for averages of field studies in ID, OR, CO, and MN. (Adapted from Stark and Westermann 2003)

nutrient status is the proportion of sand, silt, clay, and organic matter (humus). The exterior surfaces of the soil particles are predominantly negatively charged, which enable them to hold water and positively charged nutrients (cations). The measure of a soil's ability to hold and exchange cations is the Cation Exchange Capacity (CEC).

The CEC attracts positively charged elements and molecules (cations), including those listed in Table 8.1. Nearly all ammonium, potassium, calcium, and magnesium used by plants comes from the CEC exchange sites. The cationic micronutrients are also held by the CEC exchange sites; but more importantly, they are also held very tightly by organic matter complexes and chelates.

Clay and organic matter have a relatively higher surface area than sand and silt. Thus, the higher the clay and organic matter percentages in soil, the greater the water and nutrient holding capacity. Soils dominated by sand and silt tend to have low nutrient holding capacity and, as such, are relatively infertile—requiring extra care with regard to nutrient management. Ironically, potato tends to grow well in these soils as long as efficient fertilization practices are used.

Pure clay has CEC values of ~60–80 cmol/100 g soil, and pure organic matter is in excess of 200 cmol/100 g soil. By comparison, sand and silt CEC values are less than 1 cmol/100 g of soil. Due to environmental conditions, soils high in clay are usually also higher in organic matter (>3%). There is an infinite number of combinations of these various soil constituents, with CEC ranges from very sandy, low organic matter soils (CEC <10 cmol/100 g) to soils with high clay and/or organic matter (CEC >30 cmol/100 g). In general, low CEC soils require larger quantities and more frequent applications of many of the nutrients.

Besides nutrient holding capacity, there are other interactions that soil particles have with nutrient management. Soil organic matter not only holds nutrients on its CEC sites, but also contains nutrients within its structure. These nutrients are released into the soil solution as the organic matter is decomposed by microorganisms. Soil organic matter and the materials from which it is derived (crop residues, manure, etc.) contain considerable amounts of nutrients, which are particularly important sources of nitrogen, phosphorus, potassium, sulfur, zinc, and boron.

In addition to the effect of nutrient holding capacity, the mineral or inorganic portion of soil has other effects on plant nutrient availability. One major impact is related to water movement through soil as a function of pore size. Clay soils have smaller spaces between soil aggregates—resulting in reduced rates of water movement through soil (hydraulic conductivity). Conversely, water and nutrients tend to move downward rapidly in low CEC soils (assuming no physical barrier or severe compaction exists). As such, these sandy, low organic matter soils tend to be less fertile due to nutrient losses below the root zone. This is a greater problem for the negatively charged nutrients (anions), which are not held effectively by the cation exchange sites. Therefore, soluble anions (nitrate-nitrogen, sulfate-sulfur, borate, and chloride) easily leach from the active root zone—even in fertile soils—particularly when large amounts of water percolate down through the soil root zone.

Some anions (especially phosphorus) are not easily leached. This is not a function of the CEC, but is instead due to poor solubility. For example, phosphorus rapidly forms immobile precipitates with calcium, magnesium, iron, and aluminum,

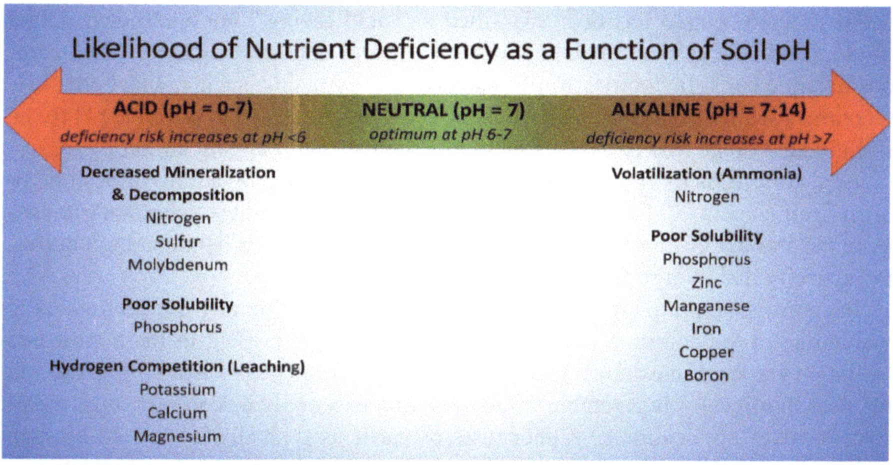

Fig. 8.2 Influence of soil pH on plant nutrient deficiencies

which form on soil particles, thereby reducing phosphorus concentrations in the soil solution.

Soil pH Most crops grow well under slightly acid to slightly alkaline conditions (pH 5.7–8.4), and potato grows well in many regions with soils having pH levels throughout this range. However, extremes at either end of this range can result in challenges for nutrient management. In most potato cropping systems, the main influence of soil pH relates to its effect on nutrient solubility and the corresponding effects on plant nutrient availability. A summary of the influence of pH on nutrient availability is shown in Fig. 8.2.

Whether a soil is acid or alkaline is primarily a function of climate—with areas receiving high rates of precipitation typically being acid and arid climates being alkaline. Water contains a percentage of dissociated hydrogen, which accumulates in soil. As hydrogen ion activity increases, the pH decreases. Hydrogen ions can dissolve minerals and increase solubility of aluminum (Al) and other nutrients (especially phosphorus, manganese, iron, zinc, copper, and boron). This solubility increase can benefit plant nutrient uptake, but very acidic conditions (pH <5.0) can increase concentrations of some elements to toxic levels, especially manganese and aluminum.

An additional effect of pH occurs as these dissolved elements chemically bond with other nutrients—resulting in a deficiency. Notably, this impacts phosphorus as it is more likely to be deficient in acid soil due to the formation of solid precipitates (aluminum, iron, manganese, phosphates, etc.) that are not very soluble and, thus, not very available for plant uptake. Molybdenum can also become deficient in acid soils. The influence of pH on the availability of potassium, calcium, and magnesium in acid soils is due principally to competition between cations and hydrogen ions for exchange sites.

Historically potato has been classified as "acid loving," but significant potato production in the U.S. occurs in the western arid and semi-arid regions with calcareous alkaline soils. In contrast to truly acid-requiring plants, such as blueberry, it is recommended for potato that extremely acidic soils be treated to raise the pH to the slightly acid range (pH 6.0–6.5), unless common scab is a significant problem.

Acid soil needs to be remedied and is readily corrected by adding limestone or other liming agents. A buffer pH test is performed by the soil test laboratory to estimate the amount of lime needed, with more required for soils having a high buffering capacity due to high CEC.

In contrast, neutralizing the pH of an alkaline soil is generally not economically viable due to the common overabundance of limestone present in these soils and irrigation waters. Limestone (calcium carbonate) has a pH of ~8 and buffers the soil pH near that level. Management of alkaline and especially calcareous soils, therefore, involves application of higher rates of nutrients typically influenced by high soil pH, especially phosphorus and some of the micronutrients—as indicated by soil testing. The fact that many of the highest potato yields in the world are produced in the alkaline soils of the Pacific Northwest is evidence that pH adjustment is not necessary for potato production on alkaline soils.

As with acid soils, phosphorus solubility in alkaline soils is poor. Additionally, manganese, iron, zinc, copper, and boron solubility is also poor in alkaline soil. As pH increases, poorly soluble precipitates of these nutrients form—rendering them less plant available.

Most alkaline soils are also calcareous, which is an abundance of free excess lime (calcium and magnesium carbonates). The mineral composition of soil, along with pH, impacts the solubility of chemical elements, with excess lime having a large impact. As soil excess lime concentration increases, the solubility and, thus, the plant availability of the nutrients impacted at alkaline pH decrease further.

Soil pH also influences soil microorganism activity. The optimum pH for most soil microorganisms is pH 6.0–7.5, with reductions in activity typically occurring outside of this range. Nutrients are stored in soil organic matter, and this supply decreases if conditions are not optimum for microbial growth. The effect of pH on nitrogen and sulfur availability is due mainly to its influence on microbial activity, although this can also impact other nutrients in the same way. Also, microbes responsible for converting atmospheric nitrogen into a plant-usable form tend to be negatively impacted when the pH is outside of the optimal range. Additionally, alkaline pH increases loss from gaseous volatilization of some nitrogen forms, particularly when ammoniacal sources of nitrogen and urea are applied to the surface of calcareous soils.

Soil Physical Properties As with the chemical and biological factors, soil physical characteristics substantially impact nutrient availability to plants. Restrictions in root growth caused by hardpans, soil compaction, or shallow soils reduce the soil volume from which nutrients can be extracted, decreasing available water and nutrient supply. Plants with limited root systems generally have stunted growth. It is notable that the combination of reduced shoot growth and a higher percentage of roots exploring the relatively more fertile topsoil can result in the concentration of some nutrients being deceptively higher compared to healthy plants.

Soil temperature also impacts root and shoot growth. Particularly, low soil temperatures early in the growing season reduce microbial degradation of organic matter and plant root activity and growth. This can result in nutrient deficiencies. This effect is particularly significant for phosphorus, which is relatively immobile in soil and requires active growth for roots to explore and find it in the soil. Banding fertilizers or applying higher rates can partially compensate for low temperature effects. Cool soil conditions also reduce the rate of nitrogen and sulfur mineralization from soil organic matter and slow the conversion of ammonium to nitrate. High soil temperatures can accelerate emergence and early-season growth, which creates a corresponding increase in nutrient demand. Nitrogen mineralization rates also increase in warm soils, which under abnormally hot weather conditions can increase soil nitrogen availability causing excessive plant nitrogen uptake. Excessive nitrogen uptake caused by over-fertilization or high nitrogen mineralization rates favors vegetative growth at the expense of tuber growth. If soil temperatures get too high (>95 °F), plant growth is significantly reduced and mineralization of soil organic matter slows. Very high soil temperatures cause problems with tuber development, particularly prior to canopy closure if the soil is dry (especially with sandy soils).

Moisture content is another important physical property of soil. In addition to impacting temperature and nutrient leaching, soil moisture directly influences plant growth and nutrient use. Excessively wet soils lack oxygen, which impacts soil chemical reactions and plant and soil organism health. Excessively dry soils limit root growth and nutrient transport to roots. These are more of an issue for regions reliant on precipitation rather than irrigation.

Pests and Pathogens Pests (such as nematodes) and diseases (such as *Rhizoctonia*, blackleg, and *Verticillium*), which attack potato root and stem tissue, can significantly reduce nutrient uptake and transport within the plant. Short rotations and nutrient deficiencies in previous potato crops can increase *Verticillium* levels, leading to greater root damage and reduced nutrient uptake. Appropriate nutrient management, especially nitrogen, helps suppress symptoms of early-die complex and early blight. Growers need to use appropriate rotations and pest/pathogen control strategies to maintain healthy roots and vascular tissues capable of optimal nutrient uptake rates. There are differences among varieties in terms of susceptibility to pests and pathogens (Fig. 8.3).

Potato Variety and Yield Potential Potential tuber yield and fertilization strategies are affected by the maturity class and the length of the growing season of the variety. Varieties that mature early, such as Russet Norkotah, have a shorter growing cycle and generally require higher rates of nutrients applied earlier in the season. Late-maturing varieties (such as Ranger Russet and Russet Burbank) may have more total leaf area, dry matter, and potential tuber yield, resulting in higher nutrient needs and a longer period of availability. Any factors, such as climate and pests, which impact yield potential will correspondingly impact nutrient needs. Fertilizer amounts, especially primary macronutrients, generally need to increase as yield potential increases. Potato yields have steadily increased during the Green

Fig. 8.3 Russet Burbank variety on left suffering from early-die complex compared to Ranger Russet variety on right. (Photo credit: BG Hopkins)

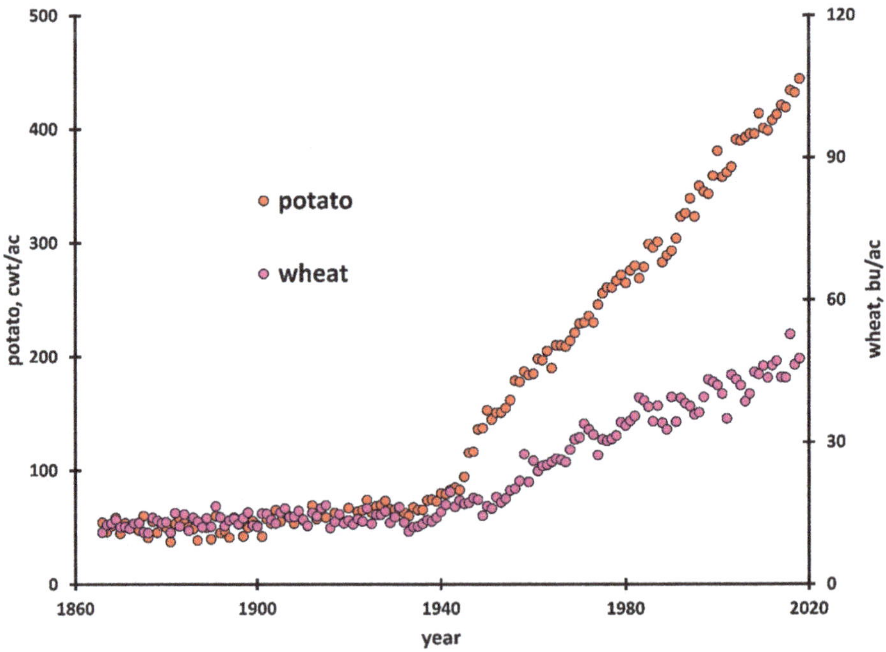

Fig. 8.4 U.S. potato yields compared to wheat. (Adapted from Hopkins and Hansen 2019)

Revolution (since about 1940) with no hint of slowing (Fig. 8.4). Thus, nutrient needs are increasing as yield potential increases.

Additionally, there can be tremendous differences across varieties in terms of nutrient needs. This is related more to the effectiveness of the root system than nutrient demand by the tissues. Although Russet Burbank is the dominant variety grown in North America, it is poor in terms of root depth and nutrient uptake efficiency. In contrast, the variety Alturas develops a deeper, more extensive root system and is far more efficient in extracting soil nutrients. The variety Shepody, which has greater root density and earlier maturity than Russet Burbank and Ranger Russet, is less responsive to nutrients, especially phosphorus, than these other varieties. Many of the newer potato varieties require less fertilizer per unit of yield produced than the older, standard varieties.

Cropping System Like many crops, potato needs to be grown in a rotation with other crop species to minimize the buildup of soilborne diseases and maintain optimal soil health. Crops grown in the field prior to potato also impact nutrient availability and uptake, which have a strong influence on vine and tuber growth.

The most recognized impact is when legumes (such as alfalfa, clover, beans, peas, peanuts, and lentils) are grown in rotation with potato. This reduces nitrogen requirements for crops grown in the field the following year. This effect can last into the second and even third year after alfalfa.

There are several reasons for this nitrogen credit phenomenon. First, the decomposing crop residues, including the root nodules, are rich in nitrogen. This is slowly released into the soil, thereby improving nitrogen availability to subsequent crops. Additionally, residues from these legume crops tend to be less fibrous than many common rotational crops (such as corn, wheat, barley, and oats) and have a lower carbon to nitrogen ratio, which results in less total nitrogen needed for microbes decomposing these residues. Finally, it is often reported that legumes grown in rotation with potato results in a healthier overall crop that has effective root and xylem transport tissues.

Legumes are not the only rotational crops that favorably impact potato nutrition. Growing about any crop between potato crops results in positive benefits. This practice reduces populations of nematodes, insects, microorganisms, and other pests. These can infect roots, decreasing nutrient use efficiency. Ideally, there should be four or more years between conventionally grown potato crops and seven or more between organically grown potatoes. Rotations are often shorter than these, but this results in increasing dependence upon pesticides and increased risk of crop failure. Although inclusion of about any species in the cropping system is beneficial, it should be noted that some crops are targeted by the same pests as potato, which has the potential to negate the positive effect of rotation. Some weeds are also hosts for potato pests and should be carefully controlled during the rotational cycle.

Additionally, the amount and type of previous crop residue impacts nitrogen availability to the potato crop. Residues that have a high carbon to nitrogen ratio (>20 parts carbon to 1 part nitrogen) result in a temporary immobilization of the soil nitrogen early in the growing season, reducing it to low levels as soil microbes scavenge available nitrogen with high populations that use the residue as a food source. For example, incorporating a large amount of wheat straw into the soil just before potato planting can result in early-season nitrogen deficiency. This effect is reduced if the straw is tilled into the soil while soil temperatures are still warm the prior year or if a majority of the straw is baled and removed (although adding organic matter to soil is generally a positive contributor to overall carbon sequestration and building fertile soils). When potato is planted into a soil with a fibrous residue from the previous crop, additional nitrogen is needed. This immobilization does not occur for residues that have a low carbon to nitrogen ratio, such as sugarbeet, cotton, and most vegetables and legumes.

Finally, using a winter cover crop can positively impacts potato nutrition. Decomposing plant residues, manure, soil organic matter, etc., continue to release nutrients into the soil even after nutrient uptake slows in the late summer. As a result, a pool of nutrients accumulates, especially nitrogen. These nutrients can be captured with a winter cover crop, which has other positive green manure effects and slows soil erosion processes.

Assessing Potato Nutrient Requirements

There are several methods used to evaluate soil nutrient status and estimate nutrient availability to potato. The most commonly used methods include soil and plant tissue analysis, as well as observation of crop development and appearance.

Soil Testing

Accurately calibrated soil tests, along with field historical records, can be used to reliably estimate lime and fertilizer needs. The accuracy of the estimate depends on how well the soil sample represents the field or area, the quality of the lab analysis, and the accuracy of the calibrations used to interpret the data. Laboratories may use different chemical extractants or instruments and may also vary in their interpretations of the results and their recommendations. Laboratories using procedures developed and calibrated for specific crop, soil, and climate variables generally provide the most accurate results. In general, accurately calibrated tests are available for many, but not all, of the nutrients (Table 8.1). Determining lime, phosphorous, and potassium needs are among the most accurately calibrated soil test parameters. Tests for nitrogen and some of the secondary and micronutrients are not as accurate and/or well-studied, but still provide useful information for estimating nutrient availability.

Soil test interpretation and subsequent recommendations are affected by multiple factors, including the: nutrient in question; extractant used; soil type; growing region; anticipated yield; and in some cases, fertilizer application method. Therefore, it is difficult to briefly summarize interpretations from various regions. For example, Table 8.2 shows the phosphorous soil test and fertilizer amount at which the lowest amount of fertilizer phosphorus is recommended for various parts of the

Table 8.2 Potato soil test phosphorous interpretations at which the lowest amount of fertilizer phosphate is recommended[a]

Location	Soil extract	Level at which a minimum recommendation is given for soil test phosphorus, ppm	Fertilizer, lb-P_2O_5/ac, minimum recommendation
Maine	Modified Morgan	25	0[b]
New York	Mehlich III	76	30[b]
Quebec	Mehlich III	Varies[c]	45
Minnesota	Bray P1	51	75
Wisconsin (sandy soils)	Bray P1	120	30
Wisconsin (non-sandy soils)	Bray P1	200	50–65[d]
Alberta	Kelowna	100	0
Idaho	Olsen sodium bicarbonate	20–30	0[e]
Oregon	Olsen sodium bicarbonate	10	150[f]
Oregon (Willamette Valley)	Olsen sodium bicarbonate	20	80[f]
Oregon (other areas)	Bray P1	40	80[f]
Washington	Olsen sodium bicarbonate	20	0[e]

[a]Adapted from Rosen et al. (2014)
[b]Higher rate for Russet Burbank
[c]P:Al ratio > 25
[d]Higher ratio for higher expected yields
[e]Varies with level of soil free lime content
[f]Increase broadcast application by 50%

Soil Sampling Procedure

An unrepresentative soil sample of a field area may be misleading, causing inappropriate fertilizer application rates. It is essential that each field be sampled using procedures that represent the majority of soils in the field or area for which a recommendation is desired. Or, preferably, multiple samples should be taken in unique zones in each field to characterize the range of soil test values present. Areas with significant variations in soil texture, color, topography, historical yields and canopy variations, and cropping and fertilization (especially manure) history should be sampled and fertilized separately from the rest of the field.

Traditional Sampling

With conventional sampling, a single set of soil test results and recommendations will be based on sample averages. The sampling guidelines in Table 8.3 are based on when the field was last tested and whether the field was responsive or non-responsive the last time it was tested.

(continued)

Table 8.3 Recommended sample intensity for relatively uniform fields

Field types	Field size, ac	Suggested sample no.
Not tested in previous 4 years and/or responsive fields	All sizes	1 sample/5–8 acres
Non-responsive fields tested in past 4 years	5–10	2
	11–25	3
	26–40	4
	41–60	5
	61–80	6
	>80	7

Take a minimum of 10 cores per sample
Adapted from Schulte et al. (2009)

A composite soil sample for a field zone consists of 10–20 soil cores (Fig. 8.5) from the topsoil taken to the depth of tillage, which in potato fields is typically about 12 in. Typically, cores are taken in a selectively random pattern, such as a zig-zag or W pattern, throughout the sampling area—avoiding field edges and other non-characteristic areas. The cores are then mixed thoroughly and sent to the laboratory (the lab needs approximately two cups of soil to perform the analysis). The soil should be transported to the lab quickly to avoid unwanted transformations (for example, warm and moist conditions result in nitrogen mineralization). The sample should be air dried or refrigerated if there is more than a one-day delay in sample delivery.

It should also be noted that procedures for nitrogen testing may vary for different regions—check with the soil testing lab for specific sampling directions.

Site-Specific Sampling

Site-specific soil sampling is typically used in conjunction with variable rate fertilization. One method is systematic grid sampling, which is typically done by taking a sample with at least eight cores at points in the field every 1.0–2.5 acres using irregular sampling positions, such as those used with a systematic, unaligned grid pattern. Grid sampling is best suited to fields expected to have significant variation present with relatively large areas in the responsive soil test range. Data are then analyzed using various statistical procedures providing an estimate of nutrient patterns within the field.

Another method is directed or zone sampling, which is typically done by collecting soil samples from 3 to 6 management zones per field. The zones are delineated using various layers of data, including aerial bare soil (Fig. 8.6) and crop canopy images (Fig. 8.7) and maps of soil survey, topography, snow and water accumulation, soil texture, color, historical yields and canopy variations, cropping/fertilization (especially manure) history, and producer experience. Depending on size and variability of each zone, a minimum of 10 soil cores from each zone are composited to produce samples representative of each zone. The samples are analyzed separately, resulting in customized fertilizer recommendations for each zone.

Fig. 8.5 Soil sampling is critical for determining residual soil nutrients and for planning a soil nutrient management program. (Photo credit: University of Idaho-top and BG Hopkins-bottom)

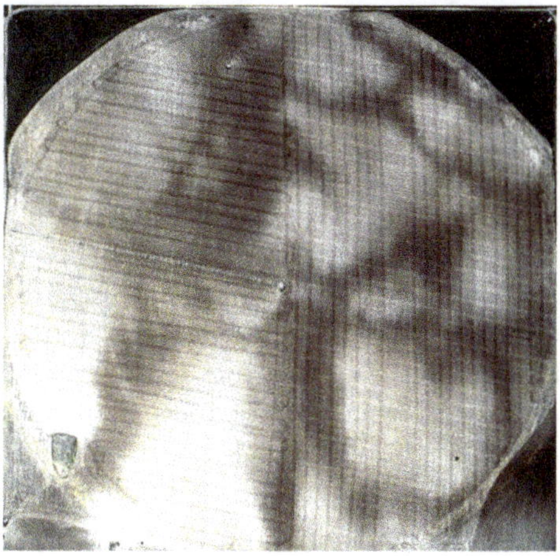

Fig. 8.6 Bare soil imagery that can be used to identify field variability (such as exposed lighter colored subsoil on eroded slopes compared to darker areas with more topsoil). (Photo credit: BG Hopkins)

Fig. 8.7 Crop canopy from overhead imagery can be used to identify field variability (such as areas with nutrient deficiencies resulting in slow canopy development and/or chlorosis). (Photo credit: BG Hopkins)

U.S. and Canada. It appears consistent that in the western U.S., where the sodium bicarbonate (Olsen) test is typically used, the recommended phosphorus rates are lowest with soils testing between 10 and 30 ppm. In the East and Midwest, however, substantially more diversity exists in the types of soil tests used and interpretation ranges. Data such as these emphasize the need to use fertilizer recommendations based on regionally calibrated research data.

Plant Tissue Analysis

Most growers use plant tissue analysis as a diagnostic tool for monitoring the nutrient status of potato and making in-season fertilizer applications. It is based on established relationships between crop yield/quality and nutrient concentrations in a standard plant part.

Factors affecting sufficiency level interpretation include plant part sampled, stage of maturity, and variety grown. Since leaves are sites of metabolic activity, the concentration of nutrients used in this activity will generally be low in these tissues. Leaf samples are effective for monitoring micronutrients, calcium, magnesium, sulfur, and total nitrogen, while petioles connecting stems with leaves are better suited for nitrate-nitrogen, phosphorus, and potassium. However, most sampling is done with either petioles or leaves but not both. Some nutrients, such as nitrogen, phosphorus, and potassium, are well studied and have research-based guidelines that can be effectively used as an in-season nutrient management tool. Other nutrients have less research data available for interpretation. For example, there is a modicum of data for micronutrient petiole interpretation; thus, using this data, although commonly done, is less reliable. In potato, the petiole of the fourth leaf from the top of the plant is generally used to determine plant nutrient status (Figs. 8.8 and 8.9).

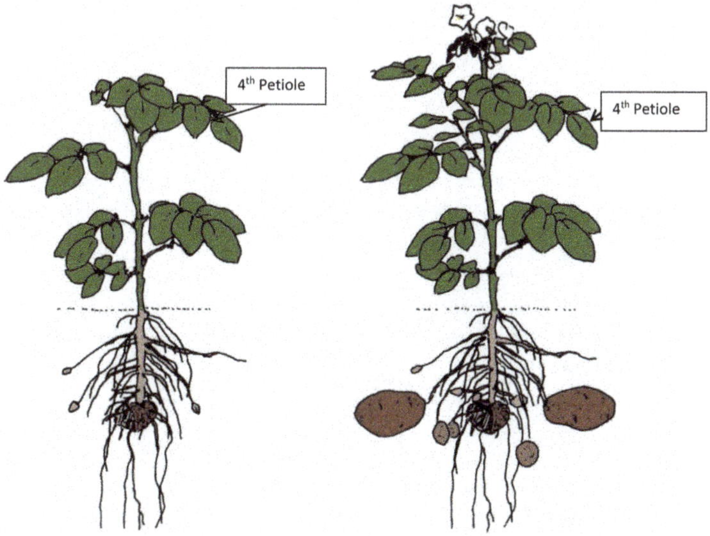

Fig. 8.8 Diagrams of vegetative shoot with petioles (**a**) and full potato plant with floral spike (**b**). The fourth petiole is used in tissue analysis for determining plant nutrient status. (Adapted with permission from Rowe 1993 and Thornton and Sieczka 1980)

Fig. 8.9 Potato leaves attached to the main stem with petioles. (Photo credit: BG Hopkins)

Nutrient concentrations in a plant tissue change with plant age. Many nutrients, including nitrogen, phosphorus, potassium, copper, zinc, and sulfur, decrease in concentration with increasing plant age—although in-season fertilization can negate or minimize the rate of decrease. However, with aging, concentrations of calcium, magnesium, boron, iron, chloride, and manganese generally increase in plant tissues.

Table 8.4 Sufficient nutrient concentrations for Russet Burbank petioles and whole leaves at early- to mid-tuber bulking

Nutrient	Sufficiency level	
	Petiole[a]	Whole leaf[b]
Total nitrogen (%)	–	3.5
Nitrate-nitrogen (ppm)	15,000	–
Phosphorous (%)	0.22	0.25
Potassium (%)	8.0	3.5
Calcium (%)	0.6	0.6
Magnesium (%)	0.3	0.25
Sulfur (%)	0.2	0.2
Zinc (ppm)	20	20
Manganese (ppm)	40	20
Copper (ppm)	4	5
Boron (ppm)	20	20

[a]Adapted from Stark and Westermann (2003)
[b]Adapted from Westermann (1993) and Schulte et al. (2000)

Petiole Sampling Procedure

Petiole sampling usually begins at late tuber initiation or early tuber bulking and is repeated every 7–10 days until nutrient uptake plateaus. As nutrient uptake in potato nearly ceases about 90–120 days after planting, sampling after this period does not allow time for meaningful corrections to be made. Of course, this date varies somewhat based on cultivar, soil, climatic, and pest conditions.

Environmental factors dramatically impact the nutrient concentration in petioles, including temperature and light intensity. Sampling at a consistent time of day and with similar weather conditions is advisable when possible. As with soil sampling, separate plant petioles can be collected from various areas or management zones differentiated by soil type, crop history, topography, or other features.

The fourth petiole should be consistently sampled, since analysis of samples taken from higher or lower positions will produce significantly different results that may not be calibrated with the nutritional status of the crop. Usually, 35–50 petioles are collected from representative areas of the field consistent with the method described previously for soil sampling. All leaflets should be stripped off the petiole immediately after sampling (Fig. 8.10), as nutrient transfer continues between petiole and leaves even after sampling. The petioles should be placed in a clean container or paper bag and shipped to arrive at the lab within 1 day after sampling. If there is a delay in shipping, the petioles should be immediately dried at 150 °F or kept cool (<40 °F) until sent to the lab. This procedure minimizes changes in nutrient forms and concentrations that can occur in warm, moist tissue samples.

Fig. 8.10 Leaves are stripped off of petioles immediately during sampling. (Photo credit: University of Idaho-left and BG Hopkins-right, used with permission)

Sufficient nutrient concentrations for Russet Burbank variety at early- to mid-bulking in the petioles without leaflets and the entire leaf (petiole and leaflets) are shown in Table 8.4. Although this is a valuable starting point, research has shown petiole nitrate-nitrogen sufficiency levels differ between variety and plant growth stage. As such, it is ideal to base fertilization on research with the variety being grown or, at a minimum, a variety with similar growth characteristics.

Although not commonly sampled, leaves are also analyzed instead of or in addition to petioles. In some cases, the sap of the petiole is tested instead of the entire petiole. The same principles as discussed above need to be employed to obtain a quality analysis, and interpretation of the results needs to be based on leaf or sap data.

Crop Symptom Observations

Nutrient deficiencies can sometimes be identified from visible symptoms (Table 8.5; Figs. 8.11, 8.12, 8.13, 8.14, and 8.15). Although visible symptoms are seen, it is more common for plants to have deficiencies without any obvious symptomology—termed "hidden hunger." Thus, by the time symptoms become noticeable it is often too late to make corrective fertilizer applications without experiencing some yield or quality loss. Accurate diagnosis is also complicated by multiple nutrient deficiency symptoms, as well as those caused by pests, pathogens, or environmental stress.

Table 8.5 Typical nutrient deficiency symptoms of potato foliage

Nutrient	Occurrence	Symptoms
Nitrogen (N)	Very common	Initially the entire plant may be light green. Younger leaves will remain green, while older leaves turn yellow (chlorosis) and may senesce. Plant will have stunted growth
Phosphorous (P)	Common	Symptoms may be indistinct with stunted, spindly plants. Leaflets can be darker green and their margins roll upward. Some purpling may occur. Leaf rolling worsens as deficiency becomes more severe
Potassium (K)	Common	Early symptoms include plant stunting and a somewhat black or uncharacteristically dark green color to leaves. Older leaves become bronzed with outer leaf margins necrotic—eventually overtaking the entire leaf. Leaflets can be small, cupped, crowded together, crinkled, and bronzed on the upper surface
Calcium (Ca)	Very rare	Deficient plants are spindly with small, upward rolling, and wrinkled leaflets. The youngest leaves near the growing point exhibit marginal chlorosis and brown spotting. The growing point may die
Magnesium (Mg)	Very rare	Initially, a slight interveinal chlorosis develops on young, fully expanded leaves with some brown spotting and brittleness that terminates as interveinal leaf scorch. Leaves near the growing point remain green
Sulfur (S)	Somewhat common	Symptoms are similar to those of nitrogen, except the yellowing appears first in the younger leaves that turn uniformly light green to yellow and may show some upward rolling
Boron (B)	Rare	Plants appear bushy with shortened internodes. Leaf blades thicken and roll upward and develop light brown edges. Eventually, growing buds die and leaf tissue darkens and collapses
Chloride (Cl)	Very rare	Young leaves are initially light green and then turn purplish bronze. Leaves curl upward and appear pebbled
Copper (Cu)	Rare	Young leaves remain green and are of normal size but exhibit a pronounced rolling and cupping. Leaf tips eventually wilt and die
Iron (Fe)	Rare	Young leaves become yellow to nearly white, usually without necrosis. Leaf veins appear greener than the interveinal tissue. Leaf tips and edges usually remain green longer than the rest of the leaf
Manganese (Mn)	Somewhat common	Younger leaves initially exhibit interveinal chlorosis and a slight upward cupping with gray and black spotting. Spot density gradually increases along the larger leaf veins, eventually producing necrotic patches
Molybdenum (Mo)	Very rare	Leaf blades become uniformly yellow similar to nitrogen and sulfur deficiencies
Nickel (Ni)	Never observed	Only recently established as an essential nutrient; however, no instances of nickel-deficient field grown crops have been documented
Zinc (Zn)	Somewhat common	The youngest leaves become narrow, erect, chlorotic, and develop tip burn and leaf cupping. Blotching and spotting of the interveinal tissues develops with severe deficiencies, and the lower leaves turn brown and die

Adapted from Kelling et al. (2001) and Westermann (1993)

Fig. 8.11 Nitrogen deficient plant on right showing chlorosis, especially with older leaves. (Adapted from Pitchay and Mikkelsen 2018)

Fig. 8.12 Nitrogen deficient leaves on right showing chlorosis and smaller size. (Adapted from Pitchay and Mikkelsen 2018)

Fig. 8.13 Nitrogen deficient plants on right show chlorosis, stunting, and slower row closure. (Photo credit: BG Hopkins)

Fig. 8.14 Stunted, phosphorus-deficient plant on left compared to healthy plant. (Photo credit: BG Hopkins)

Fig. 8.15 Potassium-deficient leaves on right showing chlorosis, smaller size, and crinkling. (Adapted from Pitchay and Mikkelsen 2018)

Efficient Fertilizer Management

Efficient fertilizer management requires an understanding of how roots operate and grow and the proper source, rate, timing, and placement of fertilizers.

Root Systems

It is imperative to understand the morphological and physiological characteristics of potato plants to manage nutrients effectively (Fig. 8.16). In contrast to other crops, potato is an *inefficient responder* when it comes to nutrient uptake. In general, potato has a shallow, poorly effective root system that is relatively less dense, less branched, and shallower than other crops (i.e., wheat has ~four-fold greater root density). Most potato roots grow in the top 2 ft of soil, with 90% of root length in the top 10 in. In addition, potato has fewer root hairs than most other crops. Root hairs comprise only about 20% of the total root mass of potato—compared to more than double that for most other crop species. This impacts nutrient use efficiency, as uptake occurs primarily through root hairs.

Furthermore, potato root systems tend to decline relatively early in the growing season when nutrients, such as phosphorus, are still being taken up by roots. This contrasts with many other species that accumulate nutrients earlier in the growing season and/or are less susceptible to disease pathogens. Because of these and possibly other plant characteristics, fertilizer recommendations and soil test sufficiency

Fig. 8.16 It is imperative to assess rooting structure and depth in order to manage nutrition. (Photo credit: BG Hopkins)

levels for potato are generally much higher than other crops. This is especially true with phosphorus, where fertilizer rates can be higher than 350 lb. P_2O_5/ac in long-season, high-yielding environments—with most other crops typically far less than this rate. Although these traits are typical for potato in general, especially Russet Burbank, there are differences among varieties.

Additionally, potato is highly susceptible to root and vascular system pathogens, partly explaining why this species is unique with respect to the higher fertilizer requirements for phosphorus and, possibly, other nutrients. Phosphorus deficiency can increase the severity of several important potato diseases, including common scab (*Streptomyces scabies*), Verticillium wilt (*Verticillium dahlia* and *Verticillium albo-atrum*), and late blight (*Phytophthora infestans*). Potato root system development mostly ceases 60–90 days after planting and begins to deteriorate. This coincides with increases in disease development while tubers are bulking rapidly and nutrient requirements are high. Although these principles are generally true for potato, there is a wide difference in disease susceptibility by cultivar. Russet Burbank is very susceptible to disease, which contributes to its problems with nutrient efficiency.

The Right Source

There are several considerations when choosing a fertilizer source, including: plant usable forms of nutrients, compatibility with soil properties, and ease and effectiveness of application. In addition, some nutrient combinations provide synergies between nutrients.

In general, fertilizer needs to be water soluble or at least have an eventual slow- or control-release form of the nutrient that will be available to plants during the growing season in a predictable manner. Unless a slow- or controlled-release pattern is desirable, the fertilizer should be at least 60% water soluble. Slow- or controlled-release materials need to be similarly water or citrate soluble within the course of a growing season. Most traditional fertilizer sources are greater than 90% soluble. If comparable fertilizers have similar solubility and chemical form of the nutrient needed, the choice of which source to use generally becomes a question of price, availability, convenience of application, and accompanying nutrients.

The primary macronutrients are the most commonly deficient and, accordingly, are the most commonly sold fertilizers—with nitrogen making up about half of worldwide fertilizer sales and phosphorus and potassium making up most of the rest. The secondary macronutrients and micronutrients make up less than 5% of annual global fertilizer sales, but are nevertheless important components of nutrient management—especially for potato. A list of the most commonly sold fertilizers is presented in Table 8.6, although there are many dozens of other fertilizer forms available and that are not shown.

There is also a wide variety of waste products used as nutrient sources, such as raw or composted manure and other animal wastes, treated biosolids, crop residues, etc. Immediate water solubility of these materials is not as relevant as it is for traditional fertilizer sources, since nutrient release is typically dependent on mineralization of these organic materials—a process that eventually breaks down the complex molecules into plant-available forms. Some of the nutrients in these fertilizer materials are present in rapidly available forms (such as all of the potassium sources) and, as such, are immediately plant available. However, a large portion of most of the nutrients are bound in organic cell structures and released slowly during decomposition. The ratio of immediate- to slow-release forms of nutrients in these materials is widely variable but should be understood when trying to match nutrient availability with nutrient uptake patterns. The slow rate of nutrient release from these materials can be an advantage, as long as the nutrients are supplied in a timely fashion coinciding with plant need. However, the slow release of nutrients from organic materials is often difficult to predict and control.

Although manure/biosolid materials can serve as very good sources of nutrients, there are potential negative effects, including presence of weed seeds, nutrient imbalances, odor, presence of toxins, cost of transportation, and compaction of soil due to heavy axle loads from manure and compost applicators. Additionally, as potato is a crop with the edible tuber in direct soil contact, it is essential that waste

Table 8.6 Common fertilizers used in potato production

Common name	Formula	Typical percentages	Additional nutrients included	Liquid or solid
Nitrogen				
Urea (U)	$CO(NH_2)_2$	46–0–0		Generally solid, but can be dissolved
Ammonium sulfate (AS)	$(NH_4)_2SO_4$	21–0–0–24(S)	Sulfur (S)	Generally solid, but can be
Ammonium nitrate (AN)	NH_4NO_3	34–0–0		Generally solid, but can be
Urea ammonium nitrate (UAN)	$CO(NH_2)_2NH_4NO_3$	28–0–0 30–0–0 32–0–0		Liquid
Phosphorus				
Monoammonium phosphate (MAP)	$H_2NH_4PO_4$	11–52–0	Nitrogen	Solid
Diammonium phosphate (DAP)	$H(NH_4)_2PO_4$	18–46–0	Nitrogen	Solid
Ammonium poly phosphate (APP)	$(NH_4PO_4)_n$	10–34–0; 11–37–0	Nitrogen	Liquid
Potassium				
Muriate of potash (MOP)	KCl	0–0–60	Chloride	Solid
Sulfate of potash (SOP)	K_2SO_4	0–0–50–18(S)	Sulfur	Solid
Potassium thiosulfate (KTS)	$K_2S_2O_3$	0–0–25–17(S)	Sulfur	Liquid

fertilizers be applied well ahead of harvest time and/or treated to minimize risk of human pathogens. In many cases there are laws governing these applications.

The popularity of the fertilizers listed in Table 8.6 is largely due to a high analysis of water-soluble nutrients, resulting in low transportation costs per unit of fertilizer compared to the typical waste products. For example, a composted manure with 1% phosphorus has transportation costs about 50 times greater per unit of phosphorus than monoammonium phosphate (MAP). For this reason, use of compost often becomes commercially unmanageable if transporting more than a few miles.

There are also commercially available fertilizers with slow- and controlled-release properties engineered to release nutrients over time. Often, these are more predictable than waste products. In some cases, release is also temperature controlled, similar to organically complexed nutrients early in the growing season when soil temperatures are cool. However, others have a time-release mechanism that is not temperature dependent. In the case of highly soluble nutrients, most notably nitrogen and sulfur, a slow- or controlled-release form can minimize nutrient loss from leaching and loss to the atmosphere. For example, use of polymer-coated urea from manufacturers with proven products of known patterns and timing of the control release is particularly effective at supplying nitrogen to plants. Doing so minimizes risk of nitrogen loss to the environment, while supplying a steady source of nitrogen during the time period when it is being taken up rapidly. Other controlled- and slow-release nitrogen sources, such as various inhibitors and long chain polymers, can also be effective.

In the case of highly insoluble nutrients, such as phosphorus and iron, fertilizer efficiency depends more on complexation in the soil than leaching or gaseous loss from the system. For example, adding immediately soluble phosphorus fertilizer results in a temporary spike in soil solution phosphorus concentration at levels exceeding chemical equilibrium constants, forcing precipitation of phosphate minerals. A slow- or controlled-release phosphorus fertilizer may minimize formation of these phosphate compounds, as the soil solution phosphorus concentration typically does not reach excessively high concentrations. In addition, the phosphorus is released gradually over time as a function of temperature and moisture, potentially resulting in increased phosphorus use efficiency.

Another approach is complexing the nutrient in some way to minimize the formation of poorly soluble precipitates in the soil. Various organic acids have been proven effective, especially in low organic matter soils. Chelated nutrients are also known to be effective for many of the secondary and micronutrients. However, not all sources are proven. Therefore, valid scientific testing proving the efficacy of a fertilizer source is vital.

Growers selling organically certified potato are required to use Organic Materials Review Institute (OMRI)-certified fertilizer. These products can be effective fertilizer sources. However, they are often relatively costly. Most traditional, low-cost fertilizers have not gone through the certification process, despite their proven safety for crops and consumers. Nevertheless, these products are not allowed if not certified organic. Some organically certified fertilizers are not as effective as traditional sources. Most notably, untreated rock phosphate is commonly promoted as an organic fertilizer, but this material is insoluble and, as such, is a poor source of plant-available phosphorus in anything but highly acidic soils.

Another consideration in making fertilizer source decisions is pH, especially for phosphorus fertilizers. For example, the pH of some fertilizers is acidic (MAP = 3.5, triple super phosphate = 1.5, and ammonium poly phosphate (APP) = 6.2) and others are alkaline (urea ammonium phosphate and diammonium phosphate (DAP) = 8.0). Despite these differences, the pH in the microsite around the fertilizer does not remain acidic or alkaline as the soil pH eventually absorbs the effect. Therefore, the uptake of phosphorus is not impacted, to a large extent, because of the difference in reaction pH among sources. Other fertilizers in a concentrated fertilizer band, particularly nitrogen, can also impact pH, a common example being the acidifying effect that occurs during nitrification. For example, APP results in reduced phosphorus uptake and potato tuber yield and quality compared to DAP on soils with a pH of about 5. The effect is likely due to the further pH reduction in the banded application with the APP, which has an acidic reaction pH, as opposed to DAP, which has an alkaline reaction pH.

Some differences have been observed for P sources containing N due to volatilization of NH_3 and/or accumulation of nitrite (NO_2^-). For example, DAP has resulted in more damage than MAP and may be more damaging to seeds and seedlings when in direct contact. High rates (≥250 lb P_2O_5 per ac) of DAP or urea ammonium phosphate placed near (≤5 cm) or in contact with potato seed piece delay emergence, reduce stand, and negatively impact yields. When growing on alkaline soil and with high rates of phosphorus, DAP is generally avoided.

Moreover, liquid vs. dry fertilizers is a source consideration. Liquid products tend to have higher transportation costs due to lower analysis because of the extra water weight. The most popular liquid fertilizer is APP, which is an effective fertilizer source. Reasons cited for choosing liquids over solid fertilizers in some circumstances are: (1) homogeneity of blends resulting in uniform application, (2) sequestering of micronutrients to aid in their uptake, (3) ease for combination applications with liquid pesticides, (4) ease of fertigation, (5) no "caking" up of solids, and (6) advantages for the fertilizer dealer in terms of safety and equipment logistics. Although these are legitimate considerations, if the same amount of dry or liquid nutrient is applied at the same distance from a root, the plant will not likely distinguish the source of the nutrients once they are in the soil. So it is generally recommended to choose a nutrient source based on pricing and convenience factors in most instances.

Furthermore, another source issue is heterogeneity of nutrient blends with dry fertilizer materials. Blends of dry products can segregate in handling, transportation, and spreading due to differences in granule shape and density. This is typically a greater problem with micronutrients applied at very low rates. Using a well-mixed liquid blend or a homogenous granule eliminates this problem.

Various fertilizer additives and/or enhanced efficiency fertilizers are frequently marketed to potato growers. These are an especially important consideration for nitrogen, as it can be easily lost from soils. It is highly beneficial to use products that match nitrogen availability in the soil with plant uptake timing. Some products are proven to be effective in this regard, including slow- and control-release materials (such as polymer-coated urea) and various inhibitors that slow conversion to forms more easily lost.

Similarly, many products claim to harness an increase in phosphorus solubility when applied in combination with organic acids. However, the sale of these additives, unlike fertilizer, is largely unregulated and products may be unreliable. Thus, buyers should work with products from reliable companies that provide independent research confirmation.

Also, it is important to evaluate products under specific soil conditions. For example, most of the work showing effective response to organic acids has been done in sandy, low organic matter soils that are relatively low in the natural forms of these acids. Before an additive is used, it should be independently verified—preferably by more than one independent scientist. They also need to be cost effective. If the products claim to enhance efficiency, the rate of the nutrient applied should be reduced unless vastly improved yields are expected. In other words, if a plant already has adequate nutrition due to high rates of fertilizer or residual nutrients in the soil, fertilizer enhancement will not likely provide any benefit.

Manure and similar materials take time to decompose. Nutrient release from these can occur over 2–3 years. These materials should, therefore, be tested and research-based recommendations applied to estimate release rates. Additionally, irrigation water is commonly high in calcium, magnesium, sulfur, and chloride—often enough to meet the entire need of the crop. Further, seed treatments and foliar fungicide sprays often contain micronutrients at rates that meet the needs of the crop.

The Right Rate

Choosing the correct fertilizer rate is imperative for successful potato production. Fertilizer calibration studies correlate yield response to soil test concentrations. Most fertilizer recommendation tables derived from such research provide a base fertilizer recommendation with an adjustment upward in rate of fertilizer for incrementally greater yield potential.

After determining the rate of fertilizer needed based on yield potential and residual nutrients in the soil, further adjustments in rate may need to be made. Determining the right rate involves factoring in all available nutrient sources discussed in the previous section, such as irrigation water, manure, previous crop, etc. Evaluation of trends in soil and tissue analysis as well as field history, including yield, help predict rate adjustments to customize fertilizer recommendations for specific field conditions. All of these factors, as well as the interactive effects of timing and placement, need to be applied against the economics of the expected fertilizer response. Depending on crop value and fertilizer cost, the ideal rate can change—even if all other factors are the same. Typical rates are shown for each nutrient later in this chapter.

The Right Placement and Timing

Several timing and placement choices exist, including: (1) pre-plant broadcast applications left on the surface or incorporated into soil, (2) concentrated bands applied near the seed piece, (3) concentrated bands applied either during the season to the surface or injected between rows, (4) in-season fertilizer applications injected into the irrigation water (fertigation), and (5) low-concentration liquid foliar sprays.

In order to make good decisions with regard to timing and placement, it is essential to understand soil nutrient mobility. Very soluble anionic nutrients are poorly held by the soil and more likely to leach below the root zone, especially in sandy soils with low CEC and rapid hydraulic conductivity rates. These include sulfate-sulfur, chloride, borate-boron, and especially nitrate-nitrogen. Additionally, nitrogen can be lost to the atmosphere as a result of various chemical and microbial processes. Therefore, it is important to time nutrient applications to coincide with plant uptake patterns using in-season applications or control/slow-release fertilizers.

In contrast, poorly soluble nutrients tend to move very slowly in soil, and roots may not encounter them in an effective way unless placement is correct. These include zinc, iron, manganese, and copper in alkaline soil and phosphorus in alkaline and acidic soil. Thus, it is more important that these relatively immobile nutrients are placed in the root's pathway. Applying concentrated bands of these and other fertilizers to the side of the seed piece has been shown to be effective in improving nutrient uptake efficiency. If these bands are placed prior to planting (Fig. 8.17), it is vital to place the band deep enough (about 3 in below seed piece depth) to avoid disrupting it with the planting process. If the concentrated band is applied with the planter, placement can be at or up to 3 in above the seed piece with equal effectiveness.

Cationic nutrients with high to moderate solubility tend to be held by the soil CEC so placement and timing is not typically as critical. These include potassium, calcium, and magnesium. Ammonium-nitrogen is also classified with this group, but this form of nitrogen converts so rapidly to nitrate that it needs to be managed as described above unless a nitrification inhibitor is used to slow this process. It is also noteworthy that some soils have high levels of potassium-fixing clays and, as such, benefit from in-season potassium application during the time of major plant uptake. Similarly, and more commonly, soils with low CEC are relatively more susceptible to leaching losses of these nutrients and also benefit from in-season application.

Fig. 8.17 Placing a concentrated band of fertilizer while "marking out" or forming rows. (Photo credit: BG Hopkins)

Finally, soils with calcium and boron deficiency are rare, but when this occurs, placement near the tuber development zone is beneficial.

Individual nutrient recommendations factoring in source, rate, placement, and timing follow.

Macronutrients

Nitrogen

Nitrogen is the nutrient with the greatest impact on potato production; thus, efficient management is essential. Nitrogen management affects vine and tuber biomass production, tuber size, grade, specific gravity, and internal and external quality. The potential for nitrogen losses to the environment are closely related to the efficiency of the nitrogen management program. For this reason, there are laws and guidelines that govern nitrogen fertilizer management in many regions.

Uptake Total nitrogen uptake for potato crops usually ranges from 150–400 lb. N/ac, depending on cultivar and yield potential of the growing region. At harvest, about 60–65% of the total plant nitrogen is contained in the tubers, while about 30% remains in the vines. During tuber bulking, potato plants require about 2.0–4.0 lb N/ac/day depending on tuber growth rate.

About 60% of the seasonal nitrogen requirement is taken up by 75 days after planting (Fig. 8.1). Consequently, adequate nitrogen must be available to the crop early in the season to allow for sufficient canopy development. Research shows about 120–160 lb N/ac from soil and fertilizer sources is required by the time the rows begin to close to provide for optimum canopy development and yield. However, excessive nitrogen availability prior to tuber initiation can delay tuber bulking by up to 2–3 weeks, reducing tuber yields by as much as 80–120 cwt/ac. Excessive early-season nitrogen can also increase susceptibility to brown center and hollow heart. Excessive late-season nitrogen applications usually reduce specific gravity and skin set, as well as increase the potential for nitrate leaching.

Nitrogen Sources Crops can acquire nitrogen from a number of sources, including: (1) the inorganic soil nitrogen forms: nitrate (NO_3) and ammonium (NH_4); (2) nitrogen mineralized from soil organic matter, previous crop residues (especially legumes), and animal wastes; (3) nitrogen present in irrigation water; and (4) fertilizers. Although potato prefers nitrogen in the nitrate form, research shows similar performance from both ammonium and nitrate.

To determine nitrogen fertilizer needs, crop nitrogen requirements and the availability of nitrogen from all sources must be assessed. Research on commercial potato fields indicates a total of about 300 lb N/ac from all sources is required to produce a 400–450 cwt/ac crop of Russet Burbank potato. Total nitrogen uptake at these sites was about 180–220 lb N/ac, indicating total nitrogen use efficiency was approximately 60–75%, which is similar to other crops. Nitrogen mineralization is highly variable, but for these fields it averaged about 60–80 lb N/ac. Applying excessive nitrogen substantially increased post-harvest soil nitrate concentrations and reduced potato yield and quality.

Nitrogen Fertilizer Recommendations Research studies were used to validate and refine the total season nitrogen fertilizer recommendations presented in Table 8.7. The recommended rates are based on total nitrogen requirements adjusted for yield potential, soil test NO_3–N and NH_4–N in the top 12 in, and the previous crop. The recommendations assume an average of 60 lb N/ac of mineralized nitrogen.

Nitrogen Fertilizer Timing The most efficient management systems for potato release nitrogen to coincide with nutrient uptake. This is often done by splitting nitrogen fertilizer applications with 50–75% of the total seasonal nitrogen supply side-dressed after emergence or applied through the irrigation system in several small applications. Several liquid nitrogen fertilizers, such as urea ammonium nitrate (UAN) or APP, can be applied by this method. Nitrogen is also a component of many other liquid fertilizers containing phosphorus, potassium, and sulfur injected through the irrigation system. When properly used, split nitrogen management can significantly increase nitrogen use efficiency and reduce nitrogen leaching

Table 8.7 Estimated total nitrogen needs for Russet Burbank potato to be supplied throughout the growing season

	Potential yield (cwt/ac)				
Soil test NO3–N	300	400	500	600	700
(0–12 in)					
ppm			(lb N/ac)		
0	200	240	280	320	360
5	180	220	260	300	340
10	160	200	240	280	320
15	140	180	220	260	300
20	120	160	200	240	280
25	100	140	180	220	260

Additional adjustments in the nitrogen recommendations include:

1. Applying an additional 15 lb N/ac for each ton of previous grain, straw, or mature corn stalk residue up to 60 lb N/ac. Mature cereal crop residues have relatively high carbon/nitrogen ratios and immobilize available nitrogen during microbial decomposition. Residues from non-cereal crops, such as sugarbeets, onions, beans, peas, and mint have lower carbon/nitrogen ratios and decompose readily without additional nitrogen applications.
2. Subtracting 80–120 lb N/ac from the recommendation following alfalfa (use the higher rate for young, pure alfalfa stands and reduce with older stands with increasing percentages of grass infestation). Subtract 40–60 lb N/ac following red clover, beans, peas, and other legumes.
3. Subtracting the nitrogen applied in the irrigation water using the formula nitrogen applied (lb N/ac) = water applied (in) × ppm NO_3–N × 0.227.
4. Subtracting the nitrogen applied in any animal manure based on analyzed samples, although be cautious about extreme variability in the manure and spreading rates.
5. Applying an additional 30–40 lb N/ac to offset nitrogen leaching losses and low soil nitrogen mineralization on fields with very sandy soils.

potential, while improving potato yield and quality. This technique is especially important in regions where precipitation is relatively more prevalent during the growing season and for soils with poor water holding capacity.

The percentage of the total anticipated nitrogen need for the season, as based on (Table 8.7), that should be applied pre-plant is as follows:

Sands	10–20%
Sandy loams	20–40%
Silt loams	25–50%

For example, a pre-plant nitrogen application for a potato crop grown on a sandy loam soil with a potential yield of 500 cwt/ac and 5 ppm soil test NO_3–N + NH_4–N, would be about 35% of the total nitrogen requirement of 260 lb N/ac (Table 8.7), which is 91 lb N/ac.

Following tuber initiation, in-season nitrogen applications should be made to maintain optimal nitrogen concentrations in the potato crop. In-season nitrogen can be applied by side-dressing, irrigation injection, foliar sprays, and/or broadcast aerially as dry fertilizer. About 40–60% of the total nitrogen requirement should be applied to the crop by 50–60 days after planting. Once tuber bulking begins, weekly crop nitrogen requirements can be estimated based on the relationship between tuber growth rates and plant nitrogen uptake. For example, an average Russet Burbank crop requires about 3 lb of nitrogen per day to maintain an average daily growth rate of 7 cwt/ac/day. Assuming 75% plant uptake efficiency for water-run nitrogen fertilizer, about 28 lb N/ac would satisfy crop nitrogen requirements for a week, while 40 lb N/ac would be adequate for 10 days without a contribution from nitrogen mineralization. Lesser amounts would be required for more frequent nitrogen applications. Adjustments to these projected rates should be based on petiole nitrate tests (Fig. 8.18).

In irrigated fields, nitrogen application frequencies range from continuous injection through center pivot and linear-move systems, especially on sandy soils, to 10–14 days intervals with set-move systems on loam soils. A common approach is to apply 20–40 lb N/ac every 7–14 days during tuber bulking, which is usually adequate to maintain petiole nitrate concentrations in the optimal range. Petiole NO_3–N analysis assists the grower in making appropriate nitrogen adjustments during the growing season.

Alternatively, using controlled- or slow-release fertilizer materials can deliver all or most of the nitrogen to coincide with plant uptake needs. For example, polymer-coated urea is applied just prior to cultivation about 2–3 weeks after planting. This effectively meters out nitrogen in a pattern similar to plant needs—minimizing the risk of losing nitrogen to the environment, but avoiding the added work of fertigation. This steady supply of nitrogen can result in equal or higher-quality tubers and yields. However, it is vital the correct fertilizer coating be used and that fertilizer be handled with care to avoid cracking the polymer coatings. It is also noteworthy that

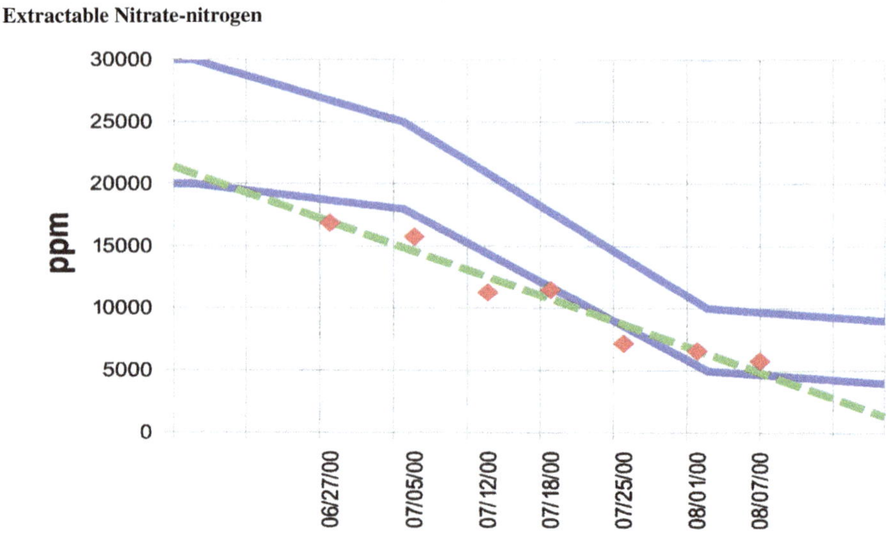

Fig. 8.18 Petiole nitrate-nitrogen concentrations help predict in-season adjustments for nitrogen fertilizer. The goal is to maintain the concentrations between the blue lines to avoid deficiency and excess

petiole nitrate concentration critical levels are 2000 ppm less for potato grown with polymer-coated urea. Other control or slow-release and urease/nitrification inhibitor products can be similarly effective. However, in some cases, it is difficult to control the release, and problems can develop if the release occurs or too late in the season.

Monitoring Nitrogen Status Relationships between tuber bulking rates and daily or weekly nitrogen requirements can be used to initially estimate in-season nitrogen applications. However, weekly petiole NO_3–N concentrations should be used to monitor actual plant nitrogen status with appropriate adjustments as needed. Soil NO_3–N concentrations in the top 18 in of soil can be used to monitor nitrogen availability. Recommended petiole and soil NO_3–N concentrations for several varieties of potato are presented for the various growth stages in Table 8.8. The combination of the two analyses provides a good indication of the effectiveness of the nitrogen management program, as well as the potential causes of decreasing petiole NO_3–N concentrations. For example, the maintenance of adequate soil NO_3–N concentrations when petiole NO_3–N concentrations are falling indicates that other factors, generally related to poor plant health inhibiting growth and/or water uptake are limiting NO_3–N uptake and/or translocation.

Following cool springs, petiole NO_3–N concentration may not provide an accurate indication of crop nitrogen status. Frequent soil fumigation also reduces the activity of nitrifying bacteria, decreasing the conversion of ammonium to nitrate and increasing the relative proportion of NH_4–N in the soil. As a result, petiole

Table 8.8 Recommended NO_3–N concentrations for soil (0–18 in) for Russet Burbank potato variety and for petioles for several potato varieties potato during different growth stages

	Vegetative (up to 44 DAP)	Tuber initiation (45–65 DAP)	Tuber bulking (66–110 DAP)	Maturation (111+ DAP)
	NO3-N, ppm			
Soil	>20	20–25	15–20	<15
Petiole NO_3–N (by variety)				
Russet Burbank		20,000–27,000	15,000–18,000	10,000–15,000
Bannock Russet		22,000–24,000	12,000–18,000	5000–8000
Clearwater Russet		21,000–26,000	15,000–23,000	11,000–19,000
Ranger Russet		24,000–27,000	17,000–20,000	10,000–15,000
Umatilla Russet		28,000–31,000	23,000–26,000	19,000–22,000
Russet Norkotah		25,000–28,000	13,000–19,000	7000–11,000
Atlantic		25,000–28,000	13,000–19,000	7000–11,000
Kennebec		25,000–28,000	13,000–19,000	7000–11,000
Shepody		20,000–22,000	12,000–16,000	9000–12,000
Snowden		23,000–25,000	15,000–19,000	7000–11,000
Superior		23,000–25,000	15,000–19,000	6000–10,000

DAP = Days After Planting

NO_3–N concentrations may be low, while plant nitrogen status is actually adequate. Measurement of soil NO_3–N and NH_4–N early in the season provides a more accurate picture of plant nitrogen availability.

Water Management Nitrogen leaching and nitrification/denitrification losses due to excessive precipitation or over-irrigation can be substantial. The potential for over-irrigation is usually greatest during the early- and late-season periods when crop water use rates are relatively low. Research conducted in southern Idaho shows over-irrigating by 20–30% over the growing season can reduce potato yield, quality, and fertilizer use efficiency and decrease net economic return. Care should be taken to closely match irrigation amounts with crop water use rates throughout the growing season while maintaining available soil water content above 65–70%, as this is vital for water and nutrient management.

Winter Cover Crops Appreciable amounts of NO_3–N can accumulate in the soil at the end of a growing season. This is attributable to reduced late-season plant nitrogen uptake efficiency, continued mineralization of soil organic matter and plant residues, and nitrogen applied in excess of that required by the crop. Planting cover crops, such as winter wheat or rye, following harvest provides an opportunity to capture some of the residual root zone nitrogen (15–30 lb N/ac) and retain it for use during the following growing season. Nitrogen absorbed by the cover crop root system is stored over winter in the plant biomass until spring when the cover crop is tilled into the soil and begins to release the stored nitrogen. Conserving soil nitrogen in this manner prevents it from leaching during the winter months and reduces potential groundwater contamination. Most importantly, winter cover crops also provide ground cover for reducing wind erosion and increasing soil organic matter content and tilth. These cover crops can be used to benefit not only the potato crop, but also the crops grown after potato, as nitrogen excesses in the soil are relatively common.

Phosphorus

Potato commonly responds to phosphorus fertilization, particularly on very acidic and calcareous soils. Adequate soil phosphorus availability is important for early crop development and tuber initiation and also enhances tuber maturity. Phosphorus deficiencies, on the other hand, significantly reduce tuber yield, shape, and size, as well as specific gravity. Phosphorus moves very slowly in soil and needs to be adequately incorporated in the plant root zone to facilitate uptake. Phosphorus is not readily leached, but can be lost in field areas prone to water runoff and soil erosion, which is a major source of water quality problems. For this reason, there are laws and guidelines that govern phosphorus fertilizer management in many regions.

Soil solution phosphorus concentrations are very low (typically 0.01–0.2 ppm) and consequently need to be constantly replenished from labile soil phosphorus sources during the growing season. Daily potato phosphorus uptake requirements are only 0.3–0.5 lb P/ac/day (0.7–1.1 lb P_2O_5/ac/day), but serious deficiencies develop if available soil phosphorus concentrations are inadequate. Soil test phosphorus concentrations are typically determined by Olsen bicarbonate, Mehlich-3, or Bray P1 extraction and are highly correlated with yield response (Bray P1 should only be used on non-calcareous soils). Other soil test phosphorus extractants are also used and can be appropriate given adequate calibration.

The primary factors used in determining potato phosphorus recommendations is the soil test phosphorus concentration and the amount of free or excess lime in alkaline soils. Research shows the presence of significant amounts of free lime (calcium carbonate) increases the precipitation of soil solution phosphorus, reducing phosphorus availability to plants. The relative reduction in soil phosphorus availability is directly proportional to the amount of free lime in the soil. Pre-plant phosphorus fertilizer recommendations for Russet Burbank are shown in Table 8.9.

Pre-Plant Fertilization Research shows that phosphorus fertilizer may be applied effectively in either fall or spring as a broadcast and/or in a concentrated band when rows are formed; i.e., during mark-out or planting. Banding generally improves uptake efficiency and early-season phosphorus availability by concentrating the fertilizer in a narrow zone near the seed piece. Banding fertilizers containing both ammonium and phosphorus is known to have a synergistic effect.

Although banding phosphorus is a best management practice, full season phosphorus availability requires that the entire root zone have ample amounts of this nutrient. Soils with low to medium soil test phosphorus concentrations should

Table 8.9 Phosphorus fertilizer recommendations for Russet Burbank potato

Soil test P (Olsen bicarbonate) (0–12 in)	% Free lime			
ppm	0	4	8	12
	(lb P_2O_5/ac)			
0	320	360	400	440
5	240	280	320	360
10	160	200	240	280
15	80	120	160	200
20	0	40	80	120
25	0	0	0	40
30	0	0	0	0

Apply an additional 40–80 lb. P_2O_5/ac as a starter at planting for soil test phosphorus levels below 35 ppm Olsen sodium bicarbonate (which is equivalent to 40–45 ppm Mehlich-3 or Bray P1, although these extractants are not recommended by the University of Idaho, and Bray P1 should not be used on calcareous soil)

Add 25 lb P_2O_5/ac for each additional 100 cwt/ac above 400 cwt/ac

receive both banded and broadcast fertilizers. Soils with moderately high soil test phosphorus levels do not need broadcast fertilizer, but may still benefit from a moderate amount of banded fertilizer (~40–80 lb P_2O_5/ac). If the soil test phosphorus level is extremely high, it is possible that no phosphorus is needed, but research shows that potato is unique among crops in responding to phosphorus fertilizers at soil test levels higher than for most other crop species. It should also be noted that high-phosphorus fertilizer rates can induce micronutrient deficiencies, such as zinc or manganese; thus, care should be taken to ensure adequate nutrition with these nutrients.

Liquid fertilizers, such as APP, and dry fertilizers, such as MAP or DAP, are equally effective phosphorus sources for potato as long as they are managed properly. Some evidence shows a slight advantage to MAP over DAP on high pH soils.

Phosphorus recommendations for Russet Burbank potato are presented in Table 8.9. As recommendations vary substantially based on extractant used and growing region, a similar table would need to be shown for each area. It is best to use locally calibrated recommendations. The rates presented in Table 8.9 represent the amounts of P_2O_5 that should be broadcast in fall or spring, in addition to what is applied in a concentrated band.

In soils with very poor phosphorus solubility, it is beneficial to use research-proven, enhanced-efficiency fertilizer products. For example, many studies show that it is beneficial to combine liquid phosphorus fertilizer in the band with humic and other organic acids, particularly when soil organic matter content is low. This practice has the potential to enhance phosphorus solubility and uptake in low organic matter soils. However, this industry is highly unregulated, and buyers should be aware of products that are scientifically unproven for the specific conditions in which the potato will be grown.

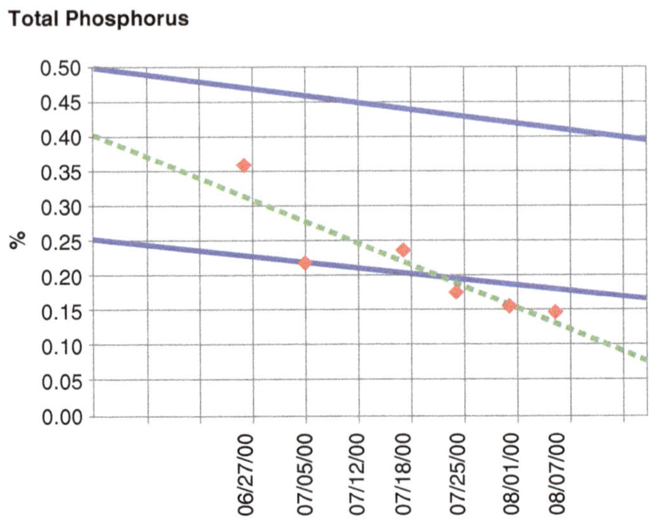

Fig. 8.19 Petiole phosphorus concentration helps predict in-season need for phosphorus fertilizer, if needed. The goal is to avoid dropping below the critical level

In-Season Application In contrast to nitrogen, foliar application or injection of phosphorus into the irrigation water is not as efficient as when the fertilizer is placed in the soil near where plant roots will concentrate. However, research shows that when petiole analysis predicts a phosphorus deficiency, fertigation can correct it, as long as active roots are near the soil surface (Fig. 8.19). Potato root density, and therefore phosphorus uptake efficiency, increases near the soil surface as plants touch across rows to close the canopy and give full shade to the soil.

Petiole phosphorus concentrations can provide a good indication of potato phosphorus status. For example, total phosphorus concentrations in the fourth petiole from the growing point for Russet Burbank should be maintained above 0.22% to provide sufficient phosphorus to satisfy both vegetative and tuber growth requirements. Petiole sampling for phosphorus should begin at or shortly after tuber initiation and continue at weekly intervals through most of the tuber bulking period. It is normal for the phosphorus concentration of the petiole to decline with time during the growing season in a somewhat linear fashion. Future petiole phosphorus concentrations may be estimated by plotting the previous test results against time and fitting a trend line to the data points. If it appears concentrations will drop below the critical value at a future date, it would be best to preemptively fertigate once with phosphorus at 40 lb P_2O_5/ac. However, it is likely this application will not be reflected in future petiole analysis. Research suggests continued applications in an effort to raise the petiole phosphorus level will not be responsive and that growers should not try to apply enough phosphorus to raise petiole phosphorus concentrations at this point in the season. Additionally, this would suggest a higher rate of pre-plant phosphorus fertilizer be applied in subsequent years to try to avoid this in-season deficiency. However, in some soils the phosphorus fixation capacity is so high that in-season fertigation is necessary.

Potassium

Potato has a relatively high potassium requirement, removing over 240 lb K/ac (289 lb K_2O/ac) in a 500 cwt/ac crop. Extractable potassium concentrations in many soils have declined over the last few decades due to crop removal. As a result, potato yield and quality responses to potassium fertilizer are common, and the intensity of potassium management programs has increased.

Potassium availability influences tuber yield and size, as well as a number of tuber quality factors, including: specific gravity, blackspot bruise susceptibility, chip and fry color, and storage quality. Potassium deficiencies decrease photosynthesis—reducing dry matter and starch formation. When potassium uptake is excessive, surplus potassium is translocated to the tubers causing increased tuber water content and decreased specific gravity.

The optimum tuber potassium concentration for maximum dry matter production is 1.8%. At this concentration, about 0.5 lb K_2O (0.6 lb K_2O) is required to produce 1 cwt of potato. A potato crop bulking at 7 cwt/ac/day requires about 3 lb. K/ac/day (3.6 lb. K_2O/ac/day) to maintain optimum dry matter production. Potassium fertilization programs should be designed to provide sufficient potassium to maintain optimum plant concentrations throughout the tuber bulking period.

Pre-Plant Fertilization Potassium fertilizer recommendations for Russet Burbank potato are presented in Table 8.10. Sodium bicarbonate-extractable potassium provides a good estimate of potassium availability in southern Idaho soils, whereas other regions are known to use different extractants to make recommendations

Table 8.10 Potassium fertilizer recommendations based on the sodium bicarbonate test for Russet Burbank potato in Idaho

	Potential yield (cwt/ac)			
Soil test K	300	400	500	600
(0–12 in)				
ppm	(lb K_2O/ac)			
25	550	600	650	700
50	450	500	550	600
75	350	400	450	500
100	250	300	350	400
125	150	200	250	300
150	50	100	150	200
175	0	0	50	100

based on crop response to added potassium. These recommendations represent the amount needed to give maximum economic yield without regard to trying to increase the soil test potassium. Research shows a soil test potassium concentration of 175 ppm via bicarbonate extraction in the top 12 in of soil is required to produce maximum potato yield and quality. Additionally, it takes about 4 lb. K_2O/ac to raise the soil test potassium concentration 1 ppm. Recommendations presented in Table 8.10 are based on these relationships and also include an adjustment for potential yield to account for differences in potassium uptake at different yield levels.

The probability of obtaining a positive yield response to potassium fertilization generally increases as the sand content of the soil increases in this order: loamy sands > sandy loams > loams > silt loams. In general, potassium source has relatively little effect on total yield, although potassium sulfate K_2SO_4 sometimes produces a slightly higher percentage of large No. 1 tubers and higher specific gravities than KCl, particularly when potassium fertilizer is applied at high rates shortly before planting. The chloride in KCl is also an essential nutrient, although excesses can be toxic. It is essential to account for the chloride in the fertilizer, as well as in the soil and irrigation water.

Although not as mobile in soil as nitrate-nitrogen, potassium is not like phosphorus in that it will move a few inches through soil. As such, it will move with the water front towards growing roots.

Banded potassium fertilizer is generally not as effective as broadcast potassium when the bulk of the seasonal potassium requirement is applied at or prior to planting, although low rates of potassium in a band have shown to be effective in some circumstances. However, potato seed pieces are high in potassium and, unlike most other crop species, the benefit from this nutrient applied in a band is relatively lower. If banded, the combined rate of potassium and nitrogen should be kept below 60 lb/ac to avoid crop damage due to fertilizer salt effects with a sliding scale of risk as the salt content of the soil increases. In other words, a soil with low salts (electrical conductivity (EC) less than about 0.3 dS/m) will allow for higher band rates than one with salt concentrations that are higher. Fertilizer rates exceeding 300 lb. K_2O/ac should have split applications to avoid yield losses. Yield reductions have been measured with spring applications of 400–600 lb. K_2O/ac. Again, this is relatively more important as salt concentrations in the soil increase (typically a greater problem in irrigated arid and semi-arid regions with saline or nearly saline waters).

In-Season Fertilization Applying all potassium pre-plant is usually more effective than applying most or all of the seasonal potassium supply via fertigation. Studies conducted in Idaho and Washington State show that applying over 50% of the seasonal potassium requirement during tuber bulking reduces tuber yield and specific gravity compared to pre-plant applications. However, if petiole analysis shows potassium deficiencies, small amounts should be applied via the irrigation system or broadcast (Fig. 8.20). In-season applications greater than 50 lb. K_2O/ac should be avoided because of the increased probability of tuber quality reductions. Studies also show there is no consistent difference in the effectiveness of potassium sources when applied through the irrigation system. Potassium fertilizer should not be applied later than 45 days before vine kill.

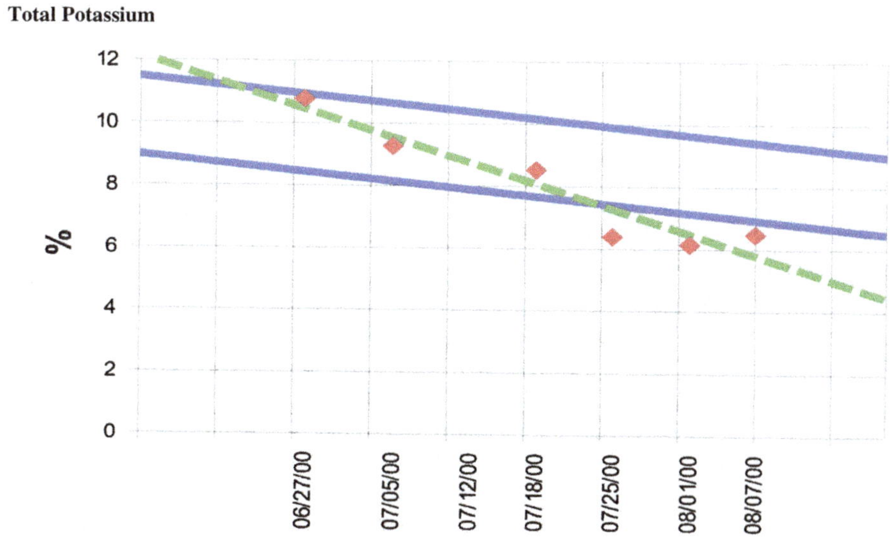

Fig. 8.20 Petiole potassium concentration helps predict in-season need for phosphorus fertilizer, if needed. The goal is to avoid dropping below the critical level

Petiole potassium concentrations generally decrease with time following tuber initiation. The rate of decrease depends on soil potassium availability and vine and tuber growth rates. Research with Russet Burbank potato shows a petiole potassium concentration of 7.0–7.5% is adequate to maintain optimal tuber growth rates and yield.

Petiole potassium concentrations respond to in-season potassium applications when plant potassium levels are deficient. However, there is a 2–3 week lag period between the time that potassium fertilizer is injected and the time petiole potassium concentrations change, although sometimes petiole potassium doesn't increase, but levels off instead of continuing downward. As a result, potassium applications should be made 15–20 days before the date petiole potassium concentrations are estimated to drop below the sufficiency level. As with phosphorus, future petiole potassium concentrations may be estimated by plotting the previous test results against time and fitting a trend line to the data points.

Sulfur

Sulfur fertilization is more commonly needed on sandy, low organic matter fields that have not received manure or similar amendments recently and where sulfate-sulfur concentrations in the irrigation water and/or precipitation are low. Sulfur is applied as a sulfate source or as elemental sulfur. Sulfate-sulfur (SO_4–S) is readily available for plant uptake, but is susceptible to leaching. Elemental sulfur, on the other hand, needs to be oxidized to SO_4–S before being taken up by plant roots. When applying elemental sulfur, there is a lag time in the conversion to SO_4–S due to the initially low activity of sulfur-oxidizing bacteria. This is particularly true for cold, wet soil conditions that further slow the oxidation process. Elemental sulfur is acidifying and should not be used on neutral or acid soils. Application to calcareous soils has been shown to be a good source of sulfur and possibly supply other benefits, but it will not reduce the overall soil pH at typical rates applied. Generally, elemental sulfur is applied at 50–100 lb./ac, and it would take several tons per acre to neutralize all of the limestone present in a typical calcareous soil. Although this is true, there is a temporary localized reduction in pH immediately around the sulfur granule, which can be used to enhance the availability of phosphorus and micronutrients when applied in the immediate proximity.

Application of sulfur fertilizer should be considered when soil SO_4–S concentrations at the 0–12 in depth are below 15 ppm and SO_4–S concentrations in the irrigation water are below 5 ppm. In general, the soil test for sulfur is less accurate than well-calibrated leaf or petiole analysis. A pre-plant application of 30–40 lb. SO_4–S/ac as ammonium sulfate, potassium sulfate, or urea-sulfuric acid should satisfy the sulfur requirement. However, potato does respond well to applications of soluble sulfur sources injected through the irrigation system during tuber growth when petiole analysis suggests deficiency. Total sulfur concentrations in petioles below 0.20% indicate a potential need for supplemental sulfur applications. As with other nutrient concentrations, trends in petiole sulfur concentrations need to be monitored on a weekly basis to provide a reliable estimate of plant sulfur status.

Calcium and Magnesium

Some tuber quality disorders, such as internal brown spot, are associated with calcium deficiencies, and both calcium and magnesium are essential for plant growth and tuber development. Although both calcium and magnesium are essential for plant growth, they are usually present in adequate amounts in calcareous, alkaline soils and irrigation waters. Although some of the calcium and magnesium in soil and irrigation water is not plant available, any that is dissolved is immediately available for plant uptake. Thus, irrigated potatoes rarely have deficiencies. Some deficiencies have been observed in very sandy soils or acid soils where supplemental applications of calcium and magnesium were needed to meet tuber growth requirements.

Exchangeable soil calcium concentrations less than 400 ppm on sandy soils and 700 ppm on heavier textured soils indicate a need for supplemental calcium, which can be met with pre-plant applications of 200 lb. Ca/ac. Magnesium deficiencies can develop at exchangeable soil magnesium levels below 100 ppm. Broadcast applications of 100 lb/ac as magnesium sulfate or potassium-magnesium sulfate, or band applications of 20 lb Mg/ac should satisfy crop requirements. Alternatively, calcium and magnesium can be applied as dolomitic lime when increases in soil pH are also desired. To improve tuber calcium uptake, fertilizer should be placed in the tuber formation zone to facilitate uptake by the stolon roots. Fertilizers such as calcium nitrate, calcium-ammonium nitrate, and calcium sulfate (gypsum) can be used to supply calcium without significantly increasing the soil pH.

Calcium applied to foliage is not translocated to the tubers, but may help satisfy some of the calcium requirement of the leaves when deficiencies develop. Immobility of calcium in alkaline soils and plants also limits the effectiveness of irrigation-applied calcium. Magnesium is also somewhat immobile in alkaline soil, but is mobile within the plants. As such, foliar sprays of magnesium sulfate can be applied to correct magnesium deficiencies when petiole concentrations are less than 0.3%.

Micronutrients

As previously discussed, micronutrients are largely supplied to plants from soil mineral and organic sources with availability strongly tied to pH. In addition, boron can be present in significant amounts in irrigation water, especially in the groundwater of arid and semi-arid regions. Critical soil micronutrient concentrations for potato are presented in Table 8.11.

Zinc

Zinc fertilization may be required on calcareous soils, particularly in areas where topsoil was removed by erosion or land leveling. Deficiencies are also sometimes seen in sands and soils with high levels of organic matter. For soils with zinc concentrations less than 1 ppm, a broadcast application of 8–10 lb. Zn/ac should be applied and incorporated prior to planting. Compared to broadcast rates, banded zinc rates can be reduced by up to 80% due to higher fertilizer uptake efficiency. Zinc fertilizer sources with high percentages of water-soluble zinc (such as $ZnSO_4$, ZnDTPA, and zinc lignosulfate) should be used to maximize uptake efficiency. Zinc will move through the phloem tissue from the leaves to the tubers; thus, foliar sprays of zinc sulfate (1 lb. Zn/ac) or chelates (0.15 lb. Zn/ac) are effective in correcting zinc deficiencies. It should be noted that some potato fungicides contain sufficient zinc to satisfy crop needs.

Table 8.11 Adequate DTPA (Zn, Mn, Fe, and Cu) and hot water (B) extractable soil micronutrient concentrations for potato

Zn	Mn	Fe	Cu	B
ppm				
>1.0	>4–8	>4.0	>0.2	>0.5

Manganese

On neutral to acid soils, manganese deficiencies are relatively rare. Deficiencies are more common on calcareous soils, and manganese should be banded or applied as a foliar spray to minimize soil fixation. Manganese can be broadcast and incorporated prior to planting at 5–10 lb. Mn/ac to correct deficiencies. Applying 5 lb. Mn/ac in a band near the seed piece has been effective in providing adequate manganese to potato when deficient. Manganese will also move from the leaves to the tubers when applied as a foliar spray. Manganese chelates and $MnSO_4$ (1 lb. Mn/ac) are effective manganese sources for foliar applications. Some potato fungicides contain sufficient manganese to satisfy crop needs.

Boron

Boron fertilizers may be needed where soils and irrigation waters have low boron concentrations. Broadcast applications of 1–2 lb. B/ac should be made prior to planting when soil test boron concentrations are below 0.5 ppm. Most nutrients can become toxic, but boron has a particularly narrow range between sufficiency and toxicity. Sodium borates and boric acid can be applied as foliar sprays. However, as with calcium, boron is not translocated from the leaves to the tubers in appreciable amounts, and benefits are seen from placement near tubers when it is deficient.

Iron

Although iron is very insoluble in alkaline soil, potato is very efficient at making the vast reserves available for uptake. Although an interpretive value is listed in Table 8.11, the soil test for iron is poorly correlated to yield response, and petiole tissues are easily contaminated with iron-rich dust. However, if petiole analysis and/or visual symptoms are present, band applications of chelated iron may be beneficial. It should be noted that most chelated forms of iron are ineffective in many soils, and a proven chelate, such as Fe-EDDHA, is needed. Iron is relatively immobile in soils and plant tissues. Consequently, multiple applications of iron sulfate or chelates to the foliage may be required to correct iron deficiencies.

Copper

Copper concentrations in mineral soils are usually sufficient for adequate plant growth; however, deficiencies have been noted in very acidic organic soils. As with boron, copper has a narrow range between where it is sufficient and toxic. Copper has intermediate mobility in plant tissue, and foliar sprays of copper sulfate or chelates applied during tuber bulking are effective in increasing petiole copper concentrations. Copper is a common component of certain potato fungicides that can provide significant amounts of this mineral to the plant.

Compatibility of Phosphorus Fertilizer with Irrigation Water (Stark and Westermann 2003)

Irrigation waters with appreciable calcium concentrations (>50 ppm) potentially form insoluble calcium phosphate deposits when liquid phosphate fertilizers are injected through irrigation systems. These deposits plug nozzles and reduce the effectiveness of irrigation systems. As the calcium concentration in the irrigation water increases, the potential for irrigation problems increases proportionately. The liquid phosphorus source, as well as the pH and bicarbonate concentration of the irrigation water, also influence the potential for calcium phosphate precipitation.

Compatibility of a phosphorus fertilizer with an irrigation water source can be tested by adding a proportional amount of the specific liquid phosphorus fertilizer to a gallon of fresh irrigation water.

For center pivot and linear move systems, the amount of fertilizer solution that can be safely added to the irrigation water can be estimated using the following relationship:

$$X = \frac{0.0283 \times F}{D \times P}$$

where:

X = teaspoons of fertilizer solution per gallon of irrigation water

F = fertilizer application rate (lb P_2O_5/ac)

D = depth of water application (inches/ac)

P = P_2O_5 concentration in the fertilizer solution (lb P_2O_5/ac)

If a fine, white precipitate forms after the calculated amount of fertilizer solution is thoroughly mixed with the irrigation water, the phosphorus fertilizer application rate should be reduced. The testing procedure should be continued until there is no visible precipitate formation.

For wheel-line, hand-line, and solid-set irrigation systems, the relationship is:

$$X = \frac{0.0283 \times F}{W \times T \times P}$$

where:
X = teaspoons of fertilizer solution per gallon of irrigation water
F = fertilizer application rate (lb P_2O_5/ac)
W = water application rate (in/h)
T = injection time (h)
P = P_2O_5 concentration in the fertilizer solution (lb P_2O_5/gal)

If precipitate forms after mixing, the phosphorus application rate should be reduced or the length of the injection period should be increased and testing repeated.

Irrigation waters with high calcium and magnesium concentrations or a high pH can be partially acidified to improve compatibility with phosphorus fertilizers. Phosphoric acid or urea phosphoric acid can be injected through the irrigation system in appropriate amounts. Alternatively, sulfuric acid or urea sulfuric acid can be injected along with ammonium polyphosphate to minimize calcium phosphate precipitation.

Achieving an irrigation water pH between 5.0 and 7.0 can substantially reduce the potential for nozzle plugging. However, the final pH should be kept above 5.0 to prevent corrosive damage to the irrigation system and nozzles.

Chloride

There is generally enough chloride present in irrigation water and/or potassium fertilizer (KCl) to provide adequate chloride for a potato crop. Deficiencies may occur when irrigation waters contain minimal chloride and when all of the potassium is applied as a non-chloride source, such as K_2SO_4 or KTS. Atmospheric deposition of chloride can also be substantial in fields in close proximity to saltwater bodies.

Molybdenum and Nickel

Although these are essential nutrients, they are needed in such minute quantities they have never been diagnosed as deficient in potato production.

Future Needs and Sustainability

World populations are growing and demand for food and other crop products is continually increasing. Agriculture developments and efficiency have improved at unprecedented rates over the past century. But further gains in production and efficiency are needed, including nutritional aspects of crop production. Increases in potato yield have outpaced all of the major grain crops and many others. These increases demand continued refinements to provide nutrients.

Additional research is needed to better predict nutritional needs, especially with recently released and future cultivars. New cultivars need to be developed that are more efficient in nutrient uptake and/or reduced nutrient needs. This is likely achieved by selecting cultivars that have deeper, more extensive root systems and/or that reach full canopy cover quickly and keep it through the entire growing season, but avoid excessive shoot growth that can be a sink for nutrients without adding any additional photosynthesis for supplying tuber growth. These new cultivars need to be equal to or superior in other important ways as well, such as storage length and efficiency. Russet Burbank continues to be the major cultivar used in the U.S. despite its poor production efficiency—including nutrition. Many cultivars have been developed that are more efficient with regard to fertilization, but these are often overlooked by end users.

Continued work also needs to be done to develop more enhanced-efficiency fertilizer products resulting in greater crop growth per unit of fertilizer applied. Although we enjoy an abundance of resources, many are limited; thus, warrant conservation. Additionally, water and air pollution from fertilizers, especially nitrogen and phosphorus, are a concern, and practices need to be adopted to increase the percentage of nutrient efficiently used and minimize the amount lost to the environment. Of course, farm sustainability is vital to society, and all of these efforts need to be made in a culture of economic sustainability for growers.

As adoption lags behind current research, growers should be encouraged to use effective products if their economic efficiency is equal to or better than traditional fertilizers. Significant work has been done to improve techniques for appropriate fertilizer rate, timing, and placement to maximize efficiency. Further work can be done, but grower adoption is equally vital.

Continued work and refinement of soil testing and petiole analysis as they correlate to yield and tuber quality is essential, with increasing yields placing greater demands on the soil to supply nutrients and as cropping systems, the environment, and cultivars change.

References

Bryson GM, Mills HA, Sasseville DN, Jones JB, Barker AV (2014) Plant analysis handbook IV. MicroMacro Publishing, Athens, GA

Hopkins BG (2015) Phosphorus in plant nutrition. In: Pilbeam DJ, Barker AV (eds) Plant nutrition handbook, 2nd edn. CRC, Boca Raton, FL, pp 65–126

Hopkins BG, Hansen NC (2019) Phosphorus management in high-yield systems. J. Environ. Qual. 48:1265–1280. https://doi.org/10.2134/jeq2019.03.0130

Kelling KA, Westermann DT, Voss RE (2001) Nutrient imbalances. In: Stevenson WR, Loria R, Franc GD, Weingartner DP (eds) Compendium of potato diseases. American Phytopathological Society, St. Paul, MN, pp 75–78

Pitchay DS, Mikkelsen RL (2018) Plant nutrition diagnostics: potato. International Plant Nutrition Institute, Peachtree Corners, GA

Rosen CJ, Kelling KA, Stark JC, Porter GA (2014) Optimizing phosphorus fertilizer management in potato production. Am J Potato Res 91(1):145–160

Rowe RD (ed) (1993) Potato health management, plant health management series. American Phytopathological Society, Eagan, MN

Schulte EE, Kelling KA, Peters JB, Combs SM (2000) Plant analysis interpretations used in the revised Wisconsin program. New horizons in soil science no. 7–2000, Department of Soil Science, University of Wisconsin-Madison

Schulte EE, Walsh LM, Kelling KA, Bundy LG, Bland WL, Woskowski RP, Peters JB, Sturgul SJ (2009) Management of Wisconsin soils (A3588). University of Wisconsin, Madison, WI

Stark JC, Westermann DT (2003) Nutrient management. In: Stark JC, Love SL (eds) Potato production systems. University of Idaho, Moscow, ID, pp 115–134

Thornton RE, Sieczka JB (1980) Commercial potato production in North America. Am Potato J 57:1–36

Westermann DT (1993) Fertility management. In: Rowe RC (ed) Potato health management. APS, St Paul, MN, pp 77–86

Westermann DT (2005) Nutritional requirements of potatoes. Am J Potato Res 82(4):301–307

Chapter 9
Disease Management

Phillip Nolte, Jeff Miller, Kasia M. Duellman, Amanda J. Gevens, and Eugenia Banks

Contents

P. Nolte (✉)
Department of Entomology, Plant Pathology and Nematology, University of Idaho, Moscow, ID, USA

J. Miller
Miller Research, Rupert, ID, USA
e-mail: jeff@millerresearch.com

K. M. Duellman
Department of Entomology, Plant Pathology and Nematology, University of Idaho, Idaho Falls, ID, USA
e-mail: kduellman@uidaho.edu

A. J. Gevens
Department of Plant Pathology, University of Wisconsin Madison, Madison, WI, USA
e-mail: gevens@wisc.edu

E. Banks
Ontario Potato Board, Elora, ON, Canada
e-mail: eugeniabanks@onpotato.ca

J. C. Stark et al. (eds.), *Potato Production Systems*,
https://doi.org/10.1007/978-3-030-39157-7_9

Introduction

Potato diseases are caused by an array of pathogens, including viruses, bacteria, phytoplasmas, and fungi. Disease can occur at any stage of potato development, from seed diseases before planting, to foliar and tuber diseases during the growing season, to storage diseases that cause rots and/or reductions in tuber quality. Some diseases are foliar only, some only attack tubers, and others can attack both. Similarly, some diseases are only found in the field, some are only seen in storage, while others occur in both locations. The fact that potatoes are vegetative propagated dictates that special emphasis be placed on seedborne diseases. A general overview of each of the important potato diseases is presented in this chapter, followed by symptom descriptions, images, economic importance, and management guidelines.

Microorganisms cause all diseases discussed in this chapter. Many plant pathologists consider physiological disorders, such as hollow heart, to be diseases, but these disorders are caused by adverse environment, nutrient imbalances, or other noninfectious factors and are discussed in Chap. 14.

The Disease Triangle

In order for disease to occur, a susceptible host plant and a pathogen that can cause disease in or on the plant must come together under environmental conditions that favor disease development. These three factors, a susceptible host, a virulent

Fig. 9.1 The disease triangle illustrating that a susceptible host and a virulent pathogen must come together in a favorable environment before infectious plant disease can occur

pathogen, and a favorable environment, make up the sides of the so-called "disease triangle" (Fig. 9.1). Most control strategies involve attempts to modify one or more of these factors to reduce the amount of disease.

The use of a variety that is resistant to a certain disease, for instance, alters the host factor. Using certified seed, produced under strict tolerances for potato viruses, is an attempt to reduce or eliminate the presence of the viral pathogen. Altering irrigation scheduling or improving drainage are examples of attempts to alter the environment. Despite growers' best efforts, however, the three factors often come together and disease occurs.

Epidemics of disease occur when a susceptible host comes into contact with an aggressive pathogen under environmental conditions extremely favorable for disease development over a wide area. Disease outbreaks such as these can be very destructive.

Many plant pathogens require a wound to gain entrance to the host plant. Others have the ability to take advantage of naturally occurring weak points or openings, such as stomates or lenticels. Still others have the ability to penetrate directly into the plant through leaves, stems, or roots.

Some, as with most of the potato viruses, require the activity of a "vector," an insect, nematode, or fungal carrier, to move to the host plant and gain entrance. Understanding how a pathogen gains entrance and infects the host plant is critical for designing and implementing a management program.

General Disease Management Strategies

Strategies for disease management are classified in several ways, but one of the most practical schemes divides management techniques into categories, such as: management by exclusion, eradication, protection, or use of resistant varieties. The most effective approaches employ all or several of these methods in combination, a strategy usually referred to as integrated pest management (IPM).

Exclusion strategies are often regulatory and involve the use of seed certification procedures, quarantines, and inspections to prevent the establishment of a pathogen in a non-infested area.

Eradication strategies involve destruction of diseased host plants as well as sanitation procedures for storage facilities, truck beds, cutters, and handling equipment.

Some pathogens can be harbored in weeds or volunteers, and the destruction of these potential disease reservoirs would be considered forms of pathogen eradication. Eradication of soilborne pathogens by fumigation also falls under this category. This strategy attempts to eliminate a pathogen once it has become established in an area. Generally, success of this approach depends on how long the pathogen has been present—the longer the time since pathogen establishment, the less likely attempts at eradication will be effective.

Protection strategies involve the use of chemical and biological control methods to allow production of a quality crop even in the presence of the pathogen under favorable environmental conditions. Protection is probably the most widely used and most familiar method of disease management. This approach relies on the establishment of some kind of "barrier" between the host plant and the pathogen inoculum source.

"Protectants" are active ingredients that interfere with the establishment of infection. To accomplish this task, the protectant must be in place before the pathogen comes into contact with the host plants. For example, most fungicides are protectants and are ineffective against an established pathogen, since in most cases, the pathogen is within the host tissues and contact between it and the chemical does not occur. A handful of fungicides are "systemic," meaning they are actually absorbed by the plant tissues and can either provide protection within or, in rare cases, attack the pathogen after it has become established in the plant tissues. A list of common fungicides is found in Table 9.1.

Use of resistant varieties offers simple and economical management of diseases if varieties with effective resistance are available. Resistance rarely means total immunity to disease, however. More commonly, infection occurs but disease progress is slowed by the use of resistant varieties, making the disease less destructive and control less expensive.

Resistant varieties developed through traditional breeding are accepted across markets at this time. Potato varieties with targeted disease resistance or tolerance developed through novel biotech approaches provide promise for sustainable management; however, acceptance across markets has been limited as yet.

Many diseases become difficult or impossible to control after a certain amount of infection has been reached. All diseases also have an "incubation period," which is the time interval between the infection event and the appearance of visible symptoms. Control measures applied during this incubation period can appear to be relatively ineffective because the undetected disease continues to develop normally, protected within the plant, and symptoms appear even after the application of an effective fungicide. Such applications will generally be effective in preventing or reducing *new* infections.

Several diseases gain entrance into the tuber by taking advantage of wounds made during harvest and handling. Elevated temperatures and high soil moisture during harvest and handling can greatly accelerate disease progress and quickly turn a minor problem into a major one. Pythium leak, Fusarium dry rot, pink rot, bacterial soft rot, and potato late blight can all take advantage of harvest wounds and elevated temperatures. Pythium leak, for example, is strongly favored by warm harvest conditions.

Table 9.1 The following fungicides and fungicide classes are commonly used in potato cropping systems (current 2018)

Mode of action	Chemical group	Generic name	Example(s)	FRAC group
Mitosis: beta-tubulin assembly	MBC fungicides (methyl benzimidazole carbamates)	Thiabendazole, thiophanate-methyl	Mertect®, Top®, Topsin®	1
NADH inhibitor	Dicarboximides	Iprodione	Rovral®, Meteor™, Nevado®	2
C14-demethylation in sterol biosynthesis	DMI or demethylation inhibitors-fungicides	Difenoconazole, metconazole, prothioconazole	Top®, Quash®, Emesto® Silver	3
RNA polymerase I	Phenylamides	Mefenoxam, metalaxyl	Ridomil®, Ultra Flourish®, MetaStar®	4
Fungal respiration, complex II	Carboxamides	Boscalid, fluopyram, flutalonil, fluxapyroxad, penflufen, penthiopyrad	Endura®, Luna® Tranquility, Moncut®, MonCoat®, Priaxor®, Emesto® Silver, Vertisan®	7
Methionine biosynthesis	AP or anilino-pyrimidines	Cyprodinil, pyrimethanil	Switch®, Scala®, Luna® Tranquility	9
Fungal respiration, complex III	Quinone outside inhibitors (QOL)	Azoxystrobin, famoxadone, pyraclostrobin	Quadris®, Satori®, Equation®, Tanos®, Headline®	11
Signal transduction, MAP protein kinase	Phenylpyrroles	Fludioxonil	Maxim®, Spirato®	12
Complex III of fungal respiration: ubiquinone reductase site, Qi site	QIL fungicides (Quinone inside inhibitors)	Cyazofamid	Ranman®	21
B-tubulin assembly in mitosis	Benzamides, thiazole carboxamide	Zoxamide ethaboxam	Gavel®, Zing!®, Elumin®	22
Unknown	Cynanoacetamideoximes	Cymoxanil	Curzate®	27
Cell membrane permeability, fatty acids (proposed)	Carbamates	Propamocarb	Previcur® Flex	28
Uncoupler of oxidative phosphorylation	2,6-dinitroanilines	Fluazinam	Omega®	29
Inhibitors of oxidative phosphorylation, ATP synthases	Tri phenyl tin compounds	Fentin hydroxide	Agri Tin®, Super Tin®	30

(continued)

Table 9.1 (continued)

Mode of action	Chemical group	Generic name	Example(s)	FRAC group
Unknown	Phosphonates	Fosetyl-Al phosphorous acid and salts	Aliette®, Phostrol®	33 (P07)
Phospholipid biosynthesis and cell wall deposition (proposed)	CAA-fungicides (carboxylic acid amides)	Dimethomorph mandipropamid	Forum®, Zampro®, Revus®	40
Delocalization of spectrin-like proteins	Benzamides	Fluopicolide	Presidio®	43
Microbial disrupters of pathogen cell membranes	Microbial (*Bacillus* spp.)	*Bacillus amyloliquefaciens*	Double Nickel®	44
Complex III: cytochrome bc1 (ubiquinone reductase) at Qo site, stigmatellin binding sub-site	Triazolo-pyrimidylamine	Ametoctradin	Zampro®	45
Lipid homeostasis and transfer/storage	Piperidinyl-thiazole-isoxazolines	Oxathiapiprolin	Orondis®	49
Multi-site activity	Inorganics	Copper compounds	Various	M1
Multi-site activity	Inorganics	Sulfur	Various	M2
Multi-site contact activity	Dithiocarbamate	Mancozeb metiram	Dithane, Koverall, Manex, Manzate, Penncozeb, Roper, Polyram	M3
Multi-site contact activity	Chloronitriles	Chlorothalonil	Bravo®, Echo®, Equus®	M5
Salicylate related	Benzo-thiadiazole	Acibenzolar-S-methyl	Actiguard®	P01

Fungicides Commonly Used in Potato Production in United States (updated February 22, 2018)

For the purposes of this guide, potato diseases have been separated into those that are caused by viruses, bacteria, and fungi or fungus-like organisms. Some of these diseases are strictly soilborne, others are tuberborne, and still others reside in volunteers, crop residues, or weed hosts.

Some diseases spread rapidly in the wind or during rainstorms, while others are relatively stationary or require an insect or other type of vector to move from plant to plant or from one area to another. Some diseases occur only in storage, some are confined to the field, and others occur in both places.

The following are descriptions and management recommendations for important potato diseases caused by viruses, bacteria, and fungi.

Diseases Caused by Viruses

Plant viruses are a unique group of pathogens that are not considered to be living organisms by many scientists because of their inability to develop outside of a living host cell. Perhaps the simplest way to describe viruses is to refer to them as “infectious, self-replicating molecules.” The typical plant virus is constructed of a nucleic acid core surrounded by a protein coat. Viruses are so small they are not visible even with the most powerful light microscopes, and observing them requires the extremely high magnifications provided only by electron microscopes. Management procedures for all viruses employ similar strategies, and these are discussed in Table 9.2.

Potato Latent Mosaic (PVX)

Causal Agent: Rigid rod virus—Potato virus X (PVX).

Inoculum Source: Infected seed potatoes or volunteer potato plants.

Exposure: Mechanical transmission, such as by plant-to-plant contact or contact with people or machinery.

Symptoms: No symptom or mild mosaic symptom.

Economic Importance: PVX is usually symptomless and is probably of little economic importance in many potato varieties, but yield losses of up to 15% have been reported in some varieties.

Vector Relationship: No vector, mechanical transmission.

Potato Mosaic (PVY)

Causal Agent: Filamentous virus—Potato virus Y (PVY).

Inoculum Source: Infected seed potatoes, volunteer potato plants, and some weed hosts.

Exposure: Primarily by an aphid vector, although some mechanical transmission may also occur.

Symptoms: PVY symptoms range from virtually none (latent), such as that with the Shepody and Norkotah varieties, to noticeable stunting and mosaic symptoms (Fig. 9.2a) observed in Russet Burbank, to severe foliar damage and even death of the entire plant as in Ranger Russet. Tuber cracking has been observed in some potato varieties infected with some strains of PVY.

Recently, different strains of PVY, known as PVYntn or "necrotic strains," due to the type of symptoms they cause in foliage or tubers, have begun to appear in North America. Some necrotic strains cause a tuber defect known as potato tuber necrotic ringspot disease (PTNRD) in some potato varieties (Fig. 9.2b, c). There is considerable variation in susceptibility to foliar and tuber symptoms across varieties.

Table 9.2 Summary of potato diseases and management options

Disease	Cause	Management options
Diseases caused by viruses		
Potato latent virus	Potato virus X (PVX)	Plant clean, certified seed
Potato mosaic	Potato viruses Y and A (PVY, PVA)	Plant clean, certified seed
Potato leafroll/ tuber net necrosis	Potato leafroll virus (PLRV)	Plant clean, certified seed. Use a systemic insecticide at planting. Scout for aphids throughout the season and use late-season aphicides
Corky ringspot	Tobacco rattle virus (TRV)	Plant certified seed. Use soil fumigation to control stubby root nematodes
Calico	Alfalfa mosaic virus (AMV)	Use certified seed. Avoid planting near alfalfa or clover
TSWV ringspot	Tomato spotted wilt virus (TSWV)	Use certified seed. Control weeds. Avoid very susceptible varieties. Monitor and spray for thrips vector
Diseases caused by bacteria and phytoplasmas		
Bacterial ring rot	*Clavibacter michiganensis* subsp. *sepedonicus*	Plant clean, certified seed. If found, eliminate all potatoes on your farm. Thoroughly clean and disinfect all potato machinery
Blackleg	*Pectobacterium atrosepticum*, other *Pectobacterium* species, *Dickeya* spp.	Plant clean, certified seed. Use limited-generation certified seed sources. Do not wash seed potatoes. Routinely clean and disinfect seed potato cutting and transport equipment. Consider using whole seed or cut seed that has been properly healed ("precutting"). Use a fungicidal seed-piece treatment. Plant potato crops in well-drained soil and avoid irrigating non-emerged fields
Bacterial soft rot	*Pectobacterium carotovorum*, other *Pectobacterium* species of *Dickeya* spp.	Avoid bruising and mechanical damage to potatoes during harvest and transport. Properly heal stored potatoes before lowering to the holding temperature. Store as cool as possible. Avoid moisture films on potatoes in bulk piles and dry potatoes as much as possible before placing them in fresh-pack containers. Use high volumes of air and a reduction in relative humidity in storages with badly infested tubers
Common scab	*Streptomyces scabies*	Avoid low soil moisture during tuber set and early bulking. On a silt loam soil, maintain soil above 75% available soil moisture. On fields where common scab is not a severe problem, high moisture alone may be sufficient control. Chemical options are very limited against this disease
Witches' broom	Phytoplasma (not named)	Separate potato fields from alternate hosts, such as alfalfa, or avoid drying down the alternate hosts late in the season
Purple top	Phytoplasma (various strains)	Monitor and spray for leafhopper vector

(continued)

Table 9.2 (continued)

Disease	Cause	Management options
Zebra chip	*Candidatus Liberibacter solanacearum*	Monitor for potato psyllids and consider implementing use of insecticides at first detection of potato psyllids; not enough information is available for a more precise action threshold
Diseases caused by fungi		
Dry rot	*Fusarium* species, including *F. sambucinum* and *F. coeruleum*	Harvest only mature potatoes with well-developed skin. Promote wound healing immediately after placement in storage by maintaining temperatures of 50–55 °F and ventilation with high humidity. Clean and disinfect all potato handling, transfer, and cutting equipment. Provide good wound healing conditions when precutting seed before planting, or plant whole seed. Use a seed piece treatment that will effectively control *Fusarium* on potato seed pieces. Plant potato seed in warm (above 50 °F), moist soil that promotes rapid plant emergence. Avoid irrigating non-emerged fields
Potato early dying and Verticillium wilt	*Verticillium dahliae*	Plant certified seed and resistant varieties. Reduce soil populations by rotating crops and include potatoes every fourth or fifth year. Apply soil fumigants. Minimize plant stress and encourage uniform, continuous plant growth with balanced fertility and optimum soil moisture levels
Early blight	*Alternaria solani*	Provide proper plant nutrition and water management, proper pest management, and avoidance of other plant stresses. Several fungicides are available that control early blight on potato foliage during the growing season, the use of which also reduces production of spores that cause tuber lesions. If early blight moves into the canopy earlier than 3 weeks before intended vine kill, use one of the labeled and effective foliar fungicides. Prevent tuber infection through timely vine killing and harvesting of mature tubers. Avoid harvesting in wet conditions or when vines are green
Brown spot/ black pit	*Alternaria alternata*	Provide proper plant nutrition and water management. Use foliar fungicides in a preventative manner. Avoid bruising tubers at harvest

(continued)

Table 9.2 (continued)

Disease	Cause	Management options
Late blight	*Phytophthora infestans*	*Foliar*: Use only clean, certified seed. If the seed is potentially infected, use a seed piece fungicide treatment that has activity against late blight. Initiate protective spray programs at row closure, and apply applications on a weekly basis for at least 4 weeks if weather is favorable for disease. Once new crop growth slows and there are no reports of late blight in the area, increase the interval between applications. Destroy cull and volunteer potatoes *Storage*: Remove soil to examine suspect tubers for late blight, and monitor the storage for bad odors or wet spots. If late blight is detected, increase the air supply and shut down humidification until tubers are dry. Avoid conditions where free moisture is likely to develop. Do not store lots with more than 5% tuber blight. Hold temperatures at 38 °F for seed and 42 °F for fresh pack
Rhizoctonia, black scurf	*Rhizoctonia solani*	Plant seed tubers that do not have sclerotia covering more than 20% of the tuber surface. Use fungicide seed treatments. Plant in warm soils to encourage rapid emergence and green tissue development
White mold	*Sclerotinia sclerotiorum*	Implement long crop rotations to non-host plants (like grain). Maintain a good fertility program and avoid excessive vine growth. Avoid prolonged canopy wetness and excessive irrigation. Use available chemical control options with applications made before the disease has become well established in the field. Ensure good coverage on the foliage and stems, as necessary
Pink rot, water rot	*Phytophthora erythroseptica*	Maintain good irrigation practices and avoid waterlogged soils. Plant less susceptible varieties. Avoid wounding tubers and high pulp temperatures during harvest. Consider treating with mefenoxam at planting or during the production season or phosphorous acid products during the growing season. Use phosphorous acid products at harvest. Mefenoxam resistance has been reported in some areas
Pythium leak, watery wound rot	*Pythium* species	Avoid wounding during harvest, and do not attempt to harvest when pulp temperatures are above 65 °F, especially if environmental conditions are not conducive for removing field heat from stored tubers. If significant infection is found in storage, use high volumes of air and a rapid cool down. As with pink rot, consider treating with mefenoxam at planting or during the production season
Powdery mildew	*Erisyphe cichoracearum*	Sprinkle irrigate fields or spray fields with strobilurin or sulfur compounds before infections are established

(continued)

Table 9.2 (continued)

Disease	Cause	Management options
Powdery scab	*Spongospora subterranea* subsp. *subterranea*	Plant clean, certified seed. Do not plant in infested soil, and do not over irrigate. Use russet cultivars, which are more resistant to powdery scab than many white- and red-skinned varieties
Silver scurf	*Helminthosporium solani*	Plant disease-free tubers. Use fungicidal seed treatments
Black dot	*Colletotrichum coccodes*	Plant certified seed. Establish 4–5 year crop rotations. Maintain uniform and continuous growth by providing balanced nutrition and optimal soil moisture conditions. Consider using a strobilurin-based fungicide. Keep storage temperatures and humidity as low as possible, and market infected potatoes as soon as possible after harvest. Disinfect storages before the next crop
Gray mold	*Botrytis cinnerea*	Use foliar fungicides effective against gray mold. Avoid over-watering

Fig. 9.2 (**a**) Mosaic virus symptom typical of PVY or PVA on potato leaf. (Photo credit: Phillip Nolte). (**b**) External symptoms of potato tuber necrotic ringspot disease (PTNRD) typical of PVYn strains. (Photo credit: Jonathan Whitworth). (**c**) Internal symptoms of PTNRD. (Photo credit: Jonathan Whitworth)

Economic Importance: Yield losses can vary greatly, depending on the variety, the number of plants infected, and when the infection took place. Total loss of the crop has been reported on various occasions, but these are rare, isolated cases. Tests conducted in Wisconsin documented losses of up to 33% in Russet Burbank when 100% of the plants were infected early in the season. Similar results were obtained from separate Oregon and Idaho experiments.

The greatest losses occur as a result of planting infected seed, though current-season spread within the field can result in significant levels of tuber necrosis in susceptible varieties infected with necrotic strains of PVY. Because of effective certification efforts to keep virus levels low, yield losses from PVY are generally not significant.

Vector Relationship: Stylet-borne (non-persistent) transmission by green peach aphid (GPA) and numerous species of non-potato colonizing aphids.

Potato Mosaic (PVA)

Causal Agent: Filamentous virus—Potato virus A (PVA).

Inoculum Source: Infected seed potatoes, volunteer potato plants, and some weed hosts.

Exposure: Primarily by an aphid vector, although some mechanical transmission may also occur.

Symptoms: PVA is very similar to PVY in structure and in behavior. Severe mosaic symptoms can occur with PVA infection, but this virus is usually latent or causes only mild symptoms in commonly grown varieties.

Economic Importance: Unknown. Losses of up to 30% have been reported in some varieties.

Vector Relationship: Styletborne (non-persistent) transmission by GPA and several species of non-potato colonizing aphids.

Potato Leafroll, Tuber Net Necrosis

Causal Agent: Spherical virus—Potato leafroll virus (PLRV).

Inoculum Source: Infected seed potatoes, volunteer potato plants, and some weed hosts.

Exposure: Persistent transmission by potato colonizing aphids.

Symptoms: PLRV causes a characteristic rolling of leaves as well as chlorosis and stunting of the potato plant (Fig. 9.3a). More important to the commercial producer is the net necrosis symptom caused by the virus in the tubers of several common varieties, with Russet Burbank often severely affected (Fig. 9.3b).

The PLRV virus only infects the vascular (phloem) tissues of the plant, and the net necrosis symptom is caused by the selective death and damage to cells in these

Fig. 9.3 (**a**) Potato plant showing stunting and rolled leaf symptoms of potato leafroll virus infection. (Photo credit: Phillip Nolte). (**b**) Potato tuber showing discolored vascular tissues or "net necrosis" symptom of potato leafroll virus infection. (Photo credit: Nora Olsen)

vital tissues. Infection by the virus may directly damage and cause the death of the vascular tissues, or the presence of the virus may make these sensitive tissues more susceptible to damage from other stresses.

Infected seed potatoes produce plants with the characteristic leafroll foliar symptoms, but daughter tubers produced by these plants normally do not develop net necrosis. However, infected plants also serve as a source of disease for spread to healthy plants. Plants infected during the season seldom show foliar symptoms, but tubers produced by these plants will often display tuber net necrosis symptoms.

Economic Importance: Plants grown from infected seed generally do not fully develop and produce virtually no usable tubers. Plants infected later in the season develop the net necrosis tuber symptom. Losses due to net necrosis range from a slight reduction in crop value to complete crop loss.

Vector Relationship: Persistent transmission by the GPA vector is by far the most important means of movement, but transmission also has been demonstrated by potato aphid and foxglove aphid.

Corky Ringspot

Causal Agent: Multiparticle rod-shaped virus—Tobacco rattle virus (TRV).

Inoculum Source: Weed hosts and nonsymptomatic crop hosts, such as corn.

Exposure: Infested nematodes feeding on roots introduce the virus into host plants.

Symptoms: Foliar symptoms are rare, but when present, range from yellow-green mottling to foliar damage that resembles 2–4D injury. Tuber symptoms range from the namesake ringspot or arc patterns on the periderm and in the flesh of the tuber

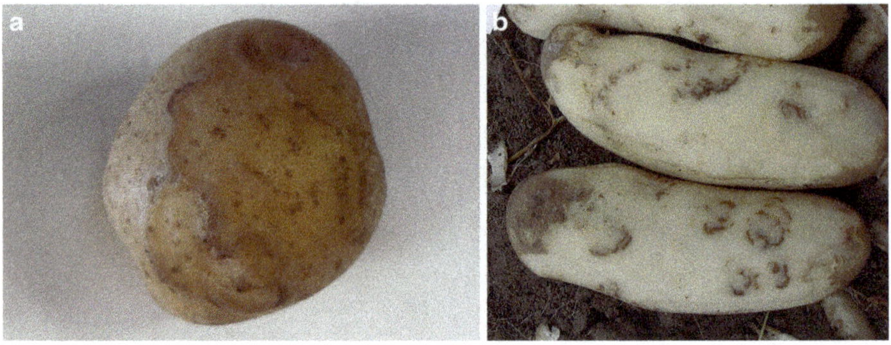

Fig. 9.4 (**a**) Tubers showing external corky ringspot symptoms caused by tobacco rattle virus (TRV). (Photo credit: Shashi Yellareddygari). (**b**) Tubers showing internal necrotic flecks and arcs resulting from TRV infection. (Photo credit: Jonathan Whitworth)

(Fig. 9.4a) and/or internal discoloration ranging from extensive discolored corky tissue to a much less conspicuous brown flecking scattered through the tuber flesh (Fig. 9.4b).

Economic Importance: Tuber discoloration can render tubers unmarketable.

Vector Relationship: Styletborne (non-persistent) transmission vectored by stubby root nematodes (*Trichodorus and Paratrichodorus* spp.)

Potato Mop Top

Causal Agent: Short rod-shaped virus—Potato mop-top virus (PMTV).

Inoculum Source: Infected seed potatoes, infested resting spores of the powdery scab plasmodiophorid pathogen vector.

Exposure: Infested powdery scab zoopores entering roots or tubers introduce the virus into host plants.

Symptoms: Foliar symptoms are characterized by shortened internodes that give infected plants the namesake "mop top" appearance in some varieties. Other foliar symptoms include yellow V-shaped or "chevron" patterns on leaves (Fig. 9.5a). Tuber symptoms include ringspot or arc patterns on the periderm and in the flesh of the tuber (Fig. 9.5b). Tuber symptoms can be difficult to differentiate from other tuber necrotic viruses, such as corky ringspot. Laboratory tests are required to confirm the identity of this virus. Symptoms are uncommon in North American production fields.

Economic Importance: Tuber discoloration can render tubers unmarketable. This virus is totally dependent on the activity of the vector to move from plant to plant. Spread is usually quite slow and transmission to daughter tubers is inefficient. Mop top is considered to be a virus of low significance in the European countries where

Fig. 9.5 (**a**) Potato leaflets showing chlorotic "chevron" pattern typical of potato mop top virus (PMTV) infection. (Photo credit Steven Johnson). (**b**) Tubers showing internal necrotic spots and arcs from infection. (Photo credit Steven Johnson)

it has occurred since the 1980s but can be a problem in very susceptible varieties grown under conditions that favor development of the powdery scab vector.

Vector Relationship: The pathogen is persistently carried by infested zoospores of the powdery scab organism *Spongospora subterranea* subsp. *subterranea*, which introduce the virus to roots and tubers when zoospores enter plant roots.

Calico

Causal Agent: Icosahedral-shaped virus—Alfalfa mosaic virus (AMV).

Inoculum Source: Alfalfa and clover, volunteer potatoes.

Exposure: Styletborne (non-persistent) transmission by GPA and numerous species of non-potato colonizing aphids.

Symptoms: Typical foliar symptoms appear as distinct yellow tissue next to green tissue in various patterns (Fig. 9.6a). The symptoms can resemble injury associated with some herbicides. Stunting and severe tuber necrosis have also been reported in some cases (Fig. 9.6b).

Economic Importance: Although not considered an important problem in the Pacific Northwest (PNW), tubers that show necrotic symptoms are unmarketable.

Vector Relationship: Stylet-borne (non-persistent) transmission vectored by various migrating aphids, such as the green peach (*Myzus persicae*), the pea aphid (*Acyrthosiphon pisum*), and others, when preferred host crops (such as alfalfa, pea, and clover) are harvested.

Fig. 9.6 (**a**) Foliar symptoms of alfalfa mosaic virus (AMV) also known as "calico virus". (Photo source unknown). (**b**) Severe tuber necrosis caused by AMV infection. (Photo source unknown)

Fig. 9.7 Tuber symptoms caused by tomato spotted wilt virus (TSWV). (Photo credit: Jonathan Whitworth)

TSWV Ringspot

Causal Agent: Membrane-bound spherical particle—Tomato spotted wilt virus (TSWV).

Inoculum Source: A wide range of crops, ornamentals, and weeds can harbor the virus.

Exposure: Persistent transmission by adult thrips.

Symptoms: Symptoms vary, but dark spots and necrotic ring spots can occur on lower and upper foliage. Tubers may show necrotic lesions on the surface that may extend to internal dead areas (Fig. 9.7).

Economic Importance: Occurrence in the field is relatively unusual, and this virus is not considered an important problem in most production areas. It can, however, be a substantial problem in greenhouses, especially since the virus can infect a wide range of vegetable and ornamental plants.

Vector Relationship: Semi-persistent transmission primarily by the Western flower thrips (*Frankliniella occidentalis*), but also onion thrips (*Thrips tabaci*) and other thrips species. Insects can only acquire the virus in the larval stage but must develop into adults before they become capable of transmitting it (semi-persistent transmission). Infected adult thrips can potentially transmit the virus for the rest of their lives (up to 30 days), but do not pass the virus on to offspring.

Virus-Vector Relationships

Each of these viruses has its own unique method of spread. An understanding of this process is vital for effective management of each disease. The simplest form of spread occurs in PVX. In this case, no insect vector is involved; instead spread occurs by contact between infected and healthy plants, also known as "mechanical spread" or "sap transmission." Simply rubbing the leaves of an infected plant onto those of a healthy plant spreads the virus.

In practice, there are probably many different means of spreading PVX, including wind-aided plant contact, machinery, and even root contact between infected and healthy plants. Humans are the most important vector because this virus may also be spread during seed cutting. One reason that mechanical transmission occurs so readily with PVX is because the virus is present in high concentrations in the plant sap.

Mechanical spread of PVY virus is also possible, but spread by aphids is much more important. The relationship between PVY and the vector is almost casual, with the virus simply adhering to the insect's mouth parts, a type of transmission referred to as styletborne or non-persistent transmission.

The most efficient vector of PVY is the GPA, but this does not mean that GPA is the most important vector. Recent investigations indicate that other aphid species—including several species of cereal aphids that do not colonize potato but land on and probe potato plants while searching for suitable grass hosts—are actually more important for spreading this virus. Not only do these non-colonizing aphids tend to move within a field, they can spread the virus to a healthy plant within minutes after acquisition from an infected plant.

Because of this "nervous feeding" habit and the fact that even the fastest-acting insecticides require several hours before aphids are killed, insecticide applications are generally NOT effective in stopping the spread of PVY by aphid vectors.

PVA spread is nearly identical to PVY. TRV spread is similar in that the virus is introduced by clinging to the mouthparts of a vector, but in this case the vector is a soil-inhabiting nematode.

PLRV spread is more complex. This virus is only spread by aphid species that colonize potato, with the GPA easily being the most important. An aphid vector is absolutely essential for spreading PLRV because mechanical transmission, like that described for both PVX and PVY, has never been demonstrated with PLRV.

The virus cannot spread by simple styletborne transmission, as is the case with PVY. Instead, the aphid must acquire the virus by feeding on a PLRV-infected plant. After acquisition, the virus must circulate from the gut of the aphid, through the circulatory system until it finally gets into the salivary glands, from which it is excreted when the aphid feeds on healthy plants. Such transmission is called circulative or persistent. The entire sequence of events from acquisition to transmission may require 24 or more hours to occur.

Unfortunately, once an aphid becomes infected with PLRV, it remains able to transmit the virus for the rest of its life. However, the "delayed reaction" between the time of acquisition and the subsequent ability to spread virus means that control of PLRV can be achieved when effective insecticides are properly used for GPA management.

The vector for TSWV is thrips, and though this insect carries the virus semi-persistently, control measures that target thrips are not considered effective at preventing virus transmission because the insecticides are unable to kill the pest rapidly enough.

Virus Management Strategies

Management of the foliar viruses starts with use of certified seed potatoes. Certification programs go to great lengths to provide commercial producers with seed potatoes free of or containing low levels of virus. Purchase of certified seed and use of the virus incidence information present on the Plant Health Certificate (see Fig. 4.11 in Chap. 4) are essential for management of PVX and PVY. Insecticides are generally not effective for managing non-persistently transmitted viruses, such as PVY. However, recent research suggests that regular applications of some insecticides or crop oils can lead to reduced PVY levels in daughter tubers, possibly due to an effect on aphid feeding behavior. For smaller plantings, use of border crops has been effective at minimizing PVY due to the aphid behavior of first probing plants at field margins before moving into a field. This probing is thought to remove virus particles from the stylets. In some regions, killing a crop early also may help minimize PVY transmission to daughter tubers, but this approach will only be effective in areas where aphid flights don't occur earlier in the season.

Use of certified seed combined with use of systemic insecticides, in-season aphid scouting, and late-season aphicide applications when aphids are present are generally effective for management of persistently transmitted viruses such as PLRV.

Management procedures for corky ringspot include use of certified seed and the use of nematicides, such as oxamyl (Vydate®) or soil fumigation. Fields contaminated with the TRV that causes corky ringspot cannot be used for seed potato production. There are few effective management tools for PMTV, and infested fields may not be suitable for potato production. Management practices to mitigate PMTV include pre- or at-planting treatments for the management of the pathogen vector, *S. subterranea*. Use of pathogen-free seed is ideal; however, most seed certification programs do not include the powdery scab pathogen in the certification process, as it's typically considered a blemish pathogen and not all populations of pathogen are viruliferous.

Diseases Caused by Bacteria

The bacteria are considerably larger in size than the viruses and are considerably more complex in their biology. However, bacteria are still small enough that higher magnifications on a light microscope are required to see them.

In contrast to the viruses, bacteria are living organisms with protoplasm, cellular membranes, and rigid cell walls. They reproduce by simple cell division and can do so extremely rapidly, doubling their population in as few as 20 min.

The three most common potato diseases caused by bacteria are bacterial ring rot, bacterial blackleg/soft rot diseases, and common scab.

Two diseases of lesser importance and infrequent appearance, witches' broom and purple top, are caused by phytoplasmas, which are bacteria-like organisms that lack a rigid cell wall and are restricted to plant phloem or their leafhopper insect vector.

A final bacterial disease that is increasing in importance in many potato producing areas is zebra chip, which is caused by the non-culturable bacterium, *Candidatus liberibacter solanacearum*. This bacterium is vectored by potato psyllids.

Bacterial Ring Rot

Causal Agent: Bacterium—*Clavibacter michiganensis* subsp. *sepedonicus* (Cms).

Inoculum Source: Usually tuberborne, although infection of healthy tubers is frequently linked to contaminated seed cutting machines, handling equipment, truck beds, and even storage facilities.

Exposure: Infected seed or contaminated machinery, equipment, and structures.

Symptoms: Foliar symptoms are first visible as "flagging," which is sudden wilt in the upper leaves on one or more stems of infected plants. More severe wilting of leaves on the symptomatic stems or on other stems of the infected plant may follow (Fig. 9.8a). In some varieties, such as Russet Burbank, a "dwarf rosette" symptom characterized by chlorosis, shortened internodes, and upright leaf growth may appear on some stems of infected plants before the development of wilt symptoms (Fig. 9.8b). Infected stems will often yield a cheesy bacterial exudate when cut and squeezed.

Tubers with internal symptoms should be sent to a laboratory to confirm the presence of this disease, because not all tubers with vascular ring discoloration or rot are infected with bacterial ring rot. Other disorders, such as freezing or *Verticillium* wilt, may also cause vascular ring discoloration, and bacterial soft rot can cause vascular ring deterioration under some circumstances.

Tuber symptoms with advanced infections are characterized by cracking of the tuber skin (Fig. 9.8c) and partial to complete breakdown of the vascular ring (Fig. 9.8d). Latent infections are not visually detectable, but they can develop over

Fig. 9.8 (**a**) Foliar wilt symptoms characteristic of bacterial ring rot (BRR) infection. (Photo credit: Phillip Nolte). (**b**) BRR "dwarf rosette" symptom. (Photo credit: Neil C. Gudmestad). (**c**) Tuber cracking caused by BRR. (Photo credit: Phillip Nolte). (**d**) Vascular ring deterioration caused by BRR. (Photo credit: Phillip Nolte). (**e**) Bacterial exudate squeezed vascular tissues of BRR-infected tuber. (Photo credit: Phillip Nolte)

time to produce visible symptoms. Infected stems and tubers often yield a cheesy bacterial exudate when squeezed (Fig. 9.8e).

Economic Importance: Tubers with severe infections are unusable and can lead to development of rotting problems in stored potatoes. Bacterial ring rot often results in substantial losses in the field and storage. Mild and latent infections often get worse in storage, progressing from mild vascular discoloration to vascular rot after a few months. For seed growers, detection of the pathogen renders the crop unsuitable for certification, since certified seed programs implement a zero tolerance for the disease and the pathogen, and thus pose a severe economic loss.

Management Strategy: Prevention of recurrence is the only practical control. Before the next growing cycle, growers need to dispose of all potatoes on the infected farm and thoroughly clean and disinfect all potato machinery and structures before bringing in new, clean, certified seed. Agreements should be made with trucking companies that haul tubers that trailers will be properly sanitized between loads, whether the truck hauls only seed potatoes or a combination of seed potatoes and commercial potatoes. The ring rot bacterium does not survive well in soil, and growing a different crop in the field for one year should be sufficient to eliminate the pathogen from soil.

Infected crops should be left in the field for 3 weeks or more after vine kill to allow severely infected tubers to completely deteriorate before harvest. It is also advisable to store potatoes as you would after a frost with high volumes of air and cool temperatures, and market as soon as possible.

Seed certification programs implement a zero tolerance for the pathogen. Sampling strategies that maximize the probability of detecting the organism in conjunction with sensitive DNA-based testing, such as polymerase chain reaction assays, are commonly used to determine eligibility of a seed lot for certification. Serological testing, such as enzyme-linked immunosorbent assays or immunofluorescent staining assays, have also been used, but validated DNA-based tests are generally considered to be more sensitive and specific.

Brown Rot

Causal Agent: Bacterium—*Ralstonia solanacearum*.

Inoculum Source: Brown rot is primarily a soilborne disease, though infected seed tubers are another important source and can be responsible for long-distance dissemination. Infected volunteers are frequently encountered where disease was present in previous crops. The bacterium moves from plant to plant infecting through contact between healthy and diseased roots. It also moves very effectively in soil water. Solanaceous crops such as pepper, eggplant, and geranium are also susceptible. Solanaceous weeds, like *Solanum dulcamara* (woody nightshade), are capable of harboring the disease.

Exposure: Bacteria in soil and irrigation water, infected seed, and volunteers.

Symptoms: Foliar wilt, stunting, and yellowing are often observed in the field. More serious is the deterioration of the vascular ring in tubers. Early tuber symptoms are a brown discoloration of the vascular ring. This discoloration progresses until the vascular ring is destroyed. Severe infection is characterized by a white, creamy exudate that oozes from the ring when tubers are cut (Fig. 9.9). Secondary invasion by soft rot bacteria often occurs, hollowing out the tuber, leaving only the outer shell. High temperatures (85–90 °F) and high soil moisture enhance symptom expression.

Brown rot and bacterial ring rot have similar symptoms.

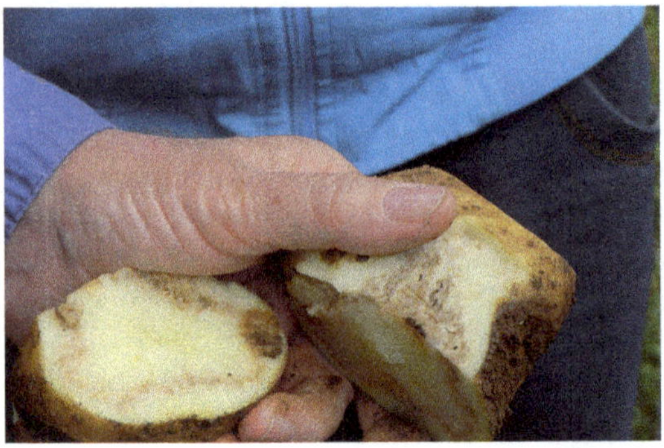

Fig. 9.9 White, creamy exudate of the brown rot bacterium oozing from the vascular ring of cut tuber. (Photo credit Charlie Higgins)

Economic Importance: Brown rot is one of the most destructive diseases of potato worldwide and total crop loss is not uncommon.

Management Strategy: Plant clean seed and disinfect cutting knives frequently. Planting whole seed bypasses this problem. Crop rotation can be effective if the alternate crops are not hosts. The disease can be spread very effectively by flood irrigation. Avoid cultivation practices that damage roots. Root knot nematode infection can greatly increase brown rot severity. Nematode management is recommended. The brown rot bacterium is not tolerant of cold temperatures, and disease is generally confined to tropical and subtropical production regions. Temperate regions are not hospitable to the bacterium, although new strains identified in Europe have greater tolerance for low temps, and spread into previously unaffected areas is possible. These strains are quarantine-level diseases in most temperate production areas. Occurrence in the U.S. is confined to the Eastern Seaboard from Maryland south down to Florida. *Suspected infections should be reported immediately*.

Blackleg-Soft Rot Diseases

Causal Agent: Bacterium—Blackleg, aerial stem rot, lenticular soft rot, and soft rot are caused by species of *Pectobacterium* and *Dickeya* (formerly in the genus *Pectobacteria*). Both groups of bacteria were once classified in the genus *Erwinia.* All have the ability to produce pectolytic enzymes. *Pectobacterium atroseptica* (Pa), *P. carotovorum* subsp. *carotovorum* (Pcc), *P. carotovorum* subsp. *brasiliensis, P. wasabiae*, *Dickeya dianthicola* (Dd), and other species of *Dickeya* can incite disease. Pa is typically associated with blackleg that is enhanced by moist, cool conditions (below 70 °F) and with soft rot in storage; Pcc is more often associated with aerial stem rot, soft rot, and lenticular soft rot (also known as pit rot). Dd has been associated with blackleg and field decay of daughter tubers under warmer conditions in eastern and southern regions of the U.S., as well as in Europe.

In storage, tuber decay may begin around areas of condensation, result from freezing injury, or begin after primary infection from other pathogens. Under the right conditions of limited airflow and high moisture, spread within a storage can be rapid. Tuber infection by Dd, or more recently *D. parmentieri,* may not be obvious if infection is latent and tubers are rapidly moved from field conditions to cool storage without extensive damage or adverse temperature or moisture conditions. Asymptomatic tubers, if purposed as seed, could cause significant emergence and decline in the subsequent production field.

Tuber decay from any of these sources can be caused by all of these bacteria; sometimes by more than one species simultaneously. Similarly, soft rot resulting from contaminated flume water and wet conditions during fresh-pack operations can be due to different species. Blackleg and soft rot bacteria may also be present within or on potatoes that have no visible signs of tuber decay.

Inoculum Source: Inoculum sources are numerous and include soil, irrigation water, rainwater, and contaminated tubers, as well as contaminated equipment such as trucks, conveyors, other handling machinery, seed cutters, and storage structures. Bacteria can enter new tubers through lenticels during the growing season, or through bruises and wounds caused by rough handling during harvest.

Stem and foliar infections also occur when aboveground portions of the potato plant are wounded and vine growth is lush. Hail damage, white mold, or late blight infection on vines, for instance, all provide a ready entrance for soft rot bacteria.

Exposure: The bacteria gain entry primarily through wounds, which include mechanical damage and foliar or tuber lesions caused by other potato diseases, such as late blight, Pythium leak, pink rot, dry rot, and white mold, or physiological damage, such as field frost or hail. Natural openings in the potato plant (e.g., tuber lenticels) are also vulnerable to infestation when conditions are right.

Symptoms: The bacteria may enter lenticels and produce lesions that appear slightly sunken, tan to brown, and as circular water-soaked areas. Brown to black pigments often develop near the margins of lesions. Soft rot pathogens also commonly infect non-healed surfaces of freshly cut seed or bruised potatoes. Moisture films on tubers enhance initiation of infection by the bacteria. As decay progresses, soft rotted tissues are cream to tan in color with a soft, slightly granular consistency (Fig. 9.10a). Infected tissues are sharply delineated from healthy ones, and the rotted tissue is easily rinsed away with water. Tubers infected with soft rot bacteria typically break down, further spreading the bacteria to adjacent tubers in soil, storage piles, or fresh-pack containers.

Blackleg in the field is commonly associated with blackened hollow cavities in the stems that may extend nearly to the growing tips of roots and vines (Fig. 9.10b). The stem-end of tubers may have external black sunken lesions and slight internal vascular discoloration. The infection typically spreads from the stem-end through the heart of the tuber.

Internally, the infected tuber flesh appears cream colored, then turns grayish and black with a mushy texture similar to bacterial soft rot (Fig. 9.10c). A dark border separates the infected flesh from the healthy flesh. Blackleg that is present in

Fig. 9.10 (**a**) Potato tuber sliced longitudinally to show stem-end soft rot. Infection in above-ground stem has progressed downward through the stolon into the tuber. (Photo credit: Phillip Nolte). (**b**) Newly emerged potato seedling with blackleg symptoms. (Photo credit: Eugenia Banks). (**c**) Blackleg-infected tuber. (Photo credit: Eugenia Banks). (**d**) Potato stem split open to show internal breakdown caused by stem soft rot. (Photo credit: Phillip Nolte)

daughter tubers in the field usually originates from infected seed pieces. The Pa bacteria commonly enter daughter tubers via stolons.

Blackleg or soft rot infected tubers have no odor during the early stages of the disease. Advanced stages of decay often develop a strong, foul odor, but this is due to the presence of other secondary bacteria that have moved into the decaying tissue. Soft rot and blackleg infected tuber tissue usually has a thick consistency and readily flows under slight pressure.

Foliar stem soft rot, not associated with blackleg, occurs in tissues with mechanical damage, such as hail, or from the action of other pathogens, such as white mold. Stem soft rot is characterized by water-soaked, green decay that later turns brown (Fig. 9.10d).

Economic Importance: Bacterial soft rots are the most serious problem in fresh potato shipments. They also cause widespread decay of potato seed pieces, often followed by blackleg, and are responsible for most of the rot in bulk potato piles in storage.

The soft rot bacteria are excellent secondary invaders, meaning that they take advantage of wounds or weaknesses in potato tubers caused by other factors. For instance, soft rot often gains entrance to tubers through lesions caused by other tuber diseases, such as late blight and dry rot. In fact, with most potato tuber diseases, secondary soft rot invasion is what is primarily responsible for serious storage problems.

Soft rot can rapidly get out of control in a storage facility and create "hot spots" that can rot previously healthy tubers. Most species of soft rot bacteria have the ability to survive with or without oxygen and thrive under low oxygen conditions either in soil or in storage.

Management Strategy (Soft Rot): Bruising and mechanical damage to potatoes should be avoided during harvest and transport, and stored potatoes should be checked before lowering storage temperatures to the holding levels. Cooler temperatures suppress pathogen activity.

Also, it is important to avoid moisture films on potatoes in bulk piles and to dry potatoes as much as possible before placing them in fresh-pack containers. Badly infested potatoes in storage may require a period of time with high volumes of air with low relative humidity for proper management.

Management Strategy (Blackleg): Growers are advised to use limited generation certified seed sources. Additionally, the following practices are strongly advised: (1) do not wash seed potatoes; (2) routinely clean and disinfest seed potato cutting and transport equipment with an approved disinfectant; and (3) precut seed and allow it to properly heal, or use small, uncut "whole seed" tubers. Seed lots infected with Dd may be grown in cooler, arid growing regions with limited to no symptom development, whereas same lots can produce significant symptoms and yield/quality losses in warmer, temperate regions.

Seed piece treatments containing ethylene bis-dithiocarbamate (EBDC), fludioxonil, or thiophanate methyl may indirectly suppress blackleg by controlling Fusarium seed piece decay, a disease caused by a fungus that creates entry points for the bacteria. Growers should plant potato crops in well-drained soil and avoid irrigating planted fields before emergence.

Common Scab

Causal Agent: Actinomycetes (filamentous bacterium)—primarily *Streptomyces scabies.* Some species strains are saprophytes, while others are antagonists against the pathogenic species. Overall, the group is highly complex.

Inoculum Source: Soil and seed.

Exposure: Most scab infections on potato occur early in the growing season after tuber initiation. The bacteria invade recently formed pores but do not invade directly through intact skin.

Symptoms: The pathogen population that causes common scab is highly variable and influenced by multiple abiotic factors, which explains the diverse symptoms that can arise. In classic infections, first appearance of this disease (generally about mid-July) begins as small, corky or netted superficial spots that may become visible on the tubers. Symptoms may become severe by the time tubers are harvested (Fig. 9.11a). After infection, a wound barrier may form a few cells below the surface. If bacterial penetration continues, a second or third barrier may be formed. When the tubers stop growing, lesions do not increase in size. Scab development occurs only during plant growth, and although disease development ceases once the tubers are harvested, the pathogen survives throughout the storage period. Another type of symptom that can develop, depending on the pathogen strain, is a deep pitted scab lesion (Fig. 9.11b).

Common scab is favored by dry soil, particularly during the early part of the growing season. Soils with a high calcium level and those rich in non-decomposed organic matter (e.g., barnyard manure) also promote the disease. Growth of most strains of the pathogen is optimum at approximately pH 7.0, while pH below 5.5 generally inhibits activity.

Economic Importance: Common scab of potato occurs in all potato production areas and can be a serious problem. This disease has no influence on total yield, unless seed potatoes have severe infection that reduces number of functional eyes. However, by severely blemishing tuber appearance, scab can drastically reduce visual attractiveness and value, especially in potatoes intended for the fresh market. In some instances, scab may predispose tubers to chewing insects that may increase tuber damage.

Management Strategy: Source seed with low to no visible common scab symptoms. Optimal soil moisture suppresses common scab. The most critical time for irrigation management is a period of 5 weeks after tuber initiation (when stolon tips swell to double their size). Tuber initiation generally occurs about 2 weeks after emergence. Therefore, the period when the potato is most susceptible to infection is about 2–7 weeks from emergence.

Fig. 9.11 (**a**) Common scab on potato tubers. Infected areas are typically brown, roughened, and irregularly-shaped. (Photo credit: Amanda Gevens). (**b**) Deep-pitted symptom of common scab. (Photo credit: Eugenia Banks)

Optimal levels of moisture should be maintained in the upper 9 in of the soil profile. The minimum moisture required for common scab suppression is about −40 kPa (centibars). On a silt loam soil, this would approximate 75% available soil moisture. With sprinkler irrigation on a silt loam soil, maintaining this moisture level may often be accomplished by irrigation at 4- or 5-day intervals with set length, depending on consumptive use. On fields where common scab is not a severe problem, high moisture alone may be sufficient control.

Some varieties have tolerance to common scab or show reduced symptoms. Chemical management can help to reduce severity of common scab symptoms and includes in-furrow treatment with pentachloronitrobenzene (PCNB) or fumigation with chloropicrin. Few other treatments have effective results in reducing inoculum or resulting disease symptoms.

Witches' Broom

Causal Agent: Phytoplasma—Not officially named.

Inoculum Source: Alternate host, alfalfa.

Exposure: Phytoplasma is moved into potato fields from weed or alfalfa hosts by the leafhopper vector.

Symptoms: Infected plants show a breakdown in apical dominance, and all or most of the buds on the plant will break dormancy. This leads to stunted, chlorotic plants that have a profusion of stems (often 20 or more) and shortened internodes, thus the namesake witches' broom.

Below ground, multiple small tubers are formed, often in chains. In some varieties a "tall" symptom may appear. The tall witches' broom symptom appears a month or so after planting. The affected plants, which have developed normally until that time, become chlorotic, and lateral buds begin to break as apical dominance declines.

Economic Importance: Plants with severe witches' broom symptoms seldom grow to more than a few inches high and are rapidly shaded by their healthy neighbors. Tubers remain too small to be picked up by harvesting equipment. When the tall symptom occurs, tubers may develop to sufficient size to be harvested but will either sprout in the ground or early in storage. This disease is generally not of significant economic importance unless a seed field becomes infected at a sufficient level as to be uncertifiable—a rare occurrence.

Management Strategy: Fortunately, the phytoplasma that causes witches' broom cannot be acquired by the leafhopper vector from infected potatoes and cannot spread from potato to potato, so the disease tends to be self-eliminating. Witches' broom usually occurs during dry years when the normal hosts for the leafhopper vector dry down early in the season, forcing the vector to seek other food sources.

Vector Relationship: Persistent transmission—leafhopper.

Purple Top

Causal Agent: Phytoplasma—several strains from different groups (16SrI, 16SrII, 16SrVI, 16SrXII, and 16SrXVIII), including aster yellows phytoplasma in some regions and the Columbia Basin potato purple top phytoplasma, also known as the beet leafhopper transmitted virescence agent (BLTVA).

Inoculum Source: In the Pacific Northwest, BLTVA can infect sugar beet and weeds, such as kochia (*Kochia scoparia*), Russian thistle (*Salsola kali*), groundsel (*Senecio vulgaris*), and shepherd's purse (*Capsella bursa-pastoris*).

Exposure: Phytoplasma is moved into potato fields from other plant hosts by the leafhopper vector.

Symptoms: Upper leaves of infected plants appear purplish and roll upward (Fig. 9.12). Infected plants may also experience stunting, bud proliferation, aerial tubers, discolored tubers, and premature death. The symptoms can resemble those associated with psyllid yellows, so confirmation via a laboratory test is recommended.

Economic Importance: Due to production of aerial tubers, yield loss can occur since carbohydrates are diverted from underground tubers to the aerial tubers; quality can also be compromised if tuber discoloration occurs. Research from Oregon State University reported up to 12% yield loss when only one beet leafhopper per plant was present.

Management Strategy: Additional work from Oregon State University suggests that early season monitoring and control of beet leafhoppers may help reduce the occurrence of this disease. Younger plants appear to be more susceptible to infection than plants that are at a later plant stage.

Fig. 9.12 Area of a field showing purple top symptoms. (Photo credit: Phillip Nolte)

Vector Relationship: The beet leafhopper (*Circulifer tenellus*) transmits BLTVA, while aster leafhoppers transmit the aster yellows phytoplasma. Phytoplasmas persist and can multiply in the bodies of their insect vector. Phytoplasma has not been found to be passed on to eggs by infected adult aster leafhoppers; whether this type of transmission occurs with BLTVA in the beet leafhopper is unknown.

Zebra Chip

Causal Agent: Phloem-restricted bacterium—*Candidatus Liberibacter solanacearum*.

Inoculum Source: Infected potato plants.

Exposure: The phloem-limited bacterium is introduced into the plant by pathogen-bearing potato psyllids as they feed.

Symptoms: Foliage of infected plants may show no symptoms, or plants may develop yellowing or a pinkish-red discoloration. Infected plants may show leaf scorch, form aerial tubers, or show bud proliferation (Fig. 9.13a). Smaller and deformed tubers that are more numerous may form. Brown stripes appear within tubers, and these become more pronounced after frying (Fig. 9.13b).

Economic Importance: The primary impact is reduction in quality due to the internal tuber discoloration and burnt taste after frying.

Management Strategy: Monitor for psyllids. Insecticides can be an effective tool for managing the vector. A number of insecticides with activity against psyllids are registered in the PNW for use on potato.

Vector Relationship: The phloem-feeding potato psyllid completes all stages of its life cycle (eggs, nymphs, and adults) primarily on potato, although it has a wide host range. Adults or nymphs can acquire the pathogen by feeding on an infected plant and they then become infective for the rest of their lives; thus, potato psyllids carry the pathogen in a persistent manner. A small percentage of eggs laid by infected adults also become carriers of the pathogen. Potato psyllids can cause damage to potatoes even in the absence of the pathogen.

Fig. 9.13 (**a**) Stunted plants with aerial tubers characteristic of zebra chip infection. (Photo credit: Erik Wenninger). (**b**) Potato chips produced from healthy (left) and zebra chip-infected tubers. (Photo credit: Erik Wenninger)

Diseases Caused by Fungi

The majority of plant diseases are caused by fungi. Fungi, with some exceptions, are still microscopic in size but are much larger and more complex than bacteria or viruses. Fungal cells are similar to the cells of higher plants and animals in that they have nuclei as well as complex cytoplasmic and cellular organelle structures. Included in this discussion are Oomycetes, the taxonomic group that includes *Phytophthora* and *Pythium*, (pathogens responsible for late blight and Pythium leak); and *Spongospora* (the pathogen responsible for powdery scab), which are no longer classified as fungi. They are included in this section because their behavior and the management procedures for them are similar to true fungi.

Most fungi are dispersed and reproduce by structures called spores. Depending on the fungus, these spores may result from sexual recombination, or are a form of asexual (clonal) reproduction. Within the fungi are many different spore types ranging from single to multicellular, pigmented or non-pigmented, and tiny to relatively large.

Many fungi produce more than one spore type, often a sexual spore and an asexual spore, but many fungi either do not have a sexual stage or the sexual stage is unimportant to the disease cycle. Quite often, spores produced as the result of sexual recombination not only give rise to new genetic types, but also improve survival or overwintering ability, which increases the initial inoculum for the following season.

Often the asexual spores are produced in great abundance after the initial infection events and are the most significant means by which the fungus spreads and multiplies. Some fungi, such as the white mold and Rhizoctonia fungi, survive and reproduce from special masses of hard, weatherproof fungal material called sclerotia.

Control measures often include the application of fungicides. Unfortunately, a number of the disease organisms have shown an ability to become resistant. Guidelines for management of fungicide resistance are presented in the Sidebar 9.1.

Sidebar 9.1: Fungicide Resistance Management

Many of the newest fungicides are specific in their activity and are environmentally and worker friendly. These attributes make the new fungicides effective and safe, but they can also make them vulnerable to shifts towards fungicide insensitivity or resistance in pathogen populations. Growers can best manage fungicide resistance by considering the following guidelines:

1. Use integrated pest management (IPM) practices.
2. Alternate fungicide modes of action.
3. Use fungicide combinations (when possible).

Use IPM Practices

An important point to be made regarding resistance management procedures is that relying on only one form of disease control is risky. The potential for disease development should be reduced by employing integrated pest management procedures for diseases whenever possible.

IPM practices combine cultural control measures with chemical controls, which eases the selection pressure on the fungal pathogens.

Alternate Fungicides

The use of a single-site mode of action fungicide, or different fungicides with the same single-site mode of action, can lead to the development of fungicide resistance in pathogen populations. Alternating fungicides can mitigate this risk. Growers often use this strategy for insecticide, herbicide, and fungicide resistance management. The simplest form of this procedure would be to alternate between fungicides that have different modes of action. Modes of action of different fungicides have been summarized by the Fungicide Resistance Action Committee (FRAC), an entity of CropLife International, which supports agricultural technology to advance sustainable agriculture. An example of alternating fungicides would be the use of boscalid (Endura®) alternated with chlorothalonil (Bravo®) for early blight management. This procedure ensures the fungal population is not repeatedly exposed to the single-site mode of action of boscalid, a succinate-dehydrogenase inhibitor (SDHI) fungicide in FRAC group code 7 by alternating with the multi-site contact fungicide chlorothalonil, which is a FRAC group code M5 and has low to no risk of pathogen population resistance development.

Use Fungicide Combinations

Another option for growers is the use of combinations or tank mixes of fungicides. The combination fungicide can have either multi- or single-site action, but the usual procedure is to use a multi-site material, such as mancozeb or chlorothalonil, as a partner with a single-site activity material, such as boscalid (Endura®) or cymoxanil (Curzate®).

Some fungicides are packaged as combinations (prepacks), and others must be tank mixed. When growers use combinations, the target pathogen must overcome more than one type of chemical activity in order to become resistant.

Dry Rot

Causal Agent: Fungus—*Fusarium sambucinum*, *Fusarium coeruleum*, and other species of *Fusarium*.

Inoculum Source: Soil and infected seed.

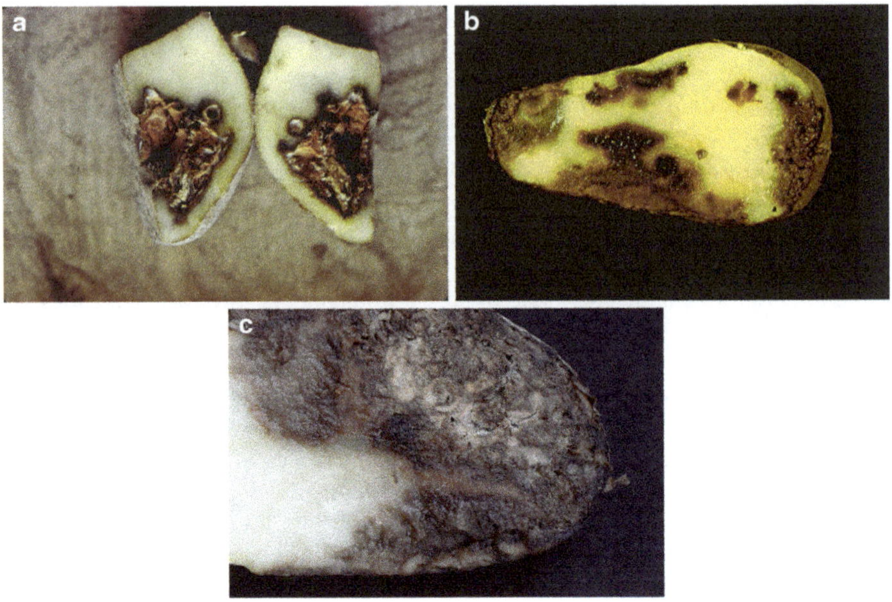

Fig. 9.14 (**a**) Advanced stages of *Fusarium* dry rot seed piece decay. (Photo credit: Phillip Nolte). (**b**) Early stages of tuber infection showing the black discoloration characteristic of *F. sambucinum* infection. (Photo credit: Phillip Nolte). (**c**) Advanced *Fusarium* dry rot infection showing brown, crumbly, rotted tissues and white fungal growth. (Photo credit: Phillip Nolte)

Exposure: These pathogens gain entry into the tuber through wounds inflicted during harvest and handling. Exposed surfaces resulting from seed cutting are especially vulnerable.

Symptoms: Infected seed-piece tissues often have a characteristic dark brown to fawn color (Fig. 9.14 a). Blackleg and soft rot bacteria often develop as a side effect of *Fusarium* seed-piece decay. In storage, the lesions of this disease appear as small, brown areas around wounds. These lesions are dry and spongy in texture and tend to form hollow cavities. As the infection enlarges, the tuber skin over the lesion collapses and wrinkles. Severely rotted tubers shrivel and become mummified.

Internal diseased areas vary in shades of brown from fawn *(F. coeruleum)* to dark chocolate *(F. sambucinum)* (Fig. 9.14b). The advancing margin of the disease is faint for lighter shades and distinct for darker shades. The decaying tissue becomes sunken, the skin wrinkles, and tufts of mold may appear at the eyes, lenticels, wounds, and stem end.

At low storage temperatures the affected tissues are usually dry and firm, sometimes powdery, but at high temperatures they are often moist and pliable. Infected tuber tissue typically appears brown and collapsed, often with a white fungal growth but other colors, including yellow, orange, or even pink, often appear in older lesions.

Dry rot does not produce a wet decay with moisture exuding from tubers as may occur with soft rot, leak, or pink rot. Tuber decay by *Fusarium* is typically dry, crumbly, brown in color, and uniquely different from most other common tuber diseases (Fig. 9.14c).

The *Fusarium* fungi do not enter potatoes through intact, healthy skin. Dry rot only infects through cuts and openings in the tuber skin caused by bruising during harvest and transport. Wounds created during the cutting of seed tubers into seed pieces are particularly vulnerable.

Fusarium often remains dormant under cold storage conditions and becomes active as the tubers are warmed up for shipment or seed use. Although dry rot does not spread from tuber to tuber during storage, it is one of the most common storage diseases.

Economic Importance: *Fusarium* dry rot is one of the most important diseases of stored potatoes and can be responsible for significant storage losses both directly and indirectly by providing an entry wound for secondary infection by bacterial soft rot. In the field, dry rot decay of seed pieces often provides an entry for bacterial soft rot, which can significantly affect stand establishment and subsequent seed performance.

Management Strategy: It is important to clean and disinfect all potato handling, transport, and cutting equipment, and plant whole seed or use a seed piece treatment that will effectively control *Fusarium* on potato seed pieces. If cutting and healing seed potatoes before planting (precutting), adequate air flow and appropriate wound healing conditions need to be maintained.

Potato seed should be planted in warm (above 50 °F), moist soil that promotes rapid plant emergence. Other precautions: (1) avoid irrigating non-emerged fields, (2) harvest only mature potatoes with well-developed skin, (3) take steps to avoid mechanical damage or bruising during harvest and transport, and (4) promote wound healing immediately after placement in storage by maintaining temperatures of 50–55 °F and ventilation with high humidity.

Early Dying and Verticillium Wilt

Causal Agent: Fungus—*Verticillium dahliae* and *V. albo-atrum* can both cause Verticillium wilt. Several other pathogens, including root lesion nematodes (*Pratylenchus penetrans* and *P. neglectus*), the soft rot bacteria (*Pectobacterium* and *Dickeya* species), the black dot fungus (*Colletotrichum coccodes*), and the early blight fungus (*Alternaria solani*) can interact with either species of *Verticillium* to cause potato early dying.

Inoculum Source: Infested soil or infected seed.

Exposure: The fungus enters through roots. Plant stress, resulting from drought, excessive water, or nutrient imbalances, will predispose the potato to early dying. Planting potatoes in soils with a previous history of *Verticillium* can also increase disease incidence.

Symptoms: Tubers that are infected with *Verticillium* may have a discolored vascular ring at the stem end. Foliar symptoms resemble a mature senescing potato

Fig. 9.15 Wilt, chlorosis, and necrosis symptoms characteristic of early dying. (Photo credit: Eugenia Banks)

plant and include uneven chlorosis, yellowing between leaf veins that may turn brown, and vascular discoloration of stem at the base (Fig. 9.15).

Symptoms will typically start on the lower leaves and may affect leaflets on only one side of the petiole or leaves on one side of the plant. As plant tissue dies the stem will often remain erect, a condition termed flagging.

Economic Importance: Early dying limits yields because the plant dies before the tubers have finished bulking.

Management Strategy: Among the steps that growers can take to prevent early dying disease are: (1) plant certified seed and resistant varieties; (2) reduce soil populations by rotating crops and include potatoes every fourth or fifth year; (3) apply soil fumigants; (4) avoid excessive irrigation, particularly early in the growing season when much of the *Verticillium* infection occurs; and (5) minimize plant stress and encourage uniform, continuous plant growth by providing a balanced fertility program and optimum soil moisture levels.

Early Blight

Causal Agent: Fungus—*Alternaria solani.*

Inoculum Source: The *Alternaria* fungi overwinter in host plant debris, soil, infected tubers, and other solanaceous plants.

Exposure: Windborne fungal spores.

Fig. 9.16 (**a**) Early blight leaf lesions. Note "concentric ring" pattern within the lesions. (Photo credit: Eugenia Banks). (**b**) Early blight stem lesions. (Photo credit: Jeffrey Miller). (**c**) Early blight tuber lesions. (Photo credit: Phillip Nolte). (**d**) Severe early blight tuber lesions on a susceptible variety after several months in storage. (Photo credit: Jeffrey Miller)

Symptoms: Early blight is characterized by dark brown to black lesions on leaves and stems. Lesions usually appear on the lower or oldest leaves first and often are first-observed before or at row closure. These lesions tend to be circular, although they may take on an angular appearance when continued lesion expansion becomes impeded by leaf veins. Typical early blight lesions contain a series of concentric rings that give them a target or "bullseye" appearance (Fig. 9.16a). These lesions are often bordered by a chlorotic zone that fades into the normal green of the surrounding healthy tissue. This chlorosis is due to the effect of alternaric acid, a toxin produced by the fungus.

As lesions expand in size, the entire leaf becomes chlorotic. The chlorotic tissue dies, turns brown, and dries out, but the leaves usually remain attached.

The first lesions can appear in late June, even before row closure, but the disease does not generally begin to spread beyond these initial infections until early- to mid-July. Stem lesions may appear later in the growing season, or may not occur at all unless the disease is particularly severe. The presence of stem lesions is an indication that early blight is building to high levels (Fig. 9.16b).

While uncommon and variety specific, under the right conditions, the early blight fungi can also infect potato tubers. Tuber lesions are typically dark brown, circular to irregular in shape, dry in integrity, and are typically superficial on most varieties (Fig. 9.16c). On Russet Burbank, lesions may be similar in color to the skin, or slightly darker.

Where the lesions penetrate the tuber flesh they are brown to black, dry, and grainy. The infected flesh is usually in a thin layer just under the visible surface lesion. Lenticel infections may show a small arc or crescent pattern. Lesions can increase in size during storage, and in advanced stages, the tubers can become shriveled (Fig. 9.16d).

Early blight lesions are quite dry and are not as prone to invasion by secondary soft rot organisms as are other tuber rots. These lesions contrast sharply against the white flesh of the tuber when the skin is peeled and typically penetrate less than one-sixteenth of an inch into the tuber flesh.

Tuber lesions can continue to develop during storage, but the lesions do not exhibit the free moisture, wet decay, or deep infection into the tuber like soft rot, leak, and water rot diseases. Early blight lesions do not have the dry, mummified appearance, visible fungal mycelium, or deep tuber penetration associated with dry rot.

Economic Importance: Early blight can be responsible for significant yield loss if infestation occurs early enough in the growing season. Early decline of potato foliage due to early blight can also result in reduction in quality and storability of tubers. The tuber early blight phase can lead to reduced crop value due to cosmetic defects for fresh pack potatoes and increased loss of raw product when peeling potatoes for processing.

Management Strategy: Sound agronomic practices, such as crop rotation in an individual field as well as across fields within proximity, proper plant nutrition and water management, proper pest management, and avoidance of other plant stresses, can greatly reduce early blight severity. While some varieties are more tolerant to early blight, most commercial varieties are moderately to highly susceptible to the disease. Several fungicides are available to control early blight on potato foliage during the growing season (e.g., chlorothalonil, succinate dehydrogenase inhibitor (SDHI), and EBDC fungicides, among others). The use of these fungicides also reduces production of the spores that cause tuber lesions.

Some growers use vine burning as a vine kill method in late summer to reduce spore counts and begin tuber maturation. Mature tubers are much more resistant to infection than immature tubers. Mature tubers must be injured during harvest for infection to occur.

Potatoes that are grown in coarse, sandy soils and are immature or harvested when wet are most susceptible to infection by early blight. Harvesting when vines are green also increases tuber infection. The fungus can infect lenticels directly if spore counts are high and moisture is present to promote spore germination.

Brown Spot/Black Pit

Causal Agent: Fungus—Primarily *Alternaria alternata*, though other small-conidia *Alternaria* species (other than *A. solani*) have also been implicated.

Inoculum Source: The *Alternaria* fungi overwinter in host plant debris, soil, infected tubers, and other solanaceous plants, as well as other agricultural and ornamental plant species.

Exposure: Windborne fungal spores. Because *A. alternata* spores are smaller, they move into potato fields from further away than *A. solani*.

Symptoms: Brown spot is characterized by dark brown to black lesions on leaves (Fig. 9.17). Stem lesions can also occur. Unlike early blight, brown spot lesions can occur on potato plants at any growth stage. These lesions tend to be smaller than early blight lesions, but may also have concentric rings that give them a target or "bullseye" appearance. As lesions expand in size, they coalesce and can cause the entire leaf to become chlorotic. The chlorotic tissue dies, turns brown, and dries out, and the leaves may drop from the plant. Brown spot can, at times, be indiscernible from early blight and, therefore, may be misidentified. Because both *Alternaria* species can develop fungicide resistance, it is important to know which disease (and species) is present, as the brown spot pathogen can become resistant to quinone outside inhibitor fungicides (QOIs), for example, much more readily than the early blight pathogen. Stem lesions have dark brown margins and can be elongated.

Black pit is the name given to tuber infections. Tuber lesions are black and sunken with definite margins and look similar to early blight tuber lesions. Black pit is a dry, grainy decay. Wounds at harvest make tubers more susceptible to black pit and symptoms develop while tubers are in storage.

Economic Importance: Brown spot is not as aggressive on potato as early blight, but the disease can be responsible for significant yield loss if infestation occurs early enough in the growing season. The tuber blight phase can lead to reduced crop value due to cosmetic defects for fresh pack potatoes and increased loss of raw

Fig. 9.17 Brown spot lesions on potato leaflets. (Photo credit: Phillip Nolte)

product when peeling potatoes for processing. Black pit, much like tuber early blight, is less common than the foliar phase of disease.

Management Strategy: Management for brown spot/black pit is similar to management for early blight. Sound agronomic practices, such as crop rotation, varietal selection, proper plant nutrition and water management, proper pest management, and avoidance of other plant stresses, can greatly reduce early blight severity. Several fungicides are available to control brown spot on potato foliage during the growing season (e.g., chlorothalonil, SDHI, and EBDC fungicides, among others). The use of these fungicides also reduces production of the spores that cause tuber lesions.

Late Blight

Causal Agent: Oomycete—*Phytophthora infestans.* New genotypes or strains of the potato late blight pathogen were found in the U.S. and Canada in the late 1980s. These newcomers rapidly went from curiosities to the dominant types of the pathogen in both countries. This story played out again during 2009–2017. During this most contemporary era of late blight epidemics, the pack of late blight pathogen strains went from a list of multiple (US–22, –23, –24, and –8) to occurrence of predominantly US–23, with sporadic US–8. The contemporary epidemics have also included tomato as a significant host, which has refocused the need for community-wide disease management.

Further, the predominance of US–23 and US–8 presents further challenges in management, as they represent each of the two mating types (akin to 'male' and 'female' in animals) A1 (US–23) and A2 (US–8), which can sexually combine to potentially form both a thick-walled, soilborne resting spore type (oospore) and offspring that can potentially be more virulent on plants or resistant to specific fungicides. Before about 1990, all isolates of late blight in the U.S. and Canada were of the A1 type, and mating was not possible. New genotypes will almost certainly continue to develop.

The rise to dominance of the new genotypes is due to characteristics that appear to give them a distinct advantage over the earlier types. Some tests have shown that the new types are more aggressive than the old, meaning that the lesions they cause get larger more quickly so that plant damage is more severe. The new types also produce abundant stem lesions on infected plants, produce a greater number of asexual and airborne spores, and have greater capacity to survive colder, longer winter conditions in northern states.

While it is fortunate that the currently predominant genotype of US–23 is sensitive to a highly effective fungicide, mefenoxam (Ridomil®, Ultraflourish®), US–8 is resistant and as US–23 continues to predominate epidemics, there is current evidence that it is developing some resistance to the phenylamide fungicides, such as mefenoxam and metalaxyl, which are single-site mode of action fungicides. Mefenoxam is a purified form of the former active ingredient metalaxyl. Metalaxyl is the best fungicide ever developed for late blight management when pathogen populations are sensitive to the fungicide, giving excellent control of both foliar and tuber phases of the disease. Mefenoxam application is also still considered an effective treatment for management of other diseases caused by Oomycetes such as Pythium (leak) and pink rot (water rot, see below).

Inoculum Source: The late blight pathogen overwinters in seed potatoes, cull piles, and volunteer tubers.

Exposure: The pathogen produces spores on infected tubers in cull piles, or it can sporulate on late blight lesions, sprouts grown from infected seed, or volunteer tubers. Although overwintering in the sexual spore stage (oospore) has not been observed in most production regions of the U.S., the new strains of the pathogen, consisting of both A1 and A2 mating types, make the occurrence of oospores and, therefore, soil survival of the pathogen possible.

Symptoms: Leaf lesions may first appear as small pale to dark green spots that may rapidly enlarge under a favorable environment to become large, brown to purplish black necrotic lesions (Fig. 9.18a). In contrast to early blight, late blight lesions tend to be circular in shape since leaf veins are not barriers to this pathogen (Fig. 9.16a) A pale green to yellow halo may also be associated with these lesions.

Fig. 9.18 (**a**) Late blight leaf lesions. Light green border may not always be present. (Photo credit: Eugenia Banks). (**b**) Underside of leaf showing late blight sporulation. (Photo credit: Eugenia Banks). (**c**) Late blight lesions on potato stems. (Photo credit: Jeffrey Miller). (**d**) Potato tubers showing late blight symptoms. (Photo credit: Eugenia Banks)

Under moist conditions (usually early in the morning), a white cottony growth may be present on the underside of infected leaves (Fig. 9.18b).

With the development of the disease, infection can spread to entire leaflets and move rapidly from the petioles to the stem, eventually killing the entire plant. Once established under favorable conditions, this disease may take down an entire field within a matter of days.

A unique, diagnostic symptom of the disease is a dark, water-soaked-like lesion that begins at the apex and then moves down the stem (Fig. 9.18c). Stem lesions can occur readily with contemporary clonal lineages or pathogen strains and are cause for concern. A single stem lesion can girdle the stem and kill everything above the lesion.

In contrast to leaf lesions, stem lesions do not dry out in hot weather, allowing the pathogen to survive environmental stresses more readily. Stems are also much more difficult to protect with protectant fungicides than are the leaves of the plant.

On tubers, a tan-brown, dry, granular rot characteristically extends into the tuber to a depth of up to an inch or more (Fig. 9.18d). The depth of the rot may vary according to length of time after infection, variety, and temperature.

Economic Importance: Late blight is the most important disease of potatoes worldwide because it is expensive to control and it can cause significant losses both in the field and in storage. The destruction of foliage can severely limit yield, especially if the disease occurs early enough and under favorable environmental conditions. Foliar disease spread and damage can be extremely rapid, with entire fields appearing to go down in a matter of days.

The pathogen can also be responsible for huge storage losses because of the tuber blight phase, which can readily occur under certain conditions. Like several of the other storage diseases discussed herein, late blight produces lesions that are readily invaded by soft rot bacteria, and it is soft rot that presents the most significant storage management problem.

Further, negative economic impact can result from late blight in seed potato systems. Incidence of the disease can take acres out of certification, and even regional exposure may influence purchase decisions and value of the seed potato crop due to potential risk of pathogen introduction.

Management Strategy: Management of late blight must be a season-long effort that starts with seed selection, potentially fungicide treatment on seed, and investment in information systems to indicate timing of disease risk to trigger preventive fungicide applications.

Many experiments indicate that most seed tubers or seed pieces infected with late blight will readily decay from soft rot after planting and are self-eliminating. Yet, seed transmission is an important means of initiating the disease in a field. Spores of the late blight pathogen are spread from infected tubers to healthy tubers during handling, cutting, and planting operations. These spores can readily infect sprouts and the freshly cut surfaces of the seed pieces.

Seed piece fungicide treatments, such as mancozeb or cymoxanil that have activity against late blight, have been shown to be effective in minimizing this type of disease spread. Growers should not use seed piece treatments in an effort to "rescue" a badly infested seed lot.

Fields planted using seed lots with known or strongly suspected late blight infestation should be placed on a regular fungicide spray program for the entire season. Other sources of late blight are cull and volunteer potatoes. These sources should be destroyed. Contemporary clonal lineages, or strains of the late blight pathogen, can also infect tomato. As such, the introduction of tomato transplants from other production regions, as well as tomato volunteers and infected tomatoes in compost or cull piles, could be potential sources of inoculum and should be carefully managed or destroyed if late blight is identified.

By far, the most common means of managing late blight involves use of foliar-applied protectants and locally systemic fungicides at regular intervals. Currently, there are several registered and effective fungicides for late blight control in potato. Best results are obtained by fungicide applications made before disease appears.

Forecasting and Early Warning Systems

Effective systems for forecasting late blight outbreaks have been established and validated for several production regions of North America. Such systems rely on examining environmental information for periods when the weather is favorable for late blight development. Recommendations for initial fungicide application and subsequent application intervals are based on these environmental conditions. The warning systems, such as Blitecast, assume presence of the late blight pathogen somewhere in the agroecosystem. While this approach is a bit conservative, it is challenging, if not impossible, to know of all potential sources of the pathogen in the landscape. Unfortunately, such forecasting systems do not seem adequate for accurate prediction of the timing of initial disease outbreak in the arid, western production regions of North America, including Idaho. New research may remedy this situation.

Storage Management for Late Blight

Symptoms usually appear in storage within the first 3–5 weeks. The progress of late blight decay all but stops at 35 °F, but exposure to these temperatures does not kill the organism. Decay is favored by free water, inadequate air flow, and warm temperatures, especially early on in storage.

Storage should be checked at least weekly for bad odors or wet spots. Tubers that are infected may have gray lesions or be partially covered with white fungal growth. It is important to periodically wash the soil from suspect tubers and check for telltale late blight symptoms.

Late blight infected tubers will have a soft, leathery skin in the affected areas. A dry, granular rot may develop under the skin that is reddish (coppery to mahogany) in color and affects the outer one-quarter to one-half in of the tuber. Also, it is important to keep in mind that infected tubers are easily invaded by soft rot and may also have co-infection with other diseases, including Pythium leak and Fusarium dry rot depending upon tuber, field, and harvest conditions. During storage loading and immediately after, humidification should be reduced or shut down until tubers are dry.

Other precautions should be taken, including: (1) take extra care to avoid conditions where free moisture is likely to develop, (2) apply continuous air to the pile, and (3) hold storage temperatures at a recommended 38 °F for seed and 42 °F for fresh pack. Tuber lots with more than 5% tuber blight are generally a poor risk for long-term storage.

Rhizoctonia/Black Scurf

Causal Agent: Fungus—*Rhizoctonia solani.* Different strains of this fungus are divided into subgroups called anastomosis groups (AGs). *R. solani* AG–3 is the primary pathogen of potato, but other AGs can also cause disease. Other anastomosis groups, such as AG 2–2 and AG–8, are primarily pathogenic on sugar beet and small grains, respectively.

Inoculum Source: Soil and infested seed.

Exposure: This fungus invades stems and sprouts, and wounds are especially vulnerable to infection. Rhizoctonia is a problem when emergence and early plant development are slowed by cold, moist conditions.

Symptoms: Stems and stolons have brown to black, sunken lesions, and the disease may cause late emergence, weak plants, and non-uniform stands (Fig. 9.19a). Sclerotia will form on the skin of tubers and will appear as dark, scurfy residues that growers term "dirt that won't wash off" (Fig. 9.19b). The sclerotia may range from pinhead-to pea-sized.

Economic Importance: Rhizoctonia may cause poor stands, weak plants, misshapen or cracked tubers, blemished tubers, and limited yields.

Management Strategy: Seed tubers that have sclerotia covering more than 20% of the tuber surface should receive in-furrow treatments or fungicide seed treatments effective against Rhizoctonia. Rapid emergence and green tissue development can be encouraged by planting in warm soils and avoiding excessive early irrigations.

White Mold

Causal Agent: Fungus—*Sclerotinia sclerotiorum.*

Inoculum Source: Sclerotia that overwinter in the soil.

Exposure: Sclerotia—survival structures produced by the white mold fungus—in the top 1–2 in. of the soil will germinate and grow into mushroom-like bodies called

Fig. 9.19 (**a**) Healthy stems (top) contrasted to those with *Rhizoctonia* infection. Note proximity of cankers to potato seed piece. (Photo credit: Phillip Nolte). (**b**) Black scurf symptom of *Rhizoctonia* on tuber skin. (Photo credit: Eugenia Banks)

apothecia. Heavy rainfall and/or irrigation favor germination. Under favorable conditions, these mushrooms eject airborne spores (ascospores) that can infect nearby plants and be dispersed by the wind throughout the entire field or even to adjacent fields. Cooler temperatures (55–70 °F) and high humidity (95–100%) favor spore germination. Wet soils and wet foliage favor disease development.

Symptoms: Once the spores come in contact with a susceptible plant, infection causes water-soaked lesions that will soon produce a white, cottony, fungal growth. Initial infection sites usually occur where stems touch the soil. These lesions can girdle the stems of the potato plant causing the end of the vine to wilt (Fig. 9.20a).

Fig. 9.20 (**a**) Potato stem girdled by white mold lesion. (Photo credit: Eugenia Banks). (**b**) Potato stem split open to reveal sclerotia of the white mold fungus. (Photo credit: Eugenia Banks)

The fungus forms survival structures—the hard, black sclerotia—inside the dried stem (Fig. 9.20b).The sclerotia can then drop back into the soil or remain in the crop debris, ready to initiate a future cycle of disease.

Economic Importance: The effect that white mold has on yield is not clearly understood. Most research trials have not shown a significant yield difference between untreated plots and plots that have been chemically treated for white mold. However, modifications in fertility and irrigation management in an attempt to maximize yields, along with many years of accumulated white mold sclerotia, may be contributing to increased disease severity.

Management Strategy: This fungus not only infects potatoes, but is pathogenic on more than 350 species of plants. Other important hosts are beans, peas, canola, and some of the common weeds associated with potato production, such as lambsquarters and pigweed. It is important to recognize other hosts for the fungus so appropriate rotation decisions can be made to help minimize the problem. Unfortunately, white mold fungal structures can survive for perhaps 6 years or more in the soil, making it a difficult and perennial problem.

White mold can also be a problem that continues to build-up in soil over time. Cultural practices include using long crop rotations, especially with non-susceptible hosts, such as grain. Good fertility management is critical to avoid excessive vine growth.

Proper irrigation management is a critical factor in dealing with potential white mold problems. Prolonged canopy wetness and excessive irrigation should be avoided whenever possible. In general, anything that can be done to reduce humidity and wetness in the canopy will help with white mold control.

Fumigation has not been demonstrated to have a significant effect on the survival of white mold sclerotia. Chemical control options are available, but applications should be made before the disease has become well established in the field. The first fungicide application should be made when the first plant trash (e.g., flower buds or dead leaves) is on the ground. This is approximately 7–10 days after row closure. Often a second application is necessary. Good coverage on the foliage and stems is necessary.

Pink Rot, Water Rot

Causal Agent: Oomycete—*Phytophthora erythroseptica.*

Inoculum Source: Overwintering oospores in the soil.

Exposure: Oospores germinate in soil and infect stolons, roots, eyes, and lenticels. During the harvest operation, infected tubers can contaminate healthy tubers, leading to development of disease in storage. Infection is favored in wet conditions, low pH, and low soil Ca.

Symptoms: Pink rot can often be found in the field before harvest and is characterized by rotted tuber tissues that turn pink after a 20- to 30-min. exposure to air (Fig. 9.21a). Another important diagnostic trait for pink rot is that the rot will usually appear to start from the stem end of the tuber and will progress through the tuber in a uniform manner, often with a nearly straight line between the healthy and the diseased portions of the tuber (Fig. 9.21b). Underground stems can also appear brown to dark-brown and exhibit a wet decay.

In a tuber that is infected only with the pink rot organism, the rotted tissues will still retain some structure and firmness, but will have a somewhat rubbery texture. Infected tissues are easily invaded by soft rot bacteria, which will turn them wet and slimy.

Economic Importance: This disease can be responsible for significant losses in the field. Infected tubers and secondary bacterial soft rot infection can lead to serious storage problems.

Management Strategy: Pink rot in the field is usually associated with overly wet or waterlogged areas. In fact, some of the highest levels of disease are often found under the wet, central area of a center-pivot irrigation system or under drains in lateral-move or solid set sprinkler systems. Good irrigation practices that maintain optimal moisture levels can go a long way toward minimizing this disease.

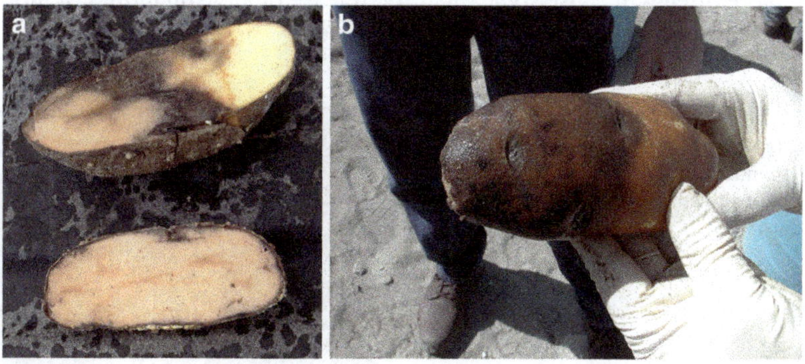

Fig. 9.21 (**a**) Tuber sliced open to reveal pink discoloration typical of pink rot infection. Color may require 10–15 min to develop. (Photo credit: Phillip Nolte). (**b**) Pink rot progresses through the tuber in a uniform manner, often with a nearly straight line between the healthy and the diseased portions of the tuber. (Photo credit: Jeffrey Miller)

Procedures to follow include: (1) plant less susceptible varieties, (2) treat with mefenoxam (at planting or during the growing season) or with phosphorous acid (during the growing season), (3) avoid wounding and high pulp temperatures (above 65 °F) during harvest, (4) treat with a phosphorous acid-based fungicide at harvest, and (5) grade out infected tubers at harvest. Be aware that mefenoxam resistance has been documented in some production areas.

Pythium Leak, Watery Wound Rot

Causal Agent: Oomycete—*Pythium* spp., usually *P. debaryanum* or *P. ultimum.*

Inoculum Source: Overwintering oospores in the soil.

Exposure: Oospores germinate in and infect tuber wounds typically made at harvest.

Symptoms: Leak is characterized by a rot that starts from an infection site on the surface of the tuber and generally rots out the entire central portion of the tuber, while leaving the portion of the tuber from the vascular ring out to the skin of the tuber intact (Fig. 9.22). This results in a condition often described as "shell rot" when the internal tissues have rotted away.

The rotted tissues are brown to black in color and may have cavities within them. The texture of the rotted tissues is soft and watery. When squeezed, diseased tissues exude a clear, watery fluid. This property is responsible for the term "watery wound rot," which is another name for leak.

Similar to pink rot, the diseased tubers can be easily invaded by bacterial soft rot. The leak pathogen invades and destroys tubers rapidly and usually shows up within the first several weeks of storage or when shipping out of the field.

Economic Importance: Pythium leak has been responsible for total losses in storage but more often will cause some tuber rot, release of large quantities of moisture, and create subsequent storage management concerns.

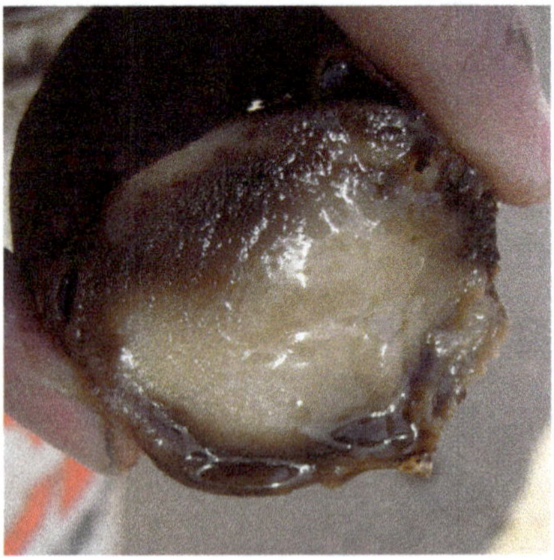

Fig. 9.22 Potato tuber showing typical symptoms of pythium infection. (Photo credit: Jonathan Whitworth)

Management Strategy: The Pythium leak pathogen invades wounds made at harvest under warm tuber pulp temperatures. It is important to avoid wounding during harvest and not attempt to harvest when pulp temperatures are above 65 °F, especially if environmental conditions are not conducive for removing field heat from stored tubers.

If significant infection is found in storage, high volumes of air and a rapid cool down are recommended. As with pink rot, treatment with mefenoxam at planting or during the production season is helpful in reducing the amount of disease. Infected tubers can also be graded out during the harvest and storage-loading operations.

Silver Scurf

Causal Agent: Fungus—*Helminthosporium solani.*

Inoculum Source: Infected seed pieces, infected tubers in storage, and infested soil.

Exposure: The scurf fungus moves from infected seed pieces to daughter tubers while the potatoes are still in the soil. Exactly when these infections take place is unknown, but daughter tubers have shown evidence of the disease as early as 6 weeks after planting.

Other research indicates that the disease spreads during periderm maturation, just before digging. This stage of the disease, which occurs while tubers are still in the ground, is referred to as the "primary infection" and results in "primary lesions." These lesions are fairly thick, prominent patches that are usually more severe on the stem end of the tuber. Field infections also occur where there are no visible symptoms.

The primary lesions and other field infections provide inoculum, in the form of fungal spores, for secondary spread of the fungus from infected to healthy potatoes within the storage facility. This cycle of infection leads to "secondary lesions," which are, individually, less severe than the primary lesions but may cause losses due to grade-off because of their high numbers.

The speed with which the silver scurf fungus spreads and establishes inside the storage facility can be greatly influenced by storage management practices, including both curing and holding conditions. The silver scurf fungus may be able to survive from the end of one storage season to the beginning of the next on materials such as wood, sheet metal, insulation (polyurethane), and even in soil from the cellar floor.

Symptoms: Lesions on infected tubers will have a smooth, gray to silvery sheen and are commonly found near the stem end. Lesions usually remain superficial, with no internal damage to the tuber. However, severe symptoms may occur where the infected cells of the periderm and underlying cortex collapse and allow moisture loss. Tuber symptoms of silver scurf and black dot can be easily confused with one another (Fig. 9.23).

Fig. 9.23 Tubers showing discolored patches on the periderm caused by silver scurf (left) contrasted with black dot (right). (Photo credit: Phillip Nolte)

Economic Importance: Silver scurf creates cosmetic defects, leading to reduced quality in fresh-marketed potatoes and increased water loss in storage. Difficulty with peeling infected tubers may reduce the acceptability of tubers for processing.

Management Strategy: Steps that growers can take to prevent silver scurf disease include: (1) plant disease-free seed tubers, (2) use effective fungicide seed treatments, (3) keep storage temperatures and humidity as low as possible, (4) monitor tubers weekly for symptoms, and (5) market infected potatoes before they go out of grade.

As previously mentioned, the silver scurf fungus can survive from the end of one storage season to the beginning of the next on a variety of surfaces; even in soil from the cellar floor. Therefore, it is important for handlers to disinfect storages before a new crop of potatoes is stored.

Black Dot

Causal Agent: Fungus—*Colletotrichum coccodes*.

Inoculum Source: Soilborne and tuberborne.

Exposure: Invades plant roots from the soil or infects leaves via wind-borne spores. Soilborne inoculum is more significant than seedborne.

Symptoms: Foliar symptoms include wilting and yellowing that begin at plant apices first, then develop later in mid to lower plant parts. Wilting symptoms are similar to those characteristic of potato early dying caused by *Verticillium* and *Fusarium* spp. Foliar lesions are small and can resemble early blight or brown spot lesions, but lack concentric rings. Symptoms on roots include lesions on the main stem, which can appear similar to Rhizoctonia lesions. Root lesions have a more uniform brown to gray color than Rhizoctonia lesions, and small, black sclerotia can be observed in black dot lesions as they age.

The disease gets its name from the abundant, black sclerotia that form on the basal stem area (Fig. 9.24). Sclerotia can also form on tubers, stolons, and roots.

Fig. 9.24 Stem showing abundant black dot sclerotia. (Photo credit: Eugenia Banks)

Severely infected plants will be reduced in size and may die early, resulting in smaller tuber sizes. At harvest, portions of infected stolons may cling to tubers. Invasion of the tuber periderm by this fungus in storage may cause irregular, rough patches on the surface, which is often confused with silver scurf (Fig. 9.23).

Economic Importance: This disease has often been considered a disease of senescence, since sclerotia on stems often appear after the potato crop is dying. Some research indicates that black dot can cause significant yield reduction. In the past, cases of black dot may have been misdiagnosed as Verticillium wilt.

Management Strategy: Black dot can be prevented by doing the following: (1) plant certified seed; (2) reduce soil populations by rotating crops, including potatoes only every fourth or fifth year; (3) use strobilurin-based or other reduced-risk fungicides with activity on ascomycete fungi, which have shown some promise in reducing plant infection by the fungus; and (4) minimize plant stress and encourage uniform, continuous plant growth by providing a balanced fertility program and optimum soil moisture levels.

Powdery Scab

Causal Agent: Protozoa—*Spongospora subterranean* f. sp. *Subterranean.*

Inoculum Source: Infected seed or infested soil.

Exposure: Spread will occur from the seed piece to progeny or neighboring tubers and can come from spores left in the soil from previous potato crops.

Symptoms: This disease causes scabby, warty lesions on the tuber surface, and the lesions fill with dark brown, powdery masses of spore balls. Tuber lesions can be confused with pitted lesions of common scab, but powdery scab lesions are often rounder and seldom as deep. As the organism pushes outward from inside the tuber,

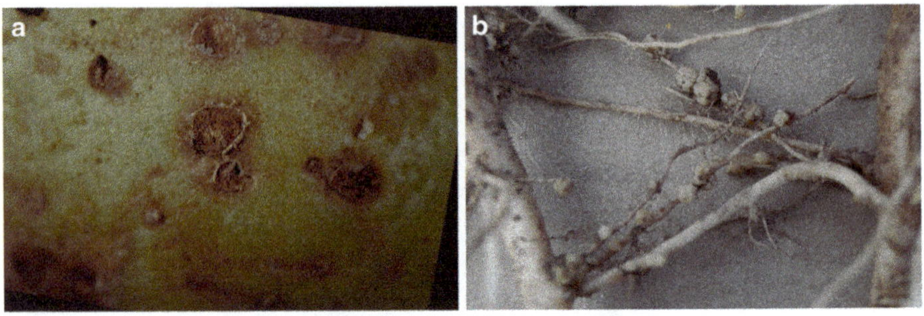

Fig. 9.25 (**a**) Powdery scab lesions on tuber. Note rolled up periderm around the lesions. (Photo credit: Phillip Nolte). (**b**) Root nodules formed by the powdery scab organism. (Photo credit: Jeffrey Miller)

the torn periderm often scrolls up around the lesions (Fig. 9.25a). The fungus also produces small, white to brown masses of galls on roots and stems (Fig. 9.25b).

Economic Importance: Powdery scab creates cosmetic defects that can lead to rejection in fresh market potatoes. This pathogen can also be a vector for PMTV.

Management Strategy: Advice to prevent powdery scab disease includes: (1) plant disease-free seed, (2) do not plant susceptible varieties in infested soil, and (3) do not overirrigate. Fluazinam fungicide can help to suppress powdery scab infection when applied in-furrow at time of planting. This fungicide use (Omega®) is provided by special state-specific registration. Infection is worse if wet soil conditions persist early in the season, so early irrigations should be avoided when possible.

The occurrence of this disease is highly dependent on environmental conditions. In general, russet varieties are relatively resistant to powdery scab, while many white- and red-skinned varieties are highly susceptible.

Gray Mold

Causal Agent: Fungus—*Botrytis cinerea*.

Inoculum Source: Spores from crop debris.

Exposure: Spores (conidia) are carried to plants by wind and water.

Symptoms: Flowers are infected first. Infected tissue will appear gray and fuzzy. The "fuzz" is a mass of spores produced by the fungus. Infected flowers fall into the lower canopy where the fungus can then spread to the lower leaves. *B. cinerea* is a weak pathogen and typically infects damaged or senescing leaves. Leaf lesions often are initiated at the leaf tip or margin where water accumulates. Lesions appear gray to dark brown and can be confused with late blight lesions (Fig. 9.26). As the fungus spreads, infected tissue will appear fuzzy from the mass of spores being produced by the fungus. When the vines are disturbed, these spores can be released into the air causing a small cloud.

Fig. 9.26 Gray mold (*Botrytis*) leaf lesions. (Photo credit: Eugenia Banks)

Economic Importance: Gray mold is generally thought to be of little economic importance. However, in some cases infections may be severe enough to cause premature senescence of potato vines, which can reduce yield.

Management Strategy: Some fungicides, which are effective against early blight and brown spot (e.g., chlorothalonil and pyrimethanil-based fungicides), are effective against gray mold. Proper irrigation management can also be helpful.

Powdery Mildew

Causal Agent: Fungus—*Erisyphe cichoracearum.*

Inoculum Source: Infested plant debris.

Exposure: Little is known about the conditions necessary for disease growth and spread; however, it is most frequently seen in surface-irrigated fields. Powdery mildews differ from most plant pathogens in that they do not require free moisture for sporulation and infection.

Symptoms: Powdery mildew first appears as brown lesions that coalesce to form streaks or stipples on the stems and petioles (Fig. 9.27). Leaf lesions may also show a superficial fungal growth that gives the lesion a powdery gray to brown color that may look, at first glance, like soil or spray residue. As the disease progresses the lower leaves will turn yellow and fall off while the rest of the plant remains erect.

Economic Importance: Vine death over large areas of a field is possible with severe yield loss caused by loss of foliage.

Management Strategy: The powdery mildew fungus is almost entirely superficial on the potato plant. Foliar applications can be made with strobilurin or sulfur compounds before infections are established.

Fig. 9.27 Powdery mildew lesions on potato stem. Note the superficial fungal growth. (Photo credit: Brad Geary)

Timeline for Disease Management Decisions

Before Planting

Losses from most diseases can be avoided by selecting and using clean, certified seed. Information on selecting a quality seed lot is contained in Chap. 7.

Early dying, caused by *Verticillium* spp., can be significantly reduced by use of soil fumigants, such as metam sodium, but they must be applied before planting, typically during the previous fall. Spring fumigation is not as good an option, but can still be effective.

Planting potato seed into spring-fumigated soils should be avoided if cool temperatures and above-average rainfall occurs within 3 weeks after the fumigant is applied. Planting too early after spring fumigation can lead to serious seed piece damage, because the fumigant will have had insufficient time or have been subjected to improper conditions for normal dispersal.

Fields that have significant root-knot nematode populations or a history of corky ringspot virus problems should probably be fumigated with a material effective against nematode pests and vectors. Fall applications are generally more effective.

In special cases, chloropicrin fumigation is utilized to manage fields with history of heavy common scab pressure. This fumigation treatment is made in the fall prior to spring potato planting. Chloropicrin treatment can also provide control of a broader range of diseases and often enhances yield in crops.

Planting

Many seed decay problems result from growers planting cut seed under soil conditions (temperature and moisture) that are unfavorable for potato growth. It is important to avoid planting under conditions where soil moisture is too high or when soil temperatures are below 45 °F or above 65 °F. Low temperatures delay wound healing in the seed, which can lead to seed decay. High temperatures can also lead to seed piece decay problems, especially if soils are excessively moist.

Fungicidal seed piece treatments can be used to manage dry rot seed decay. For fields or areas where soft rot seed decay is known to be a problem, either single drop seed or cut seed that has been healed or "suberized" should be used. A bag test can help potato producers make decisions about which fungicide to use and if a particular seed lot is a good candidate for suberizing. See Chap. 7. If producers are concerned about Rhizoctonia, they should make sure that the seed treatment is active against this fungus. In-furrow chemical treatments are also available for Rhizoctonia management. Seed-applied fungicides can also control silver scurf, black scurf, and black dot if inoculum is present on seed.

Pink rot and Pythium leak of potato are usually considered to be either late season or storage disease problems, but an in-furrow application of fungicide for management of pink rot has been shown to be effective. The same fungicides applied to foliage during the growing season, when the tubers are about a half-inch in diameter, can also aid in management of pink rot.

Growing Season

The most important disease problems encountered during the growing season are early dying, early blight, brown spot, and late blight. Early dying, early blight, and brown spot can be significantly reduced by proper fertility and proper water management. Fungicide application may be necessary for early blight and brown spot if disease begins early in the season or the variety produced is susceptible. Many white and red-skinned varieties, for instance, are susceptible to early blight and may require four or more fungicide applications to keep disease in check.

Late blight management begins with one application of protectant fungicide applied a week before row closure, followed by 4 or 5 applications applied every 7 days during the period of rapid canopy growth. Subsequent applications may be necessary if late blight is found in the area or if weather conditions are favorable for disease development. As mentioned above, late blight forecasting programs are available for many production areas.

Increasing the interval between sprays creates a greater risk for late blight infection if the disease is present in the area. Late season applications may be necessary, even after vines have begun to die, if late blight is in the field or in the area. Weather conditions during the late season tend to be both cooler and wetter, conditions that favor tuber infection, which can lead to significant rotting problems in storage.

Foliar applications of phosphorous acid-based fungicides can be effective in managing pink rot when pathogen populations are resistant to mefenoxam. These applications are initiated when the largest tubers are one-half inch in diameter and are made on 14-day intervals. This approach also provides protection for tubers against late blight.

Another disease likely to be encountered is white mold, which can be reduced by fungicide applications. This fungus reproduces by means of a small, funnel-shaped mushroom, and growers should apply fungicide as soon as the mushrooms appear. This generally occurs just before row closure.

Vine Kill and Harvest Activities

Many storage problems begin with damage to tubers that occur during harvest and handling activities. Careful attention to bruise prevention during this critical time can greatly reduce or eliminate disease problems in storage.

Vine killing is usually necessary for maturation of the tuber skin, a process called "skin set." Skin set usually takes anywhere from 14 to 21 days after the vines have been killed. Usual recommendations are to wait at least 18 days after killing vines to begin harvesting.

A mature skin provides an excellent protective barrier for the potato tuber and is vital to long-term storage. Even if the skin is properly set, some damage to the tubers is inevitable. Quality losses in storage caused by bruising and disease invasion of wounds that occur during harvest are greatly reduced when harvest and handling equipment are in good repair and properly operated.

Several important storage diseases require, or are strongly favored by, a wound to infect tubers. The most common of these are Pythium leak, pink rot, early blight, black pit, late blight, dry rot, and soft rot. In many cases if there is no wound, there is no disease. Harvesting of tubers with pulp temperatures below 42 °F or above 65 °F should also be avoided.

Post-harvest applications of fungicides and disinfestants can be made to prevent pathogens from causing infections through wounds. Applications of phosphorous acid-based products are effective in protecting against pink rot and late blight, as well as in limiting silver scurf.

Storage

Wounds that occur during harvest must be given the opportunity to heal. For this reason, potato tubers should be held under conditions favorable for wound healing to occur after they have been placed in storage. The healing process is similar to skin set and requires a 14–21 days period to complete. The terms “curing” and “suberizing” are synonyms for wound healing.

Proper storage conditions for healing are included in Chap. 17. Several diseases can cause problems in storage, but the management techniques for each of them are the same: reduce storage temperatures, increase airflow, and monitor disease progress frequently.

An estimate of how much disease is going into storage can help in making sound storage management decisions. Often marketing the potatoes as soon as possible is the only option to avoid significant losses.

Chapter 10
Nematode Management

Saad L. Hafez, Sundararaj Palanisamy, and Ann E. MacGuidwin

Contents

S. L. Hafez (✉)
Department of Entomology, Plant Pathology and Nematology, University of Idaho, Parma, ID, USA
e-mail: shafez@uidaho.edu

S. Palanisamy
Department of Zoology, Bharathiar University, Coimbatore, Tamil Nadu, India

A. E. MacGuidwin
Department of Plant Pathology, University of Wisconsin-Madison, Madison, WI, USA
e-mail: aem@plantpath.wisc.edu

J. C. Stark et al. (eds.), *Potato Production Systems*,
https://doi.org/10.1007/978-3-030-39157-7_10

Introduction

Potatoes, the fourth most important food crop in the world, yield more edible energy production per hectare (2.47 ac) per day than any other crop. Potatoes rank third among world food crops for protein production per hectare per day. Nematodes are one of the major limiting factors for potato production worldwide. Nematode infection affects plant and tuber growth and tuber quality, resulting in economic loss to the industry. Nematodes, the unseen enemies of potato growers, are microscopic worm-like animals, able to move between soil particles, between folded leaves of plant buds, in the open spaces of leaves and stems, or in plant tissues. The word "nematode" literally means "thread-like."

The typical nematode is spindle-shaped, unsegmented, and bilaterally symmetrical. These soft-bodied animals rely on soil water, so their movement in the soil is influenced by the thickness of the water films surrounding soil particles. They are inactive in dry conditions and disabled when soil is flooded, as they cannot swim. Species parasitic on plants possess a mouth stylet, which they use to puncture plant cells and extract cell contents. In the past, nematode damage to crops was often ignored or attributed to other causes, such as poor fertility, deficient soil moisture, or "soil exhaustion." This was partly because nematodes are too small to be seen without a microscope and partly because only limited information on their occurrence and pathogenicity was available.

History of Potato Nematodes in the U.S.

Discovery of root-knot nematode (RKN) in the U.S. dates back to 1889 when *Meloidogyne arenaria* was found in Florida. Considering the economic importance of RKN, Idaho potato growers are advised to test their soil for *M. hapla* before planting. Stubby-root nematode, *Paratrichodorus allius*, (SRN) was found in an Oregon in 1963. Later, it was demonstrated that *P. allius* is able to transmit the tobacco rattle virus (TRV) that causes the corky ringspot disease (CRS) of potato. The earliest record of *Pratylenchus* on potatoes in the U.S. was by Cobb (1917), who found pustules over the surface of potato tubers caused by *P. penetrans*. The

Fig. 10.1 External and internal symptoms of tubers infected by the Columbia root-knot nematode. (Photo credit: Kathy Merrifield)

potato-rot nematode, *Ditylenchus destructor*, (PRN) was found in Wisconsin in 1953, following reports in 1945 of infestations from six farms near Aberdeen, ID. More than 68 species of plant parasitic nematodes belonging to 24 genera are associated with potato fields from different parts of the world. Among them, the top five genera for potato in the U.S. are RKN (*Meloidogyne* spp.), SRN (*Trichodorus* and *Paratrichodorus* spp.), root-lesion nematodes (*Pratylenchus* spp.) (RLN), the PRN (*Ditylenchus* spp.), and potato-cyst nematodes (*Globodera pallida* and *G. rostochiensis*).

Root-Knot Nematodes

Root-knot nematodes (*Meloidogyne* spp.) (RKN) are recognized as a major nematode pest on potatoes, especially in sandy soils. Their common name comes from the knotted swellings, referred to as galls, on roots. Galls may contain one to several adult root-knot females. Damaged roots are less able to obtain soil nutrients, and symptoms appear as nitrogen or micronutrient deficiencies. Plants may wilt easily, especially in warm weather, due to root damage even though soil moisture may be adequate. Nematode feeding causes enlargement or bumps in the outer layers of the tubers and internal discoloration (Fig. 10.1) reducing their marketability for fresh packing or processing.

Occurrence and Host Range

There are many species of RKN, but only two are common on potatoes in Idaho and Eastern Oregon; the Columbia root-knot nematode (*M. chitwoodi*) (CRKN), and Northern root-knot nematode (*M. hapla*) (NRKN). The CRKN was first described on potato in Quincy, WA, and later in Iron County, UT. It now is found in the Western

U.S., Mexico, and Europe but has not been reported from Canada or potato production regions in the Midwest or Eastern U.S., where NRKN is the predominant root-knot species. The host range of root-knot nematodes is wide and includes most crops commonly grown in rotation with potatoes in Idaho, Eastern Oregon, and Washington. Two variants of the CRKN are reported based on their selective reproduction on carrots and alfalfa, and both can be distinguished from the NRKN by their ability to reproduce on maize. Pathogenicity studies of CRKN under controlled conditions showed that maize is a better rotation crop than wheat, barley, or oats for the susceptible potato crop in the Pacific Northwest (PNW). Further studies revealed that the CRKN reproduced on 53 of 68 plant species tested under greenhouse conditions. The NRKN reproduced on 21 of 32 weeds and has an extensive host range that excludes corn and small grains.

Biology and Epidemiology

The RNK life cycle has six stages: egg, first-, second-, third-, and fourth-stage juveniles, and the adult stage (Fig. 10.2). First-stage juveniles molt to second stage while still in the egg, and second-stage juveniles emerge from eggs and invade potato roots (Fig. 10.3). Inside the root, the nematodes become sedentary with their heads near the developing vascular system where they feed, increase in size, and induce cell alterations that result in the formation of galls. They undergo three additional molts and mature as adults within the potato root in as few as 20–25 days.

Females swell into a pear-like shape (Fig. 10.4) and lay 50–1000 eggs in a gelatinous matrix outside their bodies. The males emerge from the root after the fourth molt and remain worm like. Depending on climatic conditions, several generations can occur in one growing season. RKN survive winters as eggs or as second-stage juveniles in soil or in plant tissue.

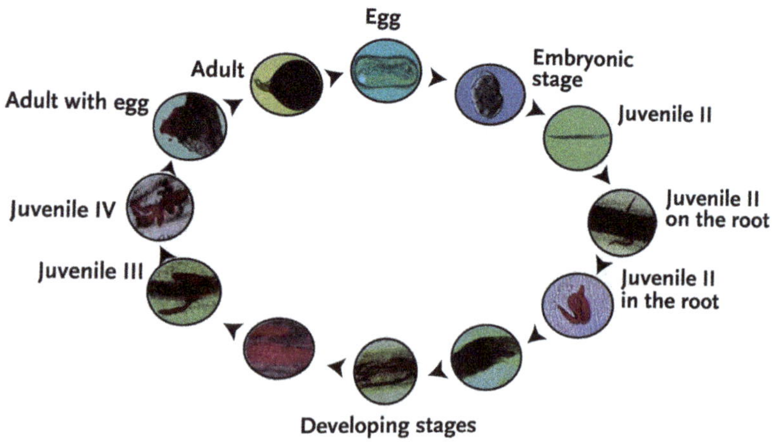

Fig. 10.2 Root-knot nematode life cycle. (Photo credit: Saad Hafez)

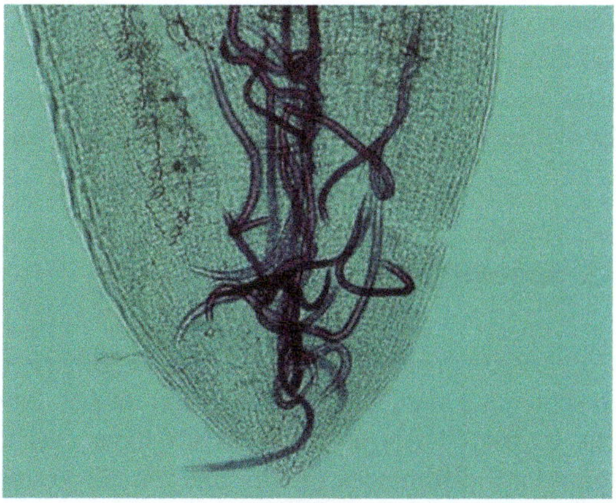

Fig. 10.3 Root infected with second-stage root-knot juveniles. (Photo credit: Jon Eisenback)

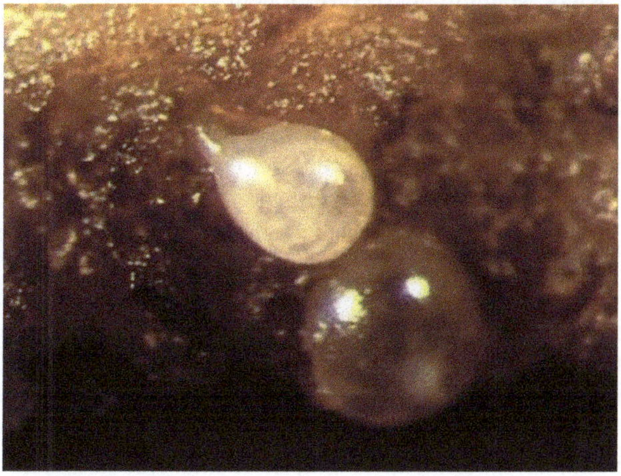

Fig. 10.4 Root-knot females dissected from galled root. (Photo credit: Jon Eisenback)

Symptoms and Disease Incidence

Plants infected with RKN develop galls of varying sizes and shapes on their roots and brown lesions in tubers. Gall shape ranges from almost spherical to irregular. Several factors influence the severity of damage caused by RKN. Typical symptoms, usually not seen until later in the growing season, include stunting, yellowing of the foliage, wilting in the presence of adequate soil moisture (particularly on warm days), or general nutrient and micronutrient deficiencies. Severely infected and galled plants may also suffer from secondary pathogens that cause tubers to rot in the field, making harvest difficult and increasing losses. At the field scale, foliar symptoms range from localized patches to large regions, but are never uniform within a field. Tubers may be damaged even when there are no foliar symptoms.

Generally, the CRKN causes more damage on potatoes than the NRKN. Yield reductions induced by RKN can range from slight to over 50%, with 25% reductions commonly reported. Total losses are possible in severely infested fields because of the loss of tuber marketability. Damage from the NRKN may be most severe after alfalfa hay crops or vegetables and during years with warm spring temperatures. Cooler seasons may delay infection, and less injury may occur.

Management Strategies

Crop rotation is the cornerstone of managing pest nematodes and should be included in schemes aimed at RKN. It is important to have information on the species present, as the CRKN and the NRKN have overlapping host ranges but with some important differences. Corn and small grains are hosts for CRKN, but not NRKN. Carrot and lucerne are hosts for NRKN, but not for some populations of CRKN. Maintaining records of nematode test results and cropping histories are useful in devising rotations that suppress RKN reproduction.

Pesticides are a valuable tool for managing RKN. If RKN is a severe economic pest, the use of nematicides or fumigants should be employed. In studies at the University of Idaho, plots infested with CRKN and treated with the systemic nematicide Vydate® yielded well. Further, it was found that the contact nematicide Mocap® applied either during the fall or spring in combination with the fumigant Vapam® significantly reduced nematode infested potatoes as compared to untreated control.

Cultural management tactics are generally not adequate to manage severe infestations of RKN, but are important for preventing the resurgence of nematode population densities to unacceptable levels. Nitrogen application significantly increased tuber yield with maximum yield at a nitrogen level of 680 lbs./ac followed by 1350 lbs./ac in Idaho. Green manure crops in the same study had no effect on yield, but other experiments conducted at microplot and field scales confirmed that rapeseed 'Humus' and oil radish '*Raphanus sativus*' reduced population densities of CRKN and increased the potato tuber yield and quality under Idaho conditions. Further, it was confirmed that addition of the bacterium, *Bacillus megaterium*, along with rapeseed 'Humus' or oil radish increased tuber yield and quality and suppressed populations of both CRKN and NRKN under greenhouse and microplot conditions. Adding prophos along with rapeseed considerably increased the grade 1 potato yield.

There are currently no cultivars of potato-resistant RKN suitable for production in the U.S. Research for resistance to CRKN is ongoing, and promising germplasm is in the pipeline. Resistance to CRKN, NRKN, and *M. incognita* has been identified in wild relatives of potatoes, so the eventual release of cultivars developed using traditional or transgenic approaches is anticipated.

Stubby-Root Nematodes

Stubby-root nematodes (*Paratrichodorus* spp.) (SRN) are migratory ectoparasites that live in the soil and on the surface of roots, especially root tips. They retain a worm-like shape throughout their life cycle and possess a unique curved stylet (Fig. 10.5). Their feeding decreases or stops root growth causing stunting of the root system for which they are named. These nematodes are important parasites of potatoes, not so much for the direct damage they cause but for the TRV they transmit to potatoes. This virus causes CRS disease of potato tubers and greatly decreases the quality of processed product (Fig. 10.6). Two genera of SRN, *P. allius* and *P. teres*, are the most damaging to potatoes.

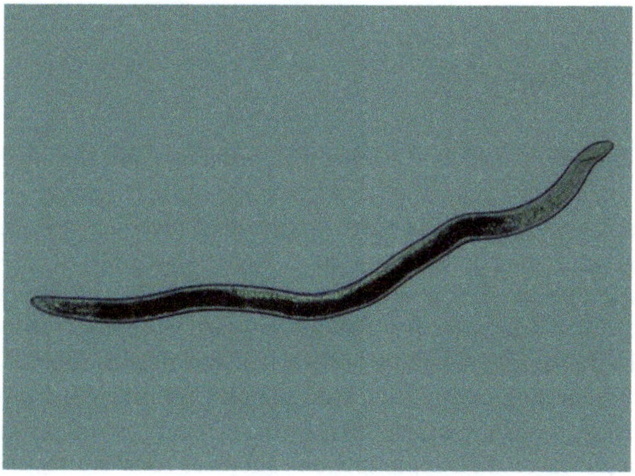

Fig. 10.5 Light microscopic view of adult stubby-root nematode. (Photo credit: Jon Eisenback)

Fig. 10.6 Internal symptoms of corky ringspot: concentric rings and diffuse spots in the flesh. (Photo credit: Saad Hafez)

Occurrence and Host Range

SRN are fairly common with a reported incidence of 22% in the PNW and 13% in Ohio. The majority of SRN infestations are not associated with TRV. About 10% of the field populations of *P. allius*, the most common species in the PNW, were found to be viruliferous in a study in 2000, suggesting that aggressive management should only be practiced in fields with a history of CRS disease. Problems with SRN have now been consistently documented in all potato regions, although the incidence of the CRS disease fluctuates.

SRN have a wide host range that includes cereal and grain crops, as well as potatoes. They are not likely to disappear from a field, but can lose the ability to cause CRS disease if plant reservoirs of TRV are eliminated.

Biology and Epidemiology

SRN are migratory ectoparasites, meaning that they are mobile during their entire life cycle and they feed on the outside of roots. The life cycle is relatively simple, as the juvenile stages resemble the adult stage, except that juveniles are smaller. Several generations can be produced within a year, so large populations of SRN can develop quickly. Damage due to SRN is related to their population density at the time of planting, but if TRV is present, population densities lower than 10 nematodes per 8 oz. soil can be damaging. Numbers of SRN can also decline rapidly after the crop is removed, so the timing of end-of-season samples for predicting disease pressure for the next potato crop is important. SRN thrive in sandy, moist soils so wetter seasons and irrigation favor their reproduction. They are sensitive to soil disturbance, relative to other pest nematodes. SRN appear to be fairly mobile, and it is thought that in many locations they survive cold winters by migrating below the frost line and going dormant.

All stages of hatched SRN can acquire TRV through feeding, but they do not retain virus particles after a molt to the next stage nor do they pass the virus to eggs during reproduction. The virus can persist in nematodes for weeks but does not multiple inside them. The population of nematodes in a field remains viruliferous for long periods of time, since all life stages of the nematode are always present, and multiple generations overlap throughout the year.

Symptoms and Disease Incidence

When SRN infestations are severe, the above-ground symptoms resemble symptoms of other nematodes, including poor growth, yellowing, and stunting. SRN transmit TRV directly to potato tubers during tuberization causing virus symptoms, so most often their presence is suspected when symptoms of CRS become prevalent. Virus infection blemishes tubers and renders them unmarketable. The most common blemish is dark-brown necrotic tissue in the tuber flesh in the form of circles, arcs, or diffuse spots (Fig. 10.6). This disease can easily be mistaken for heat necrosis if the predominant symptoms are spots. Tubers infected with TRV may become irregular in shape during early stages of growth and show bud-end folding or cracking at harvest (Fig. 10.7a). Occasionally, external symptoms appear, showing as arc-shaped lesions followed by brown concentric rings on the skin (Fig. 10.7b). Damage by SRN is greater in wetter seasons. This disease is of major concern to potato growers, as tuber injury of 5–10% may result in rejection of the harvest of an entire field.

Management Strategies

SRN are difficult to control because of their seasonal mobility in the soil. This nematode is highly sensitive to changes in soil moisture and temperature, which drive their movement up and down in the soil profile. SRN can reside at soil depths of more than 40 in, requiring increased rates of soil fumigants and shank injection. Products that are highly mobile in the soil profile and able to reach nematodes as they migrate can be effective in managing SRN, but should not be used in areas with high water tables. Nematicides with systemic activity are effective in reducing the CRS disease. Field trials showed the lowest incidence of CRS disease was achieved with a combination of systemic nematicides applied in-furrow at planting followed by foliar applications. A combination of nematicides and soil fumigants is generally recommended for fields with a high incidence of CRS disease. The high cost of this approach underscores the need for potato varieties resistant to either the TRV or nematodes.

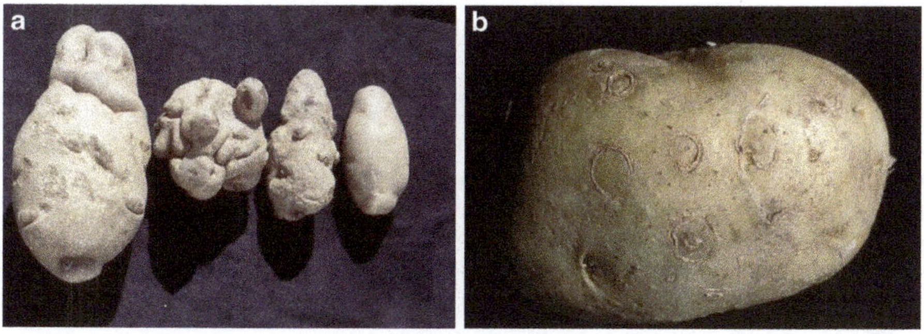

Fig. 10.7 External symptoms of corky ringspot include irregular shape and cracking (**a**) and arc-shaped lesions in the tuber skin (**b**). (Photo credit: Saad Hafez)

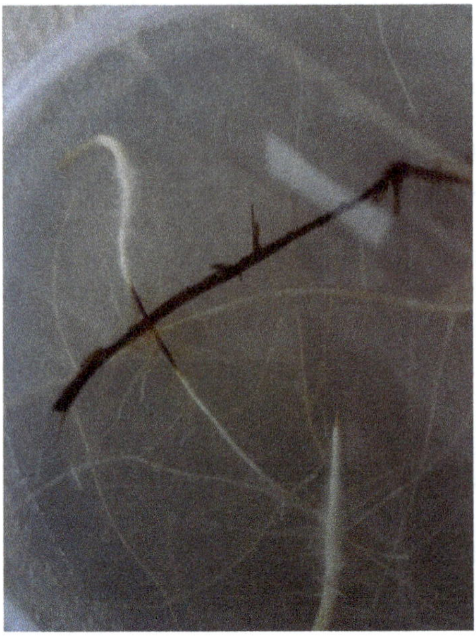

Fig. 10.8 Roots growing in a Petri dish showing lesions caused by the root-lesion nematode, *Pratylenchus penetrans*. (Photo credit: Ann MacGuidwin)

Root-Lesion Nematodes

Root-lesion nematodes (*Pratylenchus* spp.) (RLN) are the most common nematode potato pest. There are over 100 species of RLN, so named for the dark lesions they cause on roots (Fig. 10.8). The most common species of concern for potatoes in the U.S. are *Pratylenchus penetrans*, *P. neglectus*, *P. scribneri*, and *P. crenatus*, although almost every species has a wide host range and can feed on potatoes. RLN reduce yield at high population densities. Some species, most notably *P. penetrans*, interact with *V. dahliae* to cause potato early dying (PED) disease even when nematode population densities are too low to cause disease by themselves.

Biology and Epidemiology

All life cycle stages of RLN, other than the egg, can infect roots. Nematodes enter roots at all points, but tend to prefer the region immediately behind the growing tips (Fig. 10.9). They can also feed on roots without entering. The youngest juvenile stage, J2, is often observed in culture feeding on root hairs (Fig. 10.10). Lesion nematodes are migratory endoparasites, meaning they move inside the roots and continue to exit and enter roots throughout their lives.

The host range of RLN is extremely broad and includes most rotation crops, including cover crops. Each species has preferred hosts, which can influence which species predominates when infestations are mixed. Some species, such as *P. penetrans*, have a stronger interaction with *Verticillium* for PED disease than others.

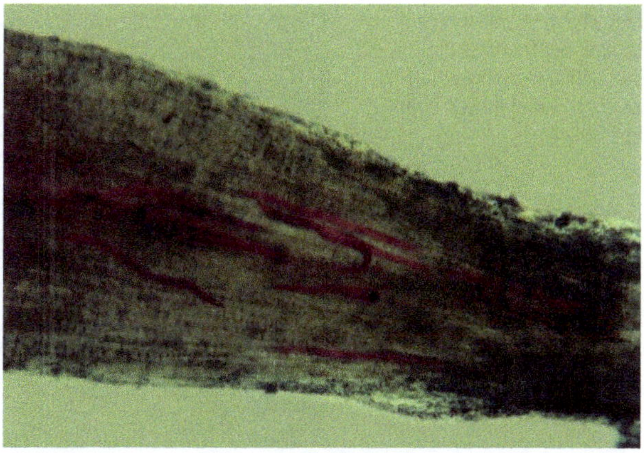

Fig. 10.9 Root-lesion nematodes within the root cortex. (Photo credit: Jon Eisenback)

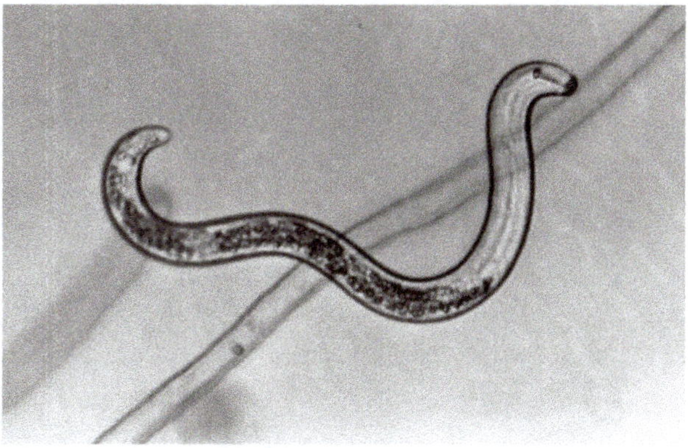

Fig. 10.10 Second-stage juvenile root-lesion nematode feeding on a root hair in a Petri-dish culture. (Photo credit: The Ann MacGuidwin lab [Dan Wixted])

The feeding behavior of RLN can be separated into phases of probing, cell penetration by the stylet, salivation, and food ingestion for brief and extended periods of time, depending on whether the nematode is eating its migration path through roots or feeding in place.

Symptoms and Disease Incidence

Symptoms of RLN damage are reddish to dark-brown lesions in the root cortex (Fig. 10.11). Lesions coalesce, turn black, and are often invaded by soil microorganisms, which can cause weakened root systems, reduced water and nutrient uptake, loss of plant vigor, and ultimately yield reduction. High population densities of RLN can cause stunting of potatoes that is visible before flowering as a slight lag in canopy closure. The distribution of RLN in the field is patchy. Shallow lesions can occur on tubers.

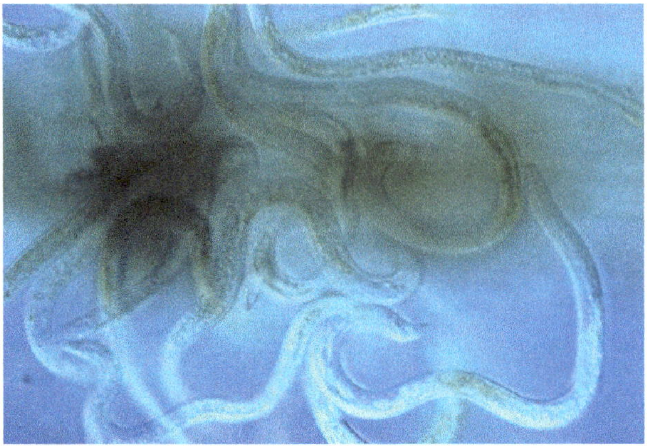

Fig. 10.11 Root-lesion nematodes feeding on a root cause browning and lesions. (Photo credit: Jon Eisenback)

Fig. 10.12 Symptoms of the potato early dying disease in field microplots inoculated with the root-lesion nematode *P. penetrans* and the fungus *V. dahliae*. (Photo credit: Ann MacGuidwin)

In addition to directly causing damage to the potato crop, *P. penetrans* and *P. neglectus* interact with *Verticillium dahliae* to cause the PED disease. Premature death and yield reduction of 30% (40–85 cwt/ac) due to PED disease have been documented in Idaho, New York, Wisconsin, Ohio, and Canada. Potato plants with PED disease are characterized by stunted growth, chlorotic foliage, deterioration of roots, premature senescence, and reduced yields (Fig. 10.12). The primary cause of PED disease is the soilborne fungus *V. dahliae*, but the fungus has a synergistic interaction with *P. penetrans*, such that very low population densities of the fungus and nematode cause a high level of disease when present together.

Fig. 10.13 Field plot without (**a**) and with (**b**) a nematicide application. (Photo credit: Saad Hafez)

Observations that nematicides reduce PED disease support the importance of a *Pratylenchus* and *Verticillium* interaction (Fig. 10.13), and formal experiments in field microplots and greenhouses have confirmed the interaction. In one study, up to 50% yield losses of Superior potatoes resulted from initial populations as low as one *P. penetrans* per cubic in of soil and 30 microsclerotia of *V. dahliae* per oz. of soil.

The interaction between the two pathogens is thought to involve modification of host plant physiology. The fact that some species of *Pratylenchus*—and not others—interact with *Verticillium*, even though all can enter and injure the roots while feeding, suggests more than root-wounding is involved. The interaction between *Verticillium* and *P. penetrans* was observed when the pathogens were physically separated in split root studies. A study using immunostaining techniques revealed that *V. dahliae* primarily infects root tips and is not associated with feeding sites or wounds caused by either *P. penetrans* or *P. crenatus*. Increased vascular colonization by *V. dahliae* of both roots and stems was observed when *P. penetrans* was present, suggesting that *P. penetrans* allows *V. dahliae* to escape or outgrow the defense responses of potato plants.

Management Strategies

Managing RLN should be a consideration throughout the entire crop rotation. All crops are hosts, but seed treatments are available for some, which can help keep nematode population densities in check. Crops harvested early enough to allow a period of fallow or the planting of select cover crops can help reduce nematode pressure. Green manure crops of oil radish, barley, velvet bean, and buckwheat prior to potatoes resulted in a decline of RKN and CRKN population densities and an increase in tuber yield compared to fallow in Idaho. Maximum yield followed barley, while minimum soil and root population of both nematode species was observed in velvet bean plots. Cover crops had less impact on *P. penetrans* in Canada and Wisconsin, with many being even better hosts than potatoes for nematode reproduction.

Fumigation is the industry standard for reducing RLN because it also is effective for *V. dahliae* and consequently, PED disease. Following the fumigant label for temperature and soil moisture conditions is important for killing RLN, as laboratory studies have shown that uptake of nematicidal compounds is greater when nematodes are active. Unlike RKN and SRN, there is no evidence that RLN migrate down in the soil profile to overwinter. Rather, they tend to shelter in roots so tillage to break and expose roots prior to fumigation can increase nematode mortality.

Nematicides can be helpful in managing high population densities of RLN and mitigating yield loss. Growers that fumigate should sample for nematodes prior to planting to see if additional means of control are needed. Research at the University of Idaho showed a systemic nematicide applied at planting significantly reduced RLN populations and increased tuber yield. Years of research and grower experience along the Snake River Plain of Southern Idaho have also demonstrated that nematicide applications often suppress PED disease with an accompanying yield response. Nematicides can prolong the interval between fumigation events once an RLN infestation has been reduced to moderate levels.

Planting a green manure crop before potatoes has produced mixed results. In Idaho, green manure crops of oil radish and barley decreased RLN population densities and increased tuber yield, particularly when combined with a nematicide. Green manure crops and the practice of biofumigation has been less effective in regions where *P. penetrans* predominates. Brassicas, such as mustards, rapeseed, and forage radish are excellent hosts for this species, so population densities increase while the cover crop is growing. Green manure cover crops kill *P. penetrans*, but if the timing or conditions of incorporation are not optimal, a considerable residual population of nematodes can remain in the field—sometimes more nematodes than were present before the cover crop was planted.

Potato-Rot Nematode

The potato-rot nematode, *Ditylenchus destructor*, (PRN) is a regulated pest for the movement of potato tubers. It enters fields when infected potato seed, bulbs, or corms are planted. The PRN can leave fields in the same way so populations wax and wane, often to the point of being undetectable through routine soil sampling. As indicated by its common name, this nematode can spread from tuber to tuber in storage, so detection efforts are focused on surveillance of potatoes during harvest.

Occurrence and Host Range

The PRN has a scattered distribution throughout the U.S. It is a zero-tolerance nematode for the international export of potatoes and in many potato seed certification programs. Potatoes are the primary host, but some fungi and more than 100 plants support reproduction of this pest including iris, tulips, garlic, hops, and snap bean.

Biology and Epidemiology

The PRN is a migratory endoparasite with all life stages, except eggs, moving freely between soil and underground plant organs. *D. destructor* can live its entire life cycle within a potato tuber or underground organs of other hosts. The nematodes enter potato tubers through *lenticles* on the skin near eyes. At first, nematodes exist singly or in small numbers in the tissue just beneath the skin of the tubers, and small white lesions are present during early- and mid-season tuber formation. Once established in a field, PRN can survive on fungi or weed hosts. Detecting PRN in soil is difficult, especially when most nematodes are in the egg stage. Not enough is known about the life cycle to recommend optimal sampling strategies. There are cases where the PRN was considered to be eradicated, only to resurge in the first potato crop following 20 years of weedy fallow.

Symptoms and Disease Incidence

Infection by PRN starts as small, whitish regions in the cortex, detected by removing the peel. Lesions coalesce in severely infected tubers, and the affected tissue darkens gradually through grayish to dark brown color. The tuber skin may remain intact but is papery thin and may crack as a result of stress. The effect of nematodes manifests at harvest or storage when infected tubers rot. Diseased tubers are readily recognized in advanced stages of infection by a dry rot (Fig. 10.14), eventually colonized by fungi. Incubation of infected tissue in water reveals scores of PRN. The distribution of infected tubers in a field is highly patchy. It is common for the infected areas of the field to be free of PRN during the next potato crop, as most nematodes are removed when tubers are harvested. It can take years for the nematodes that remain to build to detectable levels, so any field with a history of PRN should alert diagnostic labs to be vigilant for this pest when testing for nematodes.

Fig. 10.14 Potato infected with the potato-rot nematode, *D. destructor*. (Photo credit: Ann MacGuidwin)

Management Strategies

Difficulty detecting PRN in soil constrains a proactive management approach. Healthy potato seed and the strategy of avoidance are the primary means to avoid infestation with this pest. Fields with a history of PRN should be avoided for potatoes if possible, even if the last potato crop showed no symptoms of the disease. If this is not possible, crop rotation should be lengthened and the potato crop monitored for symptoms, particularly at harvest. Infected tubers tend to be small with evident cracking. Fumigation is recommended for the crop subsequent to detecting PRN.

Pale Potato-Cyst Nematode

The potato-cyst nematode, *Globodera rostochiensis*, (PCN) was described 1882. In 1973, this species was divided into two species: *G. pallida* (Stone) Behrens, the pale (or white) potato-cyst nematode (PPCN) and *G. rostochiensis* (Wollenweber) Behrens, the golden (or yellow) potato-cyst nematode (GPCN). The GPCN was first reported in the U.S. in 1941 on potatoes in Long Island, NY, and the first report of PPCN in the U.S. was in 2006 on potatoes in Idaho. Both species are of regulatory concern, and land infested with these nematodes is quarantined for potato production in the U.S.

Occurrence and Host Range

The PPCN was detected in northern Bingham County in Idaho in 2006 during routine sampling from an area known for fresh market and processed potatoes. The field is not within an Idaho Seed Potato Crop Management area.

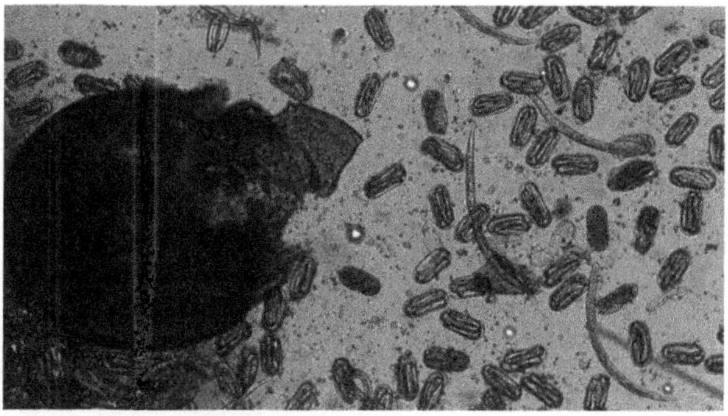

Fig. 10.15 Broken cyst and eggs of the pale potato-cyst nematode, *G. pallida*. (Photo credit: Louise-Marie Dandurand)

The PPCN, *G. pallida*, is a major pest of potato crops in cool-temperate areas. Its host range is restricted primarily to plants within the potato family including tomatoes, eggplant, and some weeds. The PPCN is widely distributed in many potato-growing regions throughout the world. In North America, the nematode is also known to be present in Canada. *G. pallida* is considered as one of the most specialized, successful plant parasitic nematodes because of its high reproductive potential and ability to tolerate adverse environmental conditions. It spreads in soil inadvertently carried by animals, plants, and humans and becomes invasive under the right conditions.

Biology and Epidemiology

The life cycle of both potato cyst nematodes is similar to RKN except that after death the body of female nematodes hardens into a cyst full of living eggs that never left the mother's body (Fig. 10.15). The eggs can remain unhatched for many years, protected not only by the eggshell but also by the durable cyst wall. Cysts can be transported in water runoff, blowing soil, or soil clinging to animals or machines. Cysts can pass through the digestive system of birds, with no harm to the eggs inside. It can take several years or more after introduction of the pest into a field before populations are high enough to cause noticeable crop injury and yield reduction.

In general, a single cyst contains up to 400 eggs, each of which contains a fully formed juvenile (Fig. 10.15). Under unfavorable conditions, when suitable hosts are absent they become dormant. During that time, potato cyst nematodes reduce their metabolic rate to a minimum and remain viable for many years. When potatoes or another suitable host plant is planted, eggs respond to root exudates and hatch. The rate of egg hatch is influenced by soil type and temperature.

Symptoms and Disease Incidence

Infestations by PPCN may display as patches of poor growth. Affected potato plants may exhibit yellowing, wilting, or death of foliage—none of which has been observed in Idaho potato fields. Severity of these symptoms is more visible when

Fig. 10.16 Pale potato-cyst nematode, *G. pallida*, on potato tubers. (Photo credit: Florida Division of Plant Industry, Florida Department of Agriculture and Consumer Services, Bugwood.org)

infestation levels are already high. Even without visible above-ground symptoms, PPCN can cause significant yield loss. Many factors influence the damage and symptoms caused by the PPCN, including site characteristics, cultivars, and weather. The nematodes themselves are visible on potato roots and tubers when they reach the adult female stage. The females break through the root surface and appear as pinhead-size spheres with a pearly sheen (Fig. 10.16).

Management Strategies

Since both species of PCN are alien to the U.S., the most important step in management is to have fields tested periodically for nematodes and be vigilant for the presence of cysts on potato roots and tubers. If an infestation of either PPCN or GPCN is confirmed, the primary focus is to contain the spread and reduce the population density of nematodes as quickly as possible. A variety of measures should be employed including fumigation, leaving the field free of potatoes for at least 6 years, using only certified seed, and planting resistant or partially resistant cultivars when potato production is resumed. Repeated use of the same resistant cultivar should be avoided, as overuse of resistant cultivars in England and Wales has allowed for the buildup of PCN.

General Strategies for Managing Nematodes

Green Manures

Oilseed rape (*Brassica napus*), mustard (*B. juncea*), and radish (*Raphanus sativus*) have shown excellent results for (CRKN) control when used as a green manure crop (Fig. 10.17), as well as sorghum-sudangrass (*Sorghum bicolor*). Some varieties of

Fig. 10.17 Mustard crop during the early growth stage. The crop was chopped and incorporated as a green manure to reduce population densities of the Columbia root-knot nematode, *M. chitwoodi*. (Photo credit: Saad Hafez)

these crops are poor hosts for CRKN and deplete the soil reservoir of nematodes because nematodes infect but do not reproduce, essentially trapping nematodes in roots. Many varieties release chemicals harmful to nematodes as they decompose, and all encourage a high level of microbial activity after incorporation that can be antagonistic to plant parasitic nematodes. Other benefits of using green manure trap crops include:

- Increased yields of subsequent potato crops
- Improved soil tilth and water-holding capacity
- Reduced nitrogen leaching into groundwater
- Weed suppression
- Reduced soil erosion by wind and water
- Potential suppression of seedling diseases

To effectively reduce CRKN populations, green manure crops require at least 8 wks growth and can be planted either in early spring or late summer or after grain harvest between the last week in July and the last week in August for the PNW.

Green manure crops should be managed as carefully as main crops to obtain maximum benefits. The following factors should be considered: (1) planting date; (2) field preparation; (3) fertilization; (4) green manure variety, seeding rate, and sowing options; (5) irrigation; (6) weed control; and (7) incorporation of crop residues.

Planting Date (Southern Idaho)

To achieve optimum results, green manure trap crops should be planted as early as possible either in spring (first 2 weeks of March) or late summer (last week of July through the second week of August). An 8–10 week growing period with soil temperatures above 60 °F is critical for effective nematode control. The length of the necessary growing period may be shortened in the near future, as new varieties that mature earlier are released. Crops in the Brassica (Brassicaceae) family grow best in cool conditions, so they may be slow to establish when planted midseason when temperatures are warm.

Field Preparation

To obtain optimum results, fields with prior crop residue should be prepared by removing the straw (baling, burning residue, or chopping) or incorporating the residue into the soil immediately after harvest. If time permits, irrigation can be used to germinate volunteer grain and weeds that can be disked. Controlling volunteers and weeds will also help achieve optimal nematode control. Soil should be loosened deep enough to allow dense root penetration and optimal aeration for egg hatching. This can be achieved by disking two or three times. Standard seedbed preparation is recommended.

Fertilization

Better nematode control is obtained when a minimum of 50 lbs. of nitrogen per acre is applied to the green manure. This rate is recommended since nitrogen aids in the decomposition of straw and enhances green manure establishment.

Green Manure Variety, Seeding Rate, and Sowing Strategy

Choosing the appropriate variety of green manure is important because the level of resistance varies among different varieties. The recommended seeding rate is 10–25 lbs./ac. A dense seeding rate reduces competition with weeds, but can also inhibit growth of the crop. Sowing options include using a grain drill and packer or a fertilizer spreader. If the seed is mixed with fertilizer in a fan spreader truck, a light harrowing is necessary to cover the seed.

Irrigation

Adequate soil moisture is important for green manure crop root establishment. Good irrigation practices will also help to maintain aeration in the soil and enhance germination. The number of irrigation events depends on the soil type, profile, and weather, but a minimum of four irrigation events is usually recommended.

Weed Control

Many commonly occurring weeds are hosts for plant parasitic nematodes and can provide additional host species for reproduction during the cropping season. These same weeds act as hosts during seasons when potatoes are not the current rotation crop, ensuring that parasitic nematodes will be available to infect future crops of potatoes. Appropriate weed control measures are recommended. See Chap. 12.

Incorporation of Crop Residues

To prevent seed production, the green manure crop should be incorporated at flowering (best for the chemical effect) or pod formation stage. Care should be taken to prevent soil moisture loss during this period, as the reactions responsible for chemicals toxic to nematodes proceed faster when fields are at field capacity for moisture. To incorporate, the green manure should first be chopped, turned, and then plowed under to mix the green foliage and roots with the soil. Plowing alone is not recommended because it results in uneven distribution of the green matter. Disking two or three times, plowing, ripping, and harrowing are recommended.

Considerations

Factors that diminish the efficacy of green manures include conditions that contribute to poor stand or growth, such as poor seedbed preparation, under-fertilization, and weeds. It is also important to consider unintended consequences of this tactic, the most serious being the buildup of other nematode pests, such as RLN or SRN.

Chemical Control

Two classes of chemical nematicides are available for the management of nematodes in potato fields: (1) nonfumigants and (2) fumigants.

Nonfumigants are nonvolatile nematicides that kill the nematodes either by contact action (e.g., Mocap®, Velum®, Majestine®) or systemic action (e.g., Vydate®, Movento®). New products are entering the market every year.

Fumigants are volatile chemicals, which under typical field conditions, exist as gases or are converted into gases. They are distributed through the soil principally as vapors, and their ability to control the pathogens varies depending on their chemical properties. True fumigants are nematicides that are injected into the soil as gases or liquids (e.g., 1,3–D, methyl bromide, and Chloropicrin®). Non-true fumigants, such as metam-sodium (Vapam®), release methylisothiocyanate (MITC) after application in the soil. The metam-sodium products, which are applied as a liquid, are active in the liquid state and finally, as chemical conversion continues, in the gaseous state.

Advantages of Fumigation

Fumigation not only suppresses nematodes but can provide numerous secondary advantages. It can help suppress disease-causing organisms, such as *V. dahliae*, deep-rooted perennial weeds, and soil-inhabiting insect pests. The optimum time for soil fumigation is in early fall. Spring fumigation is possible, but it is not as effective. Soil fumigation is an exacting procedure that must be done properly if satisfactory results are to be obtained.

Factors Influencing the Effectiveness of Soil Fumigation

Major factors that influence the effectiveness of soil fumigation include:

- Movement of fumigants
- Soil preparation
- Soil moisture
- Soil temperature
- Soil depth
- Application rate
- Time of application

Movement of Fumigants

Soil type, soil temperature, soil moisture, organic matter content, pH, and certain soil cations can affect the rate of chemical conversion, distance moved, and the rate of movement depending on the type of fumigant used. Clay soils tend to restrict movement of all types of fumigants. Although clay soils slow the conversion of Telone II® to the gas phase, they actually increase the rate of metam-sodium conversion to MITC, the liquid biocidal state. As clay soil temperature is increased, the rate of conversion of both Telone II® to the gas state and metam-sodium to MITC increases. High organic matter tends to restrict movement and tie up fumigants, sometimes necessitating the use of higher rates. As soil pH increases so, too, does the rate at which metam-sodium converts to MITC, and the presence of copper (Cu), iron (Fe), and manganese (Mn) soil cations can speed the conversion of metam-sodium to MITC. Sandy soils, however, contain large pores that are less likely to be blocked by excess moisture or compaction; but a surface seal, necessary to prevent rapid loss of vapor, is more difficult with such coarse-textured soils.

Soil Preparation

Proper soil preparation is vital to maximize product performance. The soil should be free of clods to facilitate penetration and allow a good soil seal. Plant residues should be chopped and incorporated into the top 4 in of the soil profile, as heavy

Fig. 10.18 Shank chisel-type equipment to fumigate soil while dragging a roller to seal the surface. (Photo credit: Saad Hafez)

residue may interfere with a shank application or tie up the fumigant. The soil should be loosened to aid in the movement of either liquid or gas through the profile of the desired treatment zone. A disk and drag or roller after the injection equipment will effectively seal the surface and prevent the vapor from escaping too rapidly (Fig. 10.18). The surface seal is important to increase concentration of the vapor in the top soil layers. The soil should not be disturbed for at least 3 weeks, or approximately 1 day for every gal of true fumigant applied per ac.

Soil Moisture

Soil moisture is the most important factor for achieving the desired pest control through soil fumigation. Excess moisture acts as a vapor barrier to prevent proper movement. Conversely, soil that is too dry is vulnerable to fumigant escape. Soil is usually at the proper moisture level for fumigation when it barely retains its shape after being squeezed in the palm of the hand. Nematodes are most active at this level of moisture, which increases the uptake of noxious chemicals into their bodies.

Soil Temperature

In general, the most effective gas diffusion of fumigant nematicides occurs at a temperature of 70–80 °F at an 8-in depth. The soil temperature 8 in below the surface should be between 45 and 85 °F. In addition, these conditions help ensure that the target plant parasitic nematode is in a stage susceptible to the fumigant.

In general, high soil temperatures speed gaseous diffusion, thus shortening the exposure of pests to toxic fumes, resulting in poor control and more crop damage. Low soil temperatures increase the retention of the gas, thus prolonging exposure of pests to toxic fumes, resulting in more effective control and less crop damage.

Soil Depth

The depth at which volatile fumigants are applied varies with dosage, temperature, moisture, soil type, pest species to be controlled, and desired depth of control. For minimum dosage rates under optimum soil conditions, application is generally made at a depth of 6–8 in. As the dosage is increased, the depth of application should be increased. For example, Telone II® should be injected at the rate of 12–15 gal per ac at 8 in and 18–25 gal per ac at 12–15 in. The depth of application should be increased if the soil temperature is above 80 °F and soil moisture is considerably below field capacity. Fumigants should be injected at a greater depth for in-row bed applications than they would on flat, even ground. In general, row fumigation does not perform as well as broadcast application.

Application Rate

The rate of chemical application depends on the pest species present and degree of infestation. CRKN infestations and high populations of other nematodes necessitate higher rates, as do nematodes, such as SRN, which are able to go deep into the soil profile and have dormant stages.

Previous crop history should be considered with higher rates applied after a perennial crop (e.g., alfalfa). Soil type is important because lighter soils require lower rates than heavier soils. Also, deep application requires the use of higher rates, and broadcast application methods use higher rates than row applications.

Time of Application

The optimum time for soil fumigation is in the fall. Spring fumigation is possible, but the window of opportunity is short in the spring as there should be a two-week period between fumigation and planting. Field conditions affect the efficacy of fumigation so close attention to temperature, moisture, and field preparation is important.

References

Cobb NA (1917) A new parasitic nema found infesting cotton and potatoes. J Agric Res 11:27–33

Chapter 11
Insect Pests and Their Management

Erik J. Wenninger, Arash Rashed, Silvia I. Rondon, Andrei Alyokhin, and Juan Manuel Alvarez

Contents

Modified from the previous edition of *Potato Production Systems* written by Juan M. Alvarez, Robert L. Stoltz, Craig R. Baird, and Larry E. Sandvol.

E. J. Wenninger (✉)
Department of Entomology, Plant Pathology, and Nematology, Kimberly Research and Extension Center, University of Idaho, Kimberly, ID, USA
e-mail: erikw@uidaho.edu

A. Rashed
Department of Entomology, Plant Pathology, and Nematology, University of Idaho, Moscow, ID, USA
e-mail: arashed@uidaho.edu

S. I. Rondon
Department of Crop and Soil Sciences, Hermiston Agricultural Research and Extension Center, Oregon State University, Hermiston, OR, USA
e-mail: silvia.rondon@oregonstate.edu

A. Alyokhin
School of Biology and Ecology, University of Maine, Orono, ME, USA
e-mail: alyokhin@maine.edu

J. M. Alvarez
FMC, Stine Research Center, Newark, DE, USA
e-mail: juan.alvarez@fmc.com

J. C. Stark et al. (eds.), *Potato Production Systems*,
https://doi.org/10.1007/978-3-030-39157-7_11

Introduction

This chapter presents an integrated pest management (IPM) guide for insect pests that affect potato production with an emphasis on the U.S. The information presented here will help you to identify insect pest infestations and to design an IPM program for managing insect pests in potatoes. The IPM approach combines cultural, biological, and other control methods with field scouting to improve efficacy of chemical controls. Potato growers who use IPM can increase profits while reducing pesticide use. An IPM approach should improve management while reducing production costs, as well as reduce health and environmental risks associated with pesticide use.

General Principles of Insect Management

The recommendations presented here are not to be used to the exclusion of other acceptable practices. Growers are encouraged to adapt these recommendations to their own needs and improve upon them while incorporating their own experience and creativity.

A successful insect integrated pest management (IPM) program relies upon an understanding of insect identification and assessment of damage, as well as the biology, ecology, and management options for each pest. Another key component to developing an IPM system is a thorough understanding of the crop and its entire agroecosystem.

IPM is a system for selection and use of pest control tactics harmoniously coordinated into a management strategy, based on cost-benefit analyses that consider the interests of and impacts on producers, society, and the environment. Below are steps involved in carrying out an IPM program for a given pest. Although this chapter is focused on management of insect pests, these principles are generally applicable to all pestiferous organisms that may be affecting your crop, including weeds, nematodes, and plant pathogens.

Steps of IPM

1. *Identify the Pest*: You can only manage a pest if you know what the pest is; therefore, the ability to identify the important pests (as well as beneficial arthropods) in your crop is essential. Unless proper pest identification is made, time and money may be wasted on ineffective management actions.
2. *Learn the Pest's Biology*: Certain IPM tactics might be more effective at controlling specific stages of the life cycle. For instance, Colorado potato beetle (CPB) larvae are more susceptible to chemical control than the adult form. Understanding the biology of each pest will help you to predict when and where pest problems might occur based on conditions (e.g., weather, crop rotation) that may favor or discourage a pest.
3. *Monitor for Pestiferous and Beneficial Arthropods*: IPM is based on preventative actions, but they are effective only if timed properly. Monitoring programs should begin before pests become a problem so that management tactics may be implemented in a timely manner. Key questions to address during monitoring include whether the pest is present or absent, how it is distributed (e.g., across the entire crop or only along the field edge), whether populations are increasing or decreasing over time, and whether beneficial natural enemies are present, since their presence might preclude the need to take a control action.
4. *Determine an Action threshold*: IPM uses action thresholds as guides to making management decisions. It is impractical or even impossible to eliminate pests completely, and it only makes sense to take a management action when justified economically. A threshold may indicate the point at which the cost of damage due to injury by a pest exceeds the cost of control. Precise thresholds have been worked out for relatively few insect pests. Often local practices and personal experience may factor into determining the appropriate threshold to use. In general, direct pests (i.e., those that cause damage directly to tubers) will have lower thresholds than indirect pests (i.e., those that attack foliage and have only indirect effects on tuber yield or quality).
5. *Choose IPM Tactics*: The ultimate goal in IPM is to suppress pests below injurious levels and to avoid pest outbreaks. IPM tactics should be selected carefully in such a way that they work in concert to achieve this goal with the least-toxic but effective methods available. In general, pesticides are used to supplement a framework of cultural and biological controls and other tactics and/or when these other tactics have failed to achieve acceptable results.
6. *Evaluate Results*: During and following the implementation of any IPM action, it is important to evaluate the effectiveness of that action. Questions to ask include whether the action prevented or managed the pest to your satisfaction, whether the action itself was satisfactory, and what changes might be made to improve results in the future.

Insecticide Resistance

An insect population is considered to exhibit resistance to an insecticide if a considerable portion of the pest population can survive an application of insecticide at rates that once killed most individuals of that population. Resistance is an inheritable trait, and repeated failures of an insecticide can be expected for subsequent generations of the pest. Repeated use of the same insecticide—or different insecticides with the same mode of action—will accelerate the development of resistance. It is important to consider that two of the most important insect pests of potatoes—green peach aphid (GPA) and CPB—are infamous for their ability to develop insecticide resistance, and management of these species is very challenging in many parts of the world.

Although resistant populations may sometimes revert to susceptibility after a failed insecticide is no longer being used, the primary goal of resistance management is preventing insecticide failure from happening in the first place. Several characteristics common in resistant populations may help in achieving this goal. In the absence of insecticides, resistant insects usually have lower reproductive success and/or higher mortality and are generally incapable of successful competition with susceptible insects. Also, the level of resistance in hybrid crosses between resistant and susceptible insects is somewhere in between that of their parents, allowing their successful control with a high label rate of insecticide. However, when such hybrids mate with each other, some of their offspring will be highly resistant. Additional complications arise when resistance to an insecticide with one mode of action confers resistance to another insecticide with a different mode of action, even where the insect has not been previously exposed to the latter insecticide (cross resistance).

Following insecticide resistance management (IRM) practices can slow or even prevent the development of resistance to insecticides, which will maintain the effectiveness of these important management tools. Many insecticide labels now have specific resistance management guidelines. The following are some general principles of IRM:

- *Maintain Thorough and Accurate Records*: Keep separate records for individual fields, documenting insecticide applications and responses of insect populations to those applications. This will help to select proper insecticides and identify resistance problems as they develop.
- *Rotate Modes of Action*: Avoid using the same product repeatedly. From an IRM standpoint, repeatedly using two different products with the same mode of action is no different from repeatedly using the same product. The ideal rotation avoids exposing consecutive generations of a pest to the same mode of action. Information on a product's mode of action can be found on its label, usually on the front page. Insecticides belonging to the same chemical class have the same mode of action and should not be rotated with each other. Since a given insecticide active ingredient is often sold under different brand names, simply switching among brands may not be a successful strategy to change the mode of action. More information about the insecticide mode of action classification can be

found in the webpage of the Insecticide Resistance Action Committee (https://www.irac-online.org/modes-of-action/).

- *Use Labeled Rates*: Lowering application rates below the recommended label rates accelerates the selection pressure of individuals that have some level of resistance, allowing incompletely resistant hybrids of resistant and susceptible insects to survive. When these breed with each other, some of their offspring will be completely resistant and capable of withstanding even high insecticide rates.
- *Apply Only When Necessary*: Applying insecticides to the most susceptible life stage and only when justified by scouting will not only improve efficacy but will save money and reduce selection pressure for resistant individuals.
- *Provide Untreated "Refuges"*: Leaving a portion of the population untreated (e.g., applying only a border spray to insects affecting one edge of the field or not applying a neonicotinoid seed treatment to 100% of your fields) will help to maintain susceptible individuals in the population.
- *Preserve Beneficial Arthropods with Selective Insecticides*: Many newer insecticidal chemistries are safer to both applicators and non-target organisms. Using these "softer" insecticides will help to preserve beneficial organisms that will reduce pest populations—even those populations with insecticide resistance.
- *Use a Combination of Control Tactics*: Insecticide resistance often may result from a simple spontaneous mutation in a single gene. In contrast, simultaneous adaptation to several management techniques (e.g., different insecticide classes, biological control, and crop rotation) requires much more complex changes, which are less likely to happen by chance in an insect population.

Biological Control

There are numerous natural enemies that attack the various arthropod pests of potatoes and, under certain conditions, may limit the need to apply additional control measures. Information regarding biological control that is specific to each pest is provided below, when available. Unfortunately, the role of natural enemies in controlling potato pests is not well understood. Work done on ground beetles in Idaho demonstrated that preserving natural enemies that are already present in potato fields could potentially reduce populations of CPB and aphids. Also, recent efforts to determine the actual diet of beneficial insects through gut content analysis and molecular techniques have started providing us with a better understanding as to what their impact might be. Here, we provide some general points with respect to biological control that are more broadly applicable across potato production systems. There are several approaches that can promote populations of beneficial arthropods within fields, including: limiting and targeting insecticide use, as well as establishing habitat for natural enemies within or adjacent to fields. The latter approach is discussed in detail in Chap. 6. For conventional potato production, the most practical approach to encourage beneficial arthropods is through the judicious use of insecticides.

Regrettably, arthropod natural enemies are highly susceptible to insecticides—often more so than the pests that these insecticides target. Broad-spectrum insecticides, such as pyrethroid and carbamate groups, are particularly toxic to beneficial insects, and applications of these insecticides may be followed by a surge in pest abundance due to severe reduction in natural enemies and the resurgence of secondary pests. Therefore, one of the best ways to conserve natural enemies within a potato field is to apply insecticides only when justified by scouting for pests. Avoiding use of pyrethroid and carbamate insecticides, especially during early and mid-season, will further contribute to conservation of natural enemies. These broad-spectrum insecticides might be used toward the end of the season because, although they will kill both pests and beneficials, the pests may not have enough time to rebound before harvest.

Chemical Control

Any mention of specific chemistries or classes of chemistries is used for informational purposes only. No endorsement of any named products is intended nor is criticism implied of similar products not mentioned. Consult with local Cooperative Extension Service offices and extension entomologists for information on registered insecticides specific to your area.

Box 11.1: Pesticide Warning
ALWAYS read and follow the instructions printed on the pesticide label. *The pesticide recommendations in this publication do not substitute for instructions on the label.* Pesticide laws and labels change frequently and may have changed since this publication was written; therefore, always check current recommendations. Some pesticides may have been withdrawn or had certain uses prohibited. Check local information and use pesticides with care.

- Do not use a pesticide unless the specific plant, animal, or other application site is specifically listed on the label.
- Store pesticides in their original containers and keep them out of the reach of children, pets, and livestock.
- To protect groundwater, when there is a choice of pesticides, the applicator should use the product least likely to leach.

Arthropod Pests of Potatoes

Proper identification of a pest is a critical step in its management. Local Cooperative Extension Service offices and extension entomologists can help with identification if a user of this manual is unable to identify an insect pest using the descriptions and photos presented here. In addition, a short section of useful bibliographic resources is presented in the Further Reading section found in the back matter of this text.

This chapter makes no specific recommendations regarding insecticide compounds, formulations, or rates. Information about insecticides labelled for use in potato may be obtained from web resources (e.g., the current *Pacific Northwest Insect Management Handbook*), university extension personnel, or crop consultants.

Ever-decreasing profit margins, as well as ever-increasing societal and governmental pressures to reduce pesticide use, mandate the need to pursue sustainable, ecologically based, and economically sound practices. Grower practices are the key to this sustainability. This chapter focuses on an IPM approach, providing information on insect identification, action thresholds, management decisions, conservation of natural enemies, and options for avoiding pesticide resistance.

Colorado Potato Beetle

Colorado potato beetle (CPB), *Leptinotarsa decemlineata*, is among the most important insect pests of potatoes in North America. They feed exclusively on solanaceous plants (nightshade family). Foliar feeding by larvae and adults reduces tuber yield and quality. See Fig. 6.8 in Chap. 6. Larvae can consume about four times more leaf mass than do adults; therefore, feeding by larvae is more damaging than feeding by adults. Leaf feeding has the greatest effect on yields if it occurs during tuber formation and while tubers are bulking (i.e., near the period of full bloom); feeding before or after this time is less damaging to yield. Potato plants can tolerate considerable defoliation without yield reduction: 30–40% defoliation from emergence until the appearance of first bloom, 10–60% during bloom, and up to 100% defoliation post-bloom after tuber bulking. Tolerance to defoliation may vary among cultivars.

CPB is a difficult pest to manage because it has a high reproductive potential and has developed resistance to almost all insecticides used against it, at least in certain parts of the world (see section "Insecticide Resistance", above). CPB's ability to develop resistance to so many insecticides may be due to several biochemical detoxification mechanisms, perhaps related to having evolved on solanaceous plants, which present high levels of toxins.

Identification

Adult CPBs are large, conspicuous insects, about 3/8 in long, with yellow rounded and convex wing covers (elytra) marked with ten black stripes (Fig. 11.1). CPB eggs are orange/yellow and laid in clusters of about 25–30 on the underside of leaves (Fig. 11.1). Larvae progress through four instars (stages), ranging in size from 1/8 to 1/2 in long. Early instar larvae are dark red with a black head. Later instar larvae have a black head and a slug-like, humped body; they are soft-skinned, brick red to orange or even pink, with two rows of black spots on each side of their bodies (Fig. 11.2).

Fig. 11.1 Adult Colorado potato beetle laying eggs. (Photo credit: Erik J. Wenninger, University of Idaho)

Fig. 11.2 Colorado potato beetle larvae on defoliated potato stems. Inset: close-up of mature larva. (Photo credits: Erik J. Wenninger, University of Idaho)

Biology

In temperate climates, adult beetles spend the winter buried 4–10 in deep in the soil. Adults emerge in the spring around the time of emergence of the first volunteer potatoes. Beetles begin feeding on volunteer potatoes or weed hosts and then disperse to a new potato (or other host) field. Recently emerged beetles either mate close to overwintering sites or disperse to new potato fields before mating. Beetles may feed for up to a week before laying eggs, depending on temperature. Females may lay 300–800 eggs over a 4–5 week period.

Eggs hatch in 4–9 days, depending on temperature. Young larvae remain close to the egg mass from which they hatched, but begin to disperse on the plant as the nursery leaf is consumed. The larvae develop through four growth stages in as little as 8–10 days. Development may take longer with lower temperatures. The fourth and final growth stage takes three times as long as the other three stages combined.

The mature larvae drop from the plant and burrow into the soil where they pupate and emerge as adults in 1–2 weeks. Adults of this new summer generation emerge and lay eggs within the same field or in adjacent fields. Beetles may exhibit one to three overlapping generations per year, depending on climate and temperature. Each year, individuals from the last adult generation disperse to overwintering sites, generally after mating and laying some eggs. However, low temperatures may cause beetles to forgo mating and egg laying before dispersing to overwintering sites.

Summer adults and overwintering adults are distinguishable by differences in their membranous hind wings. Overwintering adults have a smoky-orange cast to their hind wings with distinctive orange veins, whereas the hind wings of the summer adults are clear.

Management

Cultural Control

Cultural control strategies involve changing the way a crop is grown in order to make it less suitable for the pest and/or to enhance the ability of the crop to withstand attack. Cultural control tactics affect insect pests indirectly via the crop or crop environment; they generally are slower to act and, therefore, rarely are effective as standalone management tools. However, cultural controls are an important foundation of IPM.

- *Crop Rotation*: Crop rotation helps to delay or reduce beetle pressure. By planting new potato fields as far from last year's fields as possible (at least 330 yards through 0.5 mile is preferable), the number of beetles that are able to disperse to the new potato fields may be reduced greatly.
- *Elimination of Alternative Hosts*: Several herbaceous weeds, including those in the same family as potatoes, serve as alternative hosts for CPB (and other potato pests, including aphids). Elimination of these alternative hosts can reduce the presence of CPB in the crop.
- *Other Cultural Control Methods*: There are numerous other cultural controls that may be used to reduce damage from CPB. For example, beetle pressure may be reduced by adjusting planting date, using trap crops, or applying mulches. Though these methods may be effective, many conventional growers may find them to be too labor intensive or costly to implement. They are discussed in more detail in Chap. 6.

Physical Control

Some physical control methods (i.e., physical actions used directly against a pest) have proven effective against CPB in certain growing areas. These include the use of plastic-lined trench traps, propane flamers, and vacuums. Another method is to manipulate the habitat that beetles use for diapause in order to enhance overwintering mortality. Similar to certain cultural controls, these methods may be considered by conventional growers to be too labor intensive or costly relative to using conventional insecticides. These methods are discussed in more detail in Chap. 6.

Host Plant Resistance

Despite a long history of efforts to develop potato varieties resistant to CPB, few commercial varieties exist with both true resistance to CPB and acceptable yields. Commercially available varieties may differ in their susceptibility to defoliation by beetles; however, use of less susceptible varieties could not be used as a standalone management tactic for CPB.

Biological Control

CPB can exhibit high reproductive output, which makes reliance on natural enemies alone untenable; however, natural enemies still provide an important contribution to overall management of CPB.

A number of insects and other arthropods that are generalist predators may feed on CPB eggs and/or larvae. These include lady beetles, ground beetles, predatory stink bugs, harvestmen (daddy longlegs), and spiders. Adults of at least one species of ground beetle feed on CPB eggs and larvae. There is a species of wasp that parasitizes CPB eggs and may reduce beetle pressure considerably; however, this wasp cannot overwinter in temperate climates. There are also two species of tachinid flies that parasitize beetle larvae and adults. Beneficial insects are highly susceptible to insecticides, so the best way to preserve them is to use insecticides only when necessary to reduce pest incidence.

There is a species of pathogenic fungus, *Beauveria bassiana*, which attacks CPB (and many other insect species). It is available commercially in several different formulations that may be used as a foliar insecticide spray. It may reduce beetle populations by up to 75%.

Chemical Control

Selecting an Insecticide

Numerous insecticides are registered for use against CPB in potatoes, with registrations varying by state. Consult your local extension office to find out which insecticides are labelled for your area. In most cases, the choice of an insecticide is based on cost, effectiveness, and ease of application.

Which Pest Stage to Target

The efficacy of a foliar insecticide spray may depend, in part, upon which life stages are present in the field at the time of application. The larval stages, especially the earlier instars, tend to be the most susceptible to insecticides.

When to Apply

Insecticides should be used only when justified by scouting. As mentioned above, potato plants can tolerate some defoliation without affecting yield. In areas where CPB are a problem every year, a systemic insecticide applied at planting may be the best insecticide option. Depending on the insecticide class, and depending on the geographical location and history in controlling CPB, systemic insecticides applied at planting may provide up to 90 days of residual control. This should provide protection for much, if not all, of the first generation of beetles. Generally, if the first generation is well managed, additional controls are not needed for the second generation. However, it is important to scout fields for CPB feeding damage to detect the time at which the systemic insecticide is no longer effective against this pest and to determine whether foliar sprays are justified.

If the systemic insecticide is no longer protecting the crop when tubers are beginning to bulk (within 2 weeks of flowering), a foliar insecticide application is necessary. Where damaging infestations of CPB are not an annual occurrence, the best option for control is to scout the crop regularly after emergence and apply foliar insecticides as needed. Rotation among insecticides with different modes of action is an important component of IRM (see section "Insecticide Resistance", above).

No formal economic thresholds exist across growing areas for CPB, but defoliation by developing larvae of above 10–15% when tubers are beginning to bulk may warrant spraying. When a foliar insecticide application is needed, sprays against the first generation should be applied when many (about 80%) of the larvae have hatched, but few, if any, have completed larval development and pupated. This timing maximizes the effect on the susceptible larval stage. Most foliar sprays will have little or no effect on eggs or on pupae, which are well protected in the soil and will contribute to the second generation if allowed to reach this stage.

Where to Spray

Because infestations of CPB may be patchy, frequently occurring along the edge of part of the field, it may not be necessary to treat an entire field to achieve satisfactory control. Spot treatments of areas with high beetle activity are often sufficient. Similarly, an in-furrow border spray may be effective in interrupting field colonization by overwintered adults, particularly on rotated fields.

Aphids

Aphids feed by sucking sap from their host plants. Although their feeding alone may cause damage to potatoes by excessive removal of sap, they rarely become abundant enough to cause economic damage by direct feeding. The primary concern regarding aphids in potatoes is their ability to transmit viruses. Four species of aphids are known to colonize potatoes: *Myzus persicae* (green peach aphid, often abbreviated as GPA), *Macrosiphum euphorbiae* (potato aphid), *Aphis nasturtii* (buckthorn aphid), and *Aulacorthum solani* (foxglove aphid). Of those, GPA and potato aphid are the species that are most widespread and most commonly found in potato fields. However, there are also many different species of aphids that may transiently visit potato fields from other crops or weeds and spread virus to potatoes. Although most of them are not very efficient virus vectors, populations of some of these non-colonizing species can be extremely abundant. Thus, their importance with respect to epidemiology of the non-persistently transmitted *Potato virus Y* (PVY) generally far exceeds that of colonizing aphid species (see section "*Potato virus Y* (PVY)", below).

GPA is the most common and abundant aphid in North America. It is a European native that occurs throughout the world and feeds on over 875 species of plants; it also can transmit more than 100 viruses to various cultivated crops. GPA is considered to be one of the most difficult insect species to control primarily because of its high reproductive potential, diverse host range, and high propensity to develop resistance to insecticides.

Identification

GPA are small, pear- or teardrop-shaped insects that are wingless in the immature (nymph) stages and may or may not have wings in the adult stage. One distinctive characteristic (relative to other potato-colonizing aphids) of this species is the presence of small, inward-pointing structures located at the base of the antennae called tubercles (Fig. 11.3). Antennal tubercles are similar on both the winged and wingless forms of GPA.

Winged adult GPA are pale or bright green with a dark head and thorax (Fig. 11.4). Irregular dark patches on the abdomen are characteristic, but not unique, to this species. Wingless adults are light yellowish-green to pinkish (Fig. 11.5). Adult GPA are about 0.08–0.11 in long.

The immature forms of GPA are wingless, yellow, pinkish, or pale green and resemble wingless adults (Fig. 11.6). Eggs are laid on primary host trees (see below) and are dark green at first but later turn shiny black.

GPA and the potato aphid (Fig. 11.7) are the most common potato-colonizing species, but the buckthorn aphid (Fig. 11.8) and foxglove aphid (Fig. 11.9) may be abundant in some areas. Important characteristics to distinguish between these two species include: body size and shape, length and shape of the cornicles (tube-like structures on the abdomen), shape of the head and tubercles, and shape of the cauda (tip of the abdomen). The potato aphid is larger (3–4 mm) than the GPA, and has longer legs and cornicles. The cauda of the potato aphid is long and extends beyond the tips of the cornicles. Infestations of GPA typically begin on the leaves of the lower part of the plant, whereas potato aphid infestations may be on the upper parts or scattered over the plant. When exposed to sunlight, potato aphids rapidly move to the opposite side of the leaf, whereas GPAs generally do not move. Potato aphids are green or sometimes pink in color.

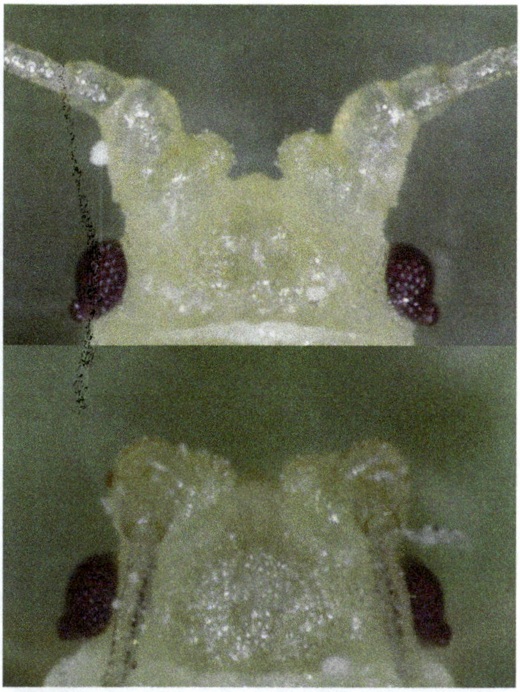

Fig. 11.3 Antennal tubercles of green peach aphid (upper) versus potato aphid (lower). (Photo credits: Erik J. Wenninger, University of Idaho)

Fig. 11.4 Winged adult green peach aphid. (Photo credit: Aaron Buzza, University of Maine)

If an aphid nymph is found in a potato field, it is almost certainly one of the relatively few species that reproduces in potatoes. Transient aphid species may occasionally deposit nymphs in potatoes, but the nymphs will not survive for long.

Not all winged aphids found in potato fields are potato-colonizing species. Many species of winged aphids that have developed on other crops or weeds may be present in potatoes. As mentioned above, these species are important because of their ability to transmit virus to potatoes, and the potential of some species to be far more abundant than the potato-colonizing species. There are dozens, if not hundreds, of such transient aphid species that may be found in potatoes; therefore, it is beyond

Fig. 11.5 Wingless adult green peach aphid. (Photo credit: Aaron Buzza, University of Maine)

Fig. 11.6 Green peach aphid nymph. (Photo credit: Aaron Buzza, University of Maine)

the scope of this chapter to present a comprehensive identification guide. However, it is worth mentioning that growers should expect movement of winged aphids from surrounding crops, including cereals, alfalfa, and soybean, often as those crops mature or are harvested. Controlling these aphids is critical to reduce transmission of PVY, especially for seed potato growers.

Biology of the GPA (Fig. 11.10)

The life cycle of aphids, in general, is unusual and complicated and includes several body forms, as well as both sexual and asexual modes of reproduction. Asexual reproduction occurs on secondary host plants during the growing season when

Fig. 11.7 (**a**) Winged adult potato aphid, (**b**) wingless adult potato aphid (n), and (**c**) potato aphid nymph. (Photo credits: Aaron Buzza, University of Maine)

females give "live" birth to female nymphs (i.e., eggs are not laid). Some aphid species may exhibit upwards of 25 overlapping generations per season.

During late summer to autumn, winged adult aphids develop. They fly to their primary/winter hosts where sexual reproduction occurs. Females then lay eggs on these primary hosts; the egg stage is the typical overwintering stage for many aphid species. However, in mild climates or during mild winters in temperate climates, GPA also may overwinter as nymphs or adults on weeds, including mallow, filaree, and several mustards. "Stem mothers" hatch from eggs that have overwintered on primary host plants. Eggs hatch in response to warm temperatures (which in some years may occur during winter months). Females hatch from these eggs, develop to the adult stage, and produce live nymphs without mating. Winged "spring migrants" later develop and fly to secondary hosts, including crops, weeds, or ornamental plants. In the case of GPA, the primary hosts are plants in the genus *Prunus*, including peach, plum, and cherry; secondary hosts include potatoes and hundreds of other species of plants. Potato aphids use rose bushes as their primary hosts and potatoes, tomatoes, ground cherry, nightshades, and many others as their secondary hosts.

Fig. 11.8 (**a**) Winged adult buckthorn aphid, (**b**) wingless adult buckthorn aphid, and (**c**) buckthorn aphid nymph. (Photo credits: Aaron Buzza, University of Maine)

Fig. 11.9 Foxglove aphid. (Photo credit: A.S. Jensen)

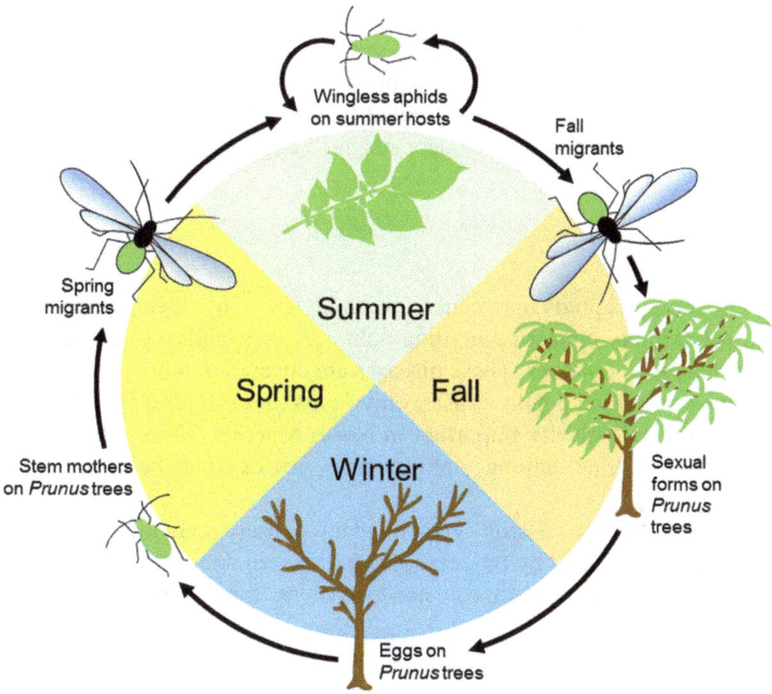

Fig. 11.10 Life cycle of green peach aphid. (Prepared by Erik J. Wenninger, University of Idaho)

Although long-distance, wind-aided flights are possible, most winged migrants likely establish colonies near the primary winter host if secondary summer host plants are present. Winged migrants may fly to multiple plants and deposit nymphs among them. When a summer host begins to mature, another winged form of aphid, known as a "summer migrant," is produced. These winged aphids are the ones that may disperse to potato fields.

Summer migrants of GPA produce nymphs that can complete development in as few as 6 days, which can, in turn, begin producing young 2–3 days later. The optimal temperature range for reproduction is about 75–80°F. Reproduction rates are lower below this optimal range and decline sharply at temperatures above 90°F. It is important to note that within the potato canopy where the aphids are reproducing, temperatures are less extreme than in the ambient air. In general, weather conditions that favor high tuber yields also favor aphid reproduction.

Each GPA female is capable of producing 30–80 nymphs over a period of 10–20 days. When population peaks result in extreme crowding, winged aphids (also summer migrants) develop and disperse throughout the field or to other fields, which may not only diffuse aphid numbers but also increase viral transmission to new plant hosts if the aphids are viruliferous. Declines in populations may be associated with periods of extremely high temperatures and with senescence of potato vines or other declines in potato vine condition. Aphids can migrate from fields planted to early maturing varieties, such as Shepody or Russet Norkotah, to late-maturing potatoes (e.g., Russet Burbank) when vines of the early maturing varieties are killed.

Cold weather in the fall and/or the lack of suitable hosts trigger the production of winged aphids ("fall migrants") that fly to the primary host. Fall GPA migrants may be female or male. Female migrants deposit nymphs on leaves of *Prunus* trees; these nymphs develop into wingless, sexual females. Male migrants mate with the wingless females. Each female then deposits 5–15 eggs on or near the axillary buds.

Aphids and Viruses

If abundant enough, aphids may cause economic losses by direct feeding damage; however, the primary concern regarding aphids is their ability to transmit viruses that cause disease in potatoes. These diseases are covered in more detail in Chap. 9. Of the aphid-transmitted potato viruses, only *Potato leafroll virus* (PLRV) and PVY are considered economically important in North America. Here, we discuss some important relationships among aphids and viruses that bear upon disease management.

Classification of transmission types of plant viruses, including those affecting potatoes, is mainly defined by the relationship between aphids and the viruses they vector based on retention times: non-persistent transmission and persistent transmission.

Non-persistent Transmission

Viruses transmitted in this way are located in the surface layer (epidermal tissues) of plant leaves. The mouthparts of the aphid (stylet) are contaminated with virus particles in the brief process of probing or feeding on the surface layer of leaves of infected plants. These viruses can then be transmitted within a few seconds when feeding on other plants. The virus particles remain on the aphid for a short time (generally less than 2 h), and the aphid must again feed on an infected plant to reacquire the virus. All aphid-borne potato viruses, except PLRV, are non-persistently transmitted. In North America, non-persistently transmitted viruses affecting potatoes include: PVY, *Potato virus A* (PVA), *Potato virus M* (PVM), *Potato virus S* (PVS), and *Alfalfa mosaic virus*. Among these viruses, PVY is the most economically important in potatoes.

Potato Virus Y (PVY)

PVY represents one of the most serious challenges facing seed potato producers but can be a major problem for commercial producers as well. PVY can severely constrain potato yield and quality, as well as hinder certification of seed potatoes. Management of PVY is complicated by the non-persistent nature of virus transmission, as well as by the ever-changing complex of different strains of the virus.

Because PVY is a non-persistent virus, it can be transmitted to potatoes by more than 50 aphid species, most of which do not normally feed and reproduce on potatoes. Aphids may acquire the virus during very short (<1 min) feeding probes and may transmit the virus to another plant within a similarly short duration with no latent period. In this way, transient aphid species that are not colonizers of potatoes may transmit PVY to potatoes during brief feeding probes before dispersing to more suitable host plants. Although GPA is the most efficient vector of PVY, many of the non-colonizing vectors may overcome their relatively low transmission efficiency by their sheer numbers. It is unlikely that any insecticide would work fast enough to kill an aphid before it can transmit PVY; however, effective insecticides can limit the spread of virus within a field by killing colonizing aphids, of which the GPA is a highly competent virus vector.

PVY can reduce yield, but certain strains of the virus also can reduce quality by causing tuber necrosis. Necrotic strains can produce a tuber defect known as potato tuber necrotic ringspot disease (PTNRD). Affected tubers show roughened rings of red or brown skin and necrosis beneath rings extending into the tuber flesh. For a detailed description of symptoms of PVY and related diseases, see Chap. 9.

Persistent Transmission

Viruses that are transmitted persistently are located in the phloem tissue of plants. Therefore, aphids must feed on the phloem (and not just probe the leaf surface) in order to acquire the virus. Virus acquisition occurs after about 20–30 min of feeding. PLRV is the only known virus in potatoes that is transmitted persistently by aphids. Only aphids that colonize potatoes will transmit a persistently transmitted virus. GPA is the most efficient vector of PLRV.

Potato Leafroll Virus (PLRV)

PLRV was once a major problem in potatoes in North America because of quality losses caused by tuber symptoms known as "net necrosis." Following the widespread adoption of early season use of systemic neonicotinoid insecticides that exhibit effective, long residual activity against aphids, the incidence of PLRV has declined dramatically. However, should the use of neonicotinoid insecticides become limited due to development of resistance (see section "Insecticide Resistance", above) or to federal regulations, this potentially devastating disease almost certainly would return to prominence quickly. For detailed description of PLRV symptoms, see Chap. 9.

Aphids acquire PLRV within about 20–30 min of feeding, and transmission (inoculation) of the virus to an uninfected host takes between 10 min and 4 h. However, after a sufficient virus load is acquired by the aphid, a "latent" period is needed before it can be transmitted to healthy plants. During the latent period, the virus moves within the aphid from the gut, into the blood, into the salivary glands, and finally into the salivary fluid. When the aphid feeds, virus particles move with the salivary fluid into the plant tissue. The time required for the virus to complete this circulative route and become transmissible can range from 8 h to 4 days. That is why some well-targeted insecticide applications can effectively reduce transmission of PLRV. Winged aphids coming into a field that are already carrying the virus can probably transmit PLRV within a few minutes. Once PLRV is acquired by an aphid, it can be transmitted for the duration of that aphid's life. The virus does not pass from the mother to its offspring.

The nature and severity of PLRV symptoms depend on several factors, such as the virus strain, the potato variety, the time and source of infection, and the environmental conditions (see Box 11.3).

Management

Cultural Control

Several cultural control options exist that can have tremendous effects on GPA, but require coordinated efforts among potato growers, fruit tree (*Prunus*) growers, homeowners, and state regulators.

- *Reducing GPA Populations on Primary Hosts*: Fields near commercial peach orchards or urban areas with backyard and abandoned peach trees usually have higher GPA populations than those in isolated areas. Removing unmanaged peach and other *Prunus* trees and spraying insecticides on commercial orchards will reduce the aphid pressure in nearby potato fields. Targeted agricultural oil applications to *Prunus* trees in the spring can also reduce aphid populations. Managing overwintering aphids on the primary hosts would not be feasible for the many non-colonizing aphid species that can transmit virus.
- *Reducing GPA Populations on Nursery Plants*: Some high-elevation seed-growing areas in the Western U.S. are too cold for survival of primary hosts of GPA; however, nursery plants and home gardens can be major sources of both vector and virus. Management of aphids on nursery plants can contribute greatly to reduction of GPA dispersing to potato fields. Aggressive local education campaigns and possibly quarantine strategies to keep infested transplants from being moved into seed areas would be required to cause appreciable reductions in GPA populations on nursery plants.
- *In-Season Management of Weed Hosts*: Early infestations of GPA commonly occur on several weeds, including species of mustards, nightshades, and ground cherries. The dozens of non-colonizing species of aphids also may feed and reproduce on myriad species of weed hosts. Winged forms produced on these weeds later may infest potatoes; high numbers may appear during a short period when one or more species of weeds dry up or mature. Management of nightshades in and around potato fields is especially important given that nightshades are excellent hosts for both GPA and potato viruses.

- *Off-Season Management of Weed Hosts*: Where the winters are mild, aphid colonies can survive the winter on host plants that maintain green growth. Even in areas where minimum temperatures are too severe for survival of certain host plants, some individual plants may remain alive near canals, springs, or even greenhouses and other heated buildings. This source of aphids is less important than the two discussed previously; however, managing these hosts should reduce potential aphid sources.
- *Management of Volunteer Potatoes*: Volunteer potato plants can serve as hosts to both colonizing aphid species as well as aphid-vectored potato viruses. Volunteer potatoes will be more abundant in areas and years with higher winter temperatures. Managing volunteers early in the spring is important in reducing these reservoir hosts.
- *Management of Aphids in Other Crops*: Although GPA is the most efficient vector of viruses in potatoes, many species of non-potato-colonizing aphids may transiently visit potato fields and transmit viruses to potatoes. Any lack of virus transmission efficiency exhibited by these species may be overcome by the sheer numbers of aphids that may disperse from adjacent fields into potatoes. This dispersal may peak when crops (e.g., cereals and alfalfa) mature or are harvested. Management of aphids in adjacent fields would require coordination with the owners of those fields and reconciled with the fact that action thresholds for aphids generally are much higher in these other crops than in potatoes.

Box 11.2: Scouting for Aphids

Winged aphids typically make short flights when dispersing to hosts; therefore, aphid infestations usually are heaviest along field margins and particularly along the margin nearest the aphid source. Potato plants and any nightshades (a preferred host for GPA) near field margins may be inspected for aphids, especially early in the season. However, any scouting should include sampling the entire field.

GPA typically infest the undersides of leaves in the lower portions of potato plants. Fields should be scouted twice per week during the season. Detection of winged aphids and timely application of insecticides will minimize the formation of colonies.

Fields should be checked for aphids at least weekly starting shortly after emergence. The most effective scouting methods include visual observation of plants and the use of beating sheets or beating trays (Fig. 11.11), especially when foliage is upright. Other sampling techniques include the use of yellow buckets filled with water with a small amount of soap (Fig. 11.12) and vacuum sampling using a “D-Vac” or a leaf blower with a vacuum attachment (Fig. 11.13). There are no well-established treatment thresholds for aphids in potatoes.

Fig. 11.11 Sampling for aphids and other foliar arthropods using a beat sheet. (Photo credit: Anastasia Stanzak, University of Idaho)

Fig. 11.12 Yellow bucket trap used for aphid sampling. A hardware cloth screen may be used on the opening of the bucket to inhibit the capture of small vertebrates. (Photo credit: Erik J. Wenninger, University of Idaho)

Biological Control

Many different species of predators, parasitoids, and pathogens affect aphid populations and, together with other natural control factors, may keep aphids below economic levels under certain conditions. The sudden decline of aphid populations that is typically observed late season is associated with several factors. Emergence of fall migrants and their dispersal to primary hosts is an important factor, as is the action of predators. Unfortunately, tolerance for aphid infestation is very low in

Fig. 11.13 Using a vacuum sampling device (i.e., leaf blower with a vacuum attachment) to sample foliar arthropods, including aphids. (Photo credit: Erik J. Wenninger, University of Idaho)

seed potato fields; therefore, natural enemies are usually not sufficient to keep aphid numbers below the level acceptable for commercial growers.

Unfortunately, many insecticides reduce or even eliminate populations of natural enemies, allowing aphid populations to increase rapidly or remain longer into the season. This may explain why high populations of aphids are sometimes observed following an insecticide application. Broad-spectrum insecticides are especially harmful to predators and parasitoids. Using insecticides that conserve beneficial insects will aid in management of aphid populations.

Several fungi in the family Entomophthoraceae are known to cause infections that kill GPA and other aphids. Unfortunately, applications of fungicides against foliar diseases, such as late blight, kill these beneficial fungi as well.

Chemical Control

The GPA is difficult to control because of its high reproductive capacity and because it has developed resistance to dozens of different insecticides representing all major insecticide classes. Most of the principles of chemical control of CPB, explained earlier, apply also to GPA.

Selecting an Insecticide

Systemic insecticides tend to work particularly well against insects that feed on plant sap, including aphids. Some systemic insecticides provide adequate control of aphids, as well as other pests, including CPBs and wireworms.

When to Apply

Aphids can be controlled effectively by applying systemic insecticides to the soil at the time of planting. Where mid- to late-season pressure from winged aphids is light (e.g., at high elevations), these applications may provide season-long protection. In higher-pressure areas, one or more foliar applications of insecticide may be necessary after midseason.

Application of foliar insecticides should begin when one to three wingless aphids per 100 leaves are detected. This is a low threshold for detection with confidence and, therefore, some growers choose to use a no-gap insecticide program (i.e., application of foliar insecticides begins just before the expected time of decline of the residual control of the systemic insecticide applied at planting.)

Insecticide applications generally cannot prevent aphids from transmitting non-persistent viruses like PVY. However, some insecticides can reduce the transmission of the persistent virus PLRV to potato plants by four factors: (1) eliminating aphid vectors before transmission occurs, (2) ceasing or inhibiting aphid feeding behavior, (3) affecting the movement and spread of vectors, and (4) reducing vector reproduction. For both viruses, insecticides can, to some extent, limit within-field spread. Winged aphids landing on treated potatoes may live from hours to days. During this period, aphids can feed and transmit viruses. However, the insecticide can kill any nymphs deposited on treated plants before they reach maturity, thus limiting further spread of virus. Such effects are less important for non-persistently transmitted viruses such as PVY (see section above), and few insecticides have shown efficacy in reducing PVY transmission.

Box 11.3: Factors Affecting Spread of Net Necrosis

1. *Aphid Growth Stage*: Some evidence indicates that efficiency of transmission varies with aphid age or stage of growth. Winged forms are obviously more important in transporting the virus into fields and moving the virus relatively long distances within fields. Wingless forms may move the virus from plant to plant, especially when crowded. Crowded aphids may either fall from an infected plant and relocate to an uninfected plant or move directly from infected to uninfected plants when foliage of adjacent plants becomes intertwined. Crowded colonies produce a high percentage of winged forms that can disperse from infected plants and move the virus short or long distances. Plants within a row next to a PLRV-infected plant have the greatest chance of becoming infected.

2. *Potato Variety*: Some varieties, although susceptible to infection, are not prone to the tuber net necrosis symptoms. Use of resistant varieties is among the most practical methods of controlling losses from PLRV. See Chap. 3.
3. *Age of Potato Plants*: GPA transmit PLRV more efficiently after feeding on a young, infected plant. Therefore, plants are most conducive to PLRV spread just after emergence and decreasingly less through the growing season. Infected plants serve as sources for aphid transmission from about the time the symptoms appear. Plants infected during about the last month of the growing season do not develop foliar symptoms; however, net necrosis readily manifests in tubers of plants that have been infected just before harvest.

   Timing of net necrosis development depends upon the time in the growing season at which plants became infected. Net necrosis symptoms usually do not occur when plants are infected before tuber initiation. If plants are infected after tubers are fully developed, net necrosis may develop after several months of storage. It is important to remember that tuber net necrosis can occur when plants are infected too late in the season to develop foliar symptoms.
4. *Numbers of Source Plants and Aphids*: Any practice that reduces the number of virus source plants will reduce the need for insecticide treatments. Severity of problems with PLRV depends on the relationship between numbers of GPA in the crop and number and distribution of virus-infected plants. PLRV-infected plants may be from infected seed or volunteer plants from a previous potato crop. Weeds can also serve as reservoirs of aphids, viruses, or in some cases reservoirs of both aphids and virus. Research in the Pacific Northwest (PNW) has shown that the presence of hairy nightshade increases GPA abundance and virus incidence in potatoes. Also, hairy nightshade appears to be a more favorable GPA host. Even the most intensive aphid management program may not prevent spread of potato viruses unless measures are also taken to reduce virus source plants. When there are few virus source plants (volunteer potatoes, weeds, or PLRV-infected seed potatoes), aphid numbers can be relatively high without causing significant loss. When the virus is abundant, the aphid population may be small, but still cause severe losses.

Potato Virus Management Recommendations

Adherence to the following steps can help reduce spread of aphid-borne viruses in potatoes.

1. Plant only clean, certified seed.
2. Plant virus-resistant varieties where possible. See Chap. 3.

3. Remove ALL volunteer potatoes. A 120-ac field with 0.05% incidence of virus contains over 900 infected plants.
4. Use a systemic insecticide at planting.
5. Soon after plant emergence, sample for aphids every 3–4 days.
6. Maintain a no-gap insecticide policy for aphid control.
7. Reduce the presence of alternative aphid and virus hosts, especially nightshade weeds.
8. Spray late-season, susceptible varieties at detection of surviving, wingless aphids, avoiding repeated use of the same insecticide mode of action. See section "Insecticide Resistance", above.
9. Control aphids on late-season, susceptible varieties until vine kill, or as close to vine kill as possible while observing pre-harvest intervals of chemicals.
10. Do not extend the growing season past normal vine kill, if possible.
11. Consider controlling aphids in fields with early maturing varieties (before winged forms develop) to prevent dispersal to late-season varieties.

Wireworms

Wireworms, the immature stages of click beetles, are becoming increasingly important pests across North America. Possible reasons include regulatory phase out of insecticides that were widely applied and had long residual activity. Another possible reason that may apply to certain areas is increased rotation with grasses used for the cattle industry. Wireworms feed upon seed pieces during the spring, which can result in weak stands, but more importantly, wireworms burrow into developing tubers causing reduced tuber quality for processing and fresh-pack use. Crop losses from wireworms may be sporadic, but considerable in some areas, and in some years up to 45% of the total potato harvest has been downgraded from a classification of U.S. No. 1 to U.S. No. 2 because of wireworm injury. Also, in the processing industry there is a zero tolerance for the presence of wireworm in the raw product because the insect is classified as foreign material.

There are approximately 885 species of wireworms in the U.S., of which 39 species are known to attack potatoes. Different species may be economically important in different parts of the country. In the Pacific Northwest (PNW), there are three species that are common. The sugar beet wireworm, *Limonius californicus*, and the Pacific Coast wireworm, *L. canus*, often are found in soils that have been under irrigation for three or more years. The Great Basin wireworm, *Selatosomus pruininus*, infests soils that have recently been brought under cultivation or that previously had been under dryland management. In the Northeastern states, economically important species include the corn wireworm complex, *Melanotus* spp., wheat wireworm, *Agriotes mancus*, eastern field wireworm, *Limonius agonus*, tobacco wireworm, *Conodoerus verspertinus*, as well as two species that do not have generally accepted common names, *Hemicrepidius decoloratus* and *Hypnoides abbreviatus*.

Identification

Wireworm larvae (Fig. 11.14) are hard-bodied, slender, cylindrical, shiny, small-legged, yellowish to brown (depending on species) "worms" that feed upon potato seed pieces, underground stems, and daughter tubers. Feeding on seed pieces (Fig. 11.15) and stems opens the plant to rotting organisms, which can result in poor or weak stands of potatoes. Wireworms also burrow into developing tubers. The holes look as if they were made by stabbing the tuber with a nail (Fig. 11.16) and after healing are lined with potato skin (Fig. 11.17). The adults, slender and brown to black (Fig. 11.18), are called click beetles because when turned on their back, they are able to right themselves with a flip into the air that makes an audible click.

Biology

Despite the economic importance of wireworms, their long life cycles and cryptic, soil-dwelling habits have contributed to the currently incomplete understanding of the biology of most species. Although a few species mature from eggs to adults in only 1 year, the life cycle of the most common wireworms in the PNW requires multiple years even under favorable conditions. Under less favorable conditions, development may take five or more years. Wireworms spend the winter in the soil either as partially grown larvae or as newly emerged adults. Adults move up to the soil surface in the spring when soil temperatures approach 50°F or above. These adults require little or no food and cause no economic damage.

Soon after emerging from the soil, the female mates and then burrows back into the soil; she lays eggs in several locations at depths of one to several inches. Infestations often are spotty because oviposition by females is not uniform, mobility of larvae is limited, and some localities are more favorable for larval development than others.

Wireworm larvae cause the most severe feeding damage during their later instars (i.e., when they are relatively large and still actively feeding). Larvae overwinter at depths of 6–24 in. A seasonal pattern of vertical movement in the soil has been

Fig. 11.14 Wireworm larva. (Photo credit: Erik J. Wenninger, University of Idaho)

Fig. 11.15 Wireworm damage to potato seed piece. (Photo credit: Erik J. Wenninger, University of Idaho)

Fig. 11.16 Tubers with feeding damage from wireworms. (Photo credit: Erik J. Wenninger, University of Idaho)

documented in many species of wireworms. During the spring, when soil temperatures exceed roughly 50°F, the larvae move toward the soil surface and begin feeding. When soil surface temperatures approach 80°F or higher later during the summer, wireworms cease feeding and move downward again. In irrigated fields with complete foliage cover, the soil surface likely does not reach this temperature. During the third or fourth season, mature larvae pupate within earthen cells. In 3–4 weeks the pupae become adults; these adults remain in the soil until the following spring. All life stages of wireworms may be present during any growing season.

Management

There are several species of wireworms that feed on potatoes, but knowledge of their individual biology is not complete enough to be able recommend specific management approaches for each. As more information is gathered on the distribution, abundance, relative importance, and biology of each species, more species-specific management strategies likely will be developed.

Fig. 11.17 Sliced tubers showing wireworm feeding damage that has healed. (Photo credit: Erik J. Wenninger, University of Idaho)

Fig. 11.18 A click beetle, the adult stage of a wireworm. (Photo credit: Erik J. Wenninger, University of Idaho)

Wireworm populations are patchily distributed, there are no robust economic thresholds, and sampling wireworms to estimate risk of damage is labor intensive. Therefore, the main approach to assessing the need for control is to look at historical problems with wireworms within each field. If past crops in a field have sustained economic damage, there may be a need to treat. Fields with a previous history of grasses for multiple years (e.g., pasture, cereals, corn, and USDA Conservation Reserve Program (CRP) land) are more likely to have wireworm problems since grasses generally are excellent hosts for the larvae and females tend to lay eggs in such fields.

USDA standards for U.S. No. 1, U.S. Commercial, and U.S. No. 2 potatoes allow only 6% external defects. This includes insect damage, as well as various physiological, mechanical, or pathological issues. If a reasonable allowance is made for defects other than wireworm damage, the limit for wireworm damage may be less than 4%. The need for application of insecticides targeting wireworm should be influenced, in part, by past occurrence of total external defects; however, if there is no field history available, baiting can give a rough indication of the need for control (see Box 11.4). If wireworms are consistently found in baits, control likely is needed.

Chemical Control

"Rescue" treatments that are applied after damage from wireworms is noticed are rarely effective. Pre-plant or at-plant insecticide applications generally work much better. Therefore, the decision to manage wireworms in potatoes using insecticides must be made before planting.

Removal of insecticides with long residual activity in the soil from the market is thought to be one reason for recent increases in wireworm problems. Wireworms can be controlled by fumigation or by using seed treatments or broadcast or band treatments of insecticides. Usually controlling wireworms in one crop of a 2–4 year rotation will reduce wireworm damage in the subsequent crops in that rotation.

Fumigants may be used to control high wireworm populations, but a combination of broadcast and band treatments may be more economical to use, depending on the rest of the pest complex that may be affecting a field. Seed treatment insecticides used to control CPB and aphids can also be effective at reducing wireworm damage, though they may not substantively reduce wireworm populations. Growers need to keep in mind that even the best insecticides will not kill all wireworms, and a small percentage of a large population can still cause economic damage.

Cultural Control

Cultural control (especially judicious crop rotation) is critical to effective wireworm management. This is partly due to the longevity of wireworms in the soil and to the limited options for "rescue" insecticide treatments against wireworms. The following are several cultural control options that can reduce wireworm problems in potatoes.

- *Crop Rotation*: Wireworms have a broad host range and may feed on many other crops and weeds. Female wireworms tend to lay more eggs in fields with grass (e.g., corn, wheat, barley, and USDA CRP land); therefore, potato crops that closely follow grasses in the rotation (especially multiple years of grasses) are more likely to have wireworm problems. Conversely, wireworms do not feed on alfalfa, so potatoes are less likely to be damaged by wireworms following an alfalfa rotation (see Leatherjackets, below). If wireworm populations are high, it may be necessary to avoid planting potatoes in a particular field. Field scouting (see Box 11.4) and consideration of field history are important parts of planning potato rotations.
- *Fallowing*: Because soil dryness can kill many wireworms, fallowing a field will reduce wireworm numbers. The success of this approach depends on the extent to which weeds are controlled and all plant material is removed from the field. Moreover, the control achieved must be weighed against the income lost from missing a crop year.
- *Late Summer Plowing*: Pupal cases of wireworms can be broken up by plowing a dry field during late summer (i.e., after cereal grains have been harvested).

Plowing during other times of the year (when pupae are not present) are less effective at killing wireworms. Any individuals that are not at the pupal stage at the time of plowing are less susceptible to this cultural management approach.

Biological Control

Known natural enemies of wireworms include birds, carabid and staphylinid beetles, entomopathogenic nematodes, and entomopathogenic fungi, such as *Beauveria* sp. and *Metarhizium* sp. Unfortunately, we do not know enough about manipulating these natural enemies to facilitate control other than limiting the use of broad-spectrum insecticides, which adversely affect beneficial arthropods.

Box 11.4: Wireworm Scouting (Baiting and Soil Sampling) (Fig. 11.19)
Detecting wireworm infestations, determining the size of a wireworm population within a field, and predicting whether a crop is at risk to damage are not easy tasks. The techniques used to assess the presence of wireworms in fields include taking soil cores or using baits. Soil sampling requires intensive labor and sampling errors are common, including failing to detect low-density populations that nonetheless will cause damage. Baiting is a useful method of determining if wireworms are present, but it provides only a rough estimate of population size and also is labor intensive. Bait systems are based on the principle that wireworms are attracted by CO_2 and other compounds given off by germinating seeds. Baits must be buried in the ground for several days at least one month before planting to determine if insecticide treatment is required.

An effective bait can be comprised of carrots, corn, or cereal seed (2–3 tablespoons) buried about 3 in below the soil surface. If corn or cereal seeds are used, they should be soaked in water overnight before deploying in the field; addition of water to the hole with seed will encourage germination and attraction. These baits should be placed in several spots throughout each field in such a way as to obtain samples that are representative of the whole field. The bait locations must be clearly marked for easy retrieval later. The more bait locations used, the better the estimate of wireworm incidence.

The bait and surrounding soil can be excavated and inspected for wireworms after several days (7–10 days is preferable because it allows for germination of any seed in baits, which enhances attractiveness). The number of wireworms recovered per bait station can be used to assess risk of damage to the subsequent crop (Table 11.1). Baits are not effective in soils that are very dry, very wet, or too cold. Additionally, if too much organic residue is present in the soil, wireworms may be feeding on this reside and will be less attracted to the bait stations. Covering the soil surface above bait stations with plastic will warm the soil and allow sampling earlier in the season when soil temperatures are otherwise too cool to promote much wireworm movement.

Fig. 11.19 Wireworm scouting using cereal seed. Hole with soaked cereal seed and water, marked with a flag. Insets: germinating cereal is visible in excavated wireworm bait trap and wireworms are found in soil. (Photo credit: Erik J. Wenninger, University of Idaho)

Table 11.1 How to interpret wireworm counts from bait stations

Average number of wireworms per bait station	Risk of economic damage (3% tuber damage)	IPM recommendation
0	Low (less than 10% chance)	Control not needed or continue sampling for greater confidence
0–0.5	Moderate (33% chance)	Continue sampling
0.5–1.0	Less than 50% chance	
1.0–2.0	Probable (more than 50% chance)	
2.0–4.0	High (75–90% chance)	Apply insecticide at planting
more than 4.0	Extreme	Do not plant potatoes

At least 25 bait stations are recommended per 30 ac
Adapted from Bechinski et al. (1994)

Much of the remainder of this chapter is focused on pests that—under most conditions—do not cause significant economic losses and/or are less geographically ubiquitous (in the U.S.) relative to CPBs, aphids, and wireworms. Biotic factors and/or abiotic environmental factors may trigger outbreaks and cause these pests to become serious problems sporadically. Prominent examples of these pests include: potato psyllids, two-spotted spider mites, leafhoppers, and armyworms/cutworms. Brief descriptions and management guides for these pests are presented below.

Potato Psyllid

The potato psyllid (*Bactericera cockerelli*) is a small insect that is related to aphids and leafhoppers that feeds on plants in a similar manner (i.e., by sucking plant sap with piercing/sucking mouthparts). Like aphids, potato psyllids may cause direct damage to plants if they occur in very high abundance. Direct feeding damage can be manifested as chlorosis and upward curling of newly emerged leaflets; this condition is known as "psyllid yellows," which purportedly is caused by salivary toxins that are injected into plant tissue during feeding. However, the primary concern regarding potato psyllids is their ability to transmit a bacterium, "*Candidatus* Liberibacter solanacearum," that causes a disease called zebra chip (ZC) in potatoes and other solanaceous crops.

ZC is a serious potato disease that has resulted in multi-million dollar losses annually in the U.S. since about 2000. Initially found in Mexico during 1994, the disease has since appeared gradually across states in the Central and Western U.S. Infected potatoes produce tubers with striped necrotic patterns that make the tubers unmarketable. Necrotic tissue in affected tubers becomes darker and more pronounced when tubers are cooked, especially as chips or fries. ZC reduces both tuber yield and quality and ultimately may kill plants.

Potato psyllids are small and may be difficult to find in potato fields. Eggs are orange/yellow and football-shaped and laid on short stalks, typically along the edges of leaves (Fig. 11.20). Nymphs go through five instars, and their body color may range from light brown to yellow to green (Fig. 11.21). Adults resemble tiny cicadas and are about 5/64–1/8 in long (Fig. 11.22). Newly emerged adults are pale green or light brown and become darker within a few hours. At least in the Pacific Northwest (PNW) potato fields, they can be distinguished from other psyllids in the adult stage by a combination of characters: a white band on the abdomen; a white margin on the top, front of the head; and clear wings with a two-branched (not three-branched) split in the wing venation toward the base of the wing (Fig. 11.23). Under optimal temperature conditions (approximately 77–86°F), potato psyllids may com-

Fig. 11.20 Potato psyllid eggs are laid on short stalks, typically along leaf margins. (Photo credit: Erik J. Wenninger, University of Idaho)

Fig. 11.21 Late-instar potato psyllid nymph and two first-instar nymphs. (Photo credit: Erik J. Wenninger, University of Idaho)

Fig. 11.22 Adult potato psyllid. (Photo credit: Erik J. Wenninger, University of Idaho)

plete one generation in 2–3 weeks. Thus, many overlapping generations may occur over the growing season.

The host range of the potato psyllid is comprised of plants in at least 20 different families, although plants in the Solanaceae (nightshade family) and Convolvulaceae (bindweed family) appear to be among the most important hosts. Once thought to overwinter only in Mexico and the Southwestern U.S. and to occur in central and northern states only via annual migration, potato psyllids are now known to overwinter on at least a few weeds (including bittersweet nightshade and matrimony vine) in lower elevation areas in the PNW. The relative importance of overwintering versus migrating populations of potato psyllids in the epidemiology of ZC remains to be clarified.

Adult potato psyllids can be monitored using yellow sticky cards, which may detect dispersing psyllids, or with sweep nets; however, presence of adults does not necessarily indicate that psyllids have begun colonizing and reproducing within a field. Plants may be inspected directly for the presence of eggs and nymphs. Eggs and nymphs will be most abundant in the top one-third of the plant. Scouting efforts should focus first along field edges, where infestations tend to begin.

Fig. 11.23 Clear forewing of a potato psyllid with three-branched wing venation toward base of wing. (Photo credit: Erik J. Wenninger, University of Idaho)

Fig. 11.24 Adult two-spotted spider mite. (Photo credit: Erik J. Wenninger, University of Idaho)

Management of ZC in potatoes currently relies heavily upon insecticides to suppress potato psyllid numbers. Resistant cultivars and cultural control options for potato psyllids and ZC are lacking. Most of the natural enemies that attack aphids also will attack potato psyllids; however, other than judicious use of insecticides, including reducing the use of broad-spectrum groups, few options exist to encourage these beneficial insects. There is evidence that pyrethroid and carbamate insecticides flare psyllid populations; the mechanism may be via reduction of natural enemies and/or promotion of egg laying by females before they die. Several other insecticide options are available for potato psyllids.

Two-Spotted Spider Mite

Two-spotted spider mites (*Tetranychus urticae*) are tiny arthropods that are more closely related to spiders than insects. Adult mites are <1/16 in long and yellowish in color with a dark spot on each side of the body (Fig. 11.24). Nymphs are similar in appearance but smaller relative to the adults, and with only six legs rather than the eight legs of adults (Fig. 11.25). Eggs are clear white spheres and are laid on leaf surfaces and other feeding areas (Fig. 11.25). Spider mites overwinter as adults in the soil or in debris in fields or along field edges. Females lay eggs on plant foliage during late spring. A female can lay 20 eggs per day and up to roughly 300 eggs during her lifetime. The female covers eggs and leaves with webbing (Fig. 11.26), which protects offspring from rain and predators. In severe infestations the leaves are adhered together with webbing. When populations become severely crowded, mites climb to the top of a plant or other vertical structure, secrete a web strand, and "balloon" or "parachute" to a new location.

Hot and dry weather is ideal for two-spotted spider mites. Infested fields are frequently located downwind from mite-infested corn, alfalfa, clover, or bean fields. Fields are also more likely to be affected by two-spotted spider mites when they are situated along dusty roads. Dusty deposits on foliage appear to be detrimental to predatory mites, which at least partly explains why such conditions favor spider mites. Spider mite infestations also appear to be more likely when plants are stressed, especially from drought. In arid growing regions, surface-irrigated fields may be particularly prone to spider mite infestations.

Plant Damage

Spider mites damage plants by puncturing the leaf tissue with their mouthparts to feed on cell contents and plant juices. They usually colonize and feed on the undersides of leaves. Injured cells and those surrounding the injury die, reducing chlorophyll content in the leaf. The injury first appears as stippling or small blotches and later turns yellow, gray, or bronze, and then brown (Fig. 11.26). These injury blotches eventually coalesce, causing the leaf to become brittle and brown. In severe infestations, browning can progress rapidly across the field.

Fig. 11.25 Various life stages of two-spotted spider mites, including an adult, several immatures, and several spherical eggs. (Photo credit: Erik J. Wenninger, University of Idaho)

Fig. 11.26 Potato leaf showing feeding damage and webbing from two-spotted spider mites. (Photo credit: Erik J. Wenninger, University of Idaho)

Management

Many predators attack spider mites, but because broad-spectrum insecticides also kill many of these beneficials, mite outbreaks often follow. Some broad-spectrum pesticides, such as pyrethroids and carbamates, tend to flare mite infestations by severely reducing predatory mite and insect populations, and perhaps through stimulating reproductive output of female mites. When chemical control is required, several effective miticides are available. Prolonged use of miticides and insecticides may result in the development of resistance in two-spotted spider mites; therefore, miticides and insecticides should be applied only when necessary.

Given that dry, dusty conditions are favorable to two-spotted spider mite infestations, sprinkler irrigation can help limit populations by washing the foliage, breaking webs, and dislodging mites.

Leafhoppers

Adult leafhoppers are wedge-shaped, generally about 1/8 in long, and yellowish to green in color, with wings held roof-like above their back (Fig. 11.27). As suggested by the name, adults tend to move by short, "hopping" flights. Nymphs are similar to adults in shape and color but have not yet developed wings. Like aphids and psyllids, leafhoppers possess piercing-sucking mouthparts that they use to penetrate plant tissue and feed on phloem sap. They may also transmit beet curly top virus (BCTV) and certain phytoplasmas. The beet leafhopper, potato leafhopper, and intermountain potato leafhopper are three species that may damage potatoes in the U.S.

The beet leafhopper, (*Circulifer tenellus*; Fig. 11.28), is the most important leafhopper in the Pacific Northwest (PNW) U.S. because it can transmit the beet leafhopper-transmitted virescence agent (BLTVA) phytoplasma, which is the pathogen responsible for the disease known as "purple top." Beet leafhoppers overwinter

Fig. 11.27 Adult *Empoasca* leafhopper. (Photo credit: A.S. Jensen)

Fig. 11.28 Adult beet leafhopper. (Photo credit: Erik J. Wenninger, University of Idaho)

in uncultivated and weedy areas, and may feed on kochia, mustards, plantains, and many other plants. The first-generation adults begin dispersing into cultivated fields during spring to summer, with the timing dependent upon when their winter/spring hosts begin to dry. They may have multiple generations during the season.

The potato leafhopper (*Empoasca fabae*) is an important potato pest in the Northeastern and Midwestern U.S. Adults are about 1/8 in long and green with white markings. Potato leafhoppers are associated with a condition in potatoes known as hopperburn, which may result from feeding by both adults and nymphs and is characterized by brown necrosis along leaflet margins distal to the feeding site (Fig. 11.29).

The intermountain potato leafhopper (*Empoasca filamenta*) is less destructive than its eastern relative, the potato leafhopper. Adults are greenish yellow with lighter-colored legs. Adults overwinter in uncultivated weedy areas surrounding fields. Females lay their eggs during the spring in stem and leaf tissues of a wide range of host plants, including beans and alfalfa. They may produce up to three overlapping generations within a season. This leafhopper does not cause hopperburn to potatoes like the eastern species. Nymphs and adults of the intermountain potato leafhopper feed on the undersurface of potato leaves and cause a speckled or white stippled appearance on the lower leaves.

Plant Damage

All species of leafhoppers are sap feeders and can weaken plants through direct feeding if they occur in great enough abundance. Direct feeding damage may result in leaf curling and growth stunting. In addition, leafhoppers may transmit plant pathogens that cause the following diseases: potato purple top wilt, Witch's broom, and BCTV.

Scouting and Management

Effective scouting methods for leafhoppers include the use of yellow sticky traps and sweep netting, as well as inspection of plants along field edges for feeding damage. Control measures specifically for the intermountain potato leafhopper are rarely warranted. Usually populations are held in check by parasitoids, predators,

Fig. 11.29 "Hopperburn" is a condition associated with potato leafhopper feeding that is characterized by brown necrosis along leaflet margins. (Photo credit: Juan Manuel Alvarez, FMC).

and fungal pathogens. Controlling weed hosts is one of the most effective cultural approaches to reduce populations of intermountain potato leafhoppers. Early in the season, within the first two months of plant emergence, is the best time to manage leafhoppers and associated diseases. Soil applications of systemic insecticides for other pests effectively control the intermountain potato leafhopper. Since hoppers frequently may move into the field from adjacent areas, multiple applications may become necessary for effective control.

Potato Tuberworm

The potato tuberworm or potato tubermoth (PTW), *Phthorimaea operculella*, is among the most important potato pests worldwide. Larval infestation of tubers renders potatoes unmarketable. There is zero tolerance for the presence of PTW larvae in raw processing products because the larvae are classified as foreign material.

PTW has four life stages: egg, larva, pupa, and adult. Adults are small moths (approximately 3/8 in long) with a wingspan of approximately ½ in (Fig. 11.30). They can be easily identified by the presence of dark spots on their forewings (two to three dots on males; and a characteristic "x" pattern on females). Both pairs of wings have fringed edges. Females lay their eggs mainly on foliage, but when foliage is not available, they will lay eggs in the soil, on plant debris, or on exposed tubers. Female moths can crawl through soil cracks or even burrow short distances through loose soil to locate tubers on which to deposit their eggs. Eggs are less than 0.02 in. diameter, spherical, translucent, and range in color from white or yellowish to light brown. Larvae usually are light brown with a characteristic brown head (Fig. 11.31). Mature larvae (approximately 3/8 in long) may be pink or greenish; larvae feed on leaves throughout the canopy, but prefer the upper foliage. PTW mines the leaves, leaving the epidermal areas on the upper and lower leaf surface intact. Mines can exhibit different shapes are not distinct

Fig. 11.30 Adult potato tuberworm. (Photo credit: OSU-Extension Experimental Station Communication (Ketchum))

Fig. 11.31 Potato tuberworm larva on a potato. (Photo credit: OSU-Extension Experimental Station Communication (Ketchum))

for this species. Extensive larval feeding results in necrotic areas on leaves that are not always visible without careful scouting. Larvae that are close to pupation drop from infested foliage to the ground and may burrow into tubers. Exposed tubers are most vulnerable to PTW damage. Ultimately, larvae spin silk cocoons and pupate on the soil surface or in debris under the plant or close to tubers. PTW pupae are smooth and brown and often are enclosed in a covering of fine sediment, which includes soil debris and their own excrement. The pupal stage can survive temperatures below freezing for more than 10 days, making this stage the most resilient to thermal extremes. Moth activity is temperature and climate dependent, and in the U.S. peak populations may occur later in the summer, overlapping with potato harvest.

Plant Damage

Damage caused by PTW can be considerable in the field and may continue during storage, especially under non-refrigerated storage conditions. Though larvae may feed on foliar tissue, damage to the tubers is more economically important. As they feed, larvae form tunnels within the tuber that become filled with frass (excrement) (Fig. 11.32). Foliar damage includes blistered leaves and mined stems; webbing on leaves also may occur, but usually only for severe PTW infestations. PTW prefers foliage over tubers; thus, the critical time for control is close to harvest when foliage is senescing or undergoing vine kill. It is unclear whether mechanical vine kill has any effect on PTW populations. Although PTW prefers potatoes, this insect has been reported to infest other solanaceous plants, including tomatoes, peppers, eggplant, tobacco, and nightshades.

Fig. 11.32 As they feed, potato tuberworm larvae form tunnels within the tuber that become filled with frass (excrement). (Photo credit: OSU-Irrigated Agricultural Entomology Program (Rondon))

Fig. 11.33 Adult potato tuberworm moths can be monitored using pheromone-baited traps like this one; high captures of adults may justify insecticide sprays to protect tubers from larval damage. (Photo credit: OSU-Irrigated Agricultural Entomology Program (Rondon))

Management

Growers in areas potentially impacted by PTW are encouraged to monitor insect numbers using pheromone traps (Fig. 11.33). Pheromones, which are concentrated quantities of the female "scent" that attract the male moths, are commercially available as lures that can be used to bait traps for monitoring. Current recommendations suggest placing at least one trap per potato field, beginning after canopy closure, changing trap liners weekly, and changing lures monthly. Treatment levels have not been established, but the University of California recommends a threshold of 15–20 moths per trap per night; Oregon State University recommends treatments if 8 moths per trap per day are captured. Moth numbers vary greatly among fields and regions; therefore, management recommendations should be based on field-specific information.

Biological Control

Parasitoid wasps, such as *Copidosoma* spp. and *Apanteles* spp., are important in PTW control in other parts of the world. A few parasitoid wasps have been collected in the U.S., but their importance in PTW management is unknown. Also unknown is the role of generalist predators, such as lady beetles, big-eyed bugs, and ground beetles. *Beauveria bassiana* and baculoviruses have also been used with some degree of success, as have nematodes such as *Hexamermis*, *Steinerma*, and *Heterorhabditis*. Fungi, viruses, and nematodes may have special requirements, such as high humidity and low solar incidence, to improve success.

Cultural Control

Cultural methods reported to reduce PTW include the following:

- *Eliminate Cull Piles and Volunteers*: Elimination of cull piles and volunteer potatoes will reduce overwintering populations, which are a source of infestations during the following year.
- *Increase Soil Moisture During and After Vine Kill*: Keeping the soil moist via frequent overhead irrigation applications reduces soil cracking, thereby reducing PTW access to tubers. Further, water-saturated soil may enhance larval mortality via reduction in soil oxygen and also may reduce mobility of larvae, thereby decreasing their ability to find tubers.
- *Minimize the Interval between Vine Kill and Harvest*: Tuber infestation increases when foliage starts to senesce. The longer dead vines remain in the field with tubers unharvested, the greater the likelihood of tuber infestation.
- *Select Less Susceptible Cultivars*: Varietal differences in susceptibility to PTW damage may be due to differential feeding by larvae and/or to adult egg-laying preferences. Varieties that set tubers deeper in the soil also are less likely to be infested. Varieties that have more foliage tend to attract more PTW.
- *Maintain Potato Hills*: Potato hills that more fully cover tubers will better protect tubers from infestation. Sandy soils are especially prone to sloughing of soil, which can expose tubers.

Chemical Control

Many insecticides are registered against PTW. Application of insecticides at and after vine kill reduces PTW populations and damage. In addition to conventional insecticides, essential oils also can provide good control.

Cutworms and Armyworms

Cutworms and armyworms are the larval stage (caterpillars) of a large group of night-flying moths (Fig. 11.34). The adult moths are brown to gray in color and may frequently be found flying around lights. Adults sometimes are referred to as miller moths because the wing scales superficially resemble flour dust when they fall off of the wings and coat clothing or other surfaces.

Cutworm and armyworm larvae generally are smooth bodied, with three pairs of legs on the thorax near the head and five pairs of fleshy leg-like appendages on the abdomen (Fig. 11.35). Some species may reach up to 2 in long as late instars. Cutworms and armyworms spend the winter as larvae or pupae in the soil. Depending on the species, they may have one or multiple generations. There are hundreds of different species, several of which may be pests in potatoes, including the following:

- *Army Cutworm (Euxoa auxiliaris)*: Gray with darker top-lateral and spiracle stripes.
- *Black Cutworm (Agrotis ipsilon)*: Gray to black on the dorsal half of the body, with an overall "greasy" appearance.
- *Pale Western Cutworm (Agrotis orthogonia)*: Pale in color without any conspicuous markings; primarily feeds on stem tissue below the surface.
- *Variegated Cutworm (Peridroma saucia)*: Grayish body with a row of yellow diamond-shaped spots on the back.
- *Spotted Cutworm (Amathes c-nigrum)*: A series of paired black oblique marks on the back that are more pronounced on the last four body segments.
- *Red-Backed Cutworm (Euxoa ochrogaster)*: Gray body with a distinctly reddish back with dark borders.
- *Armyworm (Mythimna unipuncta)*: A narrow, broken, light stripe in the middle of the back and a pale orange, mottled stripe on each side below the row of spiracles, edged with white.
- *Western Yellowstriped Armyworm (Spodoptera praefica)*: Black triangular marks are present on each side of the mid-line, bordered by white stripes (Fig. 11.36).

Fig. 11.34 Pinned specimens of adult (**a**) black cutworm and (**b**) western yellowstriped armyworm. (Photo credit: Erik J. Wenninger, University of Idaho)

Fig. 11.35 A cutworm larva; note the three pairs of "true" legs on the thorax near the head and the five pairs of fleshy leg-like appendages on the abdomen. (Photo credit: Erik J. Wenninger, University of Idaho)

Fig. 11.36 Western yellowstriped armyworm larva. (Photo credit: Jewel Brumley, USDA-ARS)

- *Bertha Armyworm (Mamestra configurata)*: Pale green, then gray, brown, or black as larvae mature; yellowish-orange stripe along each side of body (Fig. 11.37).

With the exception of western yellowstriped armyworm, which feeds during the day, the remaining species listed above feed actively at night. During the day they can be found in the soil or within soil cracks.

Plant Damage

Cutworms and armyworms may feed on plant stems, foliage, or tubers that are exposed through soil cracks. Feeding on younger shoots may result in stems being cut off at or below ground level; foliage also may be stripped off of the stem.

Fig. 11.37 Bertha armyworm larva. (Photo credit: Jewel Brumley, USDA-ARS)

Scouting and Management

Light traps and—for some species—pheromone traps can be used to monitor adult populations. Fall monitoring of adults can be used to help predict spring populations of larvae. If feeding damage is observed on foliage, larval populations may be monitored by overturning soil clumps and plant residues. Most cutworms and armyworms feed actively at night and may be found in these refuges during the day.

Cutworm and armyworm numbers usually are held in check by fungal pathogens and generalist predators commonly found in potato fields. Insecticide applications should be used only for seriously damaging infestations. If chemical control is necessary, numerous effective broadcast granular or foliar-applied insecticides are available. Foliar sprays of contact chemicals are recommended from sunset to sunrise, when larvae are feeding actively.

Some defoliation from cutworms and armyworms can be tolerated. Keeping defoliation under 10–15% generally will prevent yield loss. Weed control within fields and along field edges also aids in reducing cutworm infestations.

Flea Beetles

Flea beetles generally are very small (approximately 1/16 in long) and are so named because they use their large hind legs to jump when disturbed. They often have dark metallic bodies in the adult stage; however, species may differ in body coloration and patterning. Adults feed on foliage, and larvae feed on tubers. Several species are known to cause damage in potatoes: western flea beetle (*Epitrix subcrinita*), potato flea beetle (*E. cucumeris*), tobacco flea beetle (*E. hirtipennis*), and tuber flea beetle (*E. tuberis*; Fig. 11.38). Red headed flea beetle (*Systena frontalis*) also may cause occasional damage, but is less common than the other four species.

Fig. 11.38 Adult tuber flea beetle. (Photo credit: A.S. Jensen)

The appearance and life cycle of the western potato flea beetle is described here, though the other species have similar appearances and habits. The adult western potato flea beetle is dark metallic greenish-black in color. Western potato flea beetles seldom cause severe damage as adults; however the presence of adult leaf feeding can be interpreted as a warning sign for potential later infestation of tubers by larvae, which can burrow as far as 1/4 in under the skin.

Adult activity varies by region, but typically begins during late spring. Adults can survive by feeding on uncultivated hosts (e.g., weeds and/or natural vegetation) until potato plants emerge. Females scatter their eggs in the soil at the base of potato plants, and the eggs hatch in about 10 days. The tiny whitish larvae feed on underground stems, roots, and tubers for 3–4 weeks. One or two generations occur per year. Adult western potato flea beetles overwinter on margins of fields or along ditch banks in protected areas, such as under leaves, grass, or trash.

Plant Damage

Adult feeding of the foliar tissue is characterized as small "buckshot" holes in the leaf tissue (Fig. 11.39). Extensive feeding may result in brown discoloration of foliar tissue and plant death. Injury on tuber surfaces by larvae includes rough textures of winding trails, up to 1/16 in wide and of varying length. Internal tuber injury consists of shallow, narrow, brown feeding tunnels (Fig. 11.40). These tunnels are about 1/32 in. in diameter and up to 1/4 in deep and may occur singly or in groups. Secondary infection by fungi often fills the tunnels.

Larval damage of the tuber flea beetle can be severe because the larval tunnel may be as deep as 1/2 in directly into the tuber. The western potato flea beetle, however, burrows just beneath the skin, rarely penetrating more than 1/4 in. Internal

Fig. 11.39 "Buckshot" holes in potato leaves caused by adult flea beetle feeding. (Photo credit: A.S. Jensen)

Fig. 11.40 Tuber damage caused by tuber flea beetle. (Photo credit: Ken Gray Image Courtesy of Oregon State University)

tuber injury consists of shallow, brown feeding tunnels, often filled by secondary fungal growth.

Deep peeling during processing can help to remove flea beetle injuries. However, some lots of potatoes may be unmarketable when deep burrowing damage by the tuber flea beetle is extensive. In addition, feeding wounds from flea beetles may facilitate entry of air- or waterborne disease organisms.

Management

Soil applications of systemic insecticides and foliar applications of insecticides targeting CPB and GPA generally maintain flea beetle populations below economically injurious levels.

Blister Beetles

Blister beetles (Fig 11.41), also known as meloid beetles, are distinguished from most beetles by their short wing covers (elytra) that do not cover their abdomens. Depending on the species, their body size may range from 1/4 in to nearly 2 in. As an anti-predatory defense mechanism, adult beetles secrete a toxic liquid from their leg joints when disturbed that may cause blisters on the skin.

Five species of blister beetles commonly damage potatoes: the spotted blister beetle (*Epicauta maculata*), the striped blister beetle (*Epicauta vittata*), the ash-gray blister beetle (*Epicauta fabricii*), the Nuttall blister beetle (*Lytta nuttalli*), and the punctured blister beetle (*Epicauta puncticollis*).

Females mate and lay their eggs in the soil. Immature larvae actively search for food, feeding on grasshopper eggs and larval stages of other insects before overwintering. Adults emerge during mid-summer. In the Pacific Northwest (PNW), blister beetles may be more abundant in parts of potato fields that are adjacent to sagebrush steppe where there are more food resources for larvae. There is one generation per year.

Plant Damage

Adult beetles may inflict damage to foliage during summer; however, extensive economic loss is rare since damage is mostly patchy and localized.

Scouting and Management

Field edges should be scouted in years of heavy infestations. The beetles are strong flyers and often fly out of the area before damage is detected. If the beetles remain in the field and continue to defoliate field edges, border sprays with insecticide will

Fig. 11.41 Adult spotted blister beetle. (Photo credit: Erik J. Wenninger, University of Idaho)

help to reduce damage. If defoliation remains below 10–15%, as in most cases, controls are probably not needed.

Grasshoppers

Grasshoppers (Fig. 11.42) typically are pests of potatoes only during years when populations are high enough to result in mass dispersal from uncultivated areas. This more often occurs in Western and Central U.S. growing areas. Usually their presence is not associated with substantial damage, since populations often remain patchy and small.

The predominant species that may be damaging to crops include: the migratory grasshopper (*Melanoplus sanguinipes*), clearwinged grasshopper (*Camnula pellucida*), and the red-legged grasshopper (*Melanoplus femurrubrum*). Other species of grasshoppers may inflict sporadic damage to plants.

Female grasshoppers lay their eggs in inch-long pods just below the soil surface, generally during late summer or fall. Each female can produce multiple pods, each of which may contain several (up to about 75) eggs. Hard and uncultivated grounds are the preferred oviposition sites. Eggs also may be found on the edges of cultivated fields, along ditch banks, and in pastures and hay fields. The eggs hatch from March to June depending upon the regional climate and grasshopper species. The nymphs resemble the adults, but are smaller in size and either lack wings or have wings that do not fully cover their abdomen. Grasshoppers have one generation per year, and the nymphs become mature in summer or early fall.

Fig. 11.42 Adult grasshopper. (Photo credit: Erik J. Wenninger, University of Idaho)

Plant Damage

Both nymph and adult grasshoppers use their chewing mouthparts to feed on foliar tissues. Large populations of grasshoppers can cause heavy defoliation, and feeding may even result in transmission of certain viruses.

Scouting and Management

Grasshopper control programs are recommended only when populations become high and significant defoliation (10–15%) is observed. Spraying along just the field edge where an infestation begins is usually adequate to limit losses. Most common foliar insecticides will control grasshoppers. During outbreak years, area-wide management programs are more effective than field-by-field treatment for grasshoppers. This is because grasshoppers are strong fliers and can disperse great distances.

Leatherjacket (Crane Fly)

The adult stage of the leatherjacket (*Tipula dorsimaculata*), also known as a crane fly, is highly conspicuous because this group of insects superficially resemble very large mosquitoes (Fig. 11.43). Adults are about 1 in. in length with long, fragile legs that may break off when the insect is handled. They neither bite nor transmit pathogens. Mature larvae are grayish-brown, about 1 in long, and have characteristic fleshy anal projections. Leatherjackects are named after the leather-like appearance of their larval skin.

Leatherjackets overwinter in the soil as mature or nearly mature larvae. Adults emerge in the spring and deposit eggs in the vicinity of crop residue or other plant

Fig. 11.43 An adult crane fly on tomato. (Photo credit: Erik J. Wenninger, University of Idaho)

debris. The larvae initially feed on the decomposing plant tissue in the soil but later move to feed on developing potato tubers.

Plant Damage

Larval feeding on tubers leaves round punctures varying from shallow depressions to inch-deep holes. Severe damage to potatoes by leatherjackets is more likely to occur in a rotation that directly follows spring plowing of alfalfa. Damage may be more noticeable in low, moist, and weedy areas in the field.

Management

Management decisions should be made based on individual field history and characteristics. Cultural controls are of primary importance in managing leatherjackets, with avoidance of spring incorporation of alfalfa green manure being paramount. Maintaining effective weed control, as well as effective water management to prevent water-soaked areas, will also help with leatherjacket management. Due to the patchy and unpredictable occurrence of this pest, there are no recommended chemical controls.

Loopers

Loopers are the larval stage of a gray-brown miller moth (Fig. 11.44) that, in the adult stage, are similar in appearance to cutworm moths. The larva is smooth-skinned, greenish, and white-striped, that is about 1.5 in long when mature. When at rest or walking, the larva exhibits a characteristic looped or arched “inchworm” habit. The larvae differ from cutworms in that they have only three pairs of prolegs

Fig. 11.44 Pinned specimens of adult (**a**) cabbage looper and (**b**) alfalfa looper. (Photo credit: Erik J. Wenninger, University of Idaho)

(fleshy, leg-like appendages toward the rear of the abdomen), and the body tapers toward the rear end.

The most common loopers found in potato fields are the cabbage looper (*Trichoplusia ni*; Fig. 11.45) and the alfalfa looper (*Autographa californica*). Loopers go through two or three overlapping generations per year.

Plant Damage

Damage is inflicted at the larval stage in the form of chewed holes on leaves and ragged edges along leaf margins. Plant defoliation usually starts toward the center of the plant. Potatoes are most susceptible to damage by loopers when plants are at full bloom.

Management

Loopers seldom become a serious pest of potatoes, even though they may build up to high numbers. Damage usually occurs just after the vines have gone into senescence. Loopers often are found with cutworms and are blamed for the cutworm damage because the cutworms hide during the day. As long as defoliation remains below 10–15%, control measures are seldom warranted. Foliar sprays applied for CPB usually will control loopers.

Fig. 11.45 Cabbage looper caterpillar. (Photo credit: Jewel Brumley, USDA-ARS)

Lygus *Bugs*

Lygus bugs (including *Lygus hesperus*, *L. elisus*, and *L. lineolaris*) are about 1/4 in long, green to brown, with an obvious yellow triangle shape on the back (Fig. 11.46). Immatures (nymphs) (Fig. 11.47) are smooth, shiny, green insects that are similar in size to aphids, but tend to move rapidly when disturbed. Nymphs resemble adults but do not have fully developed wings. They may have up to four black dots toward the front of their back, plus an additional center black dot (which is a scent gland) toward the rear of the back (Fig. 11.47).

Lygus bugs are generalist plant feeders that are found on most plants and trees. Adults overwinter within debris in fields or in field margins. The insects are strong flyers and move readily from field to field. They may move into potato fields following harvest or maturation of an adjacent crop. Damage is most severe on field margins. Several overlapping generations, each about 6 weeks in length, occur each year.

Plant Damage

Damage is the result of the bugs sucking sap from buds and leaves. As they feed on plant sap, it is thought that they inject a toxin into the tissue which may kill or distort the portion of the plant where feeding occurs (Fig. 11.48). Recently, *Lygus* has been reported to carry the beet leafhopper-transmitted virescence agent (BLTVA) phytoplasma, a pathogen that is known to be transmitted by the beet leafhopper. However, the importance of *Lygus* as a vector of this pathogen remains to be clarified.

Fig. 11.46 Adult *Lygus* bug; note the obvious yellow triangle shape on the insect's back. (Photo credit: Erik J. Wenninger, University of Idaho)

Fig. 11.47 *Lygus* bug nymph; note the four black dots toward the front of the insect's back and the additional center black dot toward the rear of the insect's back. (Photo credit: Erik J. Wenninger, University of Idaho)

Fig. 11.48 Distorted leaf growth in potato from *Lygus* feeding. Inset: close-up of leaf damage. (Photo credits: Josephine Antwi, OSU-Irrigated Agricultural Entomology Program (Rondon))

Management

Due to the insect's sporadic incidence, *Lygus* bug control is rarely needed. Because *Lygus* bugs feed on a wide range of host plants, weed management in fields and around edges should help to limit populations. Damage often is not noticed until *Lygus* bugs have dispersed from the field. In the rare instance that chemical control is needed, many of the common foliar insecticides are effective, and strip spraying of field edges usually is adequate.

Thrips

Thrips are fast moving, small, cigar-shaped, yellow or brown insects (Fig. 11.49) that feed on leaves and flowers. Adults have characteristic wings, each of which consists of a thin rod that is fringed with hairs. The body shape of wingless immature thrips is similar to that of the adult, but lighter in color (Fig. 11.50). Thrips overwinter as adults in plant residue and other refuges and lay eggs in plant tissue during the spring. There may be several overlapping generations within a season. The two most damaging species in potatoes are the western flower thrips (*Frankliniella occidentalis*) and onion thrips (*Thrips tabaci*).

Fig. 11.49 Adult western flower thrips. (Photo credit: Erik J. Wenninger, University of Idaho)

Fig. 11.50 Immature thrips. (Photo credit: Erik J. Wenninger, University of Idaho)

Fig. 11.51 "Silvering" of foliage from thrips feeding damage. (Photo credit: A.S. Jensen)

Plant Damage

During the day, when thrips are most active, they puncture the plant tissue and feed on cell contents and sap causing "silvering" of the leaves, which is similar to damage that is caused by mite feeding (Fig. 11.51). Severely affected leaves will drop, and defoliated plants generally do not recover. Damage is mostly limited to the field edges adjacent to pasture, small grain, or alfalfa fields.

Management

There are no set thresholds for thrips in potatoes; however, frequent monitoring will allow for observation of population increases that may require control measures. When damage occurs, the thrips population usually has declined before the cause is discovered. However, if populations remain in potatoes and halt growth, foliar insecticides can be used for control.

Whiteflies

Adult greenhouse whiteflies (primarily *Trialeurodes vaporariorum*) are minute insects (about 0.1 in) that somewhat resemble a tiny white moth, due to their powdery wax covering and wings that are held over their back at rest (Fig. 11.52). The immature stages (Fig. 11.53) resemble scale insects, as well as the nymphal stages of the potato psyllid; misidentifications may occur by non-experts. Unlike whiteflies, later-stage nymphs of potato psyllids are fringed with a border of hairs. Wing buds are not visible on late-instar whitefly nymphs—a trait that distinguish them from potato psyllid nymphs. Behavioral differences also exist; psyllid nymphs can readily move if disturbed, whereas whitefly nymphs remain immobile. At high densities, especially toward the end of the season, whitefly presence is easily observed when potato plants are disturbed and adults fly about the plant canopy.

Plant Damage

Whiteflies feed on phloem sap and, like aphids, also produce honeydew. Honeydew can attract ants, which may negatively affect effectiveness of natural enemies in controlling the whiteflies.

Fig. 11.52 Adult whitefly. (Photo credit: Erik J. Wenninger, University of Idaho)

Fig. 11.53 Immature whitefly. Note the superficial resemblance to potato psyllid nymphs. (Photo credit: A.S. Jensen)

Management

Whiteflies rarely become a problem in potato crops. There is no economic threshold for whitefly treatment in potatoes.

White Grubs

White grubs are the larval stage of a group of beetles that, in the larval stage, are white or cream-colored, C-shaped with a brown head and three pairs of legs just behind the head (Fig. 11.54). White grubs can reach about 2 in. in length with a somewhat enlarged and transparent abdomen.

The carrot beetle (*Bothynus gibbosus*) and the ten-lined June beetle (*Polyphylla decemlineata*; Fig. 11.55) are the two white grub species that are associated with damage in potatoes. The carrot beetle has one generation per year, whereas the ten-lined June beetle spends 2 years as a grub. Adults of both beetles are not strong flyers. The name June beetle comes from the tendency to observe these beetles during May and June when they feed on leaves of trees at night and are attracted to lights. Both species are more abundant in sandy soils where grass sod or large quantities of organic matter, such as manure, have been plowed into the soil before potatoes are planted.

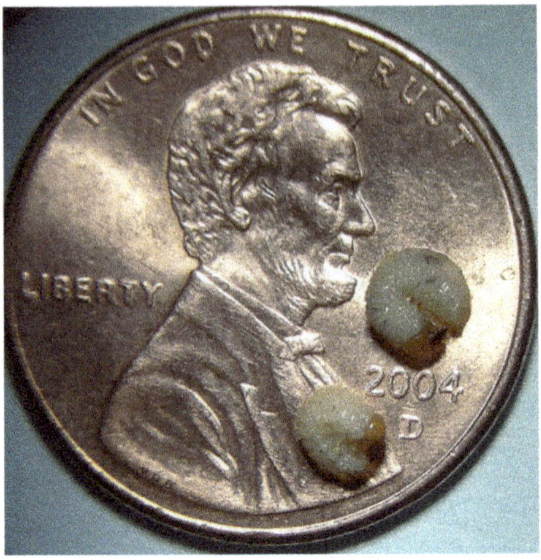

Fig. 11.54 Early instar white grub larvae collected from a potato field. (Photo credit: A.S. Jensen)

Fig. 11.55 Adult ten-lined June beetle. (Photo credit: Kansas Department of Agriculture, Bugwood.org)

Plant Damage

The larvae feed on tubers, causing cavities up to 1-in. in diameter that are rough, irregularly shaped, and wider than deep. In severe cases, more than half of the tuber may be consumed.

Management

Control of white grubs is difficult because they are found in soils with high organic matter, which tends to reduce the efficacy of soil-applied insecticides. Fall or spring tillage may reduce larval numbers by exposure to natural enemies. Insecticides used against wireworms have been somewhat effective in controlling white grubs. Good weed control also may help reduce grub damage.

Garden Symphylan

The garden symphylan (*Scutigerella immaculata*) is an occasional pest that can feed on roots and soil organic matter. However, the garden symphylan can still limit potato production in some areas.

Garden symphylans are not insects but are more primitive, centipede-like animals. The adults are small, white, soil-dwelling arthropods (Fig. 11.56) that are highly active. Symphylans move rapidly away from light; thus, visual detection of their presence must occur immediately after exposing the tubers or soil.

Symphylans lay eggs during spring or early summer in cavities in the soil, and the eggs hatch in 1–3 weeks. Under favorable conditions, a new generation develops within 60 days, and the adults may live for several years. The optimal temperature range for activity of symphylans is between 50° and 70°F. Symphylans readily move within the soil profile to stay within this temperature range.

Plant Damage

Symphylan damage usually is associated with soil containing high organic matter. Feeding on the root hairs and rootlets of potatoes prior to tuber formation may stunt plant growth. Damage to tubers may render them unmarketable; damaged tubers are characterized by the presence of tiny holes in the skin with an undercut cavity lined with hard, dark, corky tissue around each point of injury.

Management

In heavily infested soils, control measures must be thorough. Careful attention must be paid to field history for prediction of symphylan damage. If damage has not occurred in other crops, damage is expected to be minimal to potatoes. Fall fumiga-

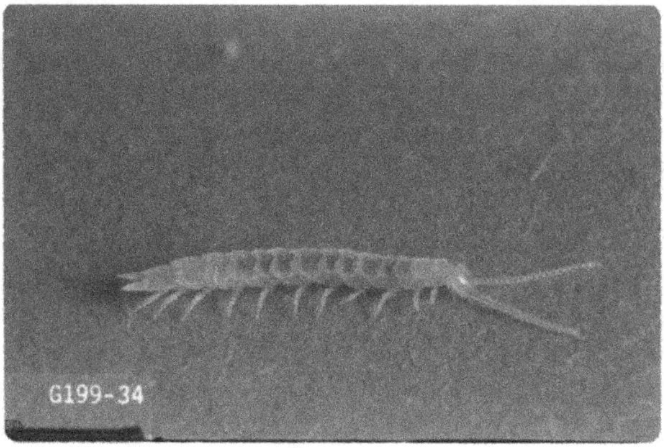

Fig. 11.56 The garden symphylan. (Photo credit: Ken Gray Image Courtesy of Oregon State University)

Fig. 11.57 European corn borer larva. (Photo credit: Frank Peairs, Colorado State University, Bugwood.org)

tion of infested areas can be effective. Insecticides can be broadcast in the spring as close to planting as possible.

European Corn Borer

The European corn borer (*Ostrinia nubilalis*) is a pest of potatoes east of the Rockies that feeds within potato stems in the larval stage. Historically a relatively common and abundant pest in potatoes, in recent years its importance has declined, possibly due to the widespread use of transgenic *Bt* corn (its primary host) and/or to its susceptibility to insecticides that are used in potatoes to manage other insects. Larvae have a brown to black head and a brownish to grayish to pinkish body with dark spots on the sides of each body segment and a faint stripe along the back (Fig. 11.57). Larvae hatch from oval, flattened, white egg clusters that are typically found on the undersides of leaves. Newly hatched larvae are 1/8 in long and mature larvae are 3/4–1 in long. Larvae, as well as pupae, may be found within the stems of their host plants, often with frass associated with the holes in the stem or branch. Adult moths are approximately ¾ in long. Males and females have slightly different colors; although both sexes have more or less straw-colored wings with dark zigzag patterns across the width of the wings, and female coloration is lighter than that of males.

Plant Damage

Like other pest species of moths, the larval stage of European corn borer is the only damaging life stage. Larvae feed within potato stems, reducing movement of nutrients through the plant tissue. This may cause wilting and leaf rolling that resembles symptoms of plant diseases. Economic reductions in yield from direct feeding damage are rare; however, European corn borer infestations are associated with higher incidence of the pathogens that cause black leg and soft rot, which may have greater effects on yield.

Management

In the absence of pathogens, 50% or more (depending on timing of infestation and variety) of stems may be infested by European corn borer without considerable reduction in tuber yield. Foliar insecticides may be used, but sprays should be applied before larvae tunnel into stems. Growing degree day models and pheromone traps are available to assist in timing of sprays.

References

Bechinski EJ, Sandvol LE, Carpenter GP, Homan HW (1994) Integrated pest management guide to wireworms in potatoes. Publication EXT 760, University of Idaho Cooperative Extension System

Chapter 12
Weed Management

Pamela J. S. Hutchinson

Contents

P. J. S. Hutchinson (✉)
Department of Plant Sciences, University of Idaho Aberdeen Research and Extension Center, University of Idaho, Aberdeen, ID, USA
e-mail: phutch@uidaho.edu

J. C. Stark et al. (eds.), *Potato Production Systems*,
https://doi.org/10.1007/978-3-030-39157-7_12

Introduction

Weeds cause losses in potato yield and tuber quality by competing for light, water, and nutrients; by interfering with harvest; and by harboring diseases, insects, and nematodes. This chapter provides a thorough treatment of potato weed management strategies. Included are discussions of weed ecology; weeds of importance in potato production; integrated weed management, including timing and targeted tank mixtures; and herbicide resistance management.

Weed Identification and Importance of Weed Species Records

The number and densities of weed species in potato fields can vary greatly from one field to the next even if fields are in close proximity and/or the weed management program, rotational crops, etc., have been similar for a number of years. Scouting and keeping a weed history record is a must for success (Fig. 12.1). Most of the herbicides currently labeled for use in potatoes have soil but no foliar activity and must be applied and incorporated into the soil before weed emergence. Therefore, unless records of the weed species present in the past are kept, choosing the most appropriate herbicide might not be possible. Only two potato herbicides that can be safely applied to potatoes after emergence have foliar activity to kill emerged broadleaf weeds. Many weeds must be small and at early growth stages for these herbicides to provide good control. As such, in-season weed identification, especially at the seedling stage, is also an integral part of successful weed management programs. Another reason proper weed identification is critical is because weed species respond differently to various control practices. For example, the herbicide, rimsulfuron (Matrix®/Prism®[1] and other trade names) can control hairy nightshade (*Solanum physalifolium* Rusby) (Fig. 12.2a–j) but not cutleaf nightshade (*Solanum triflorum* Nutt.) (Fig. 12.3a–h). The two species may look

[1]The registered trademarks of the chemicals referenced in this chapter are noted in Table 12.3. Thus, the registered trademark symbol (®) will not be repeated in the text at each occurrence. The use of a registered trademark herbicide herein is not an endorsement of that herbicide.

Fig. 12.1 Weeds compete with the potato crop for light, water, and nutrients. a) Common lambsquarters and b) a mixture of redroot pigweed, common lambsquarters, and Hairy nightshade in a potato field.

quite similar at the seedling stage, so correct identification is especially important (Fig. 12.4a–i). Hairy nightshade berries are smaller than cutleaf nightshade berries (Fig. 12.5). Other annual nightshade weeds that can be hard to control in potatoes are black (S. *nigrum* L.) and Eastern black nightshade (*S. ptychanthum* Dun.) (Figs. 12.6 and 12.7).

Weed Life Cycles

Annuals

An annual plant completes its life cycle in one year. Winter annual plants usually germinate in fall or winter, grow through spring, and produce fruit and die by mid-summer. Summer annuals germinate in spring and mature and die by fall of the same year.

Biennials

Biennials live more than one, but less than two years. The seedling usually develops vegetatively into a rosette during the first phase of growth. After a cold period, vegetative growth resumes, the plant bolts or sends up a stalk, has floral initiation and fruit set, and dies.

Fig. 12.2 (**a-g**) Hairy nightshade (*Solanum physalifolium* var. *nitidibaccatum*) plants with smooth and wavy leaf margins, **h**) stems and leaves with glandular hair, **i-j**) hairy nightshade plants and berries

Fig. 12.3 (**a–h**) Cutleaf nightshade (*Solanum triflorum*)

Perennials

Perennials live more than two years. While biennial and annual weeds rely on seeds for survival, perennials can also spread by means of creeping roots, rhizomes, stolons, tubers, or bulbils, as well as seed. If roots are injured or cut, the cut pieces may produce new plants. Creeping perennials reproduce by creeping roots, creeping above-ground stems (stolons), or creeping below-ground stems (rhizomes).

Weed Ecology

Dormancy, Seed Bank, and Germination

Weeds are successful because of their adaptability to a wide range of habitats and ability to survive, even under stressful conditions. Weed seed dormancy plays a large role in weed survival. Dormancy can be described as a physiological state in which seeds do not germinate under favorable environmental conditions. Weed seed dormancy is a problem in cropping systems because of the development of persistent weed seed banks (accumulation of viable weed seeds in the soil) and the challenge of predicting weed stand establishment in any given year and over time.

Seed banks may contain perennials, biennials, and annuals. As mentioned, perennials also can spread by means of creeping roots, rhizomes, stolons, tubers, or bulbils.

Seed densities in agricultural soils can reach as high as 95,000 seeds per square foot. Through various sources, such as contaminated crop seed; on equipment; or by wind, water, animals, and birds. The main source of weed seed in a field is through plants that escape control and produce seeds within the field. Tillage is a mechanism for vertical movement of weed seeds in soils. Chisel plowing can place seeds at a depth of 5 in. and moldboard plowing at a depth of 13 in. Over time, moldboard plowing will leave fewer

Fig. 12.4 Hairy and cutleaf nightshade plants for comparisons **a**) hairy nightshade on the left with serrated leaf margins and cutleaf nightshade with lobed leaf margins on the right, **b**) a mixture of hairy and cutleaf nightshade seedlings - the cutleaf plants generally have more linear cotyledons that those of the hairy nightshade, **c**) cutleaf and **d**) hairy nightshade seedlings for compaision with seedling plants in **b, e**) hairy nightshade plant with wavy (dentate) leaf margins, **f**) 2-leaf cutleaf nightshade seedlings, **g**) small hairy nightshade plants with highly serrated leaf margins, **h**) cutleaf nightshade plant with deeply lobed leaves, and **i**) cutleaf on the left growing adjacent to the hairy nightshade on the right

Fig. 12.4 (continued)

Fig. 12.5 Hairy nightshade plant and berries on the left and cutleaf nightshade plant and berries on the right

weed seeds in the upper 8 in. of soil than chisel plowing or no-tilling. Some studies have shown hairy nightshade emergence from seed 1–3 in. below the surface; another found that maximum emergence occurred from a 5-in. depth in soil.

Three approaches can be used to reduce weed seed bank size: (1) kill the seeds while they are in the soil, (2) stimulate germination of seeds and destroy seedlings, or (3) remove seeds before seed set. Once the seed bank has been reduced, management strategies can be incorporated that will control weed populations and require less input.

Depending upon location, conditions, and species, weed emergence can occur as early as March and as late as October (Table 12.1). Early control of weeds with

Fig. 12.6 Black nightshade (**a**) seedling; (**b**) plant with flower and berries; (**c**) berries (Photos Regents of the University of California and Phil Westra, Colorado State University, Bugwood.org)

Fig. 12.7 Eastern black nightshade (**a**-**b**) plants with flowers (**c**) leaves typically have holes from flea bettle feeding (Photo credit: Bruce Ackley, The Ohio State University, Bugwood.org)

Table 12.1 Weed emergence periods (Idaho conditions)

Weed species	Emergence period
Barnyardgrass	April–August
Canada thistle	March–October
Field bindweed	April–September
Green foxtail	April–August
Kochia	April–August
Common lambsquarters	April–August
Common mallow	April–May
Mustard sp.	August–May
Nightshade sp.	April–September
Yellow nutsedge	April–September
Redroot pigweed	April–August
Common purslane	June–July
Quackgrass	March–October
Russian thistle	May–August
Shepherd's-purse	September–May
Annual sowthistle	September–June
Wild oat	March–July

early peak germination times, such as kochia, may be critical; whereas cultivation, and/or long soil-residual or postemergence herbicides may be needed for weeds with peak germination times at or near row closure, such as redroot pigweed. At-planting herbicide applications may not last long enough to effectively control weeds that have late peak emergence. The percentage of viable seeds produced on late-germinating weeds may not be as high as those produced on weeds germinating earlier in the season.

Limiting Factors

For many weeds, light may be the limiting factor for their emergence. Seeds below the soil surface may become exposed to light during soil disturbance, such as tillage. The exposure time necessary to "trigger" germination can be as little as a fraction of a second. Crop canopies alter the light environment below the canopy and often reduce weed seed germination. Other factors affecting weed seed germination are nutrient availability, soil moisture, and soil temperature.

Weeds of Importance in Potato Production

Grasses: Annuals

Barley, volunteer (*Hordeum vulgare* L.)
Barnyardgrass (*Echinochloa crus-galli* (L.) Beauv.)
Crabgrass (*Digitaria* spp.)
Foxtail, green (*Setaria viridis* (L.) Beauv.)
Foxtail, yellow (*Setaria pumila* (Poir.) Roemer & J.A. Schultes).
Oat, volunteer (*Avena sativa* L.)
Oat, wild (*Avena fatua* L.)
Sandbur spp. (*Cenchrus* spp.)

Grasses: Perennial

Quackgrass (*Elytrigia repens* L.)

Broadleaves: Annuals

Buckwheat, wild (*Polygonum convolvulus* L.)
Cocklebur, common (*Xanthium strumarium* L.)

Box 12.1: Hairy Nightshade Correct Species Name, Varieties, and Leaf Margins

Hairy nightshade has been incorrectly referred to as *Solanum sarrachoides* Sendtner. The correct classification, however, is *S. physalifolium* Rusby. In fact, *S. Sarrachoides* Sendter and *S. physalifolium* Rusby are two morphogenetically distinct entities/species.

S. physalifolium Rusby is native to South America and according to taxonomist, Edmonds (1986) and other botanists, has two varieties. *Solanum physalifolium* Rusby var. *nitidibaccatum* (Bitter) Edmonds is the variety widely distributed to Africa, Australia, Europe, Central America, North America, and New Zealand. Leaf margins of this variety can be smooth to wavy (entire to sinuate to dentate) (Fig. 12.8a–g). Offspring of var. *nitidibaccatum* plants with smooth leaf margins can have smooth to wavy margins and vice versa. The other variety of *S. physalifolium* Rusby is var. *physalifolium*. It only has smooth leaf margins and is restricted to South America.

Dodder (*Cuscuta* spp.)
Knotweed, prostrate (*Polygonum aviculare* L.)
Kochia (*Kochia scoparia* L. Schrad).
Lambsquarters, common (*Chenopodium album* L.)
Nightshade, cutleaf (*Solanum triflorum* Nutt.)
Nightshade, black (*Solanum nigrum* L.)
Nightshade, Eastern black (*Solanum ptychanthum* Dun.)
Nightshade, hairy (*Solanum physalifolium* Rusby var. *nitidibaccatum* (Bitter) Edmonds) (Incorrectly referred to as *S. sarrachoides* Sendtner).

Palmer amaranth (*Amaranthus palmeri* S. Wats.)
Pigweed, redroot (*Amaranthus retroflexus* L.)
Prickly lettuce (*Lactuca serriola* L.)
Potato, volunteer (*Solanum tuberosum* L.)
Puncturevine (*Tribulus terrestris* L.)
Purslane, common (*Portulaca oleracea* L.)
Ragweed, common (*Ambrosia artemisiifolia* L.)
Smartweed, Pennsylvania (*Polygonum pensylvanicum* L.)
Sowthistle, annual (*Sonchus oleracea* L.)
Sunflower, common (*Helianthus annuus* L.)
Thistle, Russian (*Salsola iberica* Sennen).
Velvetleaf (*Abutilon theophrasti* Medic.)
Waterhemp (*Amaranthus tuberculatus* (Moq.) Sauer).

Fig. 12.8 (**a**–**g**) Hairy nightshade plants with smooth to wavy (entire to dentate) leaf margins

Broadleaves: Biennial and Winter Annuals

Mallow, common (*Malva neglecta* Wallr.)
Shepherd's purse (*Capsella bursa-pastoris* L.)
Flixweed (*Descurainia Sophia* L.)
Mustard, black (*Brassica nigra* L.)

Mustard, tumble (*Sisymbrium altissimum* L.)
Mustard, wild (*Brassica kaber* DC.)
Salsify, Western (*Tragopogon dubius*).

Broadleaves: Perennial

Bindweed, field (*Convolvulus arvensis* L.)
Canada thistle (*Cirsium arvense* L.)

Sedges: Perennial

Nutsedge, yellow (*Cyperus esculentus* L.)

Several weed identification guides and publications, such as *Weeds of the West*, 11th edition (Whitson 2012), are available for this most important task.

Weed Competition

Many troublesome weeds in potato cropping systems, such as hairy nightshade, can emerge as early as April and May; in other words, soon after potato planting and as potatoes are emerging. As mentioned, hairy nightshade emergence can continue through harvest in August or September. In fact, green, actively growing hairy nightshade plants can be seen in fields of potatoes naturally senescing or even after potato vine kill (Fig. 12.9a, b). Potato varieties that emerge quickly, exhibit rapid early growth, and have a dense canopy are less affected by weed interference than other varieties. Long-season potato varieties, such as Russet Burbank (RB), usually have leaf canopies that provide shading longer and later in the season which, in turn, reduces light necessary for weed seed germination. This provides more competition with weeds than early-season, less vigorous varieties such as Russet Norkotah (RN). Wisconsin weed scientists have reported that 96% or greater shading by potato is necessary for general weed suppression.

Research Results

Researchers have conducted competition and critical interference studies on weeds that affect potatoes. Russet Burbank is a late-maturing type with a large, spreading plant canopy, and RN is an early-maturing type with a small, upright canopy that often does not close between the rows in growing conditions of many potato

a b

Fig. 12.9 (**a**, **b**) Hairy nightshade plants alive and with flowers and berries in fields approximately three weeks after a vine-kill product had been applied to the potato crop

production areas. The maximum shading provided by RB during a 2-year University of Idaho hairy nightshade competition study was 87%, which occurred 6 weeks after potato emergence; whereas, maximum for RN was 61%. By 1 week later, shading had already dropped to 53 and 22%, respectively. Regardless of hairy nightshade density, which ranged from 1–100 plants per m row, biomass and berry number were 3 and 7 times greater, and seeds per sq. m were ten times greater when growing season long in RN compared with production in RB. In a University of Idaho critical interference period study, hairy nightshade, at a density of 2 per m row, growing the entire season in RN potato variety, produced 3,367 berries per sq. m. At an average of 11 seeds per berry, over 37,000 hairy nightshade seeds per sq. m could be produced in a single season. Even when hairy nightshade presence did not occur until 40 days after potato emergence in that study, as many as 2,100 berries per sq. m were produced, which would add over 23,000 hairy nightshade seeds to the weed seed bank in a single season. These seeds could begin to germinate as soon as the following spring, with some remaining with 90% or greater viability for as long as five years. Consequently, when hairy nightshade is not controlled adequately in a given potato planting year and viable seed production occurs, many of those seeds could be present to germinate and compete the next time potatoes are planted in that field even if they are grown in as long as a 4-year rotation.

In the University of Idaho hairy nightshade competition study, RN U.S. No. 1 and total tuber yields were reduced 21 and 16%, respectively, when only one hairy

Fig. 12.10 (**a–d**) Hairy nightshade competition study—one hairy nightshade per m row greenhouse grown and transplanted 1–2 leaf plant at time of potato emergence

Table 12.2 Effect of hairy nightshade competition at four densities on tuber yield of two potato varieties at Aberdeen, ID, 2004 and 2005[a]

Variety	Hairy nightshade density (per m row)	Tuber yield[b] U.S. No. 1	Tuber yield[b] Total
		% Reduction[c]	
Russet Norkotah	1	21[d]	16[d]
	2	26[d]	20[d]
	3	27[d]	25[d]
	100	48[d]	37[d]
Russet Burbank	1	6	5
	2	10[d]	9[d]
	3	11[d]	9[d]
	100	21[d]	19[d]

[a]Yield by year interaction was not significantly different so yield data were combined over years
[b]U.S. No. 1 tubers have no defects and weigh at least 4 oz (113 g). Total tuber weight includes U.S. No. 1, U.S. No. 2 (weigh at least 113 g and have 1–2 slight defects), malformed culls, and process culls (weigh less than 113 g and have no defects)
[c]Values shown are % reduction in tuber yields of weed-free control tuber yields for that variety
[d]Significantly different than the weed-free control tuber yields for that variety according to a single degree of freedom contrast performed on the means ($P < 0.05$)

nightshade plant per m row (Fig. 12.10a–d) competed season long compared with weed-free, control yields (Table 12.2). U.S. No 1 yield was reduced 26, 27, or 48% by densities of 2, 3, or 100 hairy nightshade per m row, respectively, and total tuber yield reductions ranged from 20–37% from those same weed densities. In contrast, RB U.S. No. 1 and total tuber yields in plots with one hairy nightshade plant per m row were similar to weed-free control yields. Yields in plots with 2 or 3 per m row, however, were reduced 9–11%, and when 100 per m row were present season long, RB U.S. No. 1 and total yields were reduced 21 and 19%, respectively (Table 12.2).

In a Michigan study, when 1 redroot pigweed or 1 barnyardgrass plant per m row was established at potato planting, marketable tuber yield was reduced 19–33%. Season-long interference of 14 quackgrass shoots per square ft. can reduce potato yield by 78%.

In the University of Idaho critical interference study, the periods when RN must be free of hairy nightshade, or U.S. No. 1 and total tuber yields losses of 5% or greater will occur (critical period of control), are 6–22 and 11–24 days after potato emergence, respectively. This critical weed-free period most likely begins earlier for the U.S. No. 1 than for total tuber yields, because more time is needed for tubers to develop to the size requirement for the U.S. No. 1 grade category. Mixtures of annual weeds emerging 1 week after and competing all season with potatoes have reduced tuber yields by an average of 54% compared to 16% tuber yield reduction when weeds emerged 3 weeks after potatoes. Researchers have determined that the critical weed-free period for quackgrass in potatoes depends on the level of quackgrass infestation and year. At a high level of infestation, the critical period begins before potato emergence and may last as long as 42 days after emergence. Growers may need to control quackgrass before potato emergence to prevent yield loss. On the other hand, preventing late interference is beneficial for reducing tuber perforation by quackgrass rhizomes and to facilitate harvest.

Integrated Weed Management

Integrated weed management makes use of all the cultural, mechanical, chemical, and biological tools available for weed control, rather than relying on any single control method. An integrated weed management strategy will provide the best possible weed control in potatoes and increase the chances of a successful economic, environmentally sustainable farming operation.

Cultural Controls

Cultural practices, such as growing competitive crops in the potato rotation, timely cultivation, using agronomic practices that promote vigorous crop growth, and growing a competitive potato variety, all contribute to an effective weed management program. As mentioned previously, the potato crop itself can do a good job of suppressing weed growth once the rows are closed. Maintaining vigorous potato plants that close the rows rapidly and remain healthy until vine kill will also contribute to good weed control.

Crop Rotation

A good crop rotation can disrupt weed life cycles and prevent certain species from becoming dominant. The more dissimilar the crop and weed life cycles are, the more difficult it is for a weed species to develop into a severe problem.

Including a winter annual crop, such as winter wheat or winter canola, in the rotation can reduce populations of common summer annual weeds. For example, surveys in Southeastern Idaho have shown that hairy nightshade and wild oat populations were often lower when winter wheat was included as a rotational crop with potatoes than when only spring wheat was included. Another practice is to allow weeds and volunteer crop seed to germinate and grow soon after harvesting short-season crops, such as small grains or canola. Where possible, irrigation can be used to promote growth. Weeds and volunteers emerged in the crop stubble can then be destroyed by tillage or herbicides, or frost may kill them before they flower and produce seed. Irrigation before application can also assure perennials present in the field are not stressed so that herbicides can be effective. Newly produced seed of some weed species, such as hairy nightshade, have an inherent dormancy and will not germinate until the following season at the earliest.

Diverse crop rotations allow growers to use a variety of herbicides and tillage practices that further reduce the likelihood of a particular weed species becoming dominant. A 2-, 3-, or 4-year rotation means growing potatoes in a field every other, every third, or every fourth year, respectively. Three-year or longer rotations can decrease disease, insect, and nematode populations and result in a healthier potato crop that is more competitive with weeds than one produced in a 2-year rotation.

Field Sanitation

Controlling weeds on field borders will help reduce their spread into potato fields. This is especially important when new weed species or herbicide-resistant weed populations are present on field borders. In addition, some common annual weeds are hosts for several important potato insects and diseases, so keeping field borders weed free can reduce these pest problems as well.

It is also important to clean equipment before moving from one field to the next. The most common mechanism of weed seed introduction to fields is transport by tillage and harvest equipment. Weed maps often show that new weed species appear first at the entrance to a field. Although it is time consuming, cleaning equipment with compressed air, steam, or water before moving from one field to the next can help prevent weed seed spread. Spending 20 min to clean equipment may help eliminate 20 years of weed control work.Weed seed can also enter fields through other sources, such as contaminated crop seed or by wind, water, animals, and birds.

In fields irrigated with canal water, it also is important to install screens at headgates and pumps to exclude weed seeds from irrigation water.

Using certified seed in crops grown in rotation with potatoes reduces the likelihood of introducing new weeds to a field. In areas where livestock waste is applied to fields, composting manure will reduce the potential for weed seed spread compared to applying manure directly to fields.

Cultivation

After planting potatoes, typical practice is to form a hill in the row with cultivation equipment set to throw soil out of the furrow between rows up onto the row area (Fig. 12.11a, b). This cultivation/hilling operation can reduce tuber exposure to sunlight, which lessens tuber greening. Furthermore, soils are aerated and the structure of some soils is improved; especially those high in silt and very fine sand. Soil compacted by heavy equipment and soil crusted by rain also may be loosened. Hilling should be the last cultivation in a potato field. If conducted after application of soil-active herbicides, then nontreated soil is brought up and the so called "herbicide barrier" is broken resulting in emergence of new weeds.

a b

Fig. 12.11 (**a**, **b**) Hills formed in a potato field

Fig. 12.12 (**a**, **b**) Cultivation is an important practice for controlling weeds that emerge early in the season. The size of most of the emerged weeds seen here between the potato rows are cotyledon to 2 true leaves, less than 1-in tall, so cultivation at or before this time would usually be effective at controlling these weeds

Fig. 12.13 Weeds are too large for effective cultivation

Hilling can be conducted before or after potato emergence. There are both advantages and disadvantages to using hilling alone for weed control. For instance, wind is not an issue as it can be with herbicide application, and under certain conditions, hilling may be less expensive than use of herbicides. However, timely hilling may be difficult on large acreages or in wet weather.

On the positive side, properly timed hilling can control early-germinating annual weeds, as long as they are small—2–3 true leaf stage, which is usually <1 in. (Fig. 12.12a, b).

Studies in Idaho have shown that hilling performed after potato emergence when weeds are small (0.5 in. tall) and potato plants are no taller than 4–6 in., can provide economical weed control.

Three to four-in tall weeds can sometimes still be eliminated; however, when >4 in., cultivation alone is not effective (Fig. 12.13).

On the negative side, in-row weeds will most likely not be controlled with in-furrow tillage, although soil thrown up onto tops of hills may cover small, emerged weeds enough to kill them.

- Putting more than 1 in. of soil on emerged potato plants, however, may reduce yield as a result of delayed growth.
- Large weeds initially covered with soil during cultivation can survive and re-emerge.
- If the cultivator is not set up properly, weeds may be left on side of hills. Cultivation after potato emergence may spread potato diseases, which would be especially detrimental in seed-growing areas. Late cultivation—hilling of potatoes larger than 8–10 in. tall, may damage potato foliage and cause root-pruning, resulting in reduced tuber yields and quality.

When heavy weed populations exist, multiple cultivations are sometimes performed, but weed control may still be inadequate to prevent yield loss. Multiple cultivations may cause soil compaction which, in turn, reduces aeration and potato growth. Multiple cultivations can also reduce yields. The direct effects of cultivation alone on tuber yield and quality were studied in University of Idaho weed-free experiments. U.S. No. 1 yields were 12–17% lower in plots that were hilled and cultivated twice (potato plants were 12–14 in. tall at the last cultivation) than in weed-free plots that were hilled or hilled and cultivated once when potato plants were 4–6 in. tall.

Some important points to consider when cultivating include: (1) cultivated weeds die more readily in drier than moist soil; (2) at least 1 day of lying uprooted in dry soil will kill most weeds in the 2–3 true leaf stage or smaller; otherwise in wet soils, weeds could "re-root;" and (3) less soil compaction occurs on drier soil, which minimizes yield reduction and reduces formation of clods that can interfere with the uniformity of soil- and foliar-active herbicides causing weeds to escape application (Fig. 12.14). In other words, if soil is wet, cultivation should be delayed.

Drier areas of the field should be cultivated first and irrigation scheduled to allow at least 1 full day of drying before the next irrigation or forecasted rainfall. When cultivating, enough soil needs to be thrown up by the cultivator to cover the entire hill with 1–2 in. of soil, which will kill weeds within the row. Less soil than this will not uniformly cover the hill and kill all weeds. Regardless, if more than one cultivation is planned, but a large amount of soil is needed to cover and kill the weeds at the first cultivation, the hill might be too tall to properly move the cultivator across the field during subsequent hilling operations.

Reservoir tillage creates divots in the furrow that make irrigation more efficient by slowimg water infiltration rate (Fig. 12.15a, b). When reservoir tillage is part of the management program, the equipment is set up for reservoir tillage between the potato rows and hilling the potato rows *at the same time* making it a one-pass operation (Fig. 12.16a–c).

Fig. 12.14 Cultivation was performed when soil was too wet so clods were formed that protect weeds from herbicide application

Fig. 12.15 (**a**, **b**) Divots in furrows between potato rows created by reservoir tillage

NOTE: Some refer to reservoir tillage by one of the equipment trade names, Dammer- Diker®.

Sprayers are designed to be driven across the reservoir-tillage divots/on the sides of the hills while still maintaining a level spray boom for even distribution of the herbicide (Fig. 12.17). Chemigation (or attaching a spray boom to the back of the equipment) is also an effective means of herbicide application after reservoir tillage. In this manner, herbicides are applied after the last tillage operation in the field and the herbicide barrier remains intact.

Fig. 12.16 (**a**–**c**) Reservoir tillage equipment and tractor in potato fields before potato emergence

Fig. 12.17 Herbicide application with a sprayer that can maintain a level boom across the field after reservoir tillage

Chemical Controls

Herbicides can be effective tools for weed management in potatoes. Developing an effective chemical weed-control program requires careful consideration of factors, such as the weed species present in the field, soil characteristics, tillage and irrigation practices, and crop rotation. *In other words, a herbicide program should be customized for a given field rather than using the same program for all fields.* Combining cultural and mechanical practices with appropriate herbicides gives more effective weed control than relying solely on herbicides. Appropriate tank mixtures or sequential application of herbicides with different mechanisms of action can provide control of the multiple weed species present in a given field. Just as important is using different mechanism-of-action herbicide combinations that can control the *same* weed species in order to prevent or delay the development of a herbicide-resistant population of that species.

Proper herbicide application is essential for maximizing weed control while minimizing the potential for crop injury. Mid- and late-season weed control measures may be needed on short-season, small-canopy varieties such as RN, in order to decrease competition and reduce weed seed production. In contrast and as previously mentioned, a vigorous, indeterminate variety, such as RB, usually provides more shade and, thus, effectively reduces weed germination, growth, competition, and seed production once row closure occurs.

Soil active herbicides must be incorporated into the soil with rainfall and/or sprinkler irrigation. Mechanical incorporation is sometimes used but not generally recommended.

Box 12.2: Herbicides and Potato Crop Safety
Tolerance of potatoes to potato herbicides is due mainly to the ability of the plant to quickly metabolize the herbicide to non-herbicidal compounds. Hence, if potato plant growth is slowed for reasons such as stress from drought, cool conditions, pest infestations, etc., then metabolism is slowed. If metabolism is slowed, then the chance of injury increases. Injury can be temporary, and once conditions favor metabolism and other regular plant functions, the potato plants will recover and begin typical growth.

Common Terms and Abbreviations for the Use of Herbicides in Potatoes

Units of Measurement

lb = pound (lb/gal, lbs/gal)
oz = ounce (16 oz/lb)

fl oz = fluid ounce (128 fl oz/gal)
pt = pint (8 pt/gal)
qt = quart (2 pts/qt, 4 qt/gal, 946 mL/qt)
gal = gallon, (3,785 mL/gal)
v/v = volume/volume
GPA = gallons per acre

Herbicide Group Numbering

Herbicide name$^{\text{number } 1–30}$ = herbicide site of action group

Types of Formulation

DF = Dry flowable
EC = Emulsifiable concentrate
F = Flowable
G = Granule Liquid
ME = Micro-encapsulated
OD = Oil dispersion
P = Pellet
S/SL = Solution/Soluble Liquid
SC = Suspension
SE = Solution emulsion
SG = Soluble granule
WDG = Water dispersible granule
XP = Extruded paste
ZC = Suspension of microcapsules and solid fine particles

Miscellaneous

Active ingredient (ai) = ingredient in herbicide that is biological active.

Acid equivalent (ae) = the portion of a formulation that could be converted back to the parent acid. Herbicide molecules that are acid can sometimes be altered to impart some property other than herbicidal activity. For instance, when the parent acid might not be readily absorbed into a plant, it could possibly be altered to penetrate through the leaf more effectively.

Adjuvant = Substance added to the spray mix or herbicide formulation to improve herbicidal activity or application characteristics.

Surfactant = Surface active agent, a type of adjuvant which improves the dispersing, emulsifying, absorbing, spreading, sticking and/or penetrating properties of the spray mix.

AMS = Ammonium sulfate
COC = Crop oil concentrate

MSO = Methylated seed oil
NIS = Nonionic surfactant
OM = Organic matter
PHI = Preharvest interval
RUP = Restricted use pesticide
UAN = Urea ammonium nitrate
PPE = Personal protective equipment

Rainfast = Time required between application and rain for the herbicide to perform effectively

Degradation = Process by which a pesticide is broken down to simpler structures through biological or *abiotic* mechanisms. Synonyms include *breakdown* and *decomposition.*

Dissipation = Loss of pesticide residues from an environmental compartment due to degradation and transfer to another environmental compartment.

Metabolism = Chemical transformations of a herbicide in a plant that generally results in detoxification of a herbicide, but may also increase herbicide toxicity.

Herbicide Application

Timing

PPI = Pre-plant incorporated
PRE = Preemergence to crop and/or weed
POST = Postemergence to crop and/or weed
Drag-off = A harrow or similar equipment is used to "knock down" the moderate hill that is built at planting. The field is leveled so that seed pieces are closer to the warmer soil surface for relatively quicker emergence than emergence if buried 5 to 6 inches. Drag-off can also conrol emerged weeds.
Layby = An application at the last cultivation time or the equivalent

Herbicide Activity

Soil-active herbicides, also referred to as residual herbicides, can control weeds before emergence by inhibiting germination, cell division, below-ground shoot growth, and/or root growth. Some can translocate (move) to other parts of the weed. Some inhibit photosynthesis and kill the weed as soon as exposure to sunlight occurs. Soil-active herbicides applied to control weeds that have not yet emerged must be incorporated into the soil in order to be effective.

Foliar-active systemic herbicides are applied to and absorbed by the above-ground portion of the weed; e.g., leaves, stems, and tissue, and can then translocate within the plant.

Foliar-active contact herbicides are applied to the above-ground portion of emerged weeds and only affect the part of plant that encounters the herbicide spray. Absorption by and translocation within the weed does not occur or is minimal.

As such, the spray-solution volume must be great enough for the herbicide to come into contact with as much of the above-ground plant parts as possible.

Selective herbicides only have activity on certain plant species and not others.

Non-selective herbicides can kill or damage most plant species.

Application Method

Aerial or ground sprayers. In order to be effective, herbicides must be applied in the appropriate volume of water/carrier, nozzles, pressure, etc., listed on the label (Fig. 12.18).

Chemigation is application through sprinkler irrigation. Most herbicides are labeled for chemigation via overhead systems only (Fig. 12.19).

NOTE: some herbicides are not labeled for aerial and/or chemigation application.

Incorporation: Soil-active herbicides that have been applied by aerial or ground sprayers to control weeds that have not emerged should be moved into the soil (top 2 in where seeds of most weed species present in potato production areas germinate) with water via sprinkler irrigation (Fig. 12.20) and/or rainfall or mechanically with tillage equipment. Information such as incorporation water volume, equipment type, depth, and timing are designated on the herbicide label. Herbicide incorporation can be affected by factors such as soil type/texture and existing moisture in the soil or at the soil surface, as well as herbicide characteristics. Herbicide incorporation is sometimes referred to as herbicide activation.

Fig. 12.18 Aerial application on a potato field

Fig. 12.19 Center pivot for applying herbicides via chemigation

Fig. 12.20 Sprinkler incorporation of soil active herbicides applied by ground shortly after hilling and before potato emergence

Herbicide Classification System

Definitions

Mode of action (MOA): The entire sequence of events through which a herbicide kills a weed.

Mechanism of action (sometimes also called MOA): The plant function, referred to as site of action/target site affected by the herbicide or the specific location within the plant where the herbicide has activity.

Herbicide family: A group of herbicides that is named in relation to its chemical similarities. Members of the same herbicide family will have the same mode of action and typically the same mechanism (site) of action.

Since 1989, agrichemical industry representatives and weed scientists in Canada, Australia, the U.S., and other countries have been working on a herbicide classification system based on MOA. Herbicides with similar MOA are grouped into the same herbicide class (Table 12.3). The group-classification number and/or herbicide class are often used; e.g., 2,4-D is a synthetic auxin that is a Group 4 herbicide (Table 12.3). Knowing the MOA and herbicide class is the key to planning a management strategy to reduce the potential for developing herbicide-resistant weed populations. Herbicide-resistant populations develop because of the repeated use of a herbicide or herbicides with the same MOA. The herbicide does not cause the resistance mutation. More detailed herbicide resistance information is provided later in this chapter.

Herbicide class has been required on Australian labels since 1994 and is voluntarily placed on Canadian and U.S. labels by herbicide manufacturers.

Herbicides Registered for Weed Control in Potatoes

Nineteen herbicide active ingredients are currently registered for weed control in potatoes in the U.S. and Canada). Three herbicides, carfentrazone ethyl (Aim EC), glyphosate (Roundup and various trade names), and paraquat (various trade names) are non-selective, designed to "burndown," and kill all emerged plant species. Therefore, they can only be applied before potatoes emerge. Seventeen are selective herbicides that are safe to use in potatoes when correct timing, rates, etc., are followed (see below). Note: Paraquat is not used for potatoes in Canada; Fluazifop-p-butyl (Venture) is not labeled for use in U.S. potato production.

Soil-active herbicides applied preemergence by ground or aerial sprayers must be incorporated with rain or sprinkler irrigation in order to move the herbicide into the top 2-in. layer of soil where most weed seeds germinate. Incorporation of some of these herbicides does not have to occur immediately, however, since they are applied to control weeds that have not yet emerged, the best control is usually achieved when the herbicide is incorporated as soon as possible after application and before weed emergence. In addition, adequate moisture is necessary for soil-active herbicides to be in a soil-water solution where they are available for exposure to/uptake by germinating weeds.

Table 12.3 Herbicide class, mechanism of action, resistant weed species; herbicides commonly used in U.S. and Canada potato cropping systems are in bold

WSSA group number/ class	Mechanism/site of action and chemical family	Examples: generic and trade names[a]	Resistant weeds in ID, OR, WA, MN, ND, MI, WI, NY, and Canada
1	**Acetyl CoA carboxylase (ACCase) inhibitors—Aryloxyphenoxypropanoates (FOPs), Cyclohexanediones (DIMs), Phenylpyrazolines (DENs)** ACCase is important for membrane synthesis	**Fusilade®/Venture®, Poast Plus®/Poast Ultra® Select®**	Green foxtail, Italian ryegrass, large crabgrass, Persian darnel, wild oat
2	**Acetolactate synthase (ALS), also called acetohydroxyacid synthase (AHAS), inhibitors—imidazolines (Imis), sulfonylureas (SUs), sulfonylaminocarbonyl triazolinone, triazolopyrimidines** Biosynthesis of branched-chain amino acids is blocked	**Matrix®/Prism®,** Beyond®, Raptor®, Pursuit®, Express®, Harmony®, Harmony Extra®, Oust®, Accent®, Beacon®	Common chickweed, common cocklebur, common hempnettle, downy brome, Eastern black nightshade, false cleavers, field pennycress, giant foxtail, Italian ryegrass, kochia, marshelder, mayweed chamomile, Powell amaranth, prickly lettuce, redroot pigweed, Russian thistle, shepherd's-purse, small seed falseflax, spiny sowthistle, tall waterhemp, wild buckwheat, wild mustard, wild oat, yellow foxtail
3	**Microtubule assembly inhibitors—dinitroanilines (DNA's)** Cell division is inhibited; the growing points of stems and roots are affected	**Prowl H2O®, Sonalan® HFP, Treflan® HFP** Balan®, Kerb®, Surflan®	Green foxtail, wild oat
4	**Synthetic auxins—phenoxyacetic acids Benzoicacidpyridazines** Interferes with cell formation resulting in abnormal root and shoot growth; stimulates ethylene evolution	2,4-D, MCPA, Banvel®, Starane®, Stinger®, Tordon®	Kochia, prickly lettuce, wild carrot, wild mustard, yellow starthistle

Table 12.3 (continued)

5	**Photosystem II inhibitors—triazines/pyridazinones/uracils** Binds to D1 protein and blocks electron transport and stops CO2 fixation and production of energy needed for plant growth; a chain of reactions is initiated that results in membrane leakage allowing cells and cell organelles to dry and rapidly disintegrate	**Metribuzin**, atrazine, Betamix®, Velpar®, Princep®	Annual bluegrass, barnyardgrass, common groundsel, common lambsquarters, common purslane, common ragweed, Eastern black nightshade, horseweed, kochia, redroot pigweed (and other pigweed species), ladysthumb, Powell amaranth, tall waterhemp, shepherd's purse, velvetleaf, wild mustard, witchgrass, yellow foxtail
6	**Photosystem II inhibitors** (binds to same D1 protein as Groups 5 and 7 but at a different attachment site)—**benzothiadiazoles/nitriles/phenylpyridazi nes**	Basagran®, Buctril®, Tough®	Common groundsel, redroot pigweed, smooth pigweed
7	**Photosystem II inhibitors** (binds to same D1 protein as Groups 5 and 6 but at a different attachment site)—**ureas/amides**	**Linex®/Lorox®**, Direx®, Karmex®, Spike®	Annual bluegrass, common lambsquarters, common purslane, horseweed, kochia, Powell amaranth
8	**Lipid synthesis inhibitors (not ACCase)—thiocarbamates**	**Eptam®**, Far-Go®, Ro-Neet®	Wild oat
	Similar to Group 16, the primary site of absorption and action is the emerging shoot and growing point; can also cause abnormal cell development/prevent cell division in germinating seedlings		
9	**EPSP synthase inhibitors—no family name** Inhibits a key enzyme; leads to depletion of the aromatic amino acids that are needed for protein synthesis	**Roundup®**, Touchdown IQ®, Rodeo®	Common ragweed, giant ragweed, horseweed, Italian ryegrass, kochia, Palmer amaranth, Russian thistle, tall waterhemp
10	**Glutamine synthase inhibitor—phosphinic acid** Results in massive accumulation of ammonia that destroys cells and directly inhibits photosystem I and photosystem II reactions	**Rely®**, Liberty®	Horseweed, Italian ryegrass

Table 12.3 (continued)

14	**Protoporphyrinogen oxidase (PPOs) inhibitors** Blocks production of chlorophyll and heme; inhibition of PPO also results in highly reactive molecules that attack and destroy lipids and protein membranes	**Chateau®, sulfentrazone, Reflex®**	Common ragweed, tall waterhemp, wild oat
15	**Very long chain fatty acid (VLCFA) inhibitors—chloroacetamides, oxyacetamides, isoxazoline** VLCFA are important membrane components and cuticular and epicuticular waxes; crucial for cell expansion, cell proliferation or differentiation; prevents plants from desiccation; affects susceptible weeds before emergence but do not inhibit seed germination	**Dual Magnum®/Dual II Magnum®, Me-Too-Lachor®, Outlook®, Stalwart® Zidua®**, Define®, Frontier®, Lasso®	Italian ryegrass, wild oat
16	**Lipid synthesis inhibitors—benzofuranes** (not ACCase) Similar to Group 8, the primary site of absorption and action is the emerging shoot and growing point; can also cause abnormal cell development/prevent cell division in germinating seedlings	Nortron®	Annual bluegrass
22	**Photosystem I (PSI) electron diverters—bipyridiliums** Accepts electrons from PSI and reduce to form a herbicide radical; reduces other molecules to form extremely reactive molecules that readily destroy membrane lipids, chlorophyll, and disintegrated cell membranes allowing cytoplasm to leak, which leads to rapid leaf wilting and desiccation	**paraquat, Reglone® (diquat)**	Eastern black nightshade, horseweed, Virginia pepperweed

[a]To simplify information, registered trade names have been used. No endorsement of named products is intended nor is criticism implied if similar products are not mentioned
Herbicides labeled for use in potatoes are in boldface text
Resistant weed information from Heap (2018). Some weeds listed here have multiple- or cross-resistance. Multiple resistance weeds are resistant to herbicides with different mechanisms of action. Cross-resistant weeds have resistance to more than one herbicide with the same mechanism of action

Weeds controlled/suppressed by each of the soil- and foliar-active herbicides described in this section are listed in Table 12.4. Rates for many herbicides must be adjusted for soil texture, organic matter content, soil pH, weed species, potential for soil residue, and other herbicides used in a tank mixture. Information is obtained from the herbicide labels and research results.The label is the law and must be followed.

NOTE: Of the selective herbicides labeled for use in potato, only rimsulfuron (Matrix (Prism in Canada)) or metribuzin have activity on emerged broadleaf weeds. Clethodim (Select), sethoxydim (Poast Plus and Poast Ultra), and fluazifop-p-butyl

Table 12.4 Herbicide effectiveness on weeds in potatoes

Weed	Dimethenamid-P (Outlook)	EPTC (Eptam)	Flumioxazin (Chateau)	Fomesafen (Reflex)	Metribuzin (several brands)	Linuron (Linex/ Lorox)	Metolachlor (Me-Too-Lachlor, Stalwart or others)	S-metolachlor (Dual Magnum/Dual II Magnum[a])
Barley, volunteer	F-G	F-G	N	S	P	–	–	–
Barnyardgrass	G	G	N	S	F	G	G	G
Bindweed, field	P	P	–	–	P	–	–	–
Buckwheat, wild	–	F	–	–	F	–	–	–
Cocklebur, common	–	P	–	S	F	S	–	–
Crabgrass	G	G	–	S	F	G	G	G
Foxtail spp.	G	G	N	S	F	G	G	G
Knotweed, prostrate	–	G	–	–	G	–	–	–
Kochia	P-F	P-F	S	–	G	P-F	F	F
Lambsquarters, common	P-F	G	S	P	G	G	F	F
Mallow, common	–	P	–	–	G	–	F	F
Mustard spp.	–	P	S	–	G	G	–	–
Nightshade, cutleaf	F-G	F-G	S	F	P	–	F	F-G
Nightshade, black	G	G	G	G	F	–	F	F
Nightshade, Eastern black	G	G	G	G	P-F	S	F	F
Nightshade, hairy	G	G	S	F	F	F-G	F	F
Nutsedge, yellow	F	F	–	S	P	–	F-G	F
Oat, volunteer	F-G	F-G	N	S	F-G	–	–	–
Oat, wild	F-G	F-G	N	S	F-G	–	F	P-F
Pigweed spp.	G	F-G	S	G	G	G	G	G

Purslane, common	G	G	–	G	G	G	G	G
Quackgrass	–	F-G	N	N	P-F	–	–	–
Sandbur, field	P-F	G	N	–	P	–	G	G
Smartweed spp. (annual)	–	P	–	–	F	G	P	P
Sowthistle, annual	–	F	–	–	G	–	–	–
Sunflower, wild	–	P	–	–	F	–	P	P
Thistle, Canada	–	P	N	N	F	–	–	–
Thistle, Russian	–	P	–	–	G	–	P	P
Wheat, volunteer	F-G	F-G	N	S	P	–	–	–

Weed	Pendimethalin (Prowl 3.3 or H_2O)	Ethalfluralin (Sonalan HFP)	Trifluralin (Treflan HFP or others)	Rimsulfuron (Matrix or others) preemergence	Rimsulfuron (Matrix or others) postemergence	Rimsulfuron (Prism[a]) postemergence	Pyroxysulfone (Zidua)	Clethodim (Select)/ sethoxydim (Poast Plus/ fluazifop-p-butyl (Venture[a])
Barley, volunteer	–	P	P	G	G	–	–	G
Barnyardgrass	G	G	G	G	G	G	F-G	G/G/F
Bindweed, field	P	–	P	–	P	P	–	N
Buckwheat, wild	–	F-G	F	P	P	P	–	N
Cocklebur, common	–	P	P	F	F	F	–	N
Crabgrass	G	G	G	F	G	G	F-G	G /F/F
Foxtail spp.	G	G	G	G	G	G	F-G	G
Knotweed, prostrate	G	–	G	N	–	–	–	N
Kochia	G-F	F-G	F-G	G	G	F-G	P-F	N
Lambsquarters, common	G-F	F-G	F-G	F-G	P-F	S	–	N
Mallow, common	F	–	P	–	–	–	–	N

(continued)

Table 12.4 (continued)

Mustard spp.	–	P	P	G	G	–	–	N
Nightshade, cutleaf	P-F	–	P	N	N	N	–	N
Nightshade, black	P-F	F	P	G	G	–	F-G	N
Nightshade, Eastern black	P-F	F	P	G	G	–	F-G	N
Nightshade, hairy	F-P	F	P	F-G	G	F-G	F-G	N
Nutsedge, yellow	P	–	P	–	F	–	–	N
Oat, volunteer	–	G	G	F-G	G	–	–	G
Oat, wild	F-P	F-G	F	F	G	–	–	G
Pigweed spp.	G-F	G	G	G	G	–	F-G	N
Purslane, common	G	G	G	–	F	–	–	N
Quackgrass	–	P	P	N	G	G	–	G/F/G/
Sandbur, field	–	G	G	–	–	–	P	G/G/F
Smartweed spp. (annual)	F	–	P-F	–	F	–	–	N
Sowthistle, annual	P	–	P	–	–	–	–	N
Sunflower, wild	P	–	P	G	F-G	–	–	N
Thistle, Canada	–	–	P	–	F	–	–	N
Thistle, Russian	G	F-G	F-G	P	P	–	–	N
Wheat, volunteer	–	F	F	G	G	–	–	G

G good, *F* fair, *P* poor, *N* none, *S* suppression only, – no information available

Response of weeds to any of the listed herbicides may be altered by growing conditions, weed populations, type of irrigation, genetic variations, soil type, pH, OM, time of application, and application rate. Ratings may vary from season to season and from site to site. Weed control generally decreases as the season progresses

Source: 2020 PNW Weed Management Handbook, updated chart (Hutchinson 2020) and the herbicide labels

[a]Product used in Canada

(Venture (labeled for potatoes in Canada but not the U.S.), which are applied postemergence to potatoes, have activity only on emerged grassy weeds. Non-selective herbicides, glyphosate (Roundup and various trade names), paraquat (various trade names), and carfentrazone-ethyl (Aim EC and others), are labeled for use in potatoes to burndown weeds before planting (pre-plant) or after planting but before emergence. Paraquat is not used in Canada and carfentrazone-ethyl is only used pre-plant in Canada. Sequence is a pre-mix of glyphosate and s-metolachlor labeled for burndown in the U.S.

Refer to Table 12.5 for activity (soil and/or foliar) and application timing(s) of herbicides labeled for use in potatoes. Aerial applications are not allowed for some herbicides and at some locations. The following information on herbicides labeled for use in potatoes was obtained from the labels.What is shown is brief, so read and follow the herbicide labels for more specific information about each herbicide.

Dimethenamid-p (Outlook 6EC) (6 lbs ai/gal)

- Preemergence only. Soil active, only.
- Ground, chemigation, aerial, or impregnated onto dry bulk fertilizer.
- One application only.
- 12–18 fl oz/A coarse textured soils. A (0.56–0.84 lb ai/A).
- 18–21 fl oz/A medium- to fine-textured soils. (0.84–1 lb ai/A).
- Do not exceed 21 fl oz/A per season.
- For effective control after ground or aerial preemergence application, rain or sprinkler irrigation or shallow mechanical incorporation is required. Performance is best if either rain or overhead irrigation occurs within 7 days after application.

Table 12.5 Herbicides labeled for use in potato: application timing and activity

Must be applied before potatoes emerge		Can be applied both PRE and POST (herbicides are safe to emerged potatoes)		POST only
PRE-PLANT and PRE Foliar active - Burndown	PRE only Foliar and/or Soil active	Soil active only	Soil and Foliar active	Soil and Foliar active
Aim EC and others (carfentrazone-ethyl) *Only as a pre-plant burndown In Canada*	**Chateau** (flumioxazin) **Linex or Lorox** (linuron) **Outlook** and others (dimethenamid-p) *Only soil active*	**Dual Magnum** (s-metolachlor) *Labeled as* **Dual II Magnum** *in Canada*	**Matrix** and others (rimsulfuron); *Matrix is not labeled for use in Canada* (see Prism)	**Prism** (rimsulfuron) *Only for use in Canada*
glyphosate (various trade names)		**Eptam** (EPTC) **Prowl H2O** and others (pendimethalin)	***metribuzin*** (various trade names)	**Titus Pro** (co-pack of rimsulfuron and metribuzin) *Only for use in Canada*
paraquat (various trade names) NOT FOR USE IN CANADA	**Reflex** (fomesafen) *Only soil active* **Sonalan HFP** (ethalfluralin) *Only soil active*	**Treflan HFP** (trifluralin)	Foliar active ONLY CONTROLS GRASSES **Select** (clethodim)	
Sequence (pre-mix of glyphosate + s-metolachlor) (s-metolachlor only has soil activity) *Only for use in the U.S.*	***sulfentrazone*** (various trade names) **Sulfentrazone MTZ** (pre-mix of sulfentrazone + metribuzin) sold as a co-pack in Canada, **Sencor STZ**	**Me-Too-Lachlor** and others (metolachlor) **Boundary** and others (pre-mix of s-metolachlor + metribuzin)	**Poast Plus, Poast Ultra** and others (sethoxydim) **Venture** (fluazifop-p-butyl) *Only for use in Canada*	

Note: Both U.S. and Canadian trade names are shown

Some of these herbicides may be applied preemergence + postemergence or postemergence + postemergence. Follow the label for your area

- In cold or wet growing conditions, Outlook may cause delayed emergence or early-season stunting of potatoes.
- Outlook can provide effective control of nightshade spp. A tank mix partner effective on common lambsquarters should be used.
- Re-cropping restrictions: If Outlook has been applied to potatoes and the crop fails due to weather or other reasons, replanting potatoes is not recommended. Any other crop for which Outlook soil applications are registered may be planted (e.g., corn, dry bean, grain, horseradish, sorghum, and soybean).
- Rotational cropping restrictions: Fall-seeded cereal crops may be planted 4 months after Outlook application. There are no rotational crop restrictions the spring after application in the previous year's potato crop.
- Site of action: Group 15, inhibits very long chain fatty acid synthesis.
- Chemical family: Chloroacetamide.

EPTC (Eptam 7E (7 lbs ai/gal)

- Preemergence 3.5–9 pints/A. 3.5–9 pints/A (3–7.9 lb ai/A), ground apply after planting and before emergence and incorporate.
- Drag-off or early postemergence 3.5–7 pints/A (3–6 lb ai/A).
- For nutsedge control, preemergence/drag off: 7 pints/A.
- Can be applied multiple times (preemergence and/or postemergence); however, do not exceed 14 pints/A (12.25 lb ai/A) per crop year, and the preharvest interval is 45 days.
- Soil active, only-does not control emerged weeds.
- Eptam can provide effective control of nightshade spp. A tank mix partner effective on common lambsquarters should be used.
- Highly volatile, so if applied with a ground-rig rather than chemigation, it must be incorporated (sprinkler incorporation is preferred) the same day as application.
- Can be mechanically incorporated (see label), but chemigation or ground application followed by sprinkler incorporation is recommended.
- Eptam can be ground applied preemergence at drag off, or after hilling, and incorporated with spike-toothed harrows or cultivation equipment. See label for equipment-specific details.
- Eptam can be applied preemergence by chemigation after clean cultivation. In eastern Washington, Eastern Oregon, and Idaho only, ground-apply Eptam preemergence and sprinkler incorporate with irrigation.
- Re-cropping/rotational restrictions: None listed on the Eptam label.
- Site of action: Group 8, lipid synthesis inhibitor but not an ACCase inhibitor.
- Chemical family: Thiocarbamate.

Two basic characteristics of Eptam are: (1) high volatility, and (2) relatively short life in the soil.

Factors that affect the amount of EPTC vapor loss include:

- Surface soil moisture
- Time between application and incorporation
- Depth and uniformity of incorporation

- Temperature
- Wind

Vapor loss is much greater when EPTC is applied to a wet or moist soil surface than a dry surface. To reduce vapor loss, which results in poor weed control, soil surface must be free of dew and incidental moisture and dry to at least 0.5 in. deep.

EPTC is degraded by soil microorganisms and usually persists 3–5 weeks in Idaho soils. When EPTC is applied after planting, to make best use of EPTC's relatively short persistence, the best weed control will be achieved when hills are built 2–3 weeks after planting and EPTC is applied and incorporated immediately after hilling.

Some areas of the U.S. have had problems with buildup of microbial populations that rapidly degrade EPTC. The phenomenon of rapid EPTC breakdown is called "enhanced biodegradation." Enhanced biodegradation occurs most commonly when EPTC is used on the same field year after year, but it can occur even when EPTC has been used only in the previous cropping season. For instance, in a Nebraska study, EPTC persisted 3 weeks in a field with no previous EPTC use, but only 9 days in a field where EPTC had been used the previous year. Although enhanced biodegradation is a common problem in the Midwestern U.S., only a few cases have been reported in the Pacific Northwest (PNW).

Ethalfluralin (Sonalan HFP 3 lb ai/gal)

- Preemergence only. Soil active, only.
- 1.33–2.67 pints/A. (0.5–1 lb ai/A).
- Rate ranges for coarse-, medium-, and fine-textured soils: 1.33–2 pints/A, 2–2.67 pints/A, 2.67 pints/A.
- Ground or chemigation (overhead sprinkler irrigation in 0.5–1 in. water).
- Sonalan HFP can be incorporated mechanically if rainfall or irrigation does not occur within 2 days after application; mechanical tillage is not recommended, however.
- Sonalan HFP can be effective on grassy weeds, such as foxtail sp. and barnyardgrass, and broadleaf weeds, such as redroot pigweed and kochia.
- Re-cropping restrictions: If replanting is required, then plant only crops listed on the Sonalan HFP labels.
- Rotational cropping restrictions: Sugar beets may be planted no earlier than 8 months after application and a moldboard plowing operation to a depth of at least 12 in. before planting. No other rotational cropping restrictions are listed on the label for Idaho, Oregon, or Washington; however, refer to the label for special rotation restrictions in other states.
- Site of action: Group 3, microtubule assembly inhibitor.
- Chemical family: Dinitroaniline.

NOTE: University of Idaho research has shown ethalfluralin tolerance by russeted- and white-skinned, as well as specialty-type potato varieties.

Flumioxazin (Chateau, 51% WDG)

- Preemergence only (has soil and foliar activity but not safe to emerged potatoes).
- 1.5 oz/A (0.047 lb ai/A).
- Ground, chemigation (center pivot only, end guns must be turned off), and aerial.
- Chateau is a herbicide that should be applied as early as possible after planting because at least 2 in. of settled soil must cover the potato sprouts at application (and incorporation) time.
- Rain or irrigation should be as soon as possible and within 7 days after application to incorporate and activate Chateau.
- Chateau is used in potatoes for control/suppression of hairy nightshade and other nightshade sp. weeds. Other weeds are not controlled/suppressed at the rate labeled for use in potato.
- NOTE: University of Idaho research has shown many russeted- white-skinned, and specialty potato varieties are tolerant to preemergence application of flumioxazin.
- *Caution: As mentioned, application should occur as soon after planting and hilling as possible because a*t least 2 in. of settled soil must cover the vegetative part of the potato plant at application or the crop can be damaged.
- Re-cropping restrictions: Replanting potatoes is not recommended if the treated potato crop fails (e.g., due to hail or other bad weather). Soybeans may be planted immediately after the Chateau application. Field corn, sorghum, sunflower, or wheat may be planted 30 days after application.
- Rotational cropping restrictions: barley, dry and snap beans, peas, rye, and sweet corn 3 months. Alfalfa, canola, clover, oats, sugar beets, and all other crops not listed 4 months (or 8 months after application if soil is not tilled before planting and if a successful soil bioassay was performed before planting). See labels for all other crops not listed.
- Site of action: Group 14, protoporphyrinogen oxidase (Protox) inhibitor.
- Chemical family N-phenylphthalimide.

Fomesafen (Reflex 2 lb ai/gal)

- Preemergence only-Soil active, only.
- 1 pint/A (0.25 lb ai/A) do not exceed 1 pint /A of Reflex per season.
- For use in potatoes grown with overhead irrigation only.
- Refer to EPA labels for use in potatoes; Special Local Needs, 24(c) labels for individual states/locations; e.g., for use only in ID, WA, OR with Special Local Needs 24(c) labeling.
- *Caution:* Do not use Reflex in potatoes grown for seed.
- Re-cropping restrictions: If replanting is necessary, the field may be replanted to cotton, dry beans, potatoes, snap beans, or soybeans with no delay.
- Rotational cropping restrictions: Crops that may be planted immediately after applying Reflex include cotton, dry beans, potatoes, snap beans, or soybeans. Wheat can be planted 4 months after application if a successful field bioassay has been performed. The minimum rotational interval after Reflex application for all

other crops not listed on the label is 18 months after application; however, a successful field bioassay must be performed before planting these crops.

- Site of action: Group 14, protoporphyrinogen oxidase (Protox) inhibitor.
- Chemical family: Nitrophenylether.

Linuron (Linex or Lorox 4L and Other Trade Names4 lbs ai/gal)

- Preemergence only (has soil and foliar activity but is not safe to emerged potatoes).
- 1–2 pint/A 0.5–1 lb ai/A). Use the lower end of the rate range for potatoes grown on calcareous or low OM soils.
- Ground or chemigation. Do not apply by air.
- Linex/Lorox has some activity on emerged weeds; however, it must be applied before potato emergence to avoid injury.
- Linex is the linuron product labeled for potatoes grown west of the Rocky Mountains in Idaho, Washington, and Oregon. Currently, Linex can only be used in these states per Special Local Needs 24(c) labels.
- Otherwise, refer to EPA or 24(c) labels for use of Linex/Lorox in your location.
- *Linex/Lorox can control many broadleaf and grass weeds, such as pigweed sp., Palmer amaranth, common lambsquarters, barnyardgrass, and foxtail sp., and suppress/control nightshade sp., common cocklebur, and velvetleaf.*
- Re-cropping restrictions: If a crop treated with Linex/Lorox fails, any crop registered for the rate applied may be planted immediately.
- Rotational cropping restrictions: Any crop registered for the Linex/Lorox rate applied may be planted immediately. Otherwise, do not plant any other crop until 12 months after the last Linex/Lorox application. Check the label for your location.
- Site of action: Group 7, photosystem II inhibitor but different binding site than Groups 5 or 6.
- Chemical family: Urea.

Metolachlor (Me-Too-Lachlor, Stalwart and Others, 8 lbs ai/gal)

- Preemergence or postemergence-Soil active, only.
- Preemergence: 1–2 pints/A (1–2 lb ai/A). After hilling/layby: 1.67 pints/A (1.67 lb ai/A). Use lower rate on coarse soils or soils low in OM. Use higher rate on soils that are relatively fine or high in OM. Up to 2.75 pints/A can be used on soils with 6–20% OM.
- Refer to the label for application methods.
- Does not control emerged weeds.
- Sprinkler-incorporate preemergence or layby applications within 7–10 days
- *Chemigation can only be used for preemergence timing.*
- Metolachlor can provide weed control and suppression similar to that of s-metolachlor.
- Re-cropping restrictions: If a crop treated with metolachlor fails, any metolachlor-labeled crop may be planted immediately.
- See the label for rotational crop restrictions.
- Site of action: Group 15, inhibits very long chain fatty acid synthesis.
- Chemical family: Chloroacetamide.

Metribuzin (4 L or 75 DF multiple Trade Names)

- 4 lb ai/gal or 75% dry flowable.
- Preemergence: 0.5–2 pints/A metribuzin 4 or 0.33–1.3 lb/A of 75 DF (0.25–1 lb ai/A) not exceeding 0.5 lb on sandy soils.
- Postemergence: 1 pint/A metribuzin 4 or 0.67 lb/A of the DF (0.25–0.5 lb ai/A).
- Preemergence + postemergence split applications are labeled.
- Postemergence + postemergence split applications are allowed in Idaho, Oregon, and Washington. Refer to the label for rates at your location.
- Do not exceed 1 lb ai/A total metribuzin per season.
- Can control emerged weeds listed on the label. For optimum control, apply before weeds are 1-in tall.
- May be applied preemergence and/or early postemergence via chemigation: center pivot, solid set, and lateral roll system sprinklers.
- Incorporate pre-plant applications mechanically.
- Otherwise, when applied after potato planting, chemigate or incorporate via sprinkler irrigation. Do not incorporate mechanically because of the potential for decreased weed control and increased potato injury.
- NOTE: When metribuzin is applied postemergence with a ground sprayer for control of emerged weeds, overhead irrigation or rainfall within 24 h after application may decrease weed control.
- Metribuzin controls many annual broadleaf weeds, including redroot pigweed, common lambsquarters, kochia, Russian thistle, and wild mustard, but provides only fair to poor early-season suppression of hairy and cutleaf nightshade. Metribuzin also controls some annual grasses, such as wild oat, foxtail, and barnyardgrass.
- Re-cropping restrictions: refer to the label.
- Rotational cropping restrictions: Do not plant onions, lettuce, cole and Brassica crops, or cucurbits during the growing season following metribuzin applications. Do not plant sugar beets for 18 months after metribuzin application. Refer to the label for certain cereal varieties sensitive to metribuzin and other crops not listed here.
- Consult specific product labels for recommendations on timing tank mix applications.
- Site of action: Group 5, photosystem II inhibitor.
- Chemical family: Triazinone.

Metribuzin, currently sold under several trade names, is a triazinone herbicide that kills weeds by inhibiting photosynthesis, specifically photosystem II (PS II). When metribuzin is applied postemergence under cool, wet, or cloudy conditions, foliar injury can occur even on tolerant potato varieties. Symptoms usually disappear once warm, sunny conditions prevail. Injury is cosmetic and does not result in tuber yield reductions.

Typical injury symptoms include chlorosis (yellowing) of the leaf veins (Fig. 12.21) and necrosis (death). Symptoms appear on older leaves first.

Fig. 12.21 Potato veinal leaf chlorosis as a result of metribuzin injury; application was made during cool, cloudy conditions. Once conditions improve, then the metribuzin can be metabolized and no yield losses occur (some varieties are intolerant of metribuzin regardless of external conditions affecting metabolism

Potato varieties vary in their tolerance to metribuzin. As a general rule, red-skinned varieties are susceptible to injury, round-white varieties are susceptible to moderately tolerant, and russet varieties are tolerant.

Because there are exceptions, metribuzin should be applied on a new variety only after the injury response is known. Among commonly grown varieties, Russet Burbank, Ranger Russet, Russet Norkotah, Sangre (red-skinned), and Chipeta are tolerant; Yukon Gold (yellow-skinned) is moderately tolerant; Dark Red Norland is moderately sensitive; and Shepody (white-skinned) is very sensitive. See CIS 1185, "Weed Control and Potato Crop Safety with Metribuzin" (Hutchinson 2012) for further information on metribuzin potato variety tolerance.

If labeled, metribuzin should be applied preplant only on russeted or white-skinned varieties that are not early maturing. Metribuzin cannot be applied postemergence on early-maturing, smooth-skinned or white- or red-skinned varieties.

Weed populations resistant to metribuzin and possibly other herbicides with the same mechanism of action have been identified in many potato growing regions in North America. For instance, populations of metribuzin-resistant redroot pigweed and common lambsquarters have been identified in the PNW. Some of the common lambsquarters populations in Washington are resistant to other herbicides with the same mechanism of action. Metribuzin applied alone does not control these weeds. However, a tank-mixture of metribuzin plus a herbicide with a different mechanism of action or a combination of herbicides with different mechanisms of action that are effective on these species has provided broad-spectrum weed control, including the resistant species.

Pendimethalin (Prowl H_2O, 3.8 lb ai/gal; Prowl 3.3EC, 3.3 lb ai/gal; and other trade names and formulations)

- Preemergence or postemergence-Soil active, only.
- Prowl H_2O: 1.5 pints/A on coarse-textured soil (0.7 lbs ai/A); 2 pints/A (0.95 lbs ai/A) on medium-textured soil with less than 3% OM; and 3 pints/A (1.4 lbs ai/A) on medium- textured soil with more than 3% OM or on fine-textured soil.
- Prowl 3.3EC: 1.2–1.8 pints/A (0.5–0.74 lb ai/A) on coarse-textured soil; 1.8–2.4 pints/A (0.74–1 lb ai/A) on medium-textured soil with less than 3% OM; 2.4–3.6 pints/A (1–1.5 lb ai/A) on medium-textured soil with more than 3% OM or fine-textured soil with less than 3% OM; and 3.6 pints/A (1.5 lb ai/A) on fine-textured soil with more than 3% OM.
- Preemergence application: Ground, chemigation, or aerial.
- Can be applied early postemergence up to the 6-in stage of potato growth.
- Does not control emerged weeds.
- Can be chemigated postemergence.
- Preemergence ground or aerial applications must have rain, irrigation, or shallow mechanical incorporation to move herbicide into the upper soil surface where weeds germinate. Prowl H_2O/3.3 EC is most effective with adequate rain or irrigation within 7 days of application. During the shallow mechanical incorporation for preemergence-incorporated applications, take care that equipment does not damage seed pieces or elongating sprouts.
- NOTE: This herbicide is labeled for preemergence-incorporated applications before or at drag-off; however, subsequent tillage will disrupt the herbicide layer and weeds can emerge.
- Prowl H_2O/3.3 EC is used primarily for control of annual grasses and certain broadleaf weeds, such as redroot pigweed, common lambsquarters, and kochia. It provides some early-season hairy nightshade suppression.
- Re-cropping restrictions: If crop fails because of weather, any crop registered for preplant-incorporated applications of Prowl H_2O/3.3 EC may be replanted without adverse effects the same year.
- Rotational cropping restrictions: Wheat and barley may be planted 4 months after application. Do not plant sugar beets, red beets, or spinach for 12 months after application; plow soil 12-in deep before planting these crops. All other crops can be planted the following year. Refer to the label for other rotational restrictions.
- Site of action: Group 3, microtubule assembly inhibitor; inhibits cell division (mitosis) and cell elongation.
- Chemical family: Dinitroaniline.

Pyroxasulfone (Zidua 0.85 lb ai/lb product formulated as a WG)

- Preemergence only. Can injure emerged potato plants.
- May be applied aerially or by ground.
- 1.5 oz/A on coarse-textured soil; 1.5–2 oz/A on medium- or fine-textured soils.
- Do not apply through any type of irrigation.
- Must be incorporated/activated with at least 0.5 in. rainfall and/or irrigation.

- Rotational crop restrictions: Alfalfa 10 months, sugar beet 12 month, wheat 1 month, small grains other than wheat 11 months.
- Site of action: Group 15.
- Chemical family: isoxazoline.

Rimsulfuron (Matrix 25 DF, Matrix SG, Matrix FNV, Solida, or others 25% formulations)

- Preemergence or postemergence: use 1–1.5 oz/A (0.0156–0.023 lb ai/A).
- Can split preemergence + postemergence or postemergence + postemergence applications.
- Do not exceed 2.5 oz/A (0.039 lb ai/A) per year.
- **NOTE: Prism SG is the rimsulfuron product labeled for use in Canada.** It can only be applied postemergence once per season at 60 g/ha; 24 g/A, 0.86 oz/A (0.0134 lb ai/A).
- Can control emerged weeds listed on the label; best when less than 1-in. tall.
- Can be applied via chemigation.
- After preemergence ground or aerial applications, incorporated rimsulfuron in soil with 0.33–1 in. sprinkler irrigation (or a single rain) depending upon soil texture and as soon as possible or within 5 days after application.
- Follow postemergence ground or aerial applications by 0.33–1 in. of rain or sprinkler irrigation no sooner than 4 h or later than 5 days for soil activation and best weed control.
- Optimum time for cultivation after postemergence application is 7–14 days.
- Commonly grown potato varieties, including Teton Russet, have good tolerance to rimsulfuron. However, when applied postemergence during cool, cloudy conditions, rimsulfuron may temporarily yellow potato foliage and also may cause leaf malformations and stunt growth (Fig. 12.22a, b). Potatoes recover within 7–15 days. To reduce potential injury, apply only if the weather has been sunny at least 3 successive days. Avoid using COC or MSO adjuvants when potatoes are under heat stress. This injury symptom is sometimes mistaken for PVY infection.
- Do not use on seed potatoes unless permitted by a supplemental label.
- Although many weeds are susceptible to rimsulfuron, biotypes of kochia and several other weeds that are resistant to ALS-inhibitor herbicides are present in many potato production areas. These resistant biotypes are not controlled by rimsulfuron.
- Rimsulfuron controls a broad spectrum of weeds when applied either preemergence or postemergence. However, application timing affects control of some weed species. For example, common lambsquarters control can be better when rimsulfuron is applied preemergence rather than postemergence. In contrast, quackgrass and crabgrass control are better when rimsulfuron is applied postemergence rather than preemergence. See Table 12.6 for more information on advantages and limitations of various rimsulfuron application timings.
- Some common annual weeds are not effectively controlled by rimsulfuron, including cutleaf nightshade, Russian thistle, and wild buckwheat. These annuals require the use of tank mixtures for control (see HERBICIDE TANK MIXTURES AND TARGETED WEED CONTROL section).

Fig. 12.22 (**a**, **b**) Matrix applied postemergence to potatoes during cool, cloudy conditions can cause herbicide metabolism to slow resulting in mottled, yellowing of potato foliage and leaf crinkling and/or a pinched appearance to leaves

- Rimsulfuron does not control cutleaf nightshade, so it is crucial to know which nightshade species are present.
- Site of action: Group 2, acetolactate synthase (ALS) inhibitor.
- Chemical family: Sulfonylurea.

S-metolachlor (Dual Magnum, 7.62 lb ai/gal) Dual II Magnum in Canada

- Preplant incorporated: typically for yellow nutsedge control: 1–2 pt/A (0.95–1.9 lb ai/A) to the soil and incorporate into the top 3 in. before planting.
- Preemergence: 1–2 pints/A (0.95–1.9 lb ai/A), using lower rates on soils that are coarse or low in OM.
- Postemergence: 1.67 pints/A (1.6 lb ai/A) layby/after final hilling.
- NOTE: S-metolachlor is used in Canada formulated as Dual II Magnum (7.64 lb ai/gal). Refer to the label(s) to determine rates, timing, etc., for use in potatoes.
- Refer to the label(s) for application methods.
- Soil active, only. Does not control emerged weeds.
- Postemergence chemigation through center pivot is allowed.
- Postemergence application may be applied over a previous Dual Magnum application, but do not apply more than 3.6 pt/A of Dual Magnum in a single crop season.
- Both Dual Magnum and Dual II Magnum are chloroacetamide herbicide formulations that contain the same active ingredient, s-metolachlor. Dual II Magnum has a safener (benoxacor) included in the commercial formulation. Dual Magnum does not. The safener is used to protect corn from s-metolachlor injury that may

Table 12.6 Advantages and limitations of various rimsulfuron application timings

Preemergence application of rimsulfuron
Advantages
Little to no injury to potatoes
More tank-mix options than with postemergence treatments, especially when metribuzin-sensitive varieties are grown
Better common lambsquarters control than with postemergence application, especially when metribuzin-sensitive varieties are grown
When chemigated, rimsulfuron performance is more consistent with preemergence vs. postemergence application
Limitations
In a wet spring, heavy rainfall may move some soil-applied herbicide out of the weed-seed germination zone, which may allow some weeds to escape control
May not control deep germinating wild oats
Matrix applied early may break down before row closure when the competitiveness of the potato crop itself contributes substantially to weed control
Postemergence application of Rimsulfuron
Advantages
Rimsulfuron applications can be made when and where needed for weed control
Provides consistently good control of seedling hairy nightshade
Quackgrass, crabgrass, and wild oat control are better than with preemergence application
Less time for herbicide breakdown before row closure
Limitations
May cause temporary injury to potatoes, especially under stressful environmental conditions
Weather conditions may interfere with proper application timing
Fewer tank-mix options are available compared to preemergence application
Split application of Rimsulfuron (PRE + POST; POST + POST)
Advantages
Provides excellent broad-spectrum control of weeds and is particularly effective on quackgrass
Allows flexibility of postemergence control of escaped weeds
May be particularly useful for growers who build hills when planting and do not wish to cultivate further
Limitations
Higher cost due to extra rimsulfuron applied and to a second trip across field
Timing of postemergence application(s) may be difficult in reservoir tillage (Dammer-Diker®) fields
If control with first application breaks just before row closure, potatoes will be large and more susceptible to injury from the second rimsulfuron application
Greater potential for crop injury than a single preemergence or postemergence application, especially when rimsulfuron is tank-mixed with metribuzin at each application
Greater difficulty in timing the second rimsulfuron application to avoid injury associated with stressful environmental conditions, especially when rimsulfuron is tank-mixed with metribuzin at each application

occur when conditions are abnormally wet during corn germination and emergence. Either product may be used in potatoes; crop safety and weed control are similar.

- S-metolachlor is primarily used to control annual grasses, but can also control certain broadleaf weeds, such as redroot pigweed, and suppress common lambsquarters, hairy nightshade, and cutleaf nightshade. S-metolachlor can provide fair to good yellow nutsedge control.
- Application of s-metolachlor after potato emergence can cause leaf malformations.
- Refer to the label(s) for re-cropping and rotational cropping restrictions.
- Site of action: Group 15, inhibits very long chain fatty acid synthesis.
- Chemical family: Chloroacetamide.

Sulfentrazone (4 SC, 4 lb ai/gal, various trade names)

- Preemergence only (has soil and foliar activity but not safe to emerged potatoes).
- 3–8 fl oz/A (0.094–0.25 lb ai/A) depending on soil texture, percentage of OM, and pH.
- 3 fl oz/A—the lowest labeled rate—is highly recommended especially *when pH is above 7* (higher rates are allowed depending upon soil texture).
- Do not apply to soils classified as sand and with less than 1% OM.
- NOTE: Sulfentrazone is highly mobile in soils with pH of 7.5 or greater.
- Application to high pH soils, especially if coarse textured, and/or irrigation with highly alkaline water (high pH) after applying sulfentrazone, may increase the amount of herbicide available in the soil solution for uptake by the potato crop. Crop injury can occur in these situations.
- Ground, aerial, and chemigation. Make chemigation application only with center pivot, lateral move, end tow, solid set, or hand-move irrigation systems. Apply in 0.25–0.5 in. water/A. Refer to the EPA-approved labels for specific application methods/requirements.
- Performance is best with either rain or overhead irrigation within 7 days after application. If dry conditions exist for 7 days post application, incorporate sulfentrazone into the soil to a depth no more than 2 in.
- University of Idaho research has shown several potato varieties to be tolerant of sulfentrazone.
- Factors such as heavy rain after application; cool and moist conditions before row closure; stress from heat; and soil conditions such as coarse texture, low OM, or high pH, may affect crop response to the herbicide and herbicide availability, increasing risk of adverse crop response.
- Re-cropping restrictions: Any crop for which sulfentrazone is labeled.
- Rotational cropping restrictions: Sugar beet may not be planted for 36 months after application in potatoes; field corn, sorghum 10 months; sweet potato 12 months; 1 sweet corn or popcorn 18 months; canola 24 months. See label for other crops not listed here, including spring and winter small grains.
- Site of action: Group 14, protoporphyrinogen oxidase (Protox) inhibitor.
- Chemical family: Triazolinone.

Trifluralin (Treflan HFP 4 lb ai/gal; or others)

- Preemergence and/or postemergence.
- Treflan HFP applications may be split on soils with less than 2% OM in Idaho, Oregon, and Washington: 0.75 pint/A before planting + 0.75 pint/A postemergence, when potato plants have fully emerged. See the label for your location.
- Treflan HFP may only be applied preplant when tank mixed with Eptam (see labels).
- Refer to the label for application methods.
- Not for use in the State of Maine.
- Treflan HFP is primarily an annual grass herbicide that is effective on foxtail, barnyardgrass, crabgrass, and other grassy weeds. It also controls some broadleaf weeds, including redroot pigweed, common lambsquarters, and kochia. It does not control hairy nightshade or wild mustard, and only suppresses wild oat.

Formulated Pre-mixes or co-packs

Boundary 6.5EC is a formulated pre-mix of metolachlor + metribuzin (5.25 lb ai S-metolachlor + 1.25 lb ai/gal metribuzin; and other trade names).

- Preemergence: 1.5–2.9 pints/A (1 lb ai/A S-metolachlor + 0.23 lb ai/A metribuzin to 1.9 lb ai/A S-metolachlor + 0.45 lb ai/A metribuzin).
- Preemergence: ground, irrigation (center pivot irrigation equipment), or aerial.
- Postemergence (for application in center pivot irrigation water only): 1.5–2.2 pints/A (1 lb ai/A S-metolachlor + 0.23 lb ai/A metribuzin to 1.3 lb ai/A S-metolachlor + 0.34 lb ai/A metribuzin).
- Two applications permitted per year.
- Refer to label for rate ranges according to soil texture.
- The metribuzin in this product can control emerged weeds listed on the label.
- Depending on the Boundary 6.5EC rate used, tank-mixing with additional metribuzin is allowed as long as rate limits and label specifics are followed.
- Rain or irrigation is required to activate Boundary 6.5EC. In areas of low rainfall, follow a preemergence application with irrigation of 0.25–0.5 inch of water. Do not irrigate heavily immediately after application.
- Refer to the label(s) for rate ranges according to soil texture.
- Refer to the label(s) for re-cropping and rotational crop restrictions.
- Because this product contains metribuzin, potato variety tolerance is variable. Refer to the Metribuzin section in this chapter and the Boundary 6.5EC and metribuzin labels.
- Site of action: Group 15, (s-metolachlor) and Group 5: (metribuzin).
- Chemical family: chloroacetamide (s-metolachlor); triazinone (metribuzin).

Sulfentrazone + metribuzin in a formulated pre-mix: Sulfentrazone MTZ DF is an example (0.27 lb ai metribuzin + 0.18 lb ai sulfentrazone per lb of product [total of 0.45 lb active ingredient per pound of product]).

- Preemergence only (has soil and foliar activity but is not safe to emerged potatoes).
- 8.3–22.2 oz/A (0.14 lb ai/A metribuzin + 0.09 lb ai/A sulfentrazone to 0.37 lb ai/A metribuzin + 0.25 lb ai/A sulfentrazone).
- *The lowest rate of 8.3 fl oz/A is highly recommended especially when soil pH is above 7*. Higher rates are allowed depending on soil texture, pH, and percentage of OM. Refer to the label.
- Do not apply more than 22.2 oz/A Sulfentrazone MTZ DF or more than 0.25 lb ai/A sulfentrazone from any source in the 12 months after the first application.
- This product contains metribuzin, so potato variety tolerance is variable. Refer to the Metribuzin section in this chapter and metribuzin labels.
- Refer to the information on Sulfentrazone 4SC (and various trade names) and metribuzin in this section and the specific herbicide labels for other cautionary notes on use in high-pH soils and the impact of adverse weather conditions on crop injury.
- Sites of action: Groups 14 and 5.
- Chemical families: Triazolinone (Sulfentrazone) and Triazinone (metribuzin).

Sencor STZ is a co-pac k (separate container for each herbicide—not premixed) of Sencor DF and sulfentrazone (STZ) (7.5 kg + 1.95 L) and is only sold in Canada.

- Preemergence only.
- 600–800 g/ha (243–324 g/ac) of Sencor and 157–219 mL/ha (64–89 mL/ac) of STZ.
- A minimum of 1 in. of soil must cover emerging potato shoots at application.
- If application is delayed, injury may occur if potato seed pieces are germinating or located near the soil surface.
- After a hilling event, allow the soil to settle or crust before application.
- Avoid soil disturbance, including hilling, after application. If hilling is required after application, it is recommended to wait as long as possible.
- Sencor STZ requires rainfall/sprinkler incorporation within 10–14 days of application to be activated.
- Sencor STZ is recommended for soils with OM content between 1.5–6% and a pH of less than 7.8. See label for further information.
- *The lowest rate of STZ is highly recommended, especially when soil pH is greater than 7.*
- Use higher rates for longer-season potatoes, situations where longer control is required and weed infestations are heavy, and on soils with a pH less than 7.0 and OM greater than 3%.
- See label for re-cropping and rotational crop restrictions.
- Sites of action: Groups 5 and 14.

Postemergence for Grass Control Only

Do NOT chemigate these postemergence, grass-only herbicides.

Clethodim (Select 2EC, 2 lb ai/gal)

- *Clethodim only controls emerged grass weeds - Foliar active, only.*
- 6–16 fl oz/A (0.094–0.25 lb ai/A). See label for recommended rates on specific grass species and growth stages. Use the high rate under heavy grass pressure and/or when grasses have reached maximum height. Do not exceed a total of 32 fl oz/A per season.
- Re-cropping restrictions: None on the label.
- Rotational cropping restrictions: None on the label.
- Site of action: Group 1, acetyl CoA carboxylase (ACCase) inhibitor.
- Chemical family: Cyclohexanedione.

Fluazifop-p-butyl (Venture L (125 g/L [labeled for use in Canada only])

- Postemergence only; for grass control only.
- 2 L/ha with the appropriate surfactants.
- Apply using ground equipment only.
- Do not cultivate for 5 days after applying Venture L Herbicide.
- Rotational crop restriction: 12 months for all crops other than those listed on the Venture L label.
- Site of action: Group 1, acetyl CoA carboxylase (ACCase) inhibitor.
- Chemical family: aryloxyphenoxypropanoates.

Sethoxydim (Poast Plus or Poast Ultra and others)

- Postemergence only; for grass control only.
- See the label for rates, etc.
- Controls many annual grass weeds, including foxtail, barnyardgrass, volunteer grain, and wild oat. It also suppresses quackgrass, a perennial weed.
- Must always be used with a nonphytotoxic oil concentrate in order to achieve effective weed control. Poast Plus oil concentrate is safe on all potato varieties.
- Site of action: Group 1, acetyl CoA carboxylase (ACCase) inhibitor.
- Chemical family: Cyclohexanedione.

Foliar-Active "Burndown" Herbicides Labeled for Use Only Before Planting or After Planting but before Potato Emergence

Carfentrazone-ethyl (Aim EC and others)

- Up to 2 fl oz/A Aim EC (0.031 lb ai/A).
- Apply before planting or up to 24 h after potatoes have been planted. Applications must be made before the crop emerges. Only the pre-plant timing may be used in Canada.

- Application methods: Ground in a minimum of 10 gal/A or aerial in a minimum of 3 gal/A. Use higher spray volumes when there is a dense weed population or crop canopy. Apply Aim EC with a crop oil concentrate at 1% v/v or a methylated seed oil at a minimum of 1 qt/A or 1% v/v when applied in volumes of more than 20 gal/A.
- Apply to actively growing weeds not more than 4-in tall or rosettes with a 3-in diameter. Coverage is essential for good control. Tank mixes with other herbicides may increase spectrum of control.
- Re-cropping restrictions: Following an application of Aim EC, a registered crop may be planted at any time.
- Rotational cropping restrictions: A field may be rotated to a registered crop at any time after an Aim EC application. All other crops may be planted after 12 months.
- Site of action: Group 14, protoporphyrinogen oxidase (Protox) inhibitor.
- Chemical family: Triazolinone.

Glyphosate (Glyphosate Original, Roundup PowerMAX, Roundup WeatherMAX, Roundup Custom, Roundup Original Max, or Others)

- 0.38–1.1 lb ae/A for most species.
- Delay herbicide application to allow maximum weed emergence, but apply before potatoes emerge. Herbicide will not control weeds that emerge after application. In coarse or sandy soils, apply before potato sprouting to minimize crop damage. Good growing conditions enhance glyphosate activity.
- Application methods: See label for rates and gal/A of water recommended for specific species. Some glyphosate formulations require use of nonionic surfactant; see label for details on additive use. Adding 1–2% dry ammonium sulfate (AMS) by weight or 8.5–17 lb/100 gal spray mix may increase performance. The equivalent rate of AMS in a liquid formulation also may be used.
- *Caution:* Glyphosate applied after crop emergence will injure or kill potatoes. Follow all use restrictions and precautions on label.
- Site of action: Group 9, inhibits EPSP synthase.
- Chemical family: None generally accepted.

Paraquat (Gramoxone SL or Inteon, 2 lb paraquat cation/gal; Firestorm, 3 lb paraquat cation/gal; or Others) (Not for use in Canada)

- Gramoxone SL or Inteon 1–2 pints/A (0.25–0.5 lb ae/A); or Firestorm 0.7– 1.3 pints/A (0.26–0.49 lb ae/A). Do not exceed 3 applications of Firestorm per year.
- Apply after weeds emerge but before potatoes emerge. Delay application to allow maximum weed emergence, but apply no later than ground cracking, before potatoes emerge. Paraquat will not control weeds that emerge after application.
- Application methods: Ground or aerial. Add a nonionic surfactant containing 75% or more surface-active agent at 0.125% v/v (1 pint/100 gal spray mix) or crop oil concentrate at 1% v/v (1 gal/100 gal spray mix) for ground applications or 1 pint/A for aerial applications. The herbicide kills most green plant growth on contact; thus good spray coverage is essential.

- Nonselective, postemergence herbicide. Paraquat is rainfast 30 minutes after application.
- Site of action: Group 22, photosystem I electron diversion.
- Chemical family: Bipyridilium.
- *Caution:* A restricted-use herbicide (RUP). Applications after crop emerges have reduced yields because of injury. Follow all use restrictions and precautions on label. Requires special safety equipment for handling, mixing, and spraying. **Glyphosate + s-metolachlor** (Sequence) (pre-mix of 2.5 lb ae + 3 lb ai/A glyphosate + s-metolachlor).
- Rate: Do not exceed 2.5 pints/a formulated product on coarse soils, 3.75 pints/a formulated product on medium soils with less than 3% organic matter, or 4 pints/a formulated product on fine textured soils with greater than 3% organic matter.
- Do not exceed 4 pints/a of Sequence per season.
- Time: Apply anytime before or after potato planting but before emergence.
- Caution Application must be made before crop emergence. Contact with potato foliage will result in crop injury.
- Sequence may be applied up to 30 days before planting.
- Site of action (S-metolachlor) Group 15: inhibits very long chain fatty acid synthesis; (glyphosate) Group 9: inhibits EPSP synthase.
- Chemical family (S-metolachlor) Chloroacetamide; (glyphosate) none generally accepted.

Box 12.3: Nightshade Weed Control: Differences in Species and Control Levels Listed on Labels

Labels for herbicides active on nightshade weeds do not always include all of the common species found in potato production—hairy, cutleaf, black, and Eastern black nightshade. Label language includes: suppressed, controlled, partially controlled, targeted.

Chateau: Hairy, black, and Eastern black nightshade *suppressed.*
Eptam: Hairy, cutleaf, black nightshade *controlled.*
Linex/Lorox: Hairy and black nightshade are "*targeted,*" and Eastern black nightshade is *suppressed.*
Matrix: Hairy and black nightshade *partially controlled.*
Reflex: Hairy nightshade *partially controlled;* black and Eastern black nightshade *controlled.*
Sulfentrazone: Black and Eastern black nightshade *controlled.*
Zidua: Black and Eastern black nightshade *controlled.*

Herbicide Application and Tillage Timing

Using an integrated weed management strategy will provide the best weed control in potatoes. Integrated weed management makes use of all the cultural, mechanical, chemical, and biological tools available for weed control, rather than relying on any single weed control tool.

From potato planting to row closure is a busy time for growers. It's also a busy time for weeds. Applying herbicides at the proper time and linking up with tillage operations between planting and row closure is key for weed control from start to finish. Depending upon location, potato variety, and a few other factors including the weather, time from planting to emergence can be 3–4 weeks and from emergence to row closure when potatoes finally start to help control weeds with shading and competition, 4–5 weeks (Fig. 12.23).

There are different approaches to herbicide applications related to tillage timing. Three timing scenarios are described in this chapter: (1) plant, hill at planting, then apply soil-active herbicides simultaneously or immediately after planting/hilling; (2) plant, then after potato emergence, hill and apply a planned postemergence herbicide tank mix. When using this scenario, some growers make a herbicide application via chemigation and/ or perform a "drag off" before potato emergence; (3) plant, and before potato emergence, hill then apply and incorporate soil-active herbicides as soon as possible.

NOTE: The hilling operation should be the last cultivation performed before potato harvest. *Tillage after herbicides have been applied will disrupt the herbicide barrier and bring up untreated soil after which weeds may germinate and emerge. Refer to Table 12.5 for more information about herbicide timing and activity.*

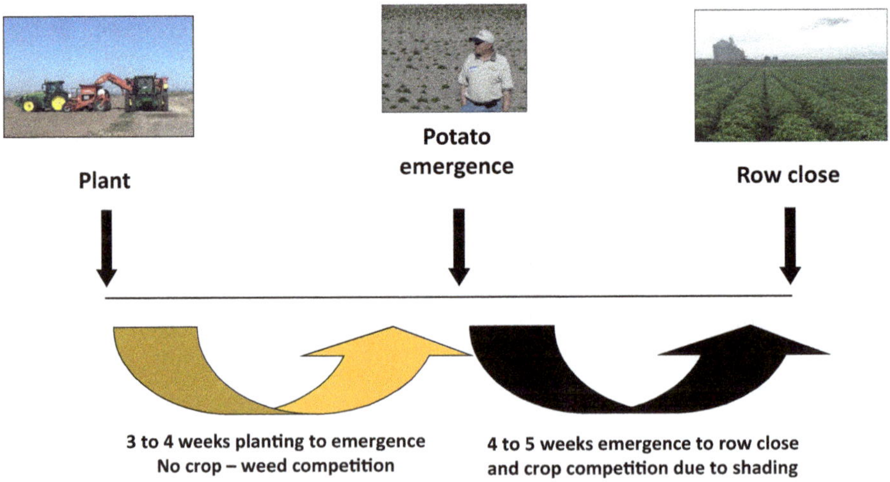

Fig. 12.23 Timeline between planting potatoes, potato emergence, and row closure

In Scenario 1, if the herbicides applied after potato emergence and hilling are soil-active and applied by ground/air, then they should be incorporated with rainfall or overhead irrigation, or chemigated if labeling allows. In Scenarios 2 and 3, pre-emergence, soil-active herbicides are chemigated or ground/aerial applied then incorporated with rainfall or overhead irrigation.

Scenario 1 Plant, hill at planting, then apply soil-active herbicides simultaneously or immediately after planting/hilling. (Fig. 12.24). All soil-active herbicides are labeled for this application timing: Chateau, Dual Magnum/Dual II Magnum, Eptam, Matrix/Prism, metribuzin (various trade names), Outlook, Prowl H_2O/3.3 EC, Reflex, Linex/Lorox, metolachlor (various trade names), Sonalan HFP, sulfentrazone (various trade names), Sulfentrazone MTZ/Sencor STZ (sulfentrazone + metribuzin), Zidua, and Treflan HFP (Table 12.5). Read labels and follow directions for correct application and incorporation methods.

NOTE: Herbicides applied at this time must be effective for at least 8–9 weeks when the potato foliage closes over the rows to provide shading and competition with weeds.

- Where labeled, growers can tank mix the soil-active herbicides with burndown herbicides glyphosate, paraquat, Sequence, and/or Aim EC if weeds are present at planting.
- These non-selective herbicides must have time to have an effect on emerged weeds before incorporation occurs for the soil-active herbicides.
- If weeds emerge after potato emergence, a herbicide application with a foliar-active herbicide safe to potatoes may be necessary.

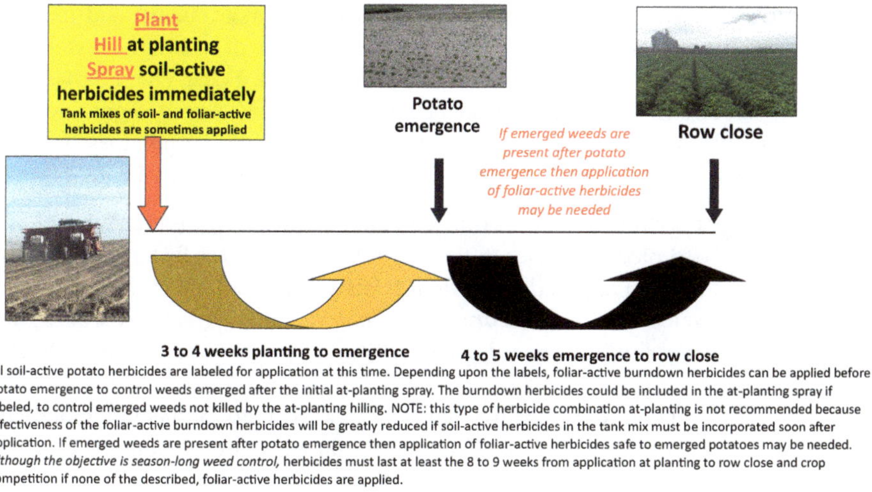

Fig. 12.24 Scenario 1: Plant, hill at planting, then apply soil-active herbicides simultaneously spray immediately after planting/hilling. If weeds emerge later, then a postemergence herbicide application including foliar active herbicides may be necessary

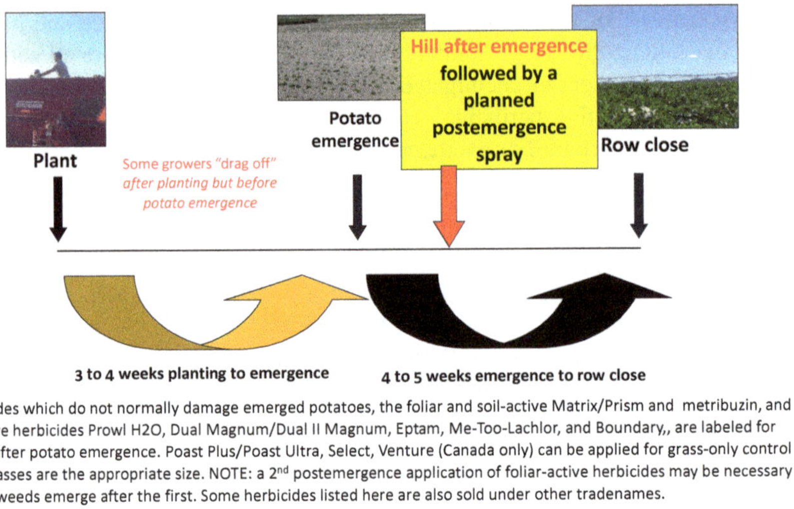

Fig. 12.25 Scenario 2: The hilling operation is not conducted until after potato emergence when the potato rows can be seen. Herbicides are applied and incorporated as soon after the hilling as possible

Scenario 2 Although not as common as Scenarios 1 and 3, in Scenario 2 potatoes are planted, but the hilling operation is not conducted until after potato emergence when the potato rows can be seen. Herbicides are applied as soon as possible after that hilling (Fig. 12.25). Consequently, only the following soil active herbicides which will not injury emerged potatoes are labeled for application after potato emergence (Table 12.5): Matrix/Prism, metribuzin, Eptam, Prowl H_2O, Dual Magnum/Dual II Magnum, Boundary (s-metolachlor + metribuzin), and metolachlor (various trade names). Only two of these herbicides, Matrix/Prism and metribuzin, have foliar activity to contol emerged weeds. Poast Plus/Poast Ultra, Select, and Venture (only in Canada) can be applied to control emerged grass weeds. Others herbicide such as Outlook can't be applied POST to the potatoes because inury can occur. A second postemergence herbicide application may be necessary.

Some growers perform a drag-off tillage to knock down the hills created at planting so that the seed piece is closer to the warm soil surface for faster germination than if buried at the 5–6 in planting depth in relatively colder soil.

The field is leveled, and *unless a precision planter with GPS features is used, the final hilling-reservoir tillage cannot occur until after potato emergence when the rows can be seen.*

The drag-off can control weeds that have already emerged. Some herbicides can even be incorporated with a shallow tillage operation with care taken not to damage the seed piece or potato shoots close to the surface.

Tillage after herbicides have been applied will disrupt the herbicide barrier. Therefore, herbicides in this scenario would be most effective if applied after the postemergence hilling-reservoir tillage.

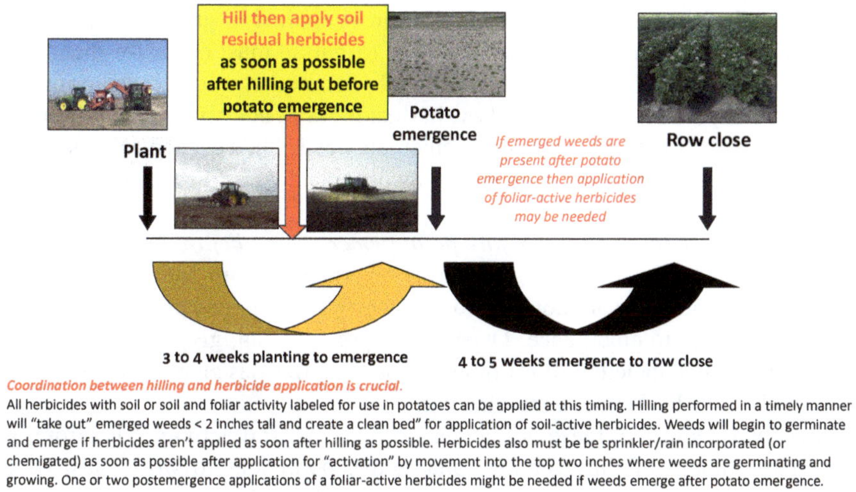

Fig. 12.26 Scenario 3: Hilling is performed before potato emergence. Soil-active herbicides are applied and incorporated immediately after hilling but before potatoes emerge. A subsequent postemergence herbicide application may be necessary

Scenario 3 Plant, then before potato emergence, hill and apply and incorporate herbicides immediately after hilling. A subsequent application of foliar-active herbicides may be necessary if weeds emerge after potato emergence (Fig. 12.26).

Hilling followed as soon as possible by herbicide application before potato emergence can be one of the most effective programs for weed control in potatoes.

- Weed control by herbicides in this scenario do not have to last as long as in the plant-hill-spray scenario.

  Hilling can be performed any time after planting and before potato emergence.

- Waiting 1–3 weeks after planting would allow for some weeds to emerge; which if small enough, will be killed with the hilling.

All herbicides labeled for use in potatoes before emergence can be applied at this time. Chateau, Dual Magnum/Dual II Magnum, Eptam, Matrix/Prism, metribuzin (various trade names), Outlook, Prowl H_2O, Reflex, Linex/Lorox, Me-Too-Lachlor (and various trade names), Sonalan HFP, sulfentrazone (various trade names), Metribuzin STZ, and Treflan HFP (Table 12.5).

As in Scenario 1, and if labeled in your area, glyphosate, paraquat, Sequence, and Aim EC can also be applied before potato emergence. These herbicides will injure/kill emerged potatoes.

- It is recommended to wait before using these burndown herbicides until after hilling and only if weeds emerge.
- Application before hilling may be necessary if weeds emerge but hilling is delayed due to conditions such as prolonged rainfall-wet soils.

NOTE: All of these soil-active herbicides can be applied by ground or via chemigation. If applied by ground, rain or sprinkler irrigation must occur for incorporation and activation in the top 2-in layer of soil where most weed seeds germinate. Although incorporation of some of these herbicides does not have to occur immediately, since they do not have activity on emerged weeds, the best control possible is usually achieved when the herbicide is incorporated as soon as possible after application.

What if weeds have emerged before potato emergence—before or after drag-off and/or hilling-reservoir tillage? "Burndown" herbicides with foliar activity only (no soil residual activity) are non-selective, destroy any emerged plant, and must be applied before potato emergence. Depending upon location, glyphosate, paraquat, and/or Aim EC are labeled for this timing in potatoes. Besides glyphosate, Sequence has the soil-active herbicide, s-metolachlor.

Other than Sequence, these herbicides do not have soil activity. Therefore, an appropriate amount of time for the foliar-active herbicides to work must occur before rainfall/sprinkler irrigation incorporation of the soil-active herbicides in the tank mixture.

Herbicide Tank Mixtures for Targeted Weed Control

An effective way to keep the integration of cultural, mechanical, and herbicides but improve control in each field is to customize the herbicide tank mixture to target the specific weed species in each field. The same weed management approach will not work for all fields because the number of weed species can vary greatly from one field to another even if they are in close proximity. Tank mixtures with different mechanism of action herbicides broaden the weed control spectrum and reduce the potential for developing herbicide-resistant weed populations. Accurate field history information is key to selecting the best tank mixtures and sequential application programs.

NOTE: Read and follow applicable "Restrictions and Limitations and Directions for Use" on all product labels in a tank mixture. The most restrictive labeling applies to use of tank mixtures.

The label is a legal document. Always read and follow instructions on the herbicide label. When tank-mixing herbicides, use the most restrictive label. Information such as rates, potato-variety sensitivity, pre-harvest intervals, and rotational crop restrictions are not provided in this chapter. Tank-mix Partner (TMP) Choice Charts can aid in developing highly successful weed control programs. Table 12.7 is an example of how to use a chart of this kind to determine various premergence-applied and incorporated herbicide tank mixtures that will control combinations of five weed species. Similar Charts can be created to determine the most effective tank mixtures for combinations of other weed species and with different tillage and applications timings.

Table 12.7 Tank Mix Partner Choice Chart to determine the season-long effectiveness of herbicides labeled for use in potatoes on five weeds: hairy nightshade, redroot pigweed, common lambsquarters, kochia, and green foxtail. H erbicides labeled for use in potatoes in the U.S. and Canada are listed in the first column. Herbicides in the same color are in the same MOA group. Not all trade names for herbicide are provided in this chart due to space limitations. Trade names specific to Canada are noted. Control levels in the chart are G, Good = 90 to 100%; F, Fair = 80 to 89%; S, Suppress = approximately 50%, PN, Poor to None = 0 to 30%, N, None = 0%. Control information for this chart was gathered from multiple sources: the herbicide label, research results, and herbicide effectiveness tables such as the ones included in this chapter. The five weeds shown are an example of species which might be present in a potato production area where the fields of interest are located

TANK MIX PARTNER CHOICE CHART					
Herbicides[a]	Weed species of interest				
	Hairy nightshade	Redroot pigweed	C. lambs-quarters	Kochia	Green foxtail
Chateau (flumioxazin)	G	G	PN	S	N
Sulfentrazone (various names)	G	G	PN	G	N
Reflex (formesafen)	F	G	PN	F	S
Outlook (dimethenamid-p)	G	G	PN	F	G
Dual Magnum/ Dual II Magnum[b] (s-metolachlor)	F	G	PN	F	G
Metolachlor (various names)	F	G	PN	F	G
Zidua (pyroxasulfone)	F	G	PN	S	F
Matrix (and others) (PRE or POST)	G	G	PN	F	F
Prism[b] (rimsulfuron (POST only)	G	F	PN	F	F
Eptam (EPTC)	G	G	S	F	G
Sonalan HFP (ethafluralin)	PN	G	F	F	G
Treflan HFP (trifluralin)	PN	G	F	F	G
Prowl H2O (and others) (pendimethalin)	S	G	G	F	G
Metribuzin (various names)	N	G	G	G	G
Linex/Lorox (linuron)	F	G	G	F	G
Boundary (and others) *s-metolachlor metribuzin*	F	G	G	F	G
Sencor STZ[b] (Canada) **Sulfentrazone MTZ** *metribuzin sulfentrazone*	G	G	G	G	G
Poast Plus or Ultra (sethoxydim) **Select** (clethodim) **Venture**[b] (fluoxifop-p-butyl)	N	N	N	N	G

s

a

Weed control goal - Best case would be to have a two- or three-way tank mix providing 90 to 100% control of all weeds present. If not possible, of course the next best would be to have at least one herbicide with 90 to 100% matched up with a herbicide that can provide 80 to 89% (F) control.

Herbicide resistance management goal during the potato crop year - Choose herbicide tank mix partners with different MOAs that have overlapping control of a weed species present i.e. a species is controlled 90 to 100% (G) with at least two different herbicide MOAs. If it is not possible to have more than one MOA controlling the same weed during the potato year, choosing herbicides with different MOAs in the rotation crops is especially important.

Using the TMP Choice Chart: Select the combination of weed species to be targeted for control. An example of two weeds to target are shown in Table 12.8: hairy nightshade and common lambsquarters. Next, note the herbicides providing 90 to 100% (G) of the two weed species in the hypothetical situation. Herbicides meeting these criteria are marked in Table 12.8 with circles, brackets, and/or arrows. As shown, the weed control goal can be met with several tank mixtures. However, since most of the herbicides that control hairy nightshade do not control common lambsquarters and vice versa, the herbicide resistance management cannot be met as easily.

Matrix + Prowl: Matrix does not control common lambsquarters (Fig. 12.27a), and Prowl H2O does not control hairy nightshade (Fig. 12.27b) at the desired level. A tank mix of Matrix and Prowl H2O applied preemergence immediately after hilling and sprinkler incorporated can provide 90 to 100% (G) season-long control of this combination of weeds (Fig. 12.27c).More than one PRE-applied tank mixture can provide 90 to 100% season-long control of hairy nightshade and common lambsquarters.

Outlook + Linex/Lorox (Fig. 12.28): Hairy nightshade is controlled 90 to 100% (G) by Outlook and 80 to 89% (F) Linex/Lorox. Common lambsquarters is controlled by Linex/Lorox 90 to 100% (G). Outlook has some activity but it is only 0 to 30% (PN).

Outlook + metribuzin is another tank mix which can satisfactorily target the combination of hairy nightshade and common lambsquarters in a potato field (Fig. 12.29a–d).NOTE: It may not be possible to customize a tank mixture with herbicides that can provide 90 to 100% control of every weed species in the field with more than one MOA. What is important is that if one herbicide does provide 90- to 100% control, then at least some control activity from the other herbicide is better than no control by that herbicide.A field that has high densities of some or all of the weeds present, however, wouldwarrant making absolutely sure that more than one herbicide providing 90 to 100% control of the same high-density weed(s) is included in the tank mixture.

University of Idaho research has shown that two-way tank mixtures of Matrix, Chateau, Eptam, Outlook, Reflex or sulfentrazone will improve preemergence hairy nightshade control compared with any of these herbicides applied alone, especially in heavily infested fields (Fig. 12.30a, b). Do not tank mix any of these herbicides which have the same MOA e.g. Chateau and Sulfentrazone are both Group 14

Table 12.8 A Tank Mix Partner Choice Chart with examples of targeted, two-way tank mixtures s for season-long control of a mixed population of hairy nightshade and common lambsquarters in a potato field. Herbicides labeled for use in potatoes in the U.S. and Canada are listed in the first column. Herbicides in the same color are in the same MOA group. Not all trade names for herbicide are provided in this chart due to space limitations. Trade names specific to Canada are noted. Control levels in the chart are G, Good = 90 to 100%; F, Fair = 80 to 89%; S, Suppress = approximately 50%, PN, Poor to None = 0 to 30%, N, None = 0%. Control information for this chart was gathered from multiple sources: the herbicide label, research results, and herbicide effectiveness charts such as the ones included in this chapter

TANK MIX PARTNER CHOICE CHART					
	Weed species of interest				
Herbicides[a]	**Hairy nightshade**	**Redroot pigweed**	**C. lambs-quarters**	**Kochia**	**Green foxtail**
Chateau (flumioxazin)	G	G	PN	S	N
Sulfentrazone (various names)	G	G	PN	G	N
Reflex (formesafen)	F	G	PN	F	S
Outlook (dimethenamid-p)	G	G	PN	F	G
Dual Magnum/ Dual II Magnum[b] (s-metolachlor)	F	G	PN	F	G
Metolachlor (various names)	F	G	PN	F	G
Zidua (pyroxasulfone)	F	G	PN	S	F
Matrix (and others) (PRE or POST)	G	G	PN	F	F
Prism[b] (rimsulfuron (POST only)	G	F	PN	F	F
Eptam (EPTC)	G	G	S	F	G
Sonalan HFP (ethafluralin)	PN	G	F	F	G
Treflan HFP (trifluralin)	PN	G	F	F	G
Prowl H2O (and others) (pendimethalin)	S	G	G	F	G
Metribuzin (various names)	N	G	G	G	G
Linex/Lorox (linuron)	F	G	G	F	G
Boundary (and others) *s-metolachlor metribuzin*	F	G	G	F	G
Sencor STZ[b] (Canada) **Sulfentrazone MTZ** *metribuzin sulfentrazone*	G	G	G	G	G
Poast Plus or Ultra (sethoxydim) Select (clethodim) Venture[b] (fluoxifop-p-butyl)	N	N	N	N	G

herbicides. Sometimes, when any of the preemergence-only herbicides are used, a postemergence application may also be needed due to heavy weed pressure, late-germinating weeds, ineffectiveness of preemergence herbicides on certain weed species, or simply as part of the herbicide program plan. Matrix and/or metribuzin

Fig. 12.27 (**a**) Matrix alone is not effective on common lambsquarters; (**b**) Prowl H_2O alone does not control hairy nightshade; (**c**) a tank mixture of Matrix and Prowl H_2O can control a mixed weed population of hairy nightshade and common lambsquarters

Fig. 12.28 A tank mix of Outlook and Linex/Lorox applied preemergence immediately after hilling and sprinkler incorporated within 24 h of application can provide season-long control of hairy nightshade, common lambsquarters, redroot pigweed, and green foxtail

are commonly used in this instance, since they can be applied postemergence to potatoes and have activity on emerged weeds. Postemergence application timing in relation to weed size is extremely important. In Fig. 12.31a, b, metribuzin was applied postemergence to the weeds and potatoes, but some of the weeds were too large at application time to be controlled. In addition, a herbicide effective on hairy nightshade was not included in the overall program.

Fig. 12.29 (**a**) A mixed population of hairy nightshade and common lambsquarters without herbicides; (**b**) metribuzin applied preemergence alone does not control hairy nightshade; (**c**) a close-up of hairy nightshade plants not controlled by metribuzin applied preemergence; (**d**) Outlook plus metribuzin applied preemergence immediately after hilling and sprinkler incorporated provides control of hairy nightshade and common lambsquarters

Solubility of Soil-Active Herbicides and Effect on Crop Injury and Weed Control

Herbicide solubility in water is often expressed as the weight of the compound that dissolves in 1 liter of water, stated as mg/L or sometimes ppm. Water with the same temperature and pH is used for this test so that solubility can be compared equitably.

The larger the value, the more soluble the herbicide is in water.

Depending upon soil characteristics, *the more soluble the herbicide, the more available it is for uptake, and the further it can move down in the soil profile*. The following lists the solubility of some of the herbicides used in potatoes from high to low:

Fig. 12.30 University of Idaho research has shown that high-density hairy nightshade infestations can be more effectively controlled when two of the "hairy nightshade herbicides," are combined in a two-way tank mixture, applied preemergence immediately after hilling and sprinkler incorporated **a**) high density population of hairy nightshade, **b**) control of a heavy infestation of hairy nightshade with Outlook + Sulfentrazone

Fig. 12.31 (**a**, **b**) Metribuzin was applied postemergence. Some of the weeds were too large at application time to be controlled. In addition, a herbicide effective on hairy nightshade was not included in the overall program

Matrix >> metribuzin, Outlook >> Dual Magnum, Eptam >> Linex, Reflex >> Chateau > Sonalan HFP, Treflan HFP, Prowl H_2O.

Solubility, herbicide action in the soil, and point of uptake by weeds not yet emerged are factors influencing potato injury. Possible injury symptoms from a given herbicide are listed in Table 12.9. For instance, Treflan HFP is mainly absorbed

Table 12.9 Potato herbicide solubility; location of plant uptake below ground; action disrupting plant germination, growth, and emergence; and possible potato injury symptoms

Herbicide trade name	Solubility in water ppm (mg/L @ 77 °F pH 7)[a]	Location of uptake by broadleaf plant *below* soil surface	Herbicide action *in Soil* to disrupt germination, growth, and/or emergence	Possible potato injury symptom(s)
Matrix	7,300	Germinating seedling	Amino acid inhibitor—Stops growth	If potato plant emerges, leaf mottling and puckering/wrinkling; growth could slowly stop
Metribuzin (multiple trade names)	1,200	Root/some shoot	Photosynthesis inhibitor—Susceptible weeds *emerge and die* immediately after emergence	If potato plant emerges, yellowing in the leaf veins (not between veins). White-skin, short-season varieties are usually more susceptible than russeted
Sulfentrazone	780[b]	Root	Cell membrane disruptor	If plants emerge, then sometimes blackening
Outlook	1,174	Below-ground shoot and some root; possibly movement w/in weed seedling before emergence	Stops growth of weed shoots, germinating seedlings	If plants emerge, then a drawstring effect on leaves and/or distorted, sometimes downward cupping leaves; stem might not recover and new stem/sprout growth below surface; can stop germination
Dual magnum	488			
Eptam	370	Absorbed as a gas by shoots; can move in germinating seedlings	Inhibits weed seedling germination	If plants emerge, then possibility of leaf distortion/puckering; possible stunting; emergence delayed
Linex	75	Root/some shoot	Photosynthesis inhibitor (*plants emerge and die quickly*)	If plants emerge, yellowing on leaf edges, overall yellowing or between-the-veins yellowing of leaves; possible stunting
Reflex	50	Root	Cell membrane disruptor	Emerged plants splashed by treated soil during intense rainfall could have burnt spots on stems, leaves; if in contact with treated, wet soil, yellowing, burnt leaves
Chateau	2			
Sonalan	1	Mainly root	Stops plant cells from dividing—Root inhibitor	Stubby, "club-foot" roots; little or no root hairs
Treflan	<1			
Prowl H_2O	<1	Roots only; *possible effect on shoots only in direct contact*	Stops cell division in roots—Cells might enlarge but do not divide	Stubby, "club-foot" roots; little or no root hairs; possible brittle stems; if somehow in contact with shoots below ground, then possible thickening

Solubility measured in pH 7 at 77 °F. *The greater the solubility value, the more soluble the herbicide* relative to the other herbicides in the table

[a]Solubility values adapted from the 10th edition of the *Herbicide Handbook, Weed Science Society of America* (Shaner 2014)

[b]When tested at pH 7.5, sulfentrazone solubility increases to 160,000 ppm

by the plant root and acts by stopping plant cells from dividing. Injury could be manifested as stubby, "club-footed" roots with little or no root hairs.

As previously shown, when herbicides are applied to a cloddy field, there is a possibility that a "blank" below and on the underside of the clod could occur. Herbicides with higher solubility will distribute more uniformly than those with low solubility. Hill fine-textured, clay soils when dry, and use other practices to prevent clods. Herbicides that are not as soluble may be more effective in sandy soils; e.g., Chateau, Linex, and Prowl H_2O. If high rainfall amounts occur after application of highly soluble herbicides, such as Matrix and metribuzin, they may leach below the weed-seed germination zone which, in turn, allows weed breaks. For instance, when rain occurs shortly after a Matrix plus Prowl H_2O mixture is applied preemergence, hairy nightshade, but not common lambsquarters, could start emerging, since the Matrix has moved down but the Prowl H_2O has stayed in the germination zone.

If breaks occur and/or the possibility of breaks exist, additional control measures usually can be taken. The following are options for a specific program whether or not herbicides have already been applied. Some herbicides and herbicide combinations can be applied sequentially, preemergence (PRE) + preemergence; PRE + postemergence (POST); POST + POST. Make sure to follow annual total amount limitations stated on the herbicide label(s) and also restrictions/allowances for sequential applications.

Scenarios of Impacts by Unusually Large Amounts of Rainfall Before or After Herbicide Application in the Spring

Rain started before any herbicide was applied; now rain has stopped.

Potatoes and weeds have not yet emerged:

- Follow the usual preemergence (PRE) herbicide program, including hilling before emergence. Wait until the soil is dry enough so that compaction will not occur.
- Adjust the irrigation incorporation amount to the low end of labeled rate, depending upon existing soil moisture.
- Do not apply Eptam to wet soils because the possibility of loss through volatility increases compared with loss when application is made to dry soils.

Potatoes have *not* yet emerged; weeds *have* emerged:

- "Burndown" herbicides with foliar activity only (no soil residual activity): Aim EC, glyphosate, and paraquat.
- Matrix and metribuzin have both foliar and soil activity and can be applied before or after potato emergence.
- Linex and Chateau are herbicides that must be applied PRE to potatoes, have soil activity, and may possibly have some foliar activity on weeds.
- Tank-mix herbicides with foliar and soil residual activity to control emerged and not-yet-emerged weeds.

Potatoes *have* emerged, weeds have *not* yet emerged:

- Herbicides with residual soil activity can be applied early postemergence (EPOST) to potatoes: Matrix, metribuzin, Dual Magnum, Eptam, and Prowl H_2O.

Both potatoes and weeds have emerged:

- Aside from Matrix and metribuzin, which can be applied POST, herbicides safe for emerged potatoes but have only soil activity, Dual Magnum, Eptam, and Prowl H_2O, can be applied early EPOST to potatoes.

Rain started after all preemergence herbicides have been applied and the last tillage operation has been performed.

Excessive rainfall occurring after herbicides have been applied and potatoes have emerged (Fig. 12.32) can cause mobile herbicides to leach below the top 2 in of soil where many weed seeds germinate.

Specific example: Matrix + metribuzin + Prowl H_2O was applied PRE on a loam soil. Hairy nightshade and common lambsquarters are present, and the potatoes may or may not have emerged:

- Hairy nightshade may germinate and emerge since Matrix may move below germination zone.
- Even though metribuzin may also move, common lambsquarters control should still be good because Prowl H_2O is still in the weed seed germination zone. For a

Fig. 12.32 Excessive rainfall occurred after herbicides have been applied and potatoes have emerged

"rescue- type" application, Matrix/Prism and metribuzin are the only herbicides that have foliar activity on emerged weeds. Care must be taken not to exceed the season maximum rates and total number of applications allowed.

Herbicide Resistance Management

As mentioned earlier, herbicide resistance is defined as the inherited ability of a plant to survive an herbicide treatment to which the original population was susceptible. Resistant plants occur naturally within a population, often in low numbers.

Herbicide-resistant populations develop because of the repeated use of an herbicide or herbicides with the same mechanism of action. The herbicide does not cause the resistance mutation. Repeated use of the herbicide allows resistant plants to survive and reproduce while susceptible plants are killed. The number of resistant plants then increases until the herbicide is no longer effective (Fig. 12.33a–e).

A weed can be resistant through any of the following mechanisms:

- Alteration at the herbicide's site of action so that the herbicide does not affect that site.
- Increased ability to metabolize or detoxify the herbicide—the rate is faster than in a susceptible biotype.
- Sequestration of the herbicide away from the site of action.
- Modification in the uptake and/or translocation of the herbicide to the target site.
- Multiple- or cross-resistance can also occur. Weeds with multiple resistance are resistant to herbicides with different active sites or modes of action. Cross-resistant weeds have resistance to more than one herbicide with the same mechanism of action.
- When and where resistance occurs depends on many factors, such as the initial frequency and fitness of the resistant biotype, herbicide history, and/or cultural practices.

Herbicide Resistance Management Strategies

Integrated weed management uses all tools available to control weeds and is important for managing herbicide-resistant weeds. The greater the variety of weed control tools used, the lower the risk of selecting resistant weeds. Some useful resistance management strategies include:

- Rotate herbicide mechanisms of action and/or crops in order to avoid or delay the onset of herbicide resistance.
- Cultivate row crops and employ different cultural practices each year to reduce the risk of developing resistant populations.

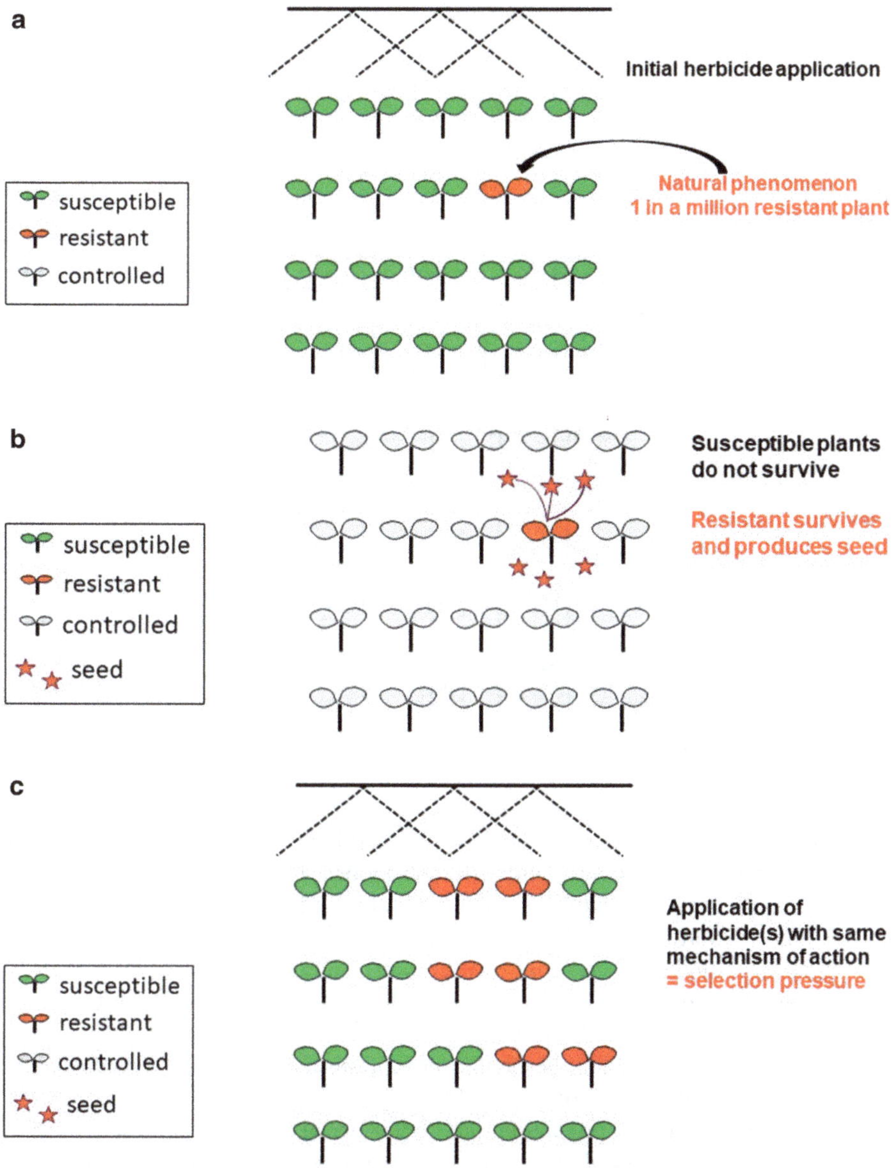

Fig. 12.33 Development of a resistant weed population through selection pressure from repeated use of the same mechanism-of-action herbicide(s). Susceptible (green) plants are killed by the herbicide. Resistant plants (red) survive and reproduce, increasing the frequency of resistant weeds in a given field for subsequent generations **a-b**). Each weed population can have a naturally occuring plant resistant to the herbicide MOA being used, perhaps at the level of one in a million. When a herbicide is applied, the susceptible plants die while the resistant plant survives and produces seed; c-e) If the same MOA is used during the same year and/or repeatedly throughout the crop rotation, then more resistant plants survive and reproduce. As a result, the weed population will eventually become reistant to herbicides with that MOA. The author greatly appreciates assistance from the University of Idaho Potato Cropping Systems Weed Science; Research Associate Brent Beutler; and Ag Field Technicians, Tina Miera, Brenda Kendall, and Tenika Trevino. Unless otherwise noted, all pictures are courtesy of the University of Idaho Potato Cropping Systems Weed Science Project

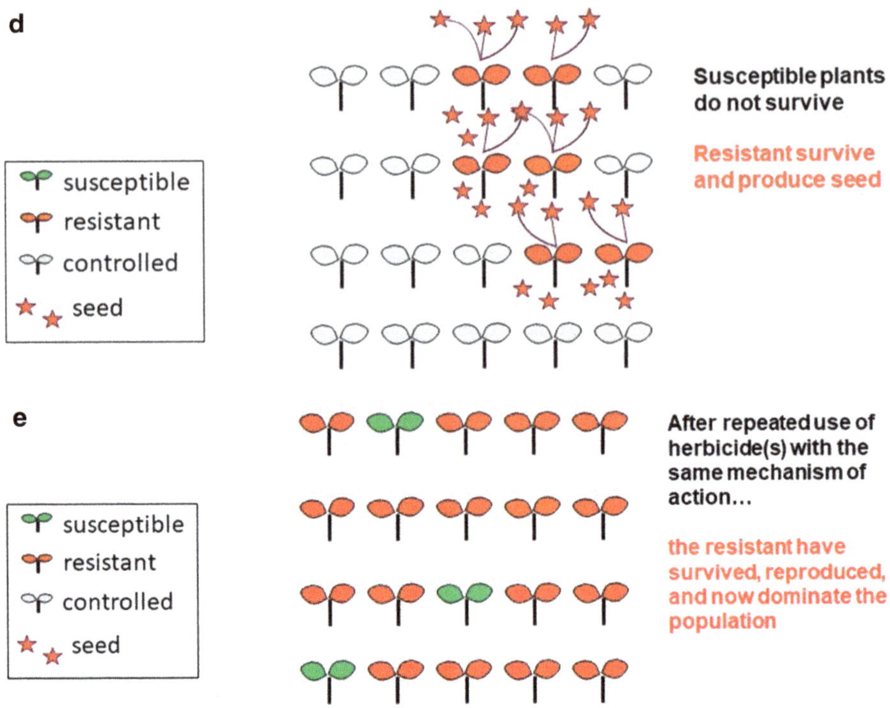

Fig. 12.33 (continued)

- Avoid sequential applications of the same herbicide within the same growing season. Tank mix herbicides with different modes of action to prevent or delay herbicide resistance, as well as to help control existing resistant weeds.
- Use tank mixtures that include herbicides with overlapping weed spectrum so that multiple modes of action are used to control the same weed. If weeds present in the field are resistant to one of the herbicides in the tank mixture, the other herbicide(s) should provide control.
- Scout fields before and after herbicide applications in order to facilitate decisions about subsequent applications.
- Make proper herbicide applications and control weed escapes as soon as possible.
- Keep herbicide and weed control histories for each field in order to track and rotate herbicide modes of action and track weed control in each field. Shifts in weed populations should be noted to guide future control strategies.
- Plant competitive crops and varieties to enhance weed control.
- Use certified seed for rotational crops because planting seed from fields contaminated with resistant weeds spreads those weeds to other fields.
- Prevent weed seed spread from field to field with sanitation methods, such as cleaning equipment before moving and screening irrigation water, when possible.

Recommendations if Resistance Is Suspected

All possibilities for poor herbicide performance, such as misapplication, environmental conditions, heavy weed pressure, inadequate coverage, sprayer skips, or inappropriate herbicide choice, should be ruled out before considering the possibility of resistance. If herbicide resistance is suspected, the field should not be resprayed with the same herbicide or herbicide class.

These questions should be asked if herbicide resistance is suspected:

- Was the same herbicide or class of herbicides applied sequentially in the same year or used year after year over the course of several years?
- Was the suspected resistant weed species controlled effectively in the past with the same herbicide(s) that is (are) not effective now?
- Is weed control good on all the other labeled weed species?

If the answer is "Yes" to one or more of these questions, it is likely that herbicide-resistant biotypes are present in the field.

Weed control by another means, such as hand weeding, cultivation, or spraying with a different mode-of-action herbicide is recommended, and seed from infected fields should not be allowed to mature. University research or extension personnel, county extension educators, or crop advisers should be informed by reporting the locations of herbicide-resistant weeds. These specialists can collect and confirm resistance in plants or seeds.

Resistance management tools available to assist growers include: PNW Bulletin No. 437, "Herbicide-Resistant Weeds and Their Management" (Mallory-Smith et al. 1999) herbicide classification charts such as the one in this chapter; and management worksheets for noting herbicide use and mode-of-action histories. Growers can customize worksheets and planning tools for particular fields and needs.

References

Edmonds JM (1986) Biosystematics of *Solanum sarrachoides* Sendtner and *S. physalifolium Rusby (S. nitidibaccatum* bitter). Bot J Linn Soc 92:1–38

Heap I (2018) The international survey of herbicide resistant weeds. www.weedscience.com. Accessed Sept 2018

Hutchinson PJS (2012) Weed control and potato crop safety with metribuzin. CIS 1185. University of Idaho Ed Comm, Moscow, p 11

Hutchinson PJS (2019) Potatoes. In: Pacific northwest 2019 weed management handbook. University of Idaho, Moscow

Mallory-Smith C, Hyslop GR, Thill DC, Morishita D (1999) Herbicide-resistant weeds and their management. University of Idaho, Moscow

Shaner DL (2014) Herbicide handbook, 10th edn. Weed Science Society of America, Champaign

Whitson TD (2012) Weeds of the West 11th ed. Western Society of Weed Science in cooperation with the Western United States Land Grant Universities Cooperative Extension Services and University of Wyoming, Las Cruces

Chapter 13
Potato Irrigation Management

Bradley A. King, Jeffrey C. Stark, and Howard Neibling

Contents

B. A. King (✉)
Northwest Irrigation and Soils Research Laboratory, USDA-ARS, Kimberly, ID, USA
e-mail: brad.king@usda.gov

J. C. Stark
Idaho Falls R&E Center, University of Idaho, Idaho Falls, ID, USA
e-mail: jstark@uidaho.edu

H. Neibling
Department of Soil and Water Systems, Kimberly Research and Extension Center, University of Idaho, Kimberly, ID, USA
e-mail: hneiblin@uidaho.edu

J. C. Stark et al. (eds.), *Potato Production Systems*,
https://doi.org/10.1007/978-3-030-39157-7_13

Introduction

Potato yield and quality are sensitive to both excess and deficit soil water. This sensitivity, coupled with a relatively shallow root zone and medium- to coarse -textured soils common in many production areas, makes economically efficient irrigated potato production challenging. Potato is grown under all types of irrigation systems worldwide, but irrigation systems capable of light, frequent, uniform water application are best. Optimum potato irrigation management requires a working knowledge of soil water relations and irrigation system characteristics. This chapter introduces both in the context of potato production in arid areas of the Pacific Northwest U.S. General guidelines and irrigation management aids are presented along with examples for implementing quantitative irrigation management of potato in an arid environment.

The Benefits of Using Irrigation in Potato Cropping Systems

Irrigation is required for profitable commercial potato production in many areas worldwide. To maximize production efficiency, soil moisture must be effectively maintained within rather narrow limits throughout the growing season. Potato is one of the most sensitive crops to both excess and deficit soil water due to its relatively shallow root system and because it is often grown on soils with low to medium water-holding capacity. These conditions necessitate that reliable irrigation systems capable of light, frequent, uniform water applications be used to optimally control soil water availability throughout the growing season. These conditions also dictate that an effective potato irrigation management program include: (1) regular monitoring of soil water content, (2) quantitative irrigation scheduling according to crop water use and soil water-holding capacity, and (3) a water supply and irrigation system capable of providing frequent, uniform water application.

The sensitivity of potato yield to irrigation management is depicted in Fig. 13.1. The results were obtained from a 1995 research study of water management practices on 45 commercial potato fields in Idaho (Stark 1996). Potato yield is reduced by both over- and under-irrigation. A mere 10% deviation from optimum water application for the growing season may begin to decrease yield. This marked response to water management is attributable to the sensitivity of potato plants to moderate water deficits and excess soil water, coupled with a very small margin for error in irrigation scheduling resulting from limited soil moisture storage in the root zone. This is due, in part, to a relatively shallow root zone. Yield reductions due to over-irrigation can be attributed to poor soil aeration, increased disease problems, and leaching of nitrogen from the shallow crop root zone. Quantitative irrigation management can increase marketable yield while reducing production costs by conserving water, energy, and nitrogen fertilizer as well as reducing potential groundwater contamination. Quantitative irrigation management, therefore, is a prerequisite for maximizing production efficiency from irrigated potato production.

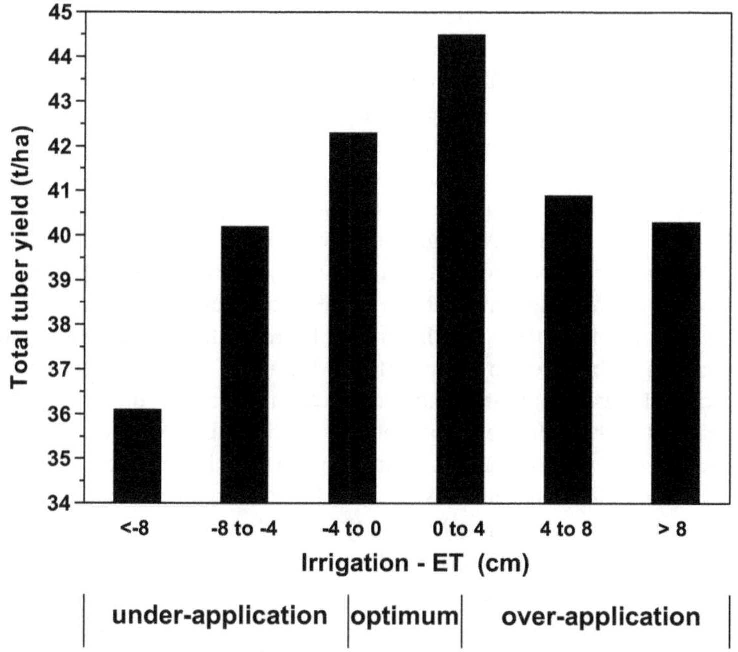

Fig. 13.1 Total tuber yield as influenced by the difference between irrigation and ET on 45 commercial potato fields in southeast Idaho

Potato Growth and Soil Water Availability

Potato root system development is relatively shallow, 18–24 in, with the majority of roots in the upper 12 in soil profile. The shallow rooting depth is largely attributable to the inability of the relatively weak root system to penetrate tillage pans or other restrictive layers. Soil compaction by field vehicle traffic can greatly restrict potato root penetration. High soil moisture content at the time of tillage operations typically increases the degree of compaction resulting from field traffic. Potato rooting depth can also be restricted by weakly cemented calcium carbonate layers in the top 2 ft of soil in arid regions, which restrict potato root penetration, but not necessarily water movement. Field determination of actual potato plant rooting depth is of primary importance in developing an effective irrigation management program.

The first physiological response of potato to water deficits is closure of the leaf stomata; the small pores in the leaf that control gas exchange between internal leaf cells and the environment. Evaporation of water from the leaves cools the plant canopy temperature below air temperature under well-watered conditions. The stomata in the leaf close under plant water deficits as a defense against further water loss. One of the first physical indications of water stress is an increase in canopy temperature because of reduced evaporative cooling of the leaves.

While stomatal closure reduces water loss through the leaves, it also reduces carbon dioxide diffusion into the leaf. This slows photosynthesis, reducing the production of photosynthetic products (starch and sugars) by the plant and their translocation from the leaves to the tubers. Potato yield and quality depend upon maximizing the steady accumulation of photosynthetic products in the tubers. When production of these products exceeds that needed for respiration and continued plant growth, they are stored in the tubers.

Another physiological response affected by plant water deficits is the expansion of leaves, stems, and tubers. Water deficits reduce plant growth by reducing the internal water pressure in plant cells (turgor pressure), which is necessary for expansion. Reduced vine and leaf growth limits total photosynthetic capacity, while reduced root development limits the plant's ability to take up water and nutrients. Water deficits also disrupt normal tuber growth patterns by reducing or temporarily stopping tuber expansion. Tuber growth resumes following relief of plant water deficits, but the disruption of the normal tuber expansion rate may result in tuber malformations such as pointed ends, dumbbells, bottlenecks, and knobs. Widely fluctuating soil moisture levels create the greatest opportunity for developing these tuber defects. Growth cracks are also associated with wide fluctuations in soil water availability and corresponding changes in tuber turgidity and volume of internal tissues.

Potato is particularly sensitive to water stress during tuber initiation and early tuber development. Water deficits at this time can substantially reduce U.S. No. 1 yields by increasing the proportion of rough, misshapen tubers. Early-season water stress can also reduce specific gravity and increase the incidence of translucent end.

Water stress during tuber bulking usually affects total tuber yield more than quality. A large photosynthetic-active leaf surface area is necessary to maintain high tuber bulking rates for extended periods. Sustaining a large photosynthetic active leaf surface area over a full growing season requires continued development of new leaves to replace older, less efficient ones. Water stress hastens leaf senescence and interrupts new leaf formation, resulting in an unrecoverable loss of tuber bulking.

Potato yield and quality are susceptible to excess soil moisture as well. Excess soil moisture from frequent or intensive irrigation or rainfall during any growth stage leaches nitrate nitrogen below the plant root zone, potentially resulting in nitrogen-deficient plants, reduced fertilizer use efficiency, and an increased hazard to groundwater. Saturation of the soil profile for more than 8–12 h can cause root damage due to a lack of oxygen required for normal respiration. Excess moisture at planting promotes seed piece decay and delayed emergence due to decreased soil temperature. Potatoes that are over-irrigated during vegetative growth and tuber initiation have a greater potential for developing brown center and hollow heart and are generally more susceptible to early die problems. Excess soil moisture can also lead to tuber quality and storage problems.

Irrigation Management

Irrigated potato production occurs over a wide range of conditions. Arid regions may have sustained hot, dry periods where irrigation provides more than 90% of the water needed for crop production. Humid regions may have short dry periods where irrigation is needed to sustain optimal soil water levels for maximum yield and quality and provides less than 10% of the water needed for crop production. Regardless of the climatic circumstances, the purpose of irrigation management is to maximize potato yield and quality by maintaining soil water content within specified limits throughout the growing season through timely, controlled water application.

Soil Water-Holding Capacity

Soil serves as the reservoir for plant nutrient and water needs. Soil has a finite capacity to hold water against gravity, which is called the water-holding capacity. A graphical representation of how water is held in soil is shown in Fig. 13.2. A given volume of soil consists of solids composed of minerals and organic matter, as well as pores, which are occupied by air and water. When soil pores are filled with water, the soil is said to be saturated (Fig. 13.2a). Under conditions of free drainage, the force of gravity will drain water from the largest pores. This free-draining water is called gravitational water, which is only available to plants during the time it is percolating through the root zone. After 12–48 hours, drainage will decrease to a negligible rate. The water content, at this point, is commonly called field capacity or upper-drained limit (Fig. 13.2b).

Water is held in the soil as a film around soil particles by molecular attraction and by water surface tension forces producing what is commonly called capillary action. Hence, water held in soil pores is called capillary water (Fig. 13.2c), which is available for plant use. As plants remove water from the soil, it is extracted from progressively smaller pores until the remaining water exists as a thin film around soil particles held tightly by molecular attraction. The molecular attraction is strong and a large amount of energy is required to remove the remaining water from the soil, so

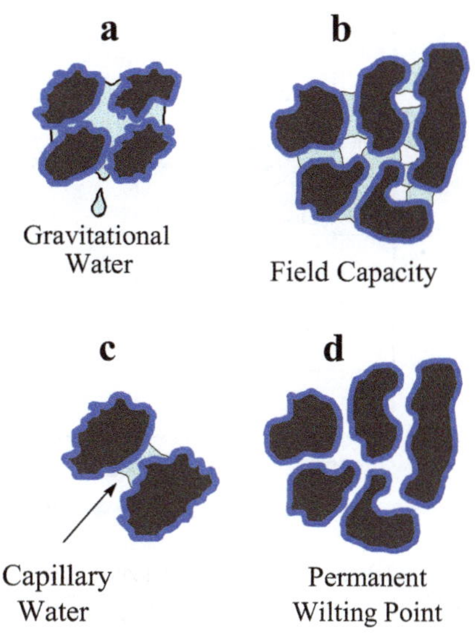

Fig. 13.2 Graphical representation of water held in soil: (**a**) near saturation, (**b**) at field capacity, (**c**) by capillary action, (**d**) at permanent wilting point

much so that plants cannot obtain water and, consequently, wilt and die. Soil water content at this point is called the permanent wilting point and is graphically illustrated in Fig. 13.2d. The volume of water held in the soil between field capacity and the permanent wilting point is called available water. Available water can also be expressed as inches of water per inch or ft. of soil depth. It is then referred to as the water-holding capacity of the soil.

Each soil has a unique relationship between soil water content and soil water energy potential called the soil water release curve. This relationship, which is highly dependent on soil texture, is shown graphically in Fig. 13.3 for four soil textures. The rather flat curve of a typical loamy sand soil indicates a narrow range in moisture content between field capacity and the permanent wilting point, indicating low water-holding capacity. In contrast, the sloping curve of the silt loam soil has a much wider range in soil moisture content between the permanent wilting point and field capacity, indicating greater water-holding capacity.

Soil moisture content is often expressed as a percentage on either a weight (gravimetric) or volumetric basis. Care must be taken to make sure which moisture content basis is being employed or measured. Conversion between the two requires knowledge of the soil's bulk density, since volumetric water content = gravimetric water content x bulk density (dry mass per unit volume). For example, if the gravimetric soil moisture of a silt loam soil with a bulk density of 1.37 g/cm^3 is 23.4%, soil moisture content on a volumetric basis is then 32.1% (23.4 × 1.37 = 32.1). Soil moisture content measured on a volumetric basis is preferred for irrigation manage-

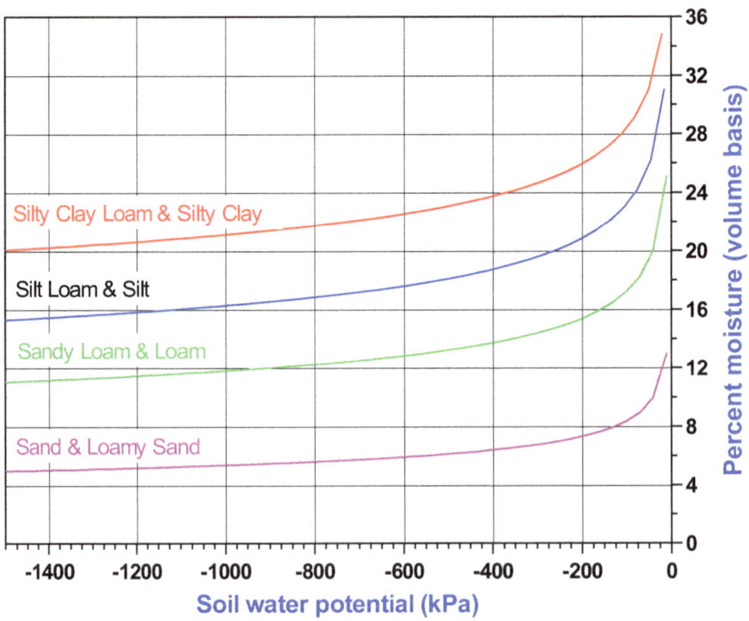

Fig. 13.3 Soil moisture release curves for four soil textures common in agriculture (100 kPa = 100 centibar = 1 bar = 14.7 psi)

Table 13.1 Soil water contents for agricultural soils

Texture class	Water content volume basis (%)							
	Field capacity		Permanent wilting point		Available water		Water-holding capacity (in/ft)	
	Average	Range	Average	Range	Average	Range	Average	Range
Sand	12	7–17	4	2–7	8	5–11	0.96	0.60–1.32
Loamy sand	14	11–19	6	3–10	8	6–12	0.96	0.72–1.44
Sandy loam	23	18–28	10	6–16	13	11–15	1.56	1.32–1.80
Loam	26	20–30	12	7–16	15	11–18	1.80	1.32–2.16
Silt loam	30	22–36	15	9–21	15	11–19	1.80	1.32–2.28
Silt	32	29–35	15	12–18	17	12–20	2.04	1.44–2.40
Silty clay loam	34	30–37	19	17–24	15	12–18	1.80	1.44–2.16
Silty clay	36	29–42	21	14–29	15	11–19	1.80	1.32–2.28
Clay	36	32–39	21	19–24	15	10–20	1.80	1.20–2.40

Adapted from Jensen et al. (1990)

ment computations because bulk density of the soil is not required. Soil moisture contents used in this publication are expressed on a volumetric basis. General soil moisture contents at critical points along with water-holding capacity for agricultural soils are given in Table 13.1. Inspection of available water listed in Table 13.1 reveals that soils having a significant portion of silt have the greatest water-holding capacity, offering the greatest flexibility in potato irrigation management.

Optimum Soil Moisture

Many field research studies have focused on determining optimum soil moisture for irrigated potato production. Most studies on the water stress-sensitive Russet Burbank variety indicate that available soil water (ASW) in the root zone (0–18 in) should be maintained above 65% to avoid yield and quality losses. In general, however, the average ASW of the root zone should be maintained between 70 and 85% during the active growth period for optimum results. In practice, ASW in the root zone will fluctuate above and below this range for short periods of time immediately before and after irrigation. This is particularly true with set-move sprinkler systems and furrow irrigation systems. Solid-set sprinkler, drip, center-pivot, and linear-move sprinkler systems allow for light, frequent irrigations and can be managed to minimize soil moisture fluctuations.

The optimal range for soil moisture at planting is about 70–80% ASW. This moisture level will provide ideal conditions for planting and early sprout development. Excessively wet soil conditions may slow soil warming and delay sprout development and emergence. Cool, wet soil conditions can increase seedpiece decay and physiological aging of seed, resulting in higher stem and tuber numbers. Excessively dry soils should be irrigated prior to planting to avoid potential seedpiece decay problems that sometime result from irrigating between planting and emergence.

During the latter part of the growing season plants begin to senesce, and crop water use rates markedly decrease. Consequently, care should be taken to adjust irrigation amounts to avoid developing excessively wet soil conditions. High soil moisture during this period can produce enlarged lenticels that provide openings for soft rot bacteria to enter the tubers. Pink rot and Pythium leak infections are also increased by excessive late-season soil moisture.

Available soil water should be allowed to decrease to about 60–65% at vine kill to provide optimal conditions for promoting tuber skin set and development of skin texture in russet potato varieties. Drier soil conditions at vine kill increase the chances of developing stem-end discoloration.

Pre-harvest irrigation should be timed to optimize soil conditions and tuber hydration levels at harvest. Tubers that have matured under relatively dry soil conditions (less than 60% ASW) will likely be dehydrated, which will increase their susceptibility to blackspot bruise. Under these conditions, fields should receive irrigation at least 1 week prior to harvest to completely rehydrate tubers. If ASW has been kept above 60% during tuber maturation, fields can be irrigated 2–3 days prior to harvest. Care should also be taken to avoid getting fields too wet at harvest because of increased potential for shatter bruise and increased soil separation and storage rot problems.

Evapotranspiration (ET)

Evapotranspiration represents the sum of water used by plants for transpiration and water loss due to evaporation from the soil surface. Evapotranspiration varies according to meteorological conditions, surface soil wetness, the stage of growth, and amount of crop cover. The meteorological parameters which affect ET are solar radiation, relative humidity, ambient air temperature, and wind speed. Since these can vary considerably from day to day, so will ET. Furthermore, seasonal ET will vary from year to year in response to yearly meteorological trends.

Daily potato ET throughout the 2015 growing season at three locations in the U.S. Pacific Northwest are shown in Fig. 13.4. Evapotranspiration is low at crop emergence and increases rapidly with crop development and increasing solar radiation and temperature into the summer months. Evapotranspiration decreases gradually as the crop begins to senesce until vine kill. Differences in the start, peak, and end of daily ET values, shown in Fig. 13.4, for the three locations are due to differences in planting and harvest dates and seasonal meteorological conditions. The dependence of ET on meteorological conditions is evident by the variation in daily ET throughout the growing season. Seasonal ET for the three locations over a

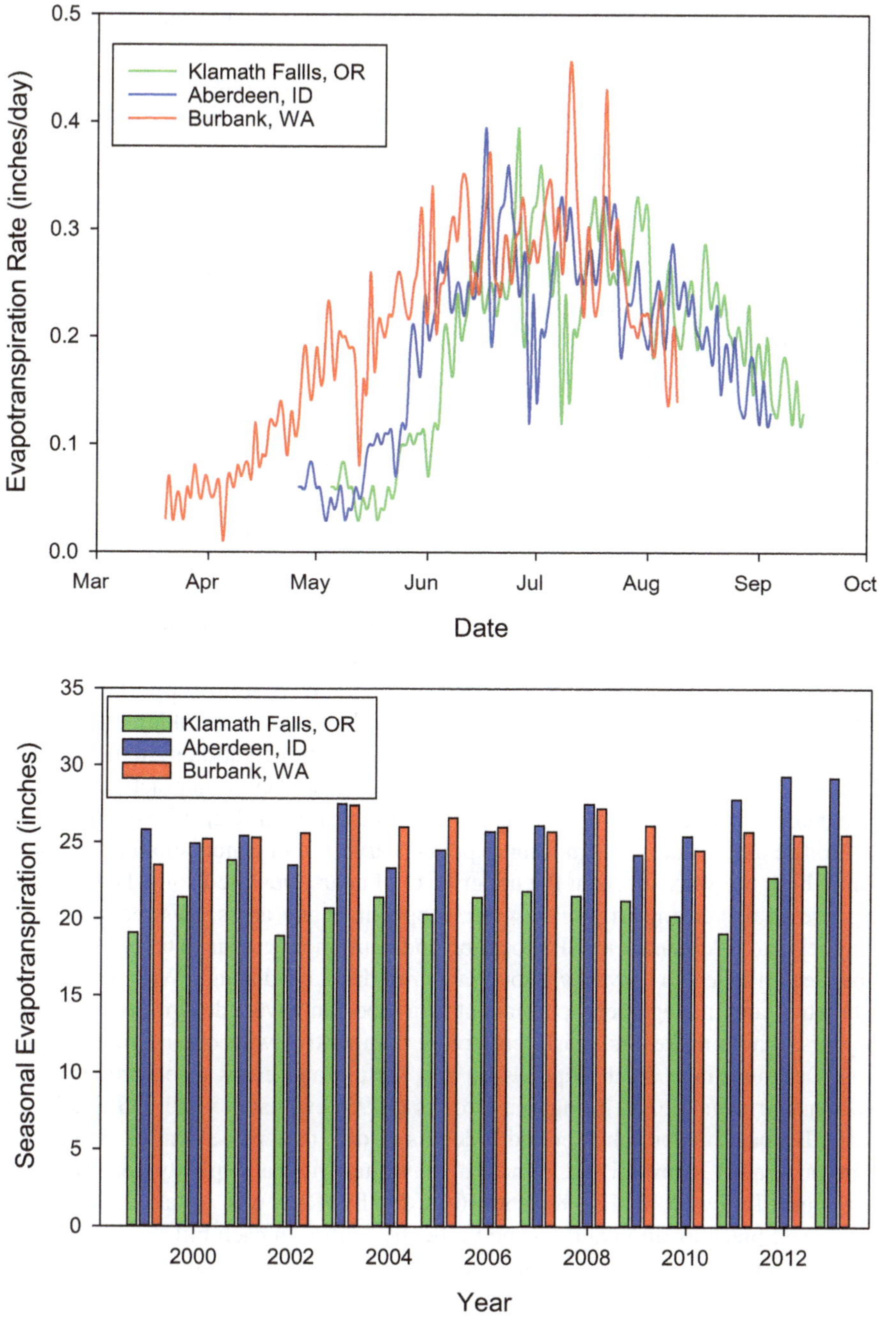

Fig. 13.4 Daily and seasonal evapotranspiration at three locations in the U.S. Pacific Northwest in 2015

15-year period is also shown in Fig. 13.4. Seasonal ET is substantially lower at Klamath Falls, OR, compared to the other two locations. Seasonal ET at Aberdeen, ID, and Burbank, WA, are similar with Burbank, WA, being greater 6 of the 15 years. The magnitude of seasonal and daily variations in ET shown in Fig. 13.4 demonstrates that an irrigation scheduling method that accounts for these variations is necessary to maximize crop production. Published daily ET values, as shown in Fig. 13.4, provide a basis upon which to develop an irrigation management program. In-field soil moisture measurement is also required to account for site-specific differences in ET resulting from differences in the type of irrigation system used; soil water-holding capacity; topography; and local meteorological conditions, such as wind and precipitation.

Irrigation Method

Potato can be grown under many types of irrigation systems; however, some are better suited than others for consistently obtaining high-quality tubers. The water sensitive nature of potato, combined with its shallow root zone, favors irrigation systems that are capable of light, frequent, and uniform water applications. Using these criteria as a basis for ranking the suitability of common irrigation methods, the order of preference from highest to lowest would be: drip, solid-set portable sprinkler, linear-move, center-pivot, side-roll sprinkler, hand-move sprinkler, and furrow. In practice, economics are often the overriding factor in irrigation system selection along with compatibility with soil type, crop rotation, and cultural practices. Buried permanent drip is expensive, incompatible with traditional deep tillage and heavy field traffic associated with traditional potato harvest, and is not suitable for maintaining high moisture levels in the upper level of coarse-textured soils. However, in response to decreasing irrigation water supply in many areas, this technology is being revisited. Advances in GPS location and equipment guidance technology now allow deep tillage and harvest without danger of drip tape damage. With proper tape installation and management, permanent drip tape can have a design life of at least 15 years, which makes its yearly cost more competitive with other irrigation systems. Optimum tape depth depends on soil texture and crops to be grown in the rotation. For example, Neibling and Brooks (1995) evaluated yield and quality of Russet Burbank under solid set and 4 depths of drip tape placement (3, 8, 12, and 16 in) on a sandy loam soil and found that maximum yield and quality occurred with 3-in depth, followed by 12-, 8-, and 16-in depths. Tape was installed above and below seed piece location and in-line vertically with it in each hill. Yield response was due, in part, to the location and shape of wetting zones as shown in Fig. 13.5. Yield and quality were highest for the 3-in placement due to more water in the active root zone. Yield and quality were lower for 12- and 16-in depths, because most of the water was applied below the active root zone due to limited capillarity water movement in the soil. Low yield and quality at the 8-in depth was due to excessive water near the seed piece and limited root development. In a more

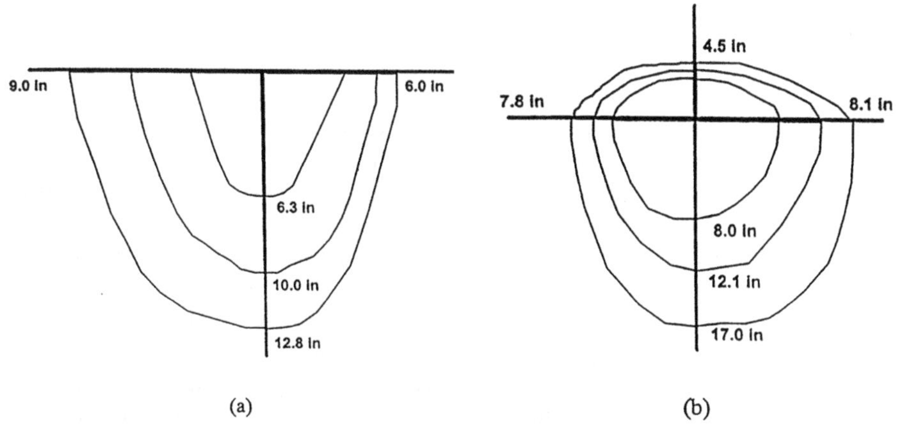

Fig. 13.5 Measured wetting pattern after 1, 2, and 3 h of irrigation at 0.18 gal per hour per emitter discharge (0.3 gpm/100 ft) in a sandy loam soil. (**a**) Tape placement on the soil surface; (**b**) Tape placement at 8-in depth

extensive study in India, Patel and Rajput (2007) studied tape depths of 0, 5, 10, 15, and 20 cm (0, 2, 4, 6, and 8 in) also in a sandy loam soil. Optimum yield was obtained at the 4-in depth in 2 of the 3 years studied and at the 6-in depth in the third year.

Alternative drip tape placement (Shock et al. 2005) evaluated multiple potato row/tape configurations on a single 72-in bed. Treatments were two rows 36-in apart with tapes directly above each, two rows 36-in apart with a drip tape offset about 7 in to the inside of each potato row, and four rows 16-in apart with one drip tape centered between each pair of rows with plants staggered in each of the paired rows. Soil texture was silt loam, and the variety was Umatilla Russet. Plant population levels were low (18,150 plants/acre) or high (24,200 plants/acre). Total yield, marketable yield, and yield of U.S. #1 was variable, with treatment in 2003 and 2004. In general, the higher plant population with four staggered rows per bed had the highest water use efficiency and the most desirable tuber size.

Temporary surface drip irrigation systems using thin-wall drip tape, flexible lay flat hose for main lines, and a portable water filtration system and pump are used on small fields. The drip tape is used a single year then recycled, with the remaining system components relocated to another site for subsequent seasons. Solid-set portable sprinkler systems are also expensive. Center-pivots are highly susceptible to excessive runoff under the outer towers unless conservation tillage practices are utilized. Side-roll and hand-move sprinklers are prone to wind skips under the windy conditions common to many arid regions. Furrow irrigation is susceptible to poor water application uniformity and excessive deep percolation and leaching. Sprinkler is the most common method of potato irrigation the U.S. Pacific Northwest, with center-pivot, side-roll, solid-set, and hand-move sprinkler being widely utilized.

Irrigation Scheduling

Effective irrigation scheduling requires regular quantitative monitoring of soil moisture and knowledge of soil water-holding capacity, crop water use, and crop rooting depth. Excess irrigation usually results from applying too much water at a given irrigation rather than from irrigating too frequently. This is particularly true for side-roll and hand-move sprinkler systems, where soil water-holding capacity and crop rooting depth are overestimated, resulting in set times that are too long and furrow irrigation in which irrigation depth is difficult to control. These assumptions lead to plant water stress when soil moisture falls below desired limits 2–3 days before irrigation and subsequent irrigation applications exceed soil water storage capacity. This characteristic problem can generally be attributed to inadequately designed systems, irrigation system equipment limitations, or improper irrigation management.

Quantitative irrigation scheduling involves maintaining a daily soil water balance that accurately accounts for water input from irrigation and rainfall and water depletion due to crop water use and leaching. Where soil salinity is not of concern, the goal is to minimize leaching during the growing season to conserve soil nutrients for maximum nutrient use efficiency by limiting irrigation depth to the amount of soil water storage available (soil water depletion). Leaching is commonly assumed to be zero when computing the soil water balance but can be substantial if the irrigation has poor application uniformity or rainfall occurs immediately following irrigation. Local daily ET estimates are assumed to represent daily soil water depletion by the crop but can differ from actual crop water use due to field-specific climatic differences from the weather station conditions used to estimate ET. Although calculation of water application based on nozzle size, spacing, and pressure usually provides a good estimate of water applied, application may be reduced by excessively hot, windy days or system maintenance issues. This technique, combined with quantitative measurements of soil moisture to adjust the computed soil water balance to actual field conditions, provides a method for determining the timing of irrigations. Computing a daily soil water balance implicitly determines the desired irrigation application depth as well. Seasonal water application is obtained by summing irrigation depths over the growing season.

Computational Steps in Quantitative Irrigation Scheduling

- Estimate field capacity and permanent wilting point based on predominate soil texture in the field using Table 13.1 as a guide.
- Estimate current crop rooting depth based on site conditions and stage of growth. Rooting depth is often assumed to increase linearly between emergence and full crop height or row closure. The maximum effective rooting depth of potato will be in the range of 18–24 in for soils without a restrictive layer. The effective

rooting depth prior to crop emergence can be assumed to be 8 in for practical purposes. As an example, if the stage of growth is 50% of maximum height or row closure and the maximum effective rooting depth is assumed to be 18 in, the effective rooting depth a can be estimated as 8 in plus 50/100 × (18–8) in = 8 + 5 in or 13 in.

- Maintain a daily soil moisture balance based on estimated values of daily ET starting at crop emergence. An initial soil moisture storage value must be assumed to start the soil water balance based on previous end-of-year irrigation history and winter precipitation or obtained by measuring soil moisture at emergence. A numerical example of quantitative irrigation scheduling is shown in Table 13.2 for two 10-day periods; one at crop emergence and one after 100% effective cover. The assumptions and calculations used to obtain the values in Table 13.2 are:
  - Loamy sand soil with 1.8 in per ft water-holding capacity (WHC)
  - Terminal rooting depth of 20 in 40 days after emergence
  - Rooting depth = 8 (in) + (20–8) (in) ∗ X/40 up to 40 days after emergence where X represents the day of concern in the range of 0 to 40. Rooting depth = 20 in more than 40 days after emergence.
  - Initial available water at 0–12 in is 70% and 12–24 in is 100%
  - Total available water (TAW) = WHC (in/ft) ∗ Root depth (in)/12
  - Initial beginning available water (in) = TAW ∗ 70%/100
  - Ending available water (in) = Beginning available (in) + Irrigation + Precipitation – ET
  - Beginning available water (in):
    - Prior to reaching final rooting depth = Previous day ending available water (in) + [Previous day TAW – Current day TAW] ∗ Initial ASW (%)/100
    - After reaching final rooting depth = Previous day ending available water (in).
- For furrow or irrigation systems with set-move sprinklers, irrigate when ASW decreases to 65–70% by applying the amount required to increase the soil moisture content to field capacity (soil water deficit). For irrigation systems designed for light, frequent irrigations, irrigate when soil water deficit is greater than the nominal application depth while maintaining ASW in the desired range.
- Periodically monitor soil moisture or soil water potential and adjust the daily soil moisture balance, if necessary, to match actual field conditions.

Web-Based Quantitative Irrigation Scheduling

A web-based method of irrigation scheduling that uses the water-balance approach, discussed above, with user-defined crop and soil conditions and daily data from a user-selected AgriMet weather station, was developed by Dr. Troy Peters at Washington State University (http://weather.wsu.edu/is/). Short videos that lead the

Table 13.2 Irrigation scheduling example for two 10-day periods

Date	Daily ET (in)	Rooting depth (in)	Total available water (in)	Beginning available water (in)	Beginning available water (%)	Irrigation depth (in)	Rainfall (in)	Ending available water (in)	Ending available water (%)	Soil water deficit (in)
5–5	0.06	8.0	1.2	0.84	70.0	0.0	0.0	0.78	65.0	0.42
5–6	0.06	8.3	1.2	0.81	65.2	0.4	0.0	1.15	92.5	0.09
5–7	0.06	8.6	1.3	1.18	91.7	0.0	0.0	1.12	87.1	0.17
5–8	0.08	8.9	1.3	1.15	86.5	0.0	0.0	1.07	80.5	0.26
5–9	0.08	9.2	1.4	1.11	80.1	0.0	0.0	1.03	74.3	0.35
5–10	0.06	9.5	1.4	1.06	74.2	0.0	0.0	1.00	70.0	0.43
5–11	0.06	9.8	1.5	1.03	70.0	0.0	0.0	0.97	65.9	0.50
5–12	0.04	10.1	1.5	1.00	66.0	0.0	0.0	0.96	63.4	0.55
5–13	0.03	10.4	1.6	0.99	63.6	0.0	0.15	1.11	71.3	0.45
5–14	0.05	10.7	1.6	1.14	71.2	0.0	0.0	1.09	68.1	0.51
5–15	0.04	11.0	1.7	1.13	68.2	0.5	0.0	1.59	96.1	0.07
7–1	0.33	20.0	3.0	2.40	80.0	0.0	0.0	2.07	69.0	0.93
7–2	0.36	20.0	3.0	2.07	69.0	0.9	0.0	2.61	87.0	0.39
7–3	0.32	20.0	3.0	2.61	87.0	0.0	0.0	2.29	76.3	0.71
7–4	0.29	20.0	3.0	2.29	76.3	0.0	0.0	2.00	66.7	1.00
7–5	0.24	20.0	3.0	2.00	66.7	0.9	0.0	2.66	88.7	0.34
7–6	0.26	20.0	3.0	2.66	88.7	0.0	0.0	2.40	80.0	0.60
7–7	0.26	20.0	3.0	2.40	80.0	0.0	0.0	2.14	71.3	0.86
7–8	0.12	20.0	3.0	2.14	71.3	0.0	0.15	2.17	72.3	0.83
7–9	0.24	20.0	3.0	2.17	72.3	0.9	0.0	2.83	94.3	0.17
7–10	0.14	20.0	3.0	2.83	94.3	0.0	0.2	2.89	96.3	0.11
7–11	0.2	20.0	3.0	2.89	96.3	0.0	0.0	2.69	89.7	0.31

user through the account and field setup processes can be found at: http://www.uidaho.edu/extension/drought/. It delivers information to any web-connected device (smartphone, tablet, laptop, etc.) in a convenient, easy-to-understand format. It has been evaluated for multiple crops, including potatoes in WA and ID. Multiple output screens are available. Two of the most useful are shown in Fig. 13.6 for a potato field at the University of Idaho Research and Extension Center.

Field Soil Water Measurement

Several methods are available to quantitatively measure soil moisture, only some are suitable for potato because of the critical threshold levels of available moisture and the limited root zone depth. Many of the methods are labor intensive and require training, experience, and expensive equipment. This requirement has led to the development of crop consulting firms specializing in irrigation management, which often provide crop nutrient and pest management services as well. A detailed discussion of soil moisture measurement methods is provided in the publication "Soil

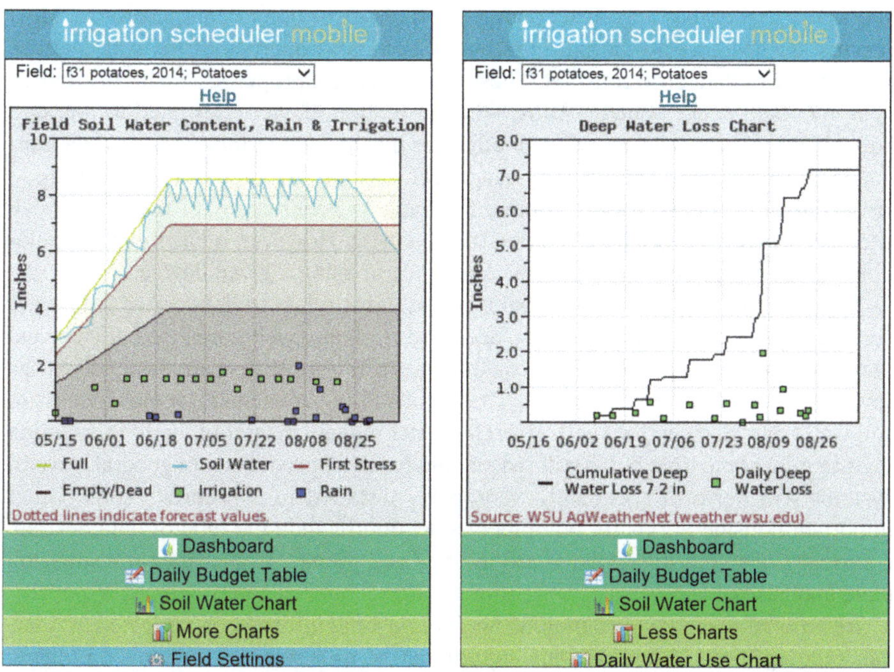

Fig. 13.6 (left) WSU Irrigation scheduler mobile output showing estimated soil water in the root zone (blue line), irrigation, and rainfall. Yellow line represents field capacity; red is MAD = 0.35. To minimize crop stress, estimated soil water should be maintained between the yellow and blue lines. (right) Estimated deep percolation. Large increases in early August due to large rainfall events

Water Monitoring and Measurement," PNW 475, University of Idaho, College of Agriculture.

Tensiometers, which measure soil water potential and soil water content indirectly, have been used to successfully monitor soil water in potato fields. Good contact between the soil and tensiometer tip is essential for proper operation. Tensiometers are often installed in the potato hill at two depths, such as 8 and 16 in below soil level. Typically, the upper tensiometer is used to track ASW within the bulk of the root zone, while the lower is used to determine whether soil water at the bottom of the root zone is increasing or decreasing over time.

The neutron probe is likely the most precise and reliable tool for soil water measurement, since it determines volumetric soil water content directly. However, licensing, training, and associated operational costs limit its use to consulting firms and large farms.

Time domain reflectometery (TDR) offers many features that make it well suited to soil water measurement in potato. However, initial equipment costs can be quite high. Other traditional instruments, such as resistance blocks, are also available and can be effectively used for water management.

Recently, many new devices have become commercially available for monitoring soil water content. Typically, they consist of two components, soil water sensors and a data logger that may also contain a cell phone or other data transmitter. Most sensors are designed to measure the bulk electrical properties of the soil; such as capacitance or dielectric constant. These bulk electrical properties are highly dependent upon soil water content. Thus, with calibration, devices designed to measure bulk electrical properties provide an effective means of determining soil water content. Soil salinity and bulk density can affect response of these devices, leading to erroneous or erratic soil water content readings. In general, any of the devices can become an effective tool for irrigation management. However, it takes experimentation and field experience to develop confidence in using a given device.

Many of the new soil water monitoring systems relate to delivery of soil water content data to a website that can be accessed from any web-connected device, and most have smartphone apps. Some data logger/cell transmitters are designed to use several different soil water sensors. These units allow convenient, remote access to soil water data and can be a very effective water management tool. Field installation does require some time but is still reasonable for most producers. In general, several systems will perform adequately if properly installed in appropriate soils. Other important considerations for equipment selection include the type of data desired, equipment cost and longevity, yearly cost of cell phone or other data transmission plan, and ease of use.

Benefits of these systems include nearly real-time information collected at intervals ranging from 30 min on some equipment, to user-selected intervals on others. Easy access to trends in water content with time, provides additional insight into soil water dynamics, and with experience, can be used to forecast potential water deficits at some soil depths unless the irrigation operation is modified. An example of one output is shown in Fig. 13.7 for the same field at the UI Kimberly Research

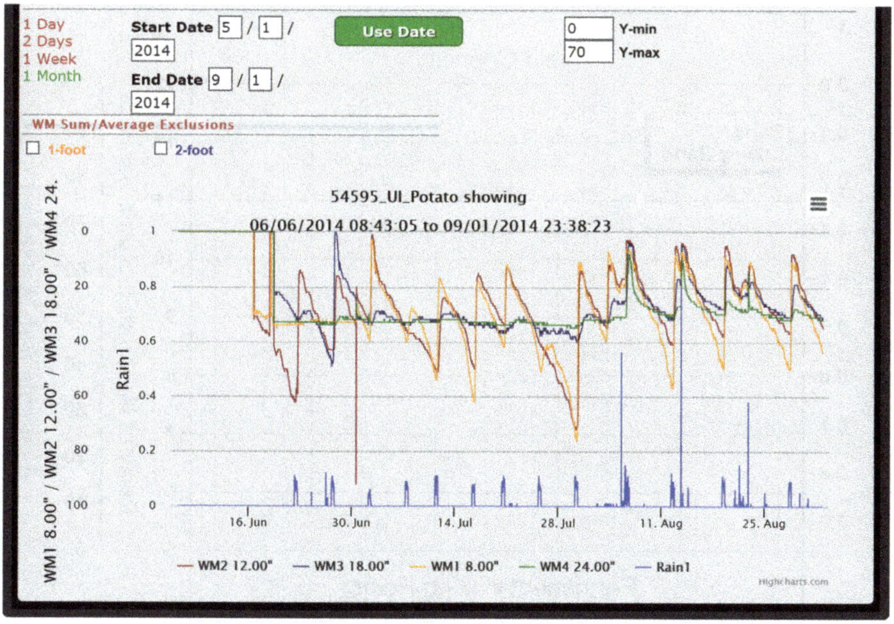

Fig. 13.7 Data logger output accessed from website. Irrigation (blue lines) applied by 10-h solid set. Watermark readings in kPa (or centibars). Threshold for irrigation is 65 kPa for this silt loam soil. Note minimal fluctuation in 18- and 24-in sensors until large August rains, showing deep percolation at that time. Note lack of soil water response on days indicated by purple. These days were hot with strong winds

and Extension Center used in Fig. 13.6. In this figure, sensors are installed at four different depths, with a shallow one for early season, then the top three for remaining season scheduling, and the deepest sensor for detection of over-irrigation. In this case, the field was well-irrigated until the first of August, with 8- and 12-in sensors indicating water content in the appropriate range rising after irrigation to about field capacity and dropping to about 65% available soil moisture before irrigation.

Soil Water-Holding Capacities for Irrigation Scheduling

For quantitative irrigation scheduling a soil water release curve is needed to relate soil water potential to volumetric soil moisture. The generalized soil water release curves shown in Figs. 13.8, 13.9, 13.10, and 13.11 can be used to relate soil water potential, volumetric soil moisture, ASW, and water depletion. These curves represent the primary soil-water relationships upon which an effective irrigation management program is developed. They allow the use of soil moisture or water potential measurements to calculate the net irrigation application amount needed to fill the

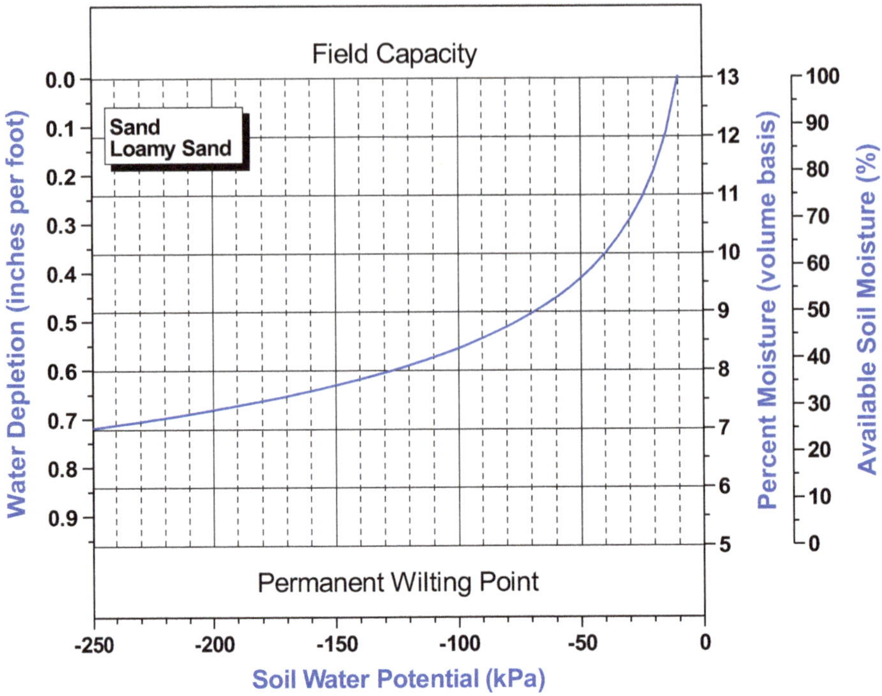

Fig. 13.8 Generalized soil moisture release curve for sand and loamy sand soils

soil water reservoir to field capacity. For example, if tensiometers show an average soil water potential of –40 kPa (centibars) in a sandy loam soil (Fig. 13.8), then available soil moisture is 62% and it's time to irrigate with a net application of 0.36 in/ft of crop root zone depth. Soil water monitoring alone can be used for irrigation scheduling if performed on a real-time basis and used to directly control an irrigation system capable of immediate response. In practice though, most field scale irrigation systems are not capable of immediate response. Thus, a soil-water balance is computed daily using estimated daily ET and forecasted daily ET to anticipate when the next irrigation should occur and the amount of water to apply. This computed soil-water balance is reconciled to actual field conditions through use of the soil water release curve, quantitative soil moisture measurements, and visual observations of the crop.

The ranges of soil water potential and volumetric soil water content corresponding to 65% available soil moisture for different soils is shown in Table 13.3. These values are obtained from the generalized soil water release curves shown in Figs. 13.8, 13.9, 13.10, and 13.11. These values are not absolute, but serve as a general guide for effective irrigation management.

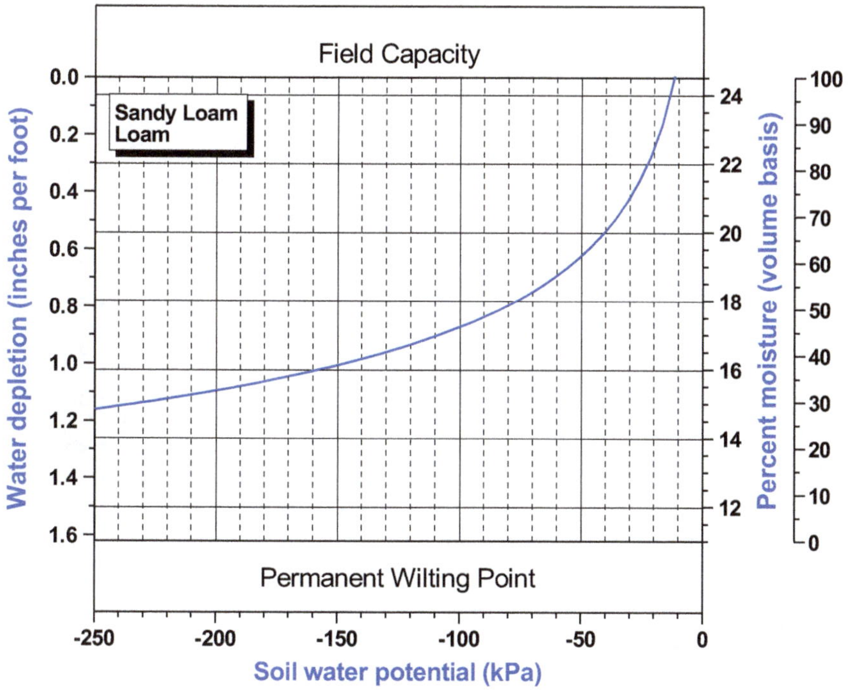

Fig. 13.9 Generalized soil moisture release curve for sandy loam and loam soil

Irrigation System Operational Parameters for Irrigation Scheduling

The primary irrigation system information needed for irrigation scheduling is net irrigation application amount, or rate of water application. For center-pivot and linear-move irrigation systems, the net application amount is dependent upon system capacity, wet run time between irrigations, and system application efficiency. For side-roll, hand-move, and solid-set sprinkler systems, the net application rate depends upon operating pressure, nozzle size, sprinkler spacing, and system application efficiency. System application efficiency is a measure of how much of the water exiting the irrigation system is stored in the crop root zone. As with all irrigation systems, some water is lost due to wind drift and evaporation under sprinkler irrigation and to deep percolation resulting from non-uniform water application. While wind drift and evaporation reduce the amount of water reaching the root zone, they also reduce the amount of water that would have been removed from the root zone in their absence. Thus, they do not represent a total loss, just less efficient irrigation. Typical irrigation system application efficiencies for Idaho are given in Table 13.4.

Irrigation efficiency values need to be applied with caution when used for irrigation scheduling purposes. This is because they result in a self-fulfilling outcome; a

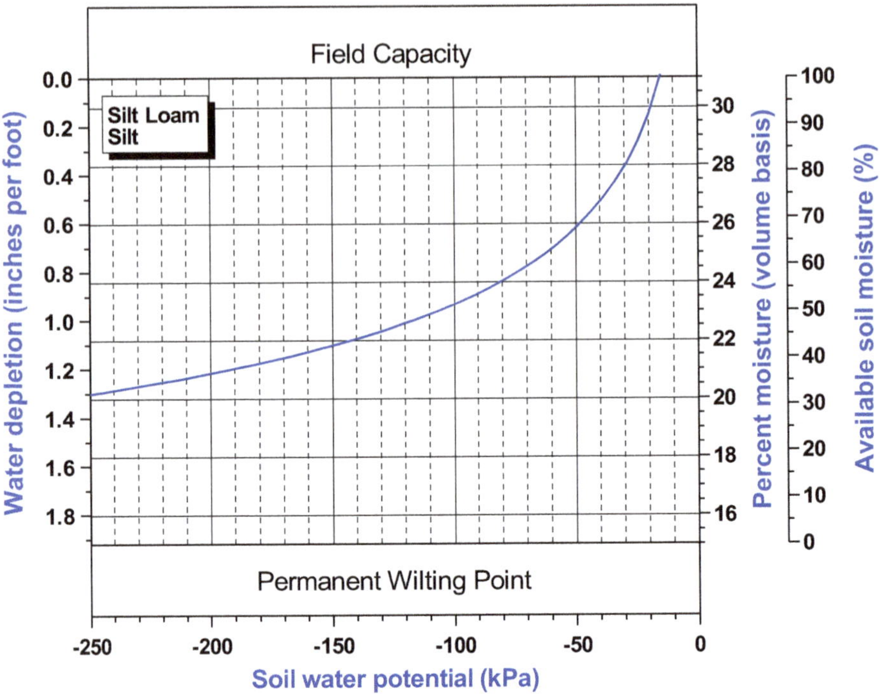

Fig. 13.10 Generalized soil moisture release curve for silt loam and silt soil

low efficiency used in irrigation scheduling calculations results in low irrigation efficiency in the field from applying excess water. The best approach is to initially use an efficiency value at the upper range of those shown in Table 13.4 or higher. As the season progresses, if soil water monitoring consistently shows less water in the soil than predicted by the soil-water balance, the assumed irrigation efficiency value can be revised downward. This further highlights the necessity of routine, consistent soil water monitoring for irrigation management.

The first step in calculating net irrigation application (desired irrigation amount) is to determine gross water application. Gross water application depth per rotation for center-pivot irrigation systems as a function of system capacity and rotation time can be obtained from the relationships presented in Fig. 13.12. System capacity in gpm/acre needed to use the curves in Fig. 13.12 can be obtained from the sprinkler application package specifications or approximated by dividing total system flow rate by the acreage irrigated. Net application depth for an 80% application efficiency can be obtained directly from right-side axis of Fig. 13.12. Net application depth for any application efficiency can be calculated as:

$$\text{Net depth} = \frac{\text{Gross depth} \times \text{Application efficiency}}{100}$$

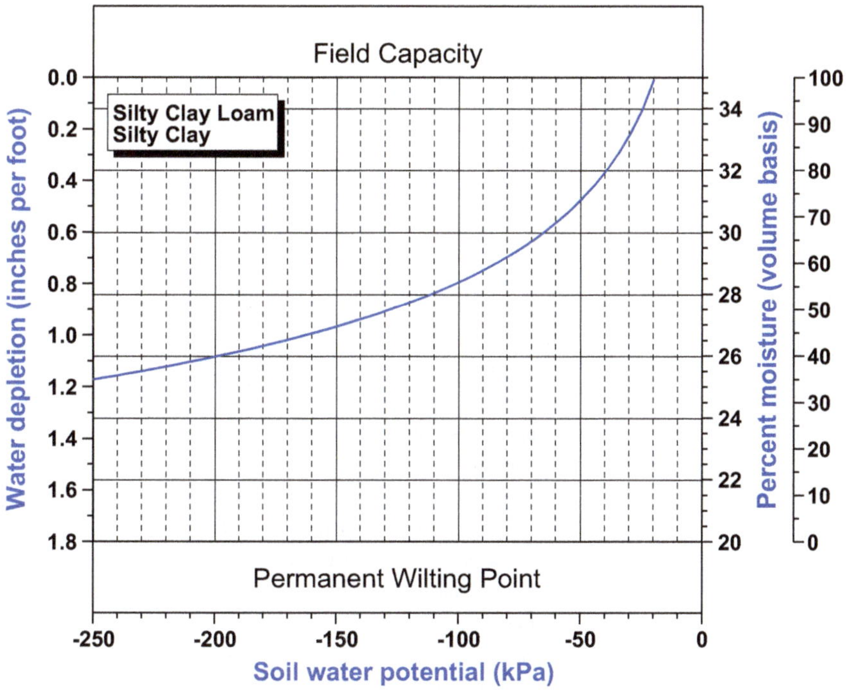

Fig. 13.11 Generalized soil moisture release curve for silty clay loam and silty clay soil

Example:

$$\text{Net depth} = \frac{0.8\,\text{in} \times 85\%}{100} = 0.68\,\text{in / rotation}$$

Gross water application rate for set-move and solid-set sprinkler systems as a function of sprinkler flow rate and spacing can be obtained from the relationships presented in Fig. 13.13. Sprinkler flow rate can be estimated from Fig. 13.14 for brass straight-bore nozzles as a function of nozzle size and pressure. Net application rate for 70% application efficiency can be obtained directly from right-side axis of Fig. 13.13. Net application rate for any application efficiency can be calculated as:

$$\text{Net application rate} = \frac{\text{Gross application rate} \times \text{Application efficiency}}{100}$$

Example:

$$\text{Net application rate} = \frac{0.22\,\text{in / h} \times 70\%}{100} = 0.154\,\text{in/h}$$

Table 13.3 Soil water potential and volumetric moisture content ranges corresponding to 65% available soil water

Soil texture	Soil water potential (kPa)	Soil water content (% by volume)
Sand, loamy sand	−25 to −35	9–12
Sandy loam, loam	−35 to −50	19–22
Silt loam, silt	−50 to −65	24–26
Silty clay loam, Silty clay	−65 to −75	29–31

Table 13.4 Typical irrigation system application efficiencies)

System type	Application efficiency (%)
Surface systems	
Furrow	35–65
Surge	50–55
Sprinkler systems[a]	
Set-move	60–75
Solid-set	60–85
High pressure center-pivot	65–80
Low pressure center-pivot	75–85
Linear-move	80–87
Micro irrigation	
Drip	90–95

Adapted from Sterling and Neibling (1994)
[a]Use lower efficiencies with larger spacing and windy conditions

Irrigation System Management

Center-Pivot Management

Center-pivot systems are sometimes designed with insufficient capacity to meet peak period daily water use. Instead, water banking is used to supply a small fraction of daily ET over the duration of the peak period. This allows for reduced system capacity that translates to reduced pump size, lower electrical demand charges, and reduced water application rates. Water banking is allowed because center-pivot systems are capable of providing light, frequent irrigations. Water banking applies to linear-move systems as well, but to a reduced extent by accounting for dry run time during repositioning. Water banking can potentially be applied to any irrigation system capable of light, frequent irrigations, such as drip and solid-set sprinkler. The degree to which water banking can be utilized is directly proportional to soil water-holding capacity and crop rooting depth. Potato grown on coarse-textured soils having water-holding capacities less than 1 in per ft. do not allow for water banking and must have a net system capacity equal to peak daily ET. For example, if peak ET is 0.34 in/day, then the net system capacity must be 6.4 gpm/acre

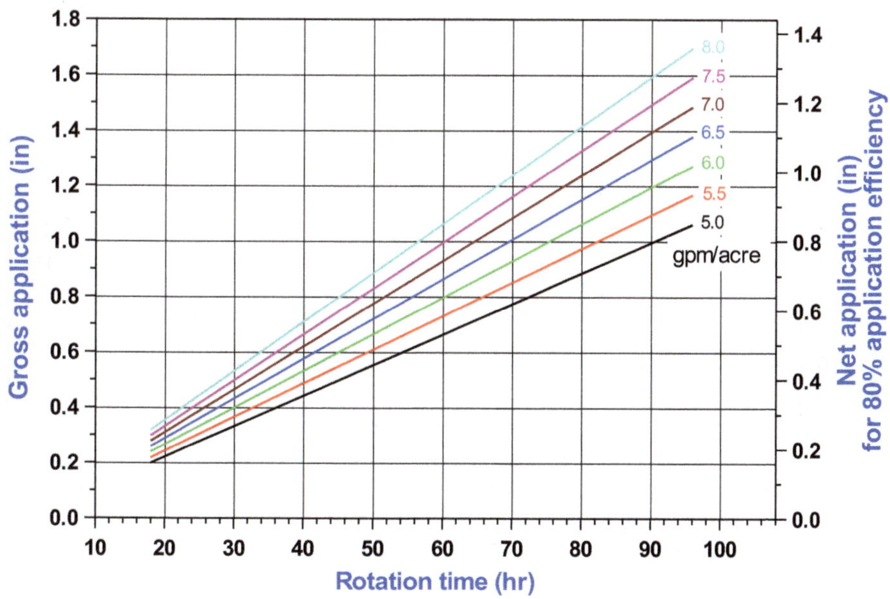

Fig. 13.12 Center-pivot application depth as a function of system capacity and rotation time (x-axis) for system capacities ranging from 8 gpm/acre to 5 gpm/acre

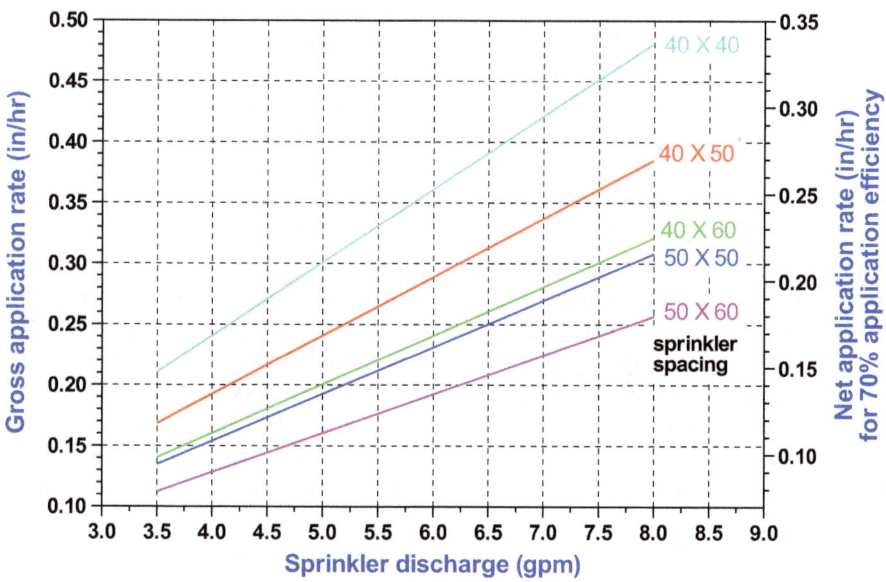

Fig. 13.13 Set-move and solid-set sprinkler application rate as a function of sprinkler discharge (x-axis) and sprinkler spacing (in ft) (lines)

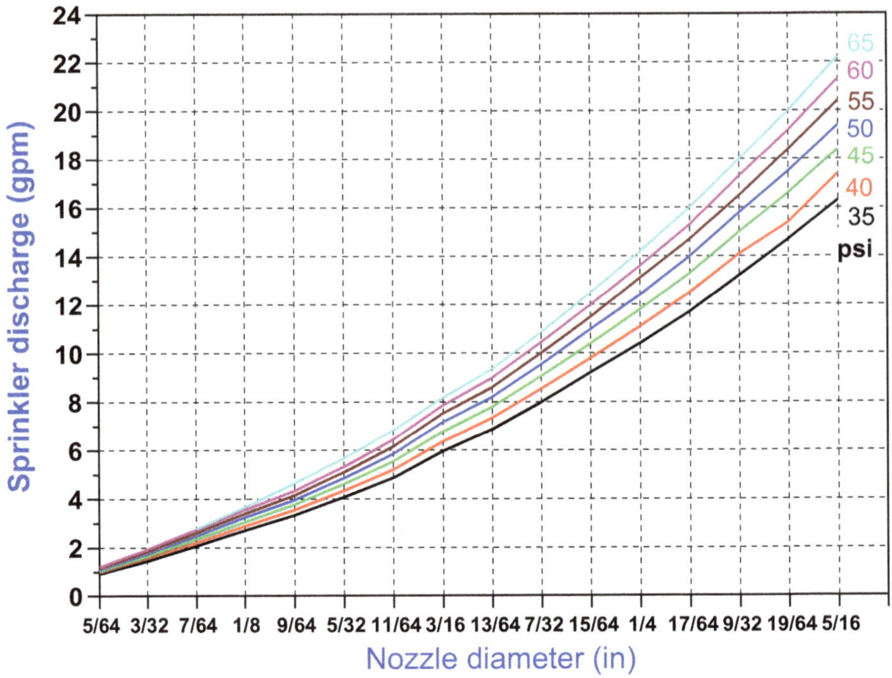

Fig. 13.14 Sprinkler discharge as a function of straight bore nozzle diameter (*x*-axis) and pressure in psi (lines)

[0.34 in/day × 18.86 (gpm/acre)/(in/day)] or a gross system capacity of 7.5 gpm/acre [6.4 (gpm/acre)/(85/100)] if application efficiency is 85%.

Center-pivot systems that utilize water banking must be managed to ensure that the soil-water reservoir is full at the beginning of the peak water use period. This requires planning and field soil moisture monitoring to the full depth of the crop root zone. Failure to do so will likely result in crop water stress near the end of the peak use period, the extent of which depends on soil and climatic conditions. The timing of the peak use period varies season to season, as does the duration of peak water use. Figure 13.15 depicts available soil moisture throughout the irrigation season for the condition where a center-pivot system is managed such that soil water is replenished to field capacity (100% ASW) early in the season (Fig. 13.15a) compared to one where soil water is replenished to only 90% ASW, either intentionally or inadvertently (Fig. 13.15b). Under both scenarios, the characteristic gradual drawdown of ASW occurs during the peak use period. However, in the second case, minimum ASW values fall below recommended limits, resulting in periodic plant water stress. When this occurs, there is no corrective course of action, as system capacity is fixed. The ultimate tuber yield and quality depends upon the season's climatic conditions, as they determine daily ET.

The natural tendency is to speed up a center-pivot system when crop water stress develops. Increasing the speed of a center-pivot produces lighter applications and

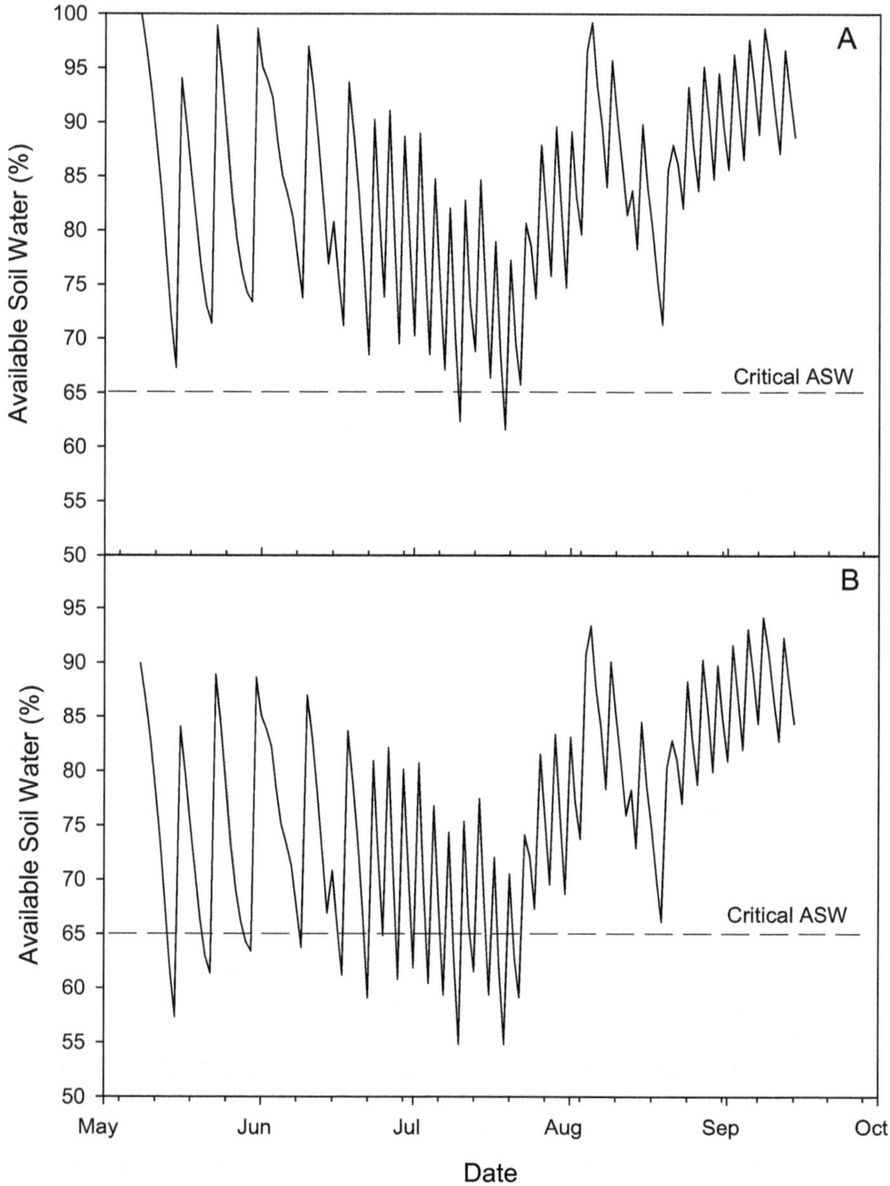

Fig. 13.15 Available soil moisture throughout the growing season for potato under center-pivot irrigation for: (**a**) 100% and (**b**) 90% available soil water (ASW) at the beginning of the peak use period

more frequent wetting of the soil and plant canopy, increasing the total amount of water lost to evaporation, and thereby decreasing the amount of water stored in the soil. Thus, system speed should remain the same or be reduced when crop water stress appears to effectively increase irrigation efficiency and stores a larger percentage of applied water in the crop root zone. However, if runoff occurs when the speed is reduced, it is better to run the pivot at the higher speed to reduce application depth.

Seemingly minor changes in application efficiency can result in a significant difference in center-pivot system performance. A 3–8% difference in application efficiency will occur between nighttime and daytime irrigation, resulting in differences in soil water storage. As a result, center-pivot speed should be adjusted such that rotation time is not a multiple of 24 h. Otherwise, areas of the field consistently watered during the daytime will have 3–8% less water stored in the soil for crop use. This small difference accumulated over time can result in water stressed areas within the field.

Conservation tillage practices, such as basin or reservoir tillage, are usually required to improve infiltration uniformity with potato under center-pivot irrigation. The hilling of potato plants causes water to concentrate in the furrow between hills under high application rates, which can result in deep percolation with water bypassing the crop root zone and/or runoff with even slight slopes. Runoff water collects in low areas causing excessive infiltration, while upslope areas have reduced infiltration and become water stressed, creating spatially differing irrigation requirements. The cumulative field scale effect is reduced yield and quality, reduced water and nutrient efficiency, and localized leaching of chemicals from the root zone. Planting potato in wide beds to minimize concentration of water in furrows has been shown to increase tuber yield and quality and reduce seasonal irrigation requirements 5–15% (King et al. 2011).

In recent years all center-pivot manufacturers and other sources have introduced equipment that can control irrigation depth on a spatial basis in a field (variable rate irrigation). Water depth can be controlled either on a pie-shaped basis by automatically adjusting system travel speed (one-dimensional control), or in irregular-shaped management zones (two-dimensional zone control), the size of which depends upon the number of individual control sections along the irrigation system lateral. In the latter case, water application depth is controlled by pulsing sprinklers on/off in a control zone using an appropriate duty cycle (timing). The additional equipment for two-dimensional control adds substantial cost to the irrigation system in terms of equipment and added maintenance. Theoretically, the sensitivity of potato to water stress suggests that tuber yield and quality can be improved and/or variability reduced field wide if water can be applied as needed everywhere in the field. Spatial differences in ASW and, hence, optimum irrigation depth, may be present initially or develop throughout the irrigation season due to spatial differences in ET and water infiltration. Spatial differences in water infiltration develop due to runoff and subsequent runon under irrigation and/or rainfall. These spatial differences in optimum irrigation depth are dynamic throughout the irrigation season making routine spatial analysis necessary to develop a dynamic prescription map to control the variable-rate irrigation system. Routine development of a prescription map is paramount to realizing improved tuber yield and quality from variable-rate irrigation systems.

Set-Move Sprinkler Management

Side-roll and solid-set sprinkler systems are normally designed to deplete soil water storage between irrigations during the peak use period. Thus, soils with greater soil water storage allow for longer irrigation intervals, resulting in reduced equipment and capital costs.

The preceding operating principal is contrary to the need to minimize soil-water fluctuations for optimum tuber yield and quality. The most typical irrigation management problems occurring with set-move sprinkler systems are irrigation intervals that are too long and excessive water applications during an irrigation set. This may be a result of overestimating soil water-holding capacity and crop rooting depth, or an insufficient number of sprinkler laterals requiring too many days to traverse the field. The maximum irrigation interval can be calculated as:

$$\text{Maximum days} = \frac{\text{Soil water} - \text{holding capacity}\left(\text{in/ft}\right) \times \text{Root zone depth}\left(\text{ft}\right) \times (1-0.65)}{\text{Peak daily ET}\left(\text{in/day}\right) + 1\,\text{day for irrigation}}$$

Maximum irrigation intervals based on a peak ET of 0.33 in/day for different soil types and root zone depths are shown in Table 13.5. Irrigation intervals exceeding 5 days during peak ET periods will likely result in ASW levels below 65%, which can adversely affect tuber yield and quality. However, in practice, irrigation intervals exceeding 5 days are not uncommon.

Field studies were conducted at Aberdeen, ID, in 1997 to evaluate the effect of irrigation frequency on potato yield and quality. Irrigation intervals of 4, 5, 6, and 7 days during tuber development were used in the study. The irrigation system was solid-set sprinkler. The soil type was a Declo silt loam with a water-holding capacity of approximately 2.2 in/ft. Water application amounts were determined based on replacement of estimated ET. Total yield, U.S. No. 1 yield, and yield of tubers >10 oz are shown in Table 13.6.

Table 13.5 Maximum irrigation interval (days) for set-move sprinkler systems based on 0.33 in/day peak ET plus 1 day irrigation time

	Root zone depth (in)				
Texture class	14	16	18	20	22
Sand, loamy sand	2.2	2.3	2.5	2.7	2.9
Sandy loam, loam	3.1	3.4	3.7	4.0	4.3
Silt loam, silt	3.4	3.7	4.0	4.4	4.7
Silty clay loam, Silty clay	3.2	3.5	3.9	4.2	4.5

An irrigation interval of 6 days resulted in the highest total yield, while the yield of U.S. No. 1 grade tubers decreased as the irrigation interval increased beyond 5 days. Yield of tubers over 10 oz. also decreased as the irrigation interval increased beyond 4 days. The results of this study show that there is an optimum irrigation interval for maximizing total yield and quality based on the water-holding capacity of the effective root zone and the rate of crop water use. This optimum interval is strongly dependent on soil texture and will be shorter for coarse-textured soils than heavier-textured soils. The percentage of large tubers is also strongly influenced by irrigation interval. Thus, irrigation management can be a useful tool in achieving tuber size goals.

Furrow Irrigation

Furrow irrigation of potato does not produce the tuber quality obtainable with other forms of irrigation, even with best achievable management practices. Water is required to traverse the field by overland flow in the furrow. The time required for the water to advance to the end of the furrow leads to greater water application at the inflow end compared to the outflow end, resulting from the difference in infiltration opportunity time. Furthermore, infiltration is a highly variable phenomena, with applications to individual plants ranging from half to twice the field average (Trout et al. 1994). Thus, furrow irrigation cannot achieve the degree of uniform water application needed to produce consistently high-quality tubers on a commercial field scale basis.

A common furrow irrigation practice for potato is to irrigate alternate furrows on successive irrigations to overcome some of the difficulty in applying small irrigation applications. Consequently, only about 15% of the soil surface is wetted, and water is expected to move upward laterally to wet the whole root zone. In the absence of a clay soil or dense soil layers, gravity causes water to move faster downward than laterally. Thus, attempts to completely wet the root zone to the top of the hill usually fail and result in excessive deep percolation losses. The lateral water distribution problem results in significant variation in soil water contents in the hill. Consequently, potato roots near the furrow experience widely varying soil water contents, while the upper portion of the hill remains dry.

Table 13.6 Influence of irrigation interval on total, U.S. No. 1, and > 10 oz. tuber yields in field studies conducted in Aberdeen, ID, 1997

	Total yield	U.S. No. 1 yield	> 10 oz. yield
Irrigation interval days	cwt/acre		
4	401	369	84
5	418	391	65
6	427	341	52
7	386	238	34

A consequence of non-uniform water distribution between and along furrows is wide variation in nitrogen availability due to both dry soil regions and leaching losses. This tends to further reduce tuber quality under furrow irrigation and reduces nutrient use efficiency.

These limitations have caused many producers to abandon furrow irrigation in favor of sprinkler irrigation. A common approach is to utilize a completely portable sprinkler irrigation system to irrigate potato, moving the system around the farm according to the crop rotation, and use furrow irrigation for the other row crops. The advantages of higher gross income and reduced risk with sprinkler irrigation are usually enough to justify the use of sprinklers for potato production. The ability to inject fertilizers and pesticides through sprinkler systems provides another significant advantage over furrow irrigation.

Irrigation Uniformity

Perfectly uniform water application is not physically or economically feasible on a field scale. Thus, some degree of variability in water application exists for all irrigation systems. The degree of water application uniformity is influenced by irrigation system type, design, and operating conditions. Regular maintenance of all irrigation systems is necessary to achieve the highest degree of uniformity throughout the life of the system. Irrigation uniformity is economically important because variations in water application, caused by worn or malfunctioning equipment, accumulate over the growing season. Portions of the field with large deviations from optimum water application rates will suffer serious losses in tuber yield and/or quality.

Summary

The primary goal of potato irrigation management is to minimize soil moisture fluctuations and maintain available soil water within the optimum range of 70–85%. Irrigation systems best suited to this task are those that are capable of light, uniform, and frequent water applications. An effective irrigation management program must include regular quantitative monitoring of soil water availability and scheduling irrigations according to crop water use, soil water-holding capacity, and crop rooting depth. Potato is more sensitive to water stress than most other crops, have relatively shallow root systems, and are commonly grown on coarse-textured soils. These conditions dictate utilization of a quantitative irrigation scheduling method for consistent, optimum economic potato production.

References

Jensen ME, Burman RD, Allen RG (1990) Evapotranspiration and irrigation water requirements. ASCE manuals and reports on engineering practice no. 70. American Society of Civil Engineers, New York, NY

King BA, Tarkalson DD, Bjorneberg DL, Taberna JP (2011) Planting system effect on yield response of russet Norkotah to irrigation and nitrogen under high intensity sprinkler irrigation. Am J Potato Res 88:121–134

Neibling WH, Brooks RH (1995) Potato production using subsurface drip irrigation: water and nitrogen management. In: Proceedings 5th international microirrigation congress, American Society of Agricultural Engineers, Orlando, FL, April 1995, pp 656–663

Patel N, Rajput TBS (2007) Effect of drip tape placement depth and irrigation level on yield of potato. Agric Water Manage 88:209–225

Shock CC, Eldridge EP, Pereira AB (2005) Planting configuration and plant population effects on drip-irrigated Umatilla russet potato yield and grade. Oregon State University Agricultural Experiment Station, special report 1062, pp 156–165

Stark JC (1996) Information management systems for on-farm potato research. Potato research and extension proposals for cooperative action. University of Idaho, College of Agriculture, pp 88–95

Sterling R, Neibling WH (1994) Final report of the water conservation task force. Idaho Department of Water Resources, Boise

Trout TJ, Kincaid DC, Westermann DT (1994) Comparison of Russet Burbank yield and quality under furrow and sprinkler irrigation. Am Potato J 71:15–28

Chapter 14
Physiological Disorders

Mike Thornton, Nora Olsen, and Xi Liang

Contents

Introduction

Potatoes are subject to a wide range of physiological disorders that are not directly caused by an infectious pest; instead they result from unfavorable environmental conditions or management practices that cause stress. The following is a discussion of common physiological problems, their causes, and management practices that will minimize their occurrence.

M. Thornton (✉)
Department of Plant Sciences, Parma Research and Extension Center, University of Idaho, Parma, ID, USA
e-mail: miket@uidaho.edu

N. Olsen
Department of Plant Sciences, Kimberly R&E Center, University of Idaho, Kimberly, ID, USA
e-mail: norao@uidaho.edu

X. Liang
Department of Plant Sciences, Aberdeen R&E Center, University of Idaho, Aberdeen, ID, USA
e-mail: xliang@uidaho.edu

J. C. Stark et al. (eds.), *Potato Production Systems*,
https://doi.org/10.1007/978-3-030-39157-7_14

Types of Disorders

Physiological disorders are also commonly referred to as abiotic (i.e., non-living), nonpathogenic, nonparasitic, or noninfectious diseases. Often a disorder is caused by the combined effects of both environmental conditions and management practices. A common characteristic of many physiological disorders is a delay from the time the stress or injury is initiated until symptoms are visible, which makes diagnosis and identification of causal factors very difficult. These disorders also often weaken the plant and allow pests to attack as a secondary consequence, which may mask the underlying causes.

Symptoms of physiological disorders are expressed in both the foliage and tubers. Tuber symptoms reduce crop value, while foliar symptoms affect productivity. Tuber disorders can be expressed in the field but can also occur in storage and can be either external or internal. External tuber disorders can reduce marketability, as well as cause reductions in processing quality and storability. Internal tuber disorders often go undetected until tubers are cut and inspected. However, these disorders can also result in significant reductions in crop quality and marketability.

Foliar Physiological Disorders

Physiological disorders of potato foliage can be caused by frost, chemicals, nutritional imbalances, hail, wind, lightning, air pollution, and waterlogged soils. These disorders are usually typified by symptom uniformity across the field or symptoms occurring in a regular pattern or topographic position within the field. In comparison, plant pathogen and insect damage can often show random distribution patterns ("hot spots") or can be associated with field edges. If visible symptoms or patterns do not provide a unique diagnosis, a plant or tuber sample should be submitted to a local extension specialist or qualified diagnostic laboratory for clinical evaluation for the presence or absence of a pest.

Frost Damage

Frost damage can occur when temperatures drop below 30 °F for any significant length of time. When injury occurs during a calm period, symptoms are usually more visible in low-lying areas.

Diagnosis

The first sign of injury is a darkening (very dark green to almost black) of the leaves. If damage is minor, the dark color will fade, leaving characteristics of leaf yellowing (chlorosis), leaf distortion, or brown, necrotic (dead) areas on the tips of the leaves

Fig. 14.1 It is common for early-season frost to injure one plant severely (right) while leaving adjoining plants untouched

on the upper part of the plant. If damage is more severe, leaves and/or stems will become water-soaked and turn dark brown to black in color, and within a few days the tissues will become desiccated and brittle (Fig. 14.1). In some cases, all foliage above ground will be killed. Any frost damage will slow plant growth, but severe frost injury will seriously delay development because plants must regrow from below-ground buds. This is especially true if freezing temperatures penetrate below the soil surface.

Management

A primary approach for prevention of frost damage is to plant potatoes late enough in the spring to minimize exposure of young plants to frost. For early planting, the use of vigorous, large seed pieces will allow the plants to recover more readily from frost injury if it does occur.

Little can be done to prevent frost damage to an emerged crop. Initiating irrigation before the frost and continuing until the temperature is above freezing can minimize damage, but this is often not practical on a large scale. If done improperly, irrigation may actually increase injury. Irrigation immediately before, but not during, the freezing temperatures will often increase symptoms.

Chemical Damage

Chemical damage or "burning" of the foliage can occur when: (1) pesticides, fertilizers, or other agricultural chemicals are applied improperly, (2) when chemicals drift onto a non-target field (Fig. 14.2), or (3) when chemical residues carry over in the soil from an application to a previous crop. Some chemical herbicides, such as glyphosate and dicamba, can carry over in seed potatoes, causing symptoms in the subsequent crop.

Fig. 14.2 Malformed leaves are symptoms of phenoxy-type herbicide drift or carryover

Diagnosis

A wide range of foliar symptoms can occur as a result of chemical damage. Growers need to systematically evaluate these symptoms in relation to crop history, chemicals applied, and environmental conditions. Some characteristics of chemical damage include leaf distortion or curling, yellowing of the leaves and stems, necrotic or brown spots on leaf margins, stunted growth, or plant death. Often the distribution of chemical damage symptoms in the field will follow a pattern consistent with the equipment or method used to apply the chemicals. Chemical carryover in seed potatoes results in a random distribution pattern in the field, usually at low levels. Some specific classes of herbicides also cause characteristic disorders in the tubers, such as multiple growth cracks, that can be used in diagnosis.

Chemical damage symptoms may often be confused with those caused by virus infections or other diseases or pests. Consequently, all pertinent chemical application information should be evaluated and related to current and previous crops in and around the affected field along with information on previous pest problems in the field and prevailing weather conditions.

Management

Effective management practices may not be available to alleviate chemical damage symptoms. In a few cases, appropriate remedial actions may be apparent when the causal factor is diagnosed. See Chaps. 9,11, and 12, for more detailed information on foliar damage symptoms related to pest problems and pesticide applications.

Nutrient Imbalances

Proper fertilization is an important part of producing a high-yielding, high-quality potato crop, but nutrient imbalances can result in foliar and tuber abnormalities. Foliar symptoms associated with improper fertilizer management are often difficult to distinguish from those caused by environmental stresses.

Diagnosis

Nutrient excesses usually cause symptoms similar to those of salt burn, such as leaf necrosis, and can best be prevented by using a proper nutrient management program. Nutrient deficiencies typically produce a wider range of symptoms, which are fairly specific for each nutrient. Information on fertilizer sources and rates, soil type, pH, salinity, soil and plant nutrient concentrations, and application rates of other chemicals can help determine if the damage is nutrient related. Petiole and/or soil samples can provide further insight into any nutrient imbalances. See Chap. 8.

Management

A proper nutrition program involving soil and plant tissue analysis will prevent most nutrient imbalance problems.

Hail Damage

On occasion, hail can cause severe crop injury, which can result in large yield and quality losses. Yield losses occur when hail damages foliage during the early part of the season while potato plants are still in the vegetative stage. Yield loss will be proportional to the timing and degree of defoliation. During early tuber bulking, minor hail damage has little effect on yield, but substantial defoliation at this stage of growth can decrease U.S. No. 1 yields, primarily as a result of the increased production of malformed tubers. Specific gravity can also be decreased by early-season defoliation. However, severe hail damage during mid tuber bulking will usually cause substantial yield and quality losses. Damage caused by hail when plants are near maturity will generally result in only minor losses in yield or quality. The impact of hail injury is generally greater for early-maturing varieties.

Diagnosis

Potato leaves will look torn and perforated after a damaging hailstorm (Fig. 14.3). In severe cases, a complete loss of foliage will occur. Damage on stems is usually superficial, producing gray lesions or pockets.

Management

Prevention measures are not available for hail damage, and subsequent management will depend on the extent of injury, developmental stage of the plants, and variety. Minor hail damage usually does not require a change in management plans. Major damage may require an adjustment in irrigation and fertility programs, usually typified by a delay in added inputs.

Fig. 14.3 Plants will be torn and perforated after a severe hailstorm; this damage will greatly reduce yield and tuber size if it occurs during tuber bulking

Fig. 14.4 Leaf tip burn is a symptom of foliar damage caused by wind, low humidity, and/or high air temperatures

Hail-damaged plants usually will not require additional foliar nutrient applications if petiole nutrient concentrations were sufficient before injury. Applications of excessive amounts of nitrogen (N) directly after severe hail damage may actually slow tuber bulking by stimulating excessive vegetative growth.

Windburn/Leaf Tip Burn

Leaf tip burns are physiological disorders that commonly occur in potato fields in arid regions. This disorder is especially prevalent when conditions are dry, warm, and windy.

Diagnosis

Symptoms are often mistaken for disease, chemical damage, fertilizer burn, or water deficits. As a result, a grower or crop adviser should carefully consider the pattern of symptom distribution in the field, as well as associated weather conditions. Foliar symptoms will typically appear in the upper canopy as brown, necrotic, irregular spots on leaves, especially on the tips and margins, but can occur on any part of the leaf (Fig. 14.4). Stems can also show wind damage as light brown lesions or pockets in the tissue.

Management

No effective management practices exist to avoid tip burn, other than making sure adequate soil moisture levels are maintained to minimize the potential for leaf dehydration.

Lightning Damage

Lightning damage produces symptoms that are similar in many respects to a "hot spot" or outbreak of a disease, such as blackleg, Rhizoctonia, or late blight.

Diagnosis

Symptoms of lightning damage will appear as circular or oval areas in the field within a few days of a thunderstorm. Affected plants initially have stems that collapse at the top and become water-soaked and black to brown in color. Characteristically, the stem pith tissues collapse and form crosshatched horizontal plates that can be seen by slicing the stem longitudinally. The damaged stems soon turn brown to a light tan as they dry. Sometimes the leaves may remain green with only the stems affected, but often the whole plant will subsequently die. Look for distribution patterns showing perfectly healthy plants next to a few dead or severely injured plants. Lightning damage often occurs in wet areas in the lowest parts of the field, where the intensity of the electrical discharge tends to be greatest.

Management

No control or management strategies exist for damage caused by lightning.

Air Pollution Damage

Air pollution injury is not a common problem in most production regions, but occasionally will cause damage to leaves. Air pollution damage is difficult to identify due to its similarity to symptoms caused by pests and other environmental stresses. Damage may be the result of high levels of chemicals in the air, such as ozone, nitrous oxide, and sulfur dioxide. Foliar injury symptoms can occur adjacent to industrial areas and major roadways.

Diagnosis

Injury will vary with pollutant type, concentration, and length of exposure, as well as with plant growth stage at the time of exposure. Symptoms vary, but generally appear as a necrotic speckled or pinpoint spotting on the leaf. This advanced injury can be confused with spider mite damage.

In diagnosing potential air pollution damage, possible sources of pollutants (factories, exhaust emissions from vehicles using a nearby highway, etc.) should be investigated near the damaged areas of the field. Prevailing wind direction and environmental conditions are other considerations in relation to the location and pattern of damage in the field. Other causes of damage that lead to similar symptoms are agricultural chemicals, weather, disease, and insects. Some varieties are more sensitive to air pollution than others.

Management

No effective field management practices are available for control of air pollution injury, other than to avoid planting potatoes in potentially polluted areas or choosing less sensitive varieties.

Waterlogged Soils

Excessively wet or flooded soil conditions that last for several days or more can damage potato foliage and tubers. The damage is primarily the result of anaerobic (oxygen deprived) conditions in the root zone that interfere with plant water and nutrient uptake.

Diagnosis

Foliage of affected plants initially becomes light green to yellow and will eventually wilt and turn brown if waterlogged conditions persist. Tubers and roots become susceptible to rot organisms under excessively wet soil conditions.

Management

It is important to ensure proper irrigation system design and management that will prevent waterlogged areas from developing in the field. Basin tillage or installing drain tiles are possible management strategies where runoff is a potential problem.

Tuber Physiological Disorders

Tuber disorders may consist of external quality damage and/or internal problems. Bruises are physiological disorders that have both internal and external symptoms. With some disorders, pathogen invasion can result, which causes additional loss of quality.

External Physiological Disorders

Malformed Shape

Malformed tubers are a result of environmental or cultural stresses, such as wide fluctuations in air and soil temperatures; water deficits; abrupt changes in nutrient availability; and defoliation due to insects, diseases, or hail. Tuber growth rates often fluctuate in response to widely varying growing conditions causing malformations, such as bottlenecks, dumbbells, pointed ends, and knobs.

Symptoms and Causes

Any stress that causes a reduction or stoppage in plant growth can cause constricted tuber growth in the bud, middle, or stem end portion of the tuber, depending upon the extent of the stress and the stage of growth at which it occurs (Fig. 14.5). For example, a tuber with a pointed bud end indicates that the stress-induced restriction in growth occurred during late tuber bulking, while pointed stem ends indicate early season stress. Growth interruptions during mid-bulking can cause dumbbell-shaped tubers. Knobby tubers are caused when secondary growth occurs at lateral eyes on the tuber due to loss of apical dominance (Fig. 14.6).

These symptoms can cause potatoes to be graded as U.S. No. 2 and can also cause sugar accumulation and specific gravity reduction in the affected area of the tuber. This sugar accumulation may make the tuber unacceptable for processing due to non-uniformity in the fry color (sugar ends).

Fig. 14.5 Malformed and misshapen tubers result from environmental or cultural stresses during critical periods of tuber growth

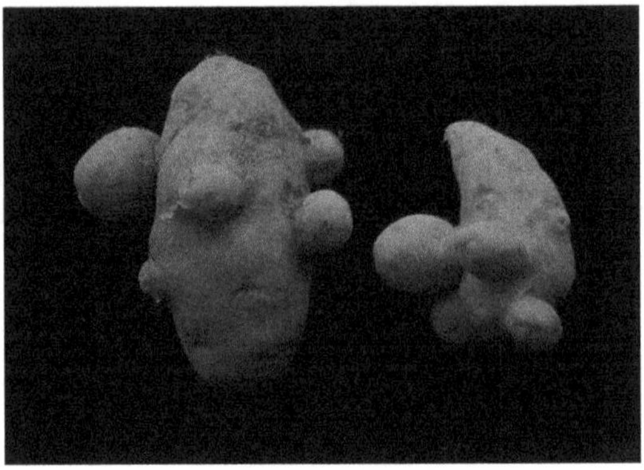

Fig. 14.6 Knobs are caused when secondary growth occurs at lateral eyes on the tuber

Management

Potatoes are particularly sensitive to stress during tuber initiation. Studies with Russet Burbank have shown that moderate to severe soil water deficits during tuber initiation can reduce U.S. No. 1 yields by 10–40%. Management approaches for preventing malformed tubers include promoting uniform growth by establishing uniform stands; avoiding large fluctuations in N availability; maintaining available soil water content above 70% and avoiding cultural practices, such as late cultivation, that may alter tuber growth patterns.

Heat Sprouts/Tuber Chaining

During periods of hot weather (85 °F and above), plants may respond by increasing top growth rather than tuber production. One consequence of this growth pattern is the tendency for stolons to remain vegetative. The result can be the development of heat sprouts and chain tubers or heat runners.

Symptoms and Causes

Heat sprouts develop when stolons continue to elongate, emerge through the soil surface, and develop into a leafy stem (Fig. 14.7). This may occur before any tubers set on the stolon, or as a result of stolons reforming from the bud end of tubers at any stage of development. Tuber chaining symptoms occur when multiple tubers develop on a single stolon (Fig. 14.8). In severe cases, tubers develop from the eye or stolon of another tuber. Heat sprouts and chain tubers are caused by renewed growth after periods of interrupted development during extended exposure to warm soil temperatures (greater than 75 °F).

Management

Prevention of these disorders includes avoidance of environmental stress and encouraging uniform vine and tuber growth by using proper planting, hilling, fertility, and irrigation practices.

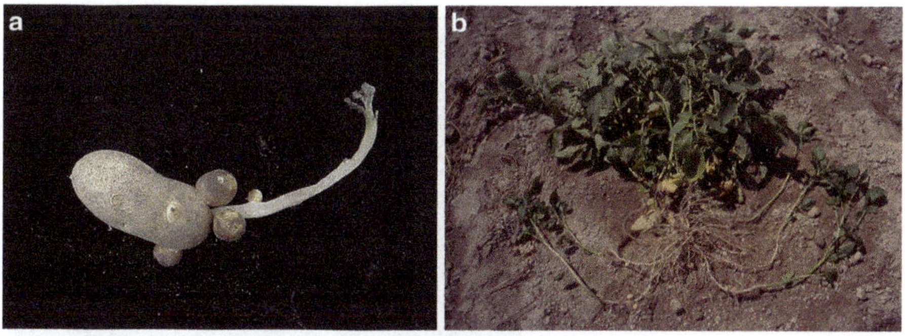

Fig. 14.7 Heat sprouts develop when stolons elongate rapidly and emerge from the soil as a leafy stem. Symptoms depend on the timing of stress. Sprouts may develop from the apical buds on a tuber (**a**) or may result from an elongating stolon that never terminates in a tuber (**b**)

Fig. 14.8 Another symptom of heat injury during growth—tuber chaining—takes two forms. Slight to moderate stress causes long stolons with multiple tubers set on side branches, while severe injury causes eyes to sprout and develop into stolons and, subsequently, new tubers

Tuber Cracking

Potato tuber cracking can take the form of growth cracks, elephant or alligator hide, skin checking, or any other symptom showing a cracked appearance of the skin. Symptoms can be either superficial or affect a major portion of the tuber.

Symptoms and Causes

Growth cracks are the most common type of cracking and are often caused by irregularities in tuber growth, especially in response to widely fluctuating water supplies (Fig. 14.9). Other factors, including virus infection and herbicide injury, can also cause tuber growth cracks and abnormal-looking tubers. Clinical diagnosis may help to identify non-management factors that cause cracking. Thumbnail cracking is caused by shallow breaks in the skin of well-hydrated tubers that have been exposed to drying conditions (see the section on shatter bruise later in this chapter). Potato varieties vary widely in their susceptibility to tuber cracking.

Fig. 14.9 Rapid tuber growth, or periods of drought stress followed by rain or irrigation, may result in growth cracks

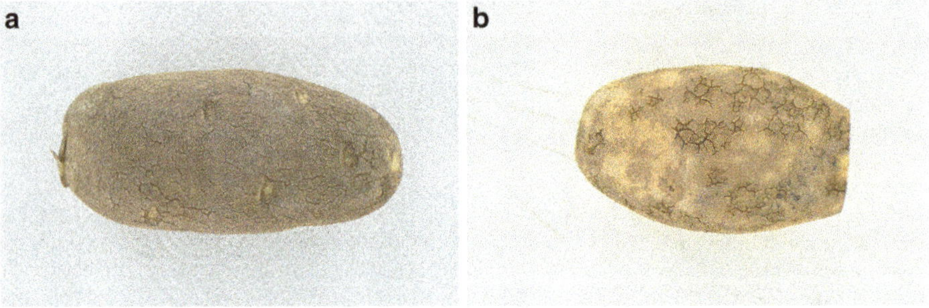

Fig. 14.10 Elephant or alligator hide is a skin-cracking defect that can show mild (**a**) or severe (**b**) symptoms

Elephant or alligator hide appears as shallow, corky cracks on the tuber skin (Fig. 14.10). The primary cause of this condition is unknown, but contributing factors may include high temperature, high soil organic matter, and excessive soil moisture and fertilization. Improper timing of maleic hydrazide applications has also been reported to cause an elephant-hide type of appearance on some varieties.

Management

Prevention of growth cracks is accomplished primarily by maintaining uniform, adequate soil moisture and nutrient levels throughout tuber bulking. Minimizing elephant hide is largely a matter of growing resistant varieties.

Feathering/Skinning

Feathering or skinning is commonly observed in immature potatoes and is a result of incomplete development of the skin layer (periderm). When tubers are immature, the skin is easily rubbed off during handling. This disorder is particularly important

Fig. 14.11 Feathering or skinning occurs during harvesting or handling of immature tubers

for fresh market potatoes because of its effect on consumer appeal. Also, the potential for weight loss and disease is considerably greater with feathered potatoes.

Symptoms and Causes

Feathering or skinning occur only upon handling. The skin of the potato is partially or completely removed, exposing the underlying tuber flesh (Fig. 14.11). If the skin is partially removed, it remains attached to the tuber but dries out and has an "onion-skin" texture.

Management

Approaches for preventing this disorder include managing irrigation, fertilizer, and vine kill practices to achieve proper maturity and skin set at harvest. Careful handling of the potatoes and proper wound healing conditions in storage are also important. High late-season N availability and excessively wet soils will delay maturity and increase susceptibility to skinning and feathering. Achieving complete vine kill 14–21 days before harvest will usually provide sufficient time to "set" the skin.

Enlarged Lenticels

Lenticels are openings on the tuber surface that allow for air exchange and can become enlarged when exposed to waterlogged soils or prolonged wet conditions.

Symptoms and Causes

Reductions in oxygen availability resulting from saturated soils or maintaining wet tuber surfaces for extended periods of time in storage or after packaging will cause lenticels to open and become enlarged. Enlarged lenticels look like small, white bumps on the surface of the tuber (Fig. 14.12). This disorder makes the tuber more susceptible to the entry of disease organisms, especially soft rot.

Fig. 14.12 Enlarged lenticels appear as small, white dots on the tuber skin

Fig. 14.13 Greening occurs when the tuber is exposed to light, resulting in the development of chlorophyll

Management

To avoid enlarged lenticels, potatoes should not be allowed to sit in saturated soils for extended periods (1–2 days), especially late in the season. Allowing soil moisture to drop to 65–70% between each irrigation is usually adequate to prevent swollen lenticels. During storage, avoid wet spots resulting from condensation, excess humidity, and temperature differences.

Greening

Light-induced formation of a green color on tuber surfaces resulting from chlorophyll accumulation is known as greening.

Symptoms and Causes

Tubers growing at or close to the soil surface may become green from direct exposure to sunlight or from light penetrating through cracks in the soil surface. This is usually an intense green color on a limited part of the tuber surface (Fig. 14.13).

Fig. 14.14 Pink eye appears as raised pink or brown water-soaked areas around the eyes

Greening also occurs from extended exposure to light in storage or on store shelves. This situation usually produces a lighter and more diffuse coloration of the entire tuber. The rate of greening is slower at lower temperatures. Thin-skinned potatoes, particularly those with white skins, are more affected because the greening is more visible.

Management

Several cultural practices can be used to minimize greening, including proper seed planting depth and hilling and rolling during vine kill to close soil cracks. In storage, potatoes should not receive prolonged exposure to light. Placing daylight blockers on outside vents helps to decrease light exposure in storage.

Pink Eye

Pink eye is a physiological disorder characterized by raised pink or brown water-soaked areas around the eyes, although the disorder may affect areas other than just the eyes. It is often accompanied with a cracking of the periderm producing a "corky patch" and providing an entry point for other pathogens, such as Pythium, soft rot, and Fusarium.

Symptoms and Causes

Small patches of disrupted periderm can be peeled to expose a brown or pink-colored tissue (Fig. 14.14). Under blue ultra-violet fluorescent light the exposed tissue will autofluorescence blue, whereas healthy tissue will not. This autofluorescence is due to higher accumulation of phenolic compounds in that area.

The pink eye disorder results from compromised skin tissue that lacks suberin, resulting in cell death. Although the cause of pink eye is unknown, higher incidence of the disorder has been associated with high soil temperatures and moisture late in the growing season during senescence or after vine death. It is also associated with compacted soils that lead to low oxygen conditions and eventual cell death within and beneath the periderm of the tuber.

Fig. 14.15 Brown center appears as brown tissue in the pith area of a tuber

Management

Maintain a healthy crop and manage to avoid "uneven" premature plant or stem death due to Verticillium or early die. Avoid excessive soil moisture when the plant does not require it, especially late in the growing season with premature vine death. Use proper tillage and crop rotation to avoid field compaction.

Internal Physiological Disorders

Brown Center/Hollow Heart

Brown center and hollow heart defects can cause serious losses in crop quality and economic return to the grower. The causes of these related disorders can be complex, and their development can occur throughout tuber bulking.

Symptoms and Causes

A brown discoloration without a visible flesh separation in the center of the tuber (pith region) is known as brown center (Fig. 14.15). The brown coloration results from the rupture of cell membranes and death of the affected cells.

This cell damage is apparently caused by inadequate nutrient and carbohydrate concentrations in the affected tuber tissue that reduce cell membrane integrity. The inadequate nutrient and carbohydrate concentrations are either the result of competition with the vines or the inability of the plant to absorb and transport sufficient quantities of nutrients to the pith area of the tubers. This disorder is also associated with periods of slow growth, followed by rapid tuber growth.

Tubers are most susceptible to brown center during the period from tuber initiation through early tuber bulking. Although this disorder does not always result in hollow heart, it is considered to be a milder form of the same defect and can even be a precursor to hollow heart.

Brown center symptoms can dissipate if subsequent tuber growth is moderate and uniform, and healthy cells grow between the damaged cells, diluting the brown

Fig. 14.16 Hollow heart is an irregular or lens-shaped cavity, of variable size, often accompanied by a brown discoloration in the tuber flesh

coloration. Research has shown that soil temperatures below 55 °F for 5–7 days during early tuber development can initiate brown center. Available soil moisture levels above 80–85% during early tuber development can also increase the incidence of brown center, possibly due to the slower warming of wet soils.

Hollow heart is characterized by the formation of an irregular cavity in the flesh of the tuber, typically surrounded with brown, discolored tissue (Fig. 14.16). The cavities can vary widely in size and form in the center of the pith or near the stem or bud ends of the tuber, depending on when the disorder develops.

Two distinct forms of hollow heart exist in potatoes; namely, early- and late-initiation types. Early-initiation hollow heart appears shortly after tuber set and is caused by the same factors that cause brown center. Rapid tuber enlargement results in the cavitation of the damaged tissue. This cavitation is often associated with a period of restricted tuber growth caused by cool soil temperatures, followed by accelerated growth rates that occur as temperatures warm up. High N availability during tuber initiation may increase the incidence of early-initiation hollow heart by producing a large canopy, which competes with tubers for nutrients and cools the soil by shading the soil surface.

Late-initiation hollow heart occurs during the latter part of the tuber-bulking period and is not usually associated with brown center. It is most commonly caused by a growth stoppage due to water or nutrient stress, followed by a return of favorable conditions and rapid tuber growth.

Management

The potential for developing brown center and hollow heart can be reduced by establishing uniform plant spacing and planting depth to encourage uniform emergence and using cultural practices that promote steady, uniform growth rates. Excessive soil moisture and N availability during tuber initiation and early development can increase the potential for brown center/hollow heart development, particularly under cool growing conditions.

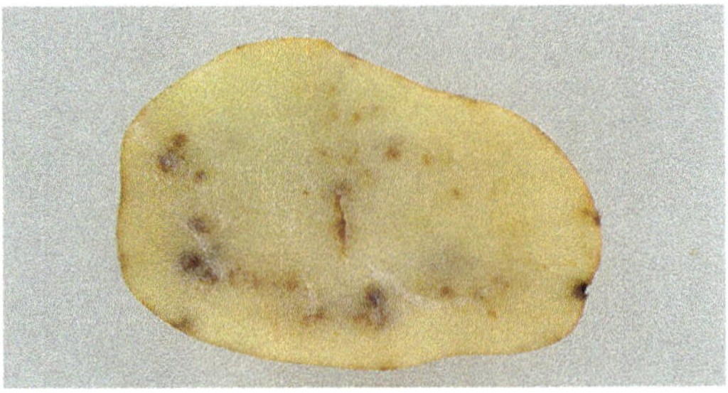

Fig. 14.17 Internal brown spot or heat necrosis symptoms can be mild or severe, and occur throughout the flesh

Surveys of growers' Russet Burbank fields during high hollow heart years have shown that N fertilization rates above 200 lb. N/acre and available soil water contents above 80–85% before row closure can significantly increase the incidence of brown center and hollow heart. Consequently, applications of N fertilizer should be moderate during early tuber development to maintain adequate tuber growth rates, while minimizing the potential for brown center and hollow heart development. Optimal rates of other nutrients should also be applied to avoid additional stresses. In addition, maintaining available soil moisture between 65 and 80% should minimize the development of these disorders, while allowing for acceptable tuber yield and quality.

Internal Necrosis

Internal necrosis is also referred to as internal brown spot (IBS) or heat necrosis.

Symptoms and Causes

This disorder can be described as small, brown, necrotic lesions or spots primarily inside the vascular ring of the tuber (Fig. 14.17). Internal necrosis differs from brown center in that it does not concentrate in the center (pith) of the tuber; rather it appears as diffuse spots distributed elsewhere in the tuber flesh. Symptoms may begin to develop shortly after tuber initiation, but more commonly, this problem becomes increasingly severe during late tuber bulking and senescence. Symptoms also tend to intensify during storage, particularly under warm conditions.

Although the factors that result in internal necrosis are not completely understood, researchers have established a relationship between the development of this disorder and lack of adequate calcium in the tuber. This may be the result of inadequate soil calcium availability, or the inability of the plant to absorb soil calcium and transport it to the tubers. Hot, dry weather, as well as high soil temperatures and fluctuating soil moisture conditions during tuber bulking, are also associated with the incidence and severity of the defect. This disorder can be more prevalent in sandier soils that have low cation exchange capacities and greater conductance of heat.

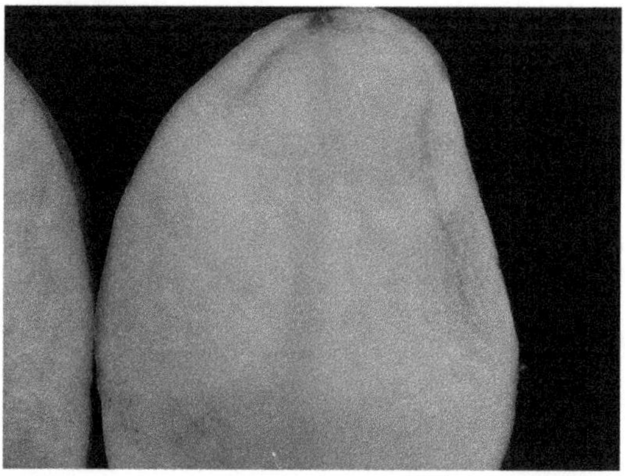

Fig. 14.18 Stem-end discoloration is a physiological problem that can be similar in appearance to net necrosis caused by the potato leafroll virus

In very susceptible varieties, such as Atlantic, there is an association between days from planting to harvest and incidence of heat necrosis. As tubers bulk and become larger, combined with warm soil and air temperatures, heat necrosis can greatly increase.

Management

Prevention of internal necrosis requires maintaining adequate soil moisture, especially during hot periods; applying adequate calcium in the tuber-forming zone in the soil (particularly in sandy or low calcium soils); and managing fertilization, irrigation, and other cultural practices to promote uniform vine and tuber growth. Harvesting the crop when mature, but before tuber size becomes excessive, can also reduce the incidence of this disorder.

Stem-End Discoloration

Several diseases, including Verticillium wilt and potato leafroll virus (PLRV), can cause vascular and/or stem-end discoloration. See Chap. 9. There is also an abiotic disorder that causes symptoms that are similar to the disease-induced vascular discoloration. This disorder is called stem-end discoloration (SED).

Symptoms and Causes

SED is typified by a shallow, brown discoloration in the vascular system (phloem and xylem) on the stem end of the tuber (Fig. 14.18). Intensity can vary with season and variety, but typically does not extend greater than 0.5–1.0 in into the tuber.

SED may or may not be visible at harvest, and may develop during storage. Laboratory analysis for PLRV will determine whether the visible symptoms are due to SED or to net necrosis caused by PLRV. Symptoms of SED will increase with time in storage but not to the extent of net necrosis.

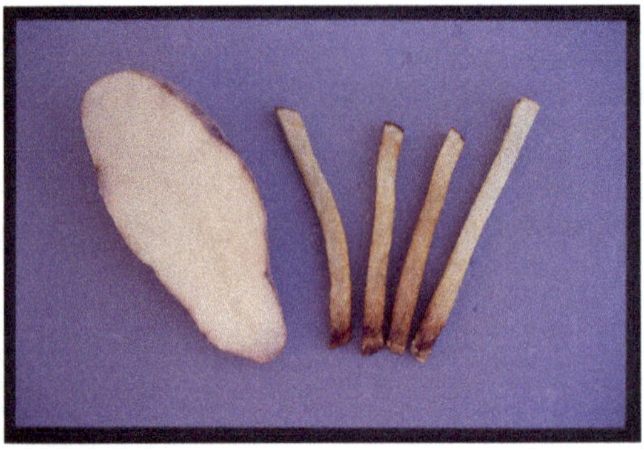

Fig. 14.19 Sugar ends are the result of stress-induced accumulation of sugar, usually in the stem end of the tuber, and are associated with translucent appearance in the raw potato and dark ends in the fried product

The causal factors of SED are not known, but several influential factors have been evaluated. Among factors previously implicated in causing SED is the rate of vine kill. Earlier research showed an increase in SED with rapid vine kill, particularly when the vines were still green and vigorously growing. However, recent research has not been able to confirm a relationship between the rate of vine kill and SED. Other studies have shown that killing water-stressed potato plants under high-temperature conditions, or a severe frost prior to plant senescence, can increase SED. But none of these factors have consistently caused SED. Many other unknown factors may be involved in the development of this physiological disorder.

Management

Practices for reducing SED are difficult to elucidate because causal factors have not been clearly identified. However, cultural practices should include avoiding vine killing when the plants are subject to moisture or temperature stress, or if the vines have not begun to senesce. Irrigation before vine kill may reduce the potential for SED development.

Sugar Ends/Translucent Ends

The physiological disorder commonly referred to as sugar ends is also known as translucent ends, glassy ends, or jelly ends. It is primarily a concern for processing potatoes and mostly affects varieties that are susceptible to a wide range of stresses, such as Russet Burbank.

Symptoms and Causes

This disorder usually shows up as a post-fry darkening of one end of a french fry, usually on the stem end of the tuber. This darkening is primarily caused by the accumulation of reducing sugars at the one end, which when fried, produces the undesirable dark color (Fig. 14.19). Often the end of the tuber that fries dark also exhibits

Fig. 14.20 Jelly end rot is an extreme expression of the sugar end disorder that often affects the stem end of heat-stressed tubers

a restriction in growth or pointed end. This darkened end will typically also have lower specific gravity and a visible "glassy" or translucent appearance. In severe cases, this disorder predisposes the tuber to subsequent tissue breakdown and the development of jelly end rot (Fig. 14.20).

Sugar ends are typically associated with periods of high air and soil temperatures and water deficits during tuber initiation and early bulking. Research has shown that high soil temperatures inhibit the conversion of sugars to starch in the tubers, thereby increasing the proportion of reducing sugars in the affected tissue. Water deficits may also interfere with the transport of sugars within the plant causing unequal distribution of reducing sugars after alleviation of the stress.

Management

Minimizing sugar end development requires avoiding moisture stress during tuber initiation and bulking, especially during early tuber development. Heat stress during this time can substantially increase the development of sugar ends. The following management checklist outlines some of the key cultural practices that can be used to minimize sugar end development:

- Choose fields that have the potential to grow a high-quality potato crop. Avoid fields with highly variable texture or topography, especially those with shallow or high-bulk density soils.
- Plant potatoes after wheat, barley, corn, or other crops that leave significant amounts of crop residue after harvest and minimize soil compaction. Shallow incorporation of crop residues can improve water infiltration and aeration and can reduce soil bulk density. Avoid planting potatoes after sugar beets or onions because of the reduced amounts of crop residue and a greater potential for soil compaction.
- Bed soil in the fall, when weather and soil conditions permit, to facilitate early planting.
- Manage irrigation to provide uniform water application and optimal water infiltration and soil water distribution throughout the field.

Fig. 14.21 Blackheart symptoms are a result of oxygen deprivation in the field or storage

- On ground of variable slope, use a properly designed sprinkler system with flow control nozzles and use basin tillage where appreciable runoff is expected.
- Monitor evapotranspiration and soil water status on a regular basis to determine crop water requirements and maintain available soil water content in the crop root zone within the optimal range. See Chap. 13.
- Use a good soil sampling and testing program and fertilize according to established guidelines for your area. Avoid excessive N applications, since high N can delay tuber bulking, thereby increasing potential exposure to heat stress.

Blackheart

Blackheart occurs in the field or in storage when the oxygen supply to tubers is insufficient to support adequate respiration rates.

Symptoms and Causes

This disorder causes distinctive symptoms typified by a dark, black, or black-blue discoloration in the center of the tuber (Fig. 14.21). In the field, this disorder is most prevalent in waterlogged soils in low-lying areas. In these areas, soil remains saturated for extended periods, thereby promoting anaerobic conditions and reducing gas exchange with the tubers. Blackheart that develops in the field is not always observed because the tubers tend to break down before harvest. Blackheart can also occur in storage if proper ventilation is not supplied to the tubers. Warmer storage temperatures increase tuber respiration rates and oxygen demands, which can accelerate blackheart development if ventilation is inadequate.

Management

In-field management for blackheart prevention includes proper irrigation scheduling and using basin tillage to minimize the accumulation of runoff water into the low spots of the field. Maintaining adequate ventilation and proper temperatures helps minimize blackheart development during potato storage and transport. Further, maintaining proper airflow through the potato pile will also minimize the potential for blackheart. In part, good airflow is the result of proper installation and operation of equipment, minimizing the amount of dirt and debris in storage, and proper distribution of air ducts.

Sidebar 14.1: Purple Pigmentation in the Tuber Flesh

Some potato varieties occasionally exhibit streaks or blotches of purple or pink coloration in the flesh. Although rare, this disorder can be a concern to the potato grower because of negative consumer response. The pigmentation is harmless, but is definitely out of the ordinary for the average potato buyer.

The pink or purple color in tuber flesh is due to anthocyanins (Fig. 14.22). This group of plant pigments is the most common source of red or purple color in nature. Red color in fall leaves, red or purple color in flowers, and the intense purple color in plants, such as table beets, are usually due to the presence of anthocyanins.

The skin color of red potatoes and red and purple flesh in novelty potatoes is also due to these same anthocyanins, demonstrating that their presence is natural and normal. Many of these pigments are used as natural food colorants, and new research has shown they have a positive impact on human nutrition as a result of antioxidant characteristics.

Upon cooking, anthocyanins in tuber flesh may fade or entirely disappear, especially if the color is initially faint. Anthocyanins can break down during baking or frying or can be leached out during boiling.

All evidence points to the fact that flesh purpling should not concern the consumer. However, even when something is understood, it is not necessarily accepted. For this reason, most growers prefer to minimize this disorder.

The causes of flesh purpling are not well understood. Variety choice is the only usable management tool. Some varieties are prone to the problem, while others are rarely affected. Flesh purpling is commonly seen in HiLite Russet, Frontier Russet, and occasionally Russet Norkotah. Pigmentation is also more common in potatoes that are subjected to certain stresses. Tubers that are close to the soil surface, especially if they show field greening, are more likely to show purple streaking in the flesh. Also, the condition has been correlated with exposure to cold temperatures, either in the field or storage.

Freezing/Chilling Injury

Potato tubers are susceptible to frost or chilling injury. Pulp temperatures below 30 °F for extended periods or colder temperatures for short periods can severely damage tubers. Widespread injury can result in total loss. See Sidebar 14.1.

Fig. 14.22 Anthocyanin in tuber flesh of the cultivar Frontier Russet

Symptoms and Causes

Potatoes that are exposed to freezing or chilling temperatures can show multiple symptoms depending upon whether they are observed frozen or thawed. Damage may occur before harvest, in transit, or in storage. Exposure to soil temperatures below 28 °F for several hours can kill the exposed tuber tissue by causing ice crystals to form in the cells. Freezing injury usually occurs on the part of the tuber closest to the soil surface, which is usually the bud end. The severity of damage will also depend upon the duration of the frost period and rate of thaw. Freezing damage is difficult to diagnose while the tubers are frozen because they show no obvious symptoms other than the surface being hard and difficult to damage. As the tubers begin to thaw, the symptoms become visible.

The first obvious sign of frost damage is free moisture (weeping) on the outside of the tuber. The next phase is cellular breakdown causing the tissue to turn brown, gray, or black. Typically, a distinct line is visible between healthy and frozen tissue. Affected tissues often break down completely and become liquid, having the appearance of a water-soaked rot. Tissues in the vascular ring are particularly sensitive to low-temperature injury.

Chilling injury can occur by exposure to temperatures between 32 and 37 °F, typically resulting in a discoloration of the internal tuber tissue (Fig. 14.23). However, cells are not frozen and tissues do not break down rapidly.

Chilling injury is not visible on the tuber surface, but internal tissues exposed to temperatures in this range can exhibit a variety of symptoms, including mahogany browning, reducing sugar accumulation and fry darkening, tissue graying upon boiling, and necrotic phloem injury. Mahogany browning is a symptom in which chilled tuber tissue turns a pinkish-red to reddish-brown or gray color.

Management

To minimize frost or freezing damage, proper hilling procedures should be used to minimize tuber exposure to freezing temperatures. If possible, potatoes should be harvested early, before potential frost exposure. In addition, potatoes should be transported only in favorable conditions or in insulated containers and storage temperatures maintained above 37 °F, accompanied by adequate ventilation.

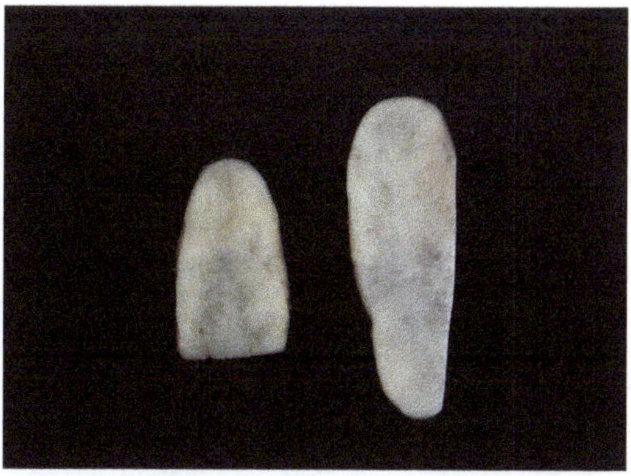

Fig. 14.23 Chilling injury showing an internal discoloration of the tuber tissue

Fig. 14.24 Internal sprouting symptoms showing inward growth of sprouts of a red-skinned variety

Internal Sprouting

Internal sprouting is a disorder that occurs in storage. As the tubers break dormancy, sprouts typically will grow away from the tuber, but because of an external growth restriction, the sprouts grow into the tuber rather than outward.

Symptoms and Causes

Internal sprouting can be caused by chemical or physical sprout inhibition. Symptoms include the growth of sprouts from eyes that penetrate back into the tuber flesh (Fig. 14.24). One physical cause of this disorder is thought to be pressure or contact by other tubers, debris, or storage walls against a sprouting eye, thus restricting outward growth and forcing the sprout to grow inward.

Another potential contributor to this problem is chemical in nature and associated with sub-optimal application of a sprout inhibitor. Internal sprouts can occur when residue levels of the sprout inhibitor are insufficient to completely inhibit growth, or when sprout inhibitors are applied after significant sprouting has begun.

Management

To prevent internal sprouting, adequate turgidity of the stored tubers should be maintained, dirt and debris should be eliminated from the stored potato pile, and the proper rate and timing of sprout inhibitor applications should be used.

Bruise Damage

Impact damage during harvest and handling or pressure damage during storage can cause blackspot, shatter, or pressure bruise. The type of tuber damage that can occur is dependent upon several factors: (1) variety; (2) physiological condition, temperature, and hydration level of the tubers; (3) size and shape of tubers; and (4) type and force of impact.

Minimizing bruising is important in the control of tuber decay, since bruises allow *Fusarium* and other organisms to enter the potato. Bruise damage can also increase weight loss in storage and diminish overall tuber quality, product recovery, contract incentives, and consumer acceptability.

Blackspot Bruise

Many physiological and physical factors affect the susceptibility of tubers to blackspot bruise. Exposure of tubers to impact forces during harvest or transport causes the cell membranes within the affected tissue to rupture and begin to leak. This damage to the cellular membranes allows the enzyme polyphenoloxidase (PPO) to come into contact with phenols (primarily tyrosine) in the intercellular space. As a result, phenols are oxidized and form the black pigment, melanin, which is responsible for the discoloration.

Symptoms and Causes

Blackspot bruises appear beneath the skin, and there are generally no external symptoms. Blackspot symptoms will typically not be fully apparent for 12–24 hours after the damage has occurred, which makes early detection of damage difficult. The damaged area first turns pink to reddish-brown, then darkens to grayish-black as the melanin begins to form (Fig. 14.25).

Factors that favor blackspot bruise include large tuber size, low tuber hydration or turgidity, high specific gravity, potassium (K) deficiency, tubers with significant curvature, and very mature tubers. Varieties differ greatly in susceptibility to blackspot bruising, primarily because of differences in cell structure, chemical composition and tuber shape.

Management

Practices for reducing blackspot bruise include selection of less-susceptible varieties; maintenance of proper tuber hydration levels between vine kill and harvest; proper fertilizer applications, especially with N and K; avoidance of early

Fig. 14.25 Blackspot bruise symptoms in a Ranger Russet tuber

Fig. 14.26 Shatter bruises, shatter cracks, and/or thumbnail cracks are external symptoms of tuber handling injury. Thumbnail cracks can occur when handling cold, turgid tubers, even with minimal impact

dying and advanced tuber maturity; and good harvest and handling procedures. See Chap. 16.

Shatter Bruise

Shatter bruise differs from blackspot bruise in that the cell walls of the tuber tissue separate due to impact. The result is development of a visible crack in the tuber surface.

Symptoms and Causes

Shatter bruise produces a cracked or shattered appearance at the point of impact due to a rupture in the tuber skin and the tissue directly underneath. Symptoms become readily apparent when the tissue dries out and separates (Fig. 14.26). Often, the margins of the damaged areas exhibit discoloration, similar to the color of blackspot-bruised tissue. Internally, the damaged tissues can have a light gray appearance that extends into the tuber. Factors that favor shatter bruise development include high tuber hydration or turgidity, cool (<45 °F) pulp temperatures, and improper handling.

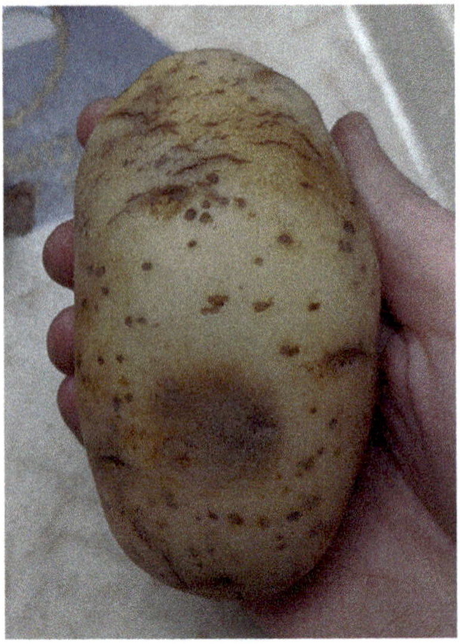

Fig. 14.27 Pressure bruise forms during storage and results in depressed areas that may or may not be discolored in the flesh underneath

Thumbnail cracks, also called air checks, are shallow, curved cracks on the tuber surface. They are caused by handling injury to cold, turgid tubers, followed by exposure to low humidity conditions. They are similar to shatter cracks in that they are due to cell wall separation. Thumbnail cracks are unique in that they are generally superficial and require minimal impact to occur. On occasion, simply lifting tubers from the soil, with an associated release of pressure, is enough to cause thumbnail cracks.

Management

Management practices for reducing shatter bruise include proper fertilization, irrigation, and pest control to allow tubers to reach proper maturity at harvest; allowing tubers adequate time to mature after vine kill; avoiding excessive soil moisture during tuber maturation; harvesting only when tuber pulp temperatures are above 45 °F; and using harvesting and handling practices that minimize tuber damage. Varieties differ greatly in susceptibility to shatter bruising; therefore, selecting a less susceptible variety can reduce incidence of this disorder.

Pressure Bruise

Pressure bruise is a disorder resulting from tissue damage due to the weight of an overlying potato pile. It is strictly a storage problem and affects otherwise healthy tubers; especially after exposure to low humidity conditions.

Symptoms and Causes

Externally, pressure bruises appear as flattened areas or indentations on the tuber surface (Fig. 14.27). If there are no internal symptoms, this is termed "pressure flattening." Internal symptoms include a gray to black discoloration in the flesh, usually darker in the vascular region. In severe cases, pressure bruise can affect one entire side of a tuber and be accompanied by tissue cavitation.

Conditions that favor pressure bruise development include dehydration of tubers coming into storage, low storage humidity (<90%), warm storage temperatures, large differentials in pile temperature, excessive potato pile height, and long storage duration.

Management

Practices to reduce pressure bruise include pre-wetting the storage floor, maintaining storage humidity above 90%, keeping storage temperatures in the optimal range, gradually cooling the storage to the final holding temperature, and not piling potatoes higher than the specifications for the storage and ventilation system.

Summary of Physiological Disorder Management

The first step in management of physiological disorders is proper diagnosis. As pointed out above, many of these disorders are easily confused with injury due to pests and diseases. Using a systematic approach that employs the following methods will help in identification of many specific disorders:

- Collect information on field history.
- Evaluate symptom distribution patterns in the field.
- Review recent and past chemical and fertilizer applications.
- Review recent weather conditions.
- Perform a clinical diagnosis to rule out possible disease pathogens.

Tuber physiological disorders can usually be diagnosed by evaluating symptoms, although some similarities exist with symptoms of a limited number of pathogens. Because tuber disorders are usually detected after harvest, a pattern in the field is more difficult to establish, and primary causes of tuber damage may be difficult to confirm.

Once a diagnosis is made, the second step is to understand the causes of the physiological disorder and how it may affect the yield, storability, and marketability of the crop. Knowledge of USDA inspection rules and individual contract specifications for external and internal tuber defects will help identify which disorders have the greatest effect on marketability and economic return.

The third step is to understand the environmental and management factors that influence the development of the disorder. Many of the foliar and tuber disorders discussed in this chapter are stress-related problems. Stress is defined as any factor (moisture, temperature, nutrient availability, etc.) above or below the optimum for plant growth and tuber production. The impacts of stress tend to be cumulative in that a longer period of stress is more damaging than a short period, and two simultaneous stresses (e.g., drought and heat) cause more damage than a single stress.

Sudden changes in conditions from one extreme to another also tend to be more damaging than a gradual change.

Since many physiological disorders in potatoes are associated with stress caused by heat and drought, knowledge of how these two factors impact vine and tuber growth is useful in management. See Chap. 2. The optimum temperatures for potato growth and development have been defined as 77 °F during the day and 54 °F at night. No growing region has those conditions every day; therefore, potatoes are subjected to some stress every growing season. At temperatures above 77 °F photosynthesis (energy production) decreases, while respiration (energy consumption) increases rapidly, doubling with every 10 °F increase. The net result is less energy in the form of carbohydrates available for transfer to tubers, and eventually tuber growth slows or even stops. Compounding this situation is the fact that high temperatures promote vine growth and also hasten the senescence rate of individual leaves. If high-temperature conditions last long enough, the plant suffers from a lack of energy production, and most of the energy that is produced is partitioned to the vines instead of the tubers. The net result is a reduced, irregular tuber growth pattern that leads to a high incidence of both external and internal disorders.

Water use by the crop is directly impacted by temperature, so it is difficult to separate heat stress from drought stress. As temperature increases, daily water use by the crop increases, and it becomes more difficult to keep soil moisture levels above the 65% threshold at which the plant begins to experience stress. See Chap. 13. The earliest observed response to drought stress is reduced stomatal conductance, which controls the exchange of water and CO_2 at the leaf surface that is necessary for regulating temperature and photosynthesis. Drought stress also reduces leaf expansion (even prior to wilting) and delays canopy development. All of these responses further reduce energy available for tuber production.

There is little growers can do to directly impact environmental conditions that cause stress. Therefore, the key to reducing stress-related disorders is not contributing to the problem by making mistakes in managing the crop. This approach starts with selecting varieties that are known to have resistance to heat and drought stress. Many of the more recently released varieties for both the fresh pack and processing markets have much better resistance to stress-related disorders compared to Russet Burbank, and as a consequence, exhibit much lower levels of defects even when the crop experiences stress during the growing season.

Another important concept is maximizing development of the root system. Potato plants are relatively susceptible to drought and heat stress due to limited ability of their roots to transport water efficiently to stems and leaves. Practices that lead to a reduction in rooting volume or depth, such as compaction and late cultivation, further increase the susceptibility to stress. Likewise, practices that lead to excessive vine development, such as over application of N fertilizer, increase water demand while having little impact on the plant's ability to take up water through the root system. The result can be an imbalance in vine and root growth that makes the crop susceptible to stress.

While excessive vine growth can increase susceptibility to stress, inadequate vine development also has negative consequences in managing stress. This is

Sidebar 14.2: Sugar-End Potatoes

One of the most severe problems resulting from sugar accumulation in potatoes takes the form of a defect called sugar ends. The name is indicative of the accumulation of high levels of sugars in one end of the tubers, usually the stem end. When fried, sugar-end potatoes produce french fries that are brown on one end, a processing defect known as dark ends (Fig. 14.28).

Generally, sugar ends are more common in years with hot temperatures or in fields that experience drought stress during the growing season. The common assumption is that the resulting stress on the foliage causes a disruption in the growth of the plant that results in interruption of sugar transport and metabolism.

Research at the University of Idaho's Kimberly Research & Extension Center, however, provides evidence that the damaging aspect of heat and drought stress is high soil temperatures, especially during tuber initiation and early bulking. Apparently, high soil temperatures have a direct disruptive effect on the biochemistry of the developing potato tuber.

In the Idaho study, four treatments were included:

- A control with ambient temperatures under natural conditions.
- Low soil moisture, allowed to fall to 50% available between irrigations.
- High soil temperature—heat cables were used to raise soil temperature 10 °F above the control.
- A combination of the low moisture and soil heating treatments.

The results of the Idaho study indicate that low moisture, by itself, impacts neither tuber specific gravity nor the percentage of sugar-end potatoes (Table 14.1). In contrast, the addition of heat to the hill had a definite negative impact on all quality parameters measured. The combination of heat and drought stress produced a slightly lower percentage of U.S. No. 1 tubers and a higher percentage of sugar ends than heat alone.

because shading from the vines is one of the few ways to reduce soil temperatures when air temperatures are high. High soil temperatures have been shown to be one of the most important factors in the incidence of external tuber defects and sugar ends. See Sidebar 14.2. Therefore, establishing a uniform stand; promoting early season growth by eliminating compaction; drought; nutrient deficiency; herbicide injury; and damage from diseases, such as Rhizoctonia stem canker, are keys to promoting adequate early-season vine development. Likewise, maintaining a healthy canopy by controlling both foliar and soilborne diseases that can lead to early plant death before vine kill will reduce exposure of tubers to large fluctuations in soil temperatures late in the season, which leads to quality defects associated with overmaturity. See Chap. 15.

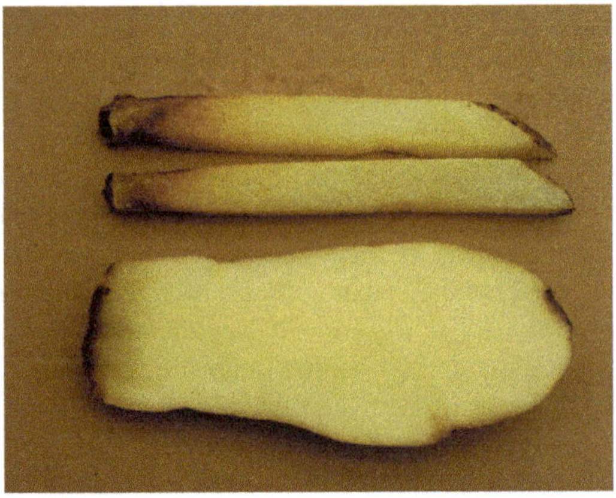

Fig. 14.28 Sugar ends are a serious disorder that impacts processing quality. A major cause of sugar ends is warm soil temperatures during early tuber bulking.

Table 14.1 Effect of soil moisture and temperature at tuberization on tuber quality and sugar ends

Treatment	Percent U.S. No. 1 tubers	Specific gravity	Percent sugar ends
Control: High moisture, ambient soil	69	1.081	4
Low moisture, ambient soil	53	1.082	0
High moisture, hot soil	39	1.068	23
Low moisture, hot soil	37	1.071	33

Source: Kleinkopf et al. (1988)

Acknowledgement Most photographs and graphics were adapted from collections of University of Idaho Extension educators, scientists, and researchers, who wrote the chapters of the first edition of this textbook.

References

Kleinkopf GE, Westermann DT, and Wille MJ (1988) Soil temperature effects on sugar end development. Proceedings of the University of Idaho Winter Commodity Schools, Moscow, ID, 20:180–181

Chapter 15
Tuber Quality

Jeffrey C. Stark, Stephen L. Love, and N. Richard Knowles

Contents

J. C. Stark (✉)
Idaho Falls R&E Center, University of Idaho, Idaho Falls, ID, USA
e-mail: jstark@uidaho.edu

S. L. Love
Aberdeen R&E Center, University of Idaho, Aberdeen, ID, USA
e-mail: slove@uidaho.edu

N. R. Knowles
Department of Horticulture, Washington State University, Pullman, WA, USA
e-mail: rknowles@wsu.edu

J. C. Stark et al. (eds.), *Potato Production Systems*,
https://doi.org/10.1007/978-3-030-39157-7_15

Introduction

Tuber size, shape, appearance, absence of diseases or defects, as well as the flavor, color, and texture of cooked or processed products all contribute to potato tuber quality. Quality may be related to visual appeal, consumer culinary preferences, or the ability to meet market specifications.

Two of the most important quality characteristics of potatoes are starch content, which impacts cooked product texture, and sugar content, which has a direct bearing on fried product color. This chapter will provide basic information on tuber sugar and starch physiology and how they respond to environmental and cultural factors. The impact of diseases and physiological disorders on quality are discussed in Chap. 9 and Chap. 14.

Processing Quality

More than half of all potatoes grown in North America are used for processing, mostly for making chips and french fries. Potatoes produced for processing must meet closely monitored specifications for starch and sugar content.

High starch content is favored by processors to ensure products have acceptable texture and keep processing costs down by limiting the amount of raw product needed, the cooking time required, and the amount of oil absorbed. Baked potato products also need high levels of starch to produce the fluffy, relatively dry texture that is preferred by most consumers.

Table 15.1 Values for converting potato specific gravity to percent dry matter and/or starch content

Tuber specific gravity	% dry matter[a]	% starch[b]
1.050	14.2	7.9
1.055	15.3	8.9
1.060	16.4	9.8
1.065	17.4	10.8
1.070	18.5	11.8
1.075	19.6	12.8
1.080	20.7	13.8
1.085	21.8	14.8
1.090	22.9	15.8
1.095	24.0	16.8
1.100	25.1	17.8
1.105	26.2	18.8
1.110	27.3	19.8

[a]Calculated from Kleinkopf et al. (1987) where % dry matter = −214.9206 + 218.1852 (specific gravity)

[b]From Hassel et al. (1997), Ohio State University Horticulture Series 666. Calculated from the Von Scheele equations where % starch = 17.565 + 199.07 (specific gravity – 1.0988)

Tuber solids make up about 20% of tuber fresh weight, comprised mainly of starch, sugars, amino acids, proteins, and minerals. Starch makes up about 70% of total tuber solids, while the other compounds are present in much smaller amounts. Starch is heavier than water, and, therefore, is the primary determinant of tuber density, which is commonly referred to as tuber specific gravity. Starch, tuber dry matter content, tuber solids content, and tuber specific gravity are terms used interchangeably when describing tuber baking and processing quality.

For most processed products, a starch content of 13% or higher, a solids or dry matter content of 20% or higher, and/or a specific gravity of 1.080 or higher is preferred (see Table 15.1 for interconversions). Tuber specific gravity is the measure of choice for estimating starch content and characterizing the processing potential of tubers. Consequently, it is a commonly used measurement when calculating quality incentives in processing contracts.

Although high solids and low sugars are essential for the processing industry, potatoes produced for other uses may have very different requirements. When boiled, potatoes with high solids content often fall apart, an undesirable characteristic known as sloughing.

Low specific gravity potatoes, typical of red varieties, for example, tend to be best for boiling and canning. Obtaining good boiling potatoes is largely a matter of choosing the proper variety, while producing good processing potatoes involves many aspects of management.

Tuber sugar content also has an important effect on the quality of processed products because of its large influence on fried product color. When exposed to high levels of heat, which is typical of the frying process, sugars (glucose and fructose) combine with amino acids (esp. asparagine) and other compounds to form the dark color and flavor we associate with "burned" food. This process is a non-enzymatic reaction known as the Maillard reaction (Fig. 15.1).

Even a low content of the reducing sugars glucose and fructose has unwanted effects on the quality of processed tuber products, such as chips and french fries. Sucrose, however, contributes little to dark color development in fried products but is still important because it is the substrate for creating more reducing sugars under the right environmental and physiological conditions. Reducing sugar accumulation in tubers varies with variety, storage temperature, physiological maturity, plant stresses, and tuber age.

Carbohydrate Production and Storage

As with all plants, potatoes use chlorophyll in leaves to trap energy from light and convert it to sugar through the process of photosynthesis. A portion of the sugars produced in the leaves is retained in the leaves and vines and utilized in respiration

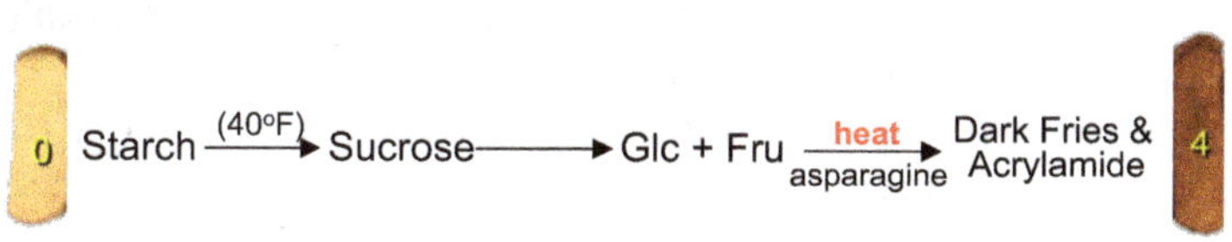

Fig. 15.1 The Maillard reaction

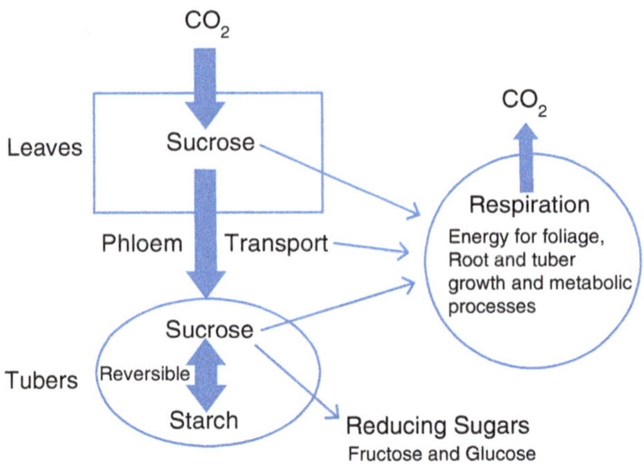

Fig. 15.2 Carbohydrate production and metabolism in potato plants

to provide energy for the plant, maintain growth, and drive critical metabolic processes (Fig. 15.2). Surplus sugars are transported to the tubers where they are either deposited in the cytoplasm for use by the tuber cells or converted into starch in the amyloplasts. Because sugars are the substrate for the production of starch in tubers, their respective biochemical pathways are closely linked.

During maturation, potato tubers accumulate nearly all of their carbohydrate reserves as starch and only a small amount as soluble sugars such as sucrose, glucose, and fructose. During storage, sugar concentrations increase at the expense of starch in response to low storage temperatures. This "cold-induced sweetening" response is observed in potatoes and other plants, with the soluble sugars functioning as cryoprotectants in the cells. The temperature-dependent conversion of starch into sugars is reversible in relatively young, healthy, non-stressed tubers, which provides an opportunity to "recondition" tubers that have undergone cold-induced sweetening. Reconditioning involves raising the storage temperature to 50 °F or above. However, as tubers age during prolonged storage, they gradually lose the ability to recondition, resulting in irreversible "senescent sweetening." As a result of starch-sugar interconversion in tubers, their starch and sugar contents are inversely correlated.

Starch and sugar content of tubers is controlled by a number of genetic factors, which can differ significantly among varieties. Since there is an optimal range for tuber starch content for fresh and processing potatoes, tuber specific gravity is an important trait to select for in potato breeding programs.

Numerous environmental and management factors also affect the concentration and proportion of sugars and starch in tubers (Fig. 15.3). For example, the processes of sugar production and transport to the tubers can be disrupted by water or heat stress, disease, or early senescence. This will usually limit starch deposition in the tubers, resulting in lower specific gravity.

Other conditions, such as high soil temperatures, may not hinder production or transport of the sugars, but may disrupt conversion to starch in the tubers. Either way, the result is nearly always a loss of quality.

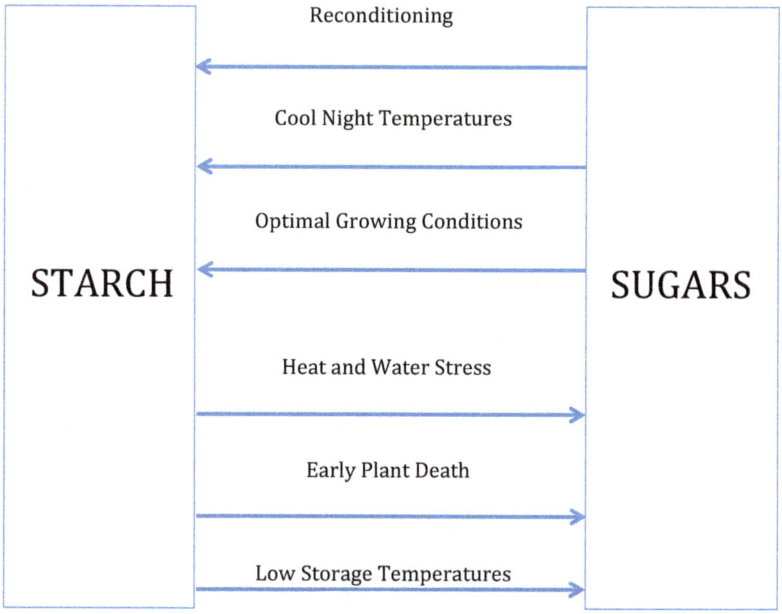

Fig. 15.3 Environmental and management factors influencing the relative proportion of sugars and starch in potato tubers

During early development, tubers are low in starch and high in sugars (Fig. 15.4). Near the end of vine growth, the tubers reach the point—known as physiological maturity—when they achieve maximum dry matter content and minimum sugar content. Specific gravity in overly mature tubers typically decreases while reducing sugar concentrations, particularly in the stem ends of tubers, increase following physiological maturity. Delaying harvest well beyond physiological maturity can result in sugar ends at harvest or continued sugar increase in the tubers during storage, resulting in loss of process quality (darker and/or non-uniform process fry color) developing early in storage, as shown in Fig. 15.4.

The general management strategy for maximizing tuber quality is to maintain good plant health through the tuber bulking period, time vine kill and harvest to correspond with physiological maturity, and employ appropriate storage temperatures and techniques with respect to the intended market.

Factors Affecting Specific Gravity

Greater consumer preference for potato products made from high specific gravity tubers is reflected in the quality incentives contained in potato processing contracts, which typically are based on tuber grade, size distribution, specific gravity, and percent defects. Consequently, a primary management goal of potato growers is to

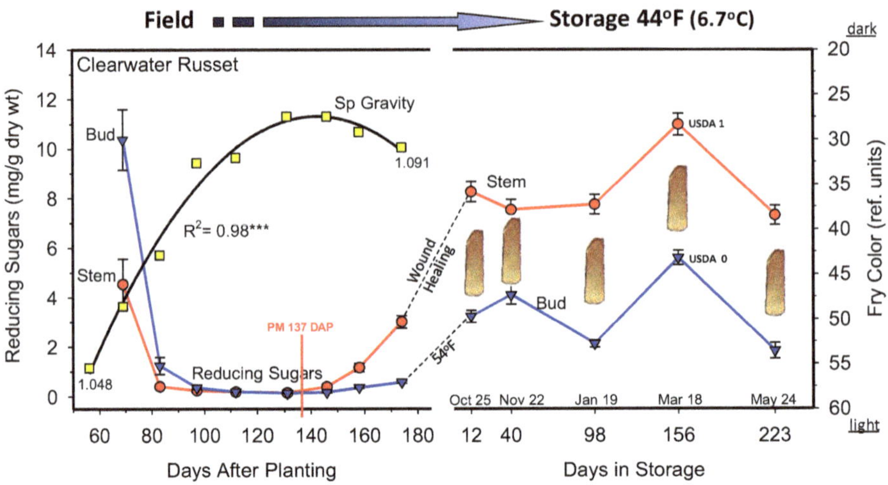

Fig. 15.4 Changes in tuber specific gravity and reducing sugar concentrations in the bud and stem ends of Clearwater Russet tubers versus days after planting (Othello, WA 2010). Tubers reached physiological maturity approximately 137 days after planting (DAP). Following harvest (174 DAP), the tubers were wound-healed at 54 °F for 12 days and then stored at 44 °F through a 224-day storage season. Fry colors were measured on the indicated dates with a Photovolt reflectance meter. (NR Knowles, Washington State University, unpublished)

produce a high-yielding crop with few external and internal defects and high specific gravity.

Given the large proportion of potatoes used for processing and the difficulty in producing a crop with good processing quality, it is important for producers to consider the environmental and cultural factors that maximize tuber specific gravity and minimize tuber sugar content. Any condition, natural or imposed, that affects crop growth can affect tuber specific gravity. Some of these factors are environmental in nature, such as climate, weather, and soil type, and are difficult to control. Others are management related, and the grower can manipulate these factors to improve tuber specific gravity and produce a higher-value crop.

Environmental Factors

Air and soil temperatures are the primary environmental factors affecting specific gravity of irrigated potatoes. Warm days (80–90 °F) and cool nights (50–60 °F) provide optimal conditions for producing high specific gravity tubers. Air temperatures greater than 90 °F can reduce net photosynthesis, thereby reducing the amount of carbohydrate that can be transported to the tubers and converted into starch.

High soil temperatures have a direct effect on tuber physiology and inhibit starch deposition. For that reason, high soil temperatures are even more detrimental to tuber quality than high air temperatures. Extended periods with soil temperatures above 70 °F can have a marked negative effect on tuber specific gravity.

Other weather conditions can also affect tuber specific gravity. High evaporative demand caused by low relative humidity, high solar radiation, and/or high wind speed can also reduce photosynthesis by causing stomata to close with an accompanying restriction in CO_2 uptake. Conversely, prolonged periods with overcast skies can reduce light intensity to levels below that required for maximum dry matter production.

Any event or condition that destroys foliage or shortens the growing season can influence tuber specific gravity. This includes serious disease or insect infestations. Hail injury during mid to late tuber bulking can cause a severe loss of tuber quality if damage to foliage is extensive. Late summer frosts can also reduce specific gravity by destroying leaves and shortening the time period available for transporting sugars to the tubers.

Obviously, weather and climate are not under the control of the potato grower. Stress resulting from natural factors cannot be eliminated. However, through proper management, the damage to tuber quality caused by environmental factors can be minimized.

Cultural Factors

Unlike environmental factors, management factors are under the direct control of the producer. Management can impact tuber specific gravity as much or more than weather or climate. Ideally, the grower's management system should produce a growing environment that allows for high photosynthetic rates over the entire season. Management decisions that can influence specific gravity involve a wide range of agronomic and pest management practices.

Choice of Variety

Potato varieties vary widely in their ability to accumulate starch in the tubers. The choice of variety is probably the most critical decision with respect to matching tuber quality with intended market. There are distinct expectations on the part of consumers for certain types of potatoes to have specific cooking qualities.

- Red-skinned varieties generally have lower specific gravity and are primarily for boiling or microwaving.
- Russet-type varieties with moderate to high specific gravity generally are best for baking and frying.
- Long-white and round-white varieties are divided into those suitable for general home use where potatoes are often boiled, and those that have high specific gravity and are intended only for making potato chips or french fries.

Before selecting a variety for processing, growers should consider market potential and quality characteristics, as well as the ability for producing potatoes with high specific gravity.

Seed Management and Planting

Producing a crop with high, consistent tuber specific gravity is dependent on having a uniform stand of vigorous plants. Growers need to design seed management programs that produce optimal plant populations and stem numbers. This will allow for efficient use of nutrients and water and help provide uniform growing conditions that will allow each plant to maximize productivity and tuber dry matter production.

Irregular stands and low plant populations, which result from planting skips, poor seed quality, or disease, can cause excessive nutrient uptake in the affected parts of a field. Excessive nitrogen uptake can delay tuber bulking and maturation and reduce specific gravity. Irregular stands can also increase variability in specific gravity, which can make processing more difficult.

Planting density and uniformity also affect tuber size distribution, which, in turn, influences specific gravity. Fields with closely spaced plants and doubles usually produce a higher proportion of small tubers compared to fields with normal or optimal spacing. In general, large tubers tend to have higher specific gravity than small tubers. This is especially true in growing areas with long seasons. Therefore, seed spacing and uniformity are important factors in maximizing both tuber size and specific gravity.

Physiological age of seed can affect the developmental rate of the crop, as well as yield and specific gravity. Potato plants grown from physiologically young seed typically begin to bulk later in the growing season than those produced from physiologically old seed. With respect to tuber specific gravity, this delayed response can be either positive or negative, depending on the production area and climate.

In areas with relatively short growing seasons, specific gravity of late-developing tubers produced from physiologically young plants may be reduced due to inadequate time for sugars to be transported to the tubers and converted to starch. By comparison, plants produced from physiologically old seed may die too early to reach maximum tuber solids content. Physiologically old plants also bulk and mature sooner and utilize soil nutrients earlier in the growing season. Therefore, appropriate adjustments in the nutrient management program need to be made to maximize quality.

Nutrient Management

Optimum plant nutrient concentrations are essential for maintaining high vine and tuber growth rates over the entire growing season. However, applying excessive amounts of nitrogen and potassium can decrease specific gravity.

Nitrogen (N)

When other growing conditions are favorable, increasing nitrogen availability up to the optimum level increases U.S. No. 1 yield and average tuber size with minimal reductions in specific gravity. However, excessive N rates stimulate vine and root growth and delay tuber bulking and maturation. In areas with short growing seasons, delayed tuber growth and dry matter accumulation can reduce specific gravity. High N rates also increase the concentration of soluble organic N compounds, further reducing specific gravity.

Potassium (K)

Potassium nutrition is also an important factor in determining tuber specific gravity as a result of its influence on starch synthesis and water content. Starch synthesis and specific gravity increase with increasing K concentration up to an optimum tuber concentration of about 1.8%. However at higher K concentrations, specific gravity decreases as tubers begin to absorb more water due to the osmotic effects of increased tissue salt concentrations.

The effects of high tuber K concentrations on water absorption and reduced specific gravity are generally greatest at the bud end. This response is also more pronounced when fertilizing at high rates with KCl than with K_2SO_4 due to the greater salt effect of KCl.

Large amounts of K fertilizer applied during tuber bulking are more detrimental with respect to effects on specific gravity than the same amounts applied preplant. The negative effects of in-season K applications on specific gravity are greatest when applications of more than 50 lbs. K_2O/ac are made late in the season. Growers, therefore, should avoid late-season K applications when specific gravity is a concern if petiole K concentrations are at adequate levels.

Phosphorus (P)

Phosphorus tends to increase starch synthesis when applied at increasing rates up to the optimum, but in contrast with N, it hastens rather than delays maturity. Phosphorus-deficient potato plants typically produce tubers with lower specific gravity compared to those with adequate P nutrition. As a result, adequate P nutrition is an essential factor in obtaining high specific gravity.

To a certain extent, phosphorus can counteract the negative effects of high N rates on specific gravity (Fig. 15.5). However, applying optimal rates of all required nutrients is the best approach for obtaining both high yield and high specific gravity.

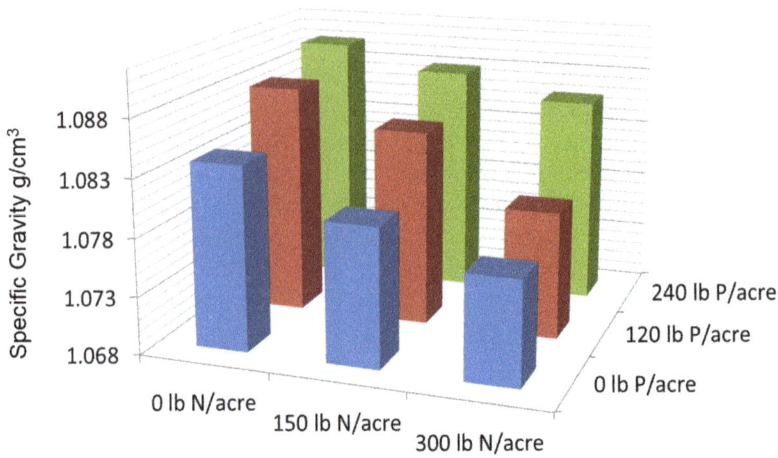

Fig. 15.5 Specific gravity of Russet Burbank potato tubers as influenced by N and P fertilization rates. (Stark 1985)

Irrigation

Water stress during early- to mid-season tuber development tends to decrease specific gravity due to reduced photosynthesis, particularly when accompanied by high temperatures. Late season water deficits, on the other hand, partially dehydrate tubers creating what might be viewed as an artificial increase in specific gravity that is due to lower tuber water content. However, late season water stress increases both the level of reducing sugars in tubers and susceptibility to blackspot bruise.

To promote high specific gravity, available soil water content should be maintained above 65% throughout the tuber growth period until just before vine kill. Soil moisture after vine kill should remain above 60% to minimize tuber dehydration.

Soil Conditions, Tillage, and Cultivation

Potato root system development can be greatly affected by soil physical conditions. Medium-textured soils, such as sandy loams, loams, and silt loams, generally produce potatoes with higher specific gravity than very sandy or heavy clay soils. Well-managed loam soils have good water-holding and nutrient-supplying characteristics that allow for high rates of growth and tuber dry matter production.

Tillage, planting, or cultivation practices that increase soil compaction and hardpan development can increase plant water stress, restrict root and tuber growth, and decrease tuber dry matter accumulation. In addition, cultivation practices, such as late weed tillage that increase root pruning, can increase plant water stress and reduce tuber quality.

Length of Tuber Growth Period

Because specific gravity is related to maturity, tubers that are harvested early often have lower specific gravities than those allowed to develop for the entire growing season. Early planting can increase the number of days that can potentially contribute to tuber starch deposition if environmental conditions are conducive to early season growth and development.

Disease Management

Diseases, such as Verticillium wilt, early blight, and late blight, which shorten the length of the tuber growth period and damage foliage can also affect specific gravity. Crop rotations and cultural and pest management practices that help suppress diseases can extend the period of tuber dry matter accumulation and increase specific gravity. However, disease suppression should involve season-long processes, rather than quick-fix strategies that can detrimentally affect crop growth. For example, growers should avoid late season N applications for the purpose of slowing disease progression because of the negative effects of N on specific gravity.

Vine Kill and Harvest Management

Several factors associated with vine kill and harvest can influence the specific gravity of harvested tubers. Tubers that remain in the soil after vine death can actually lose dry matter as starch is converted back to sugars for use in respiration. This is especially evident when soil temperatures are high. This problem can be minimized by timely harvest and placing the tubers in storage where conditions can be controlled.

Killing green vines while the root system is still actively absorbing water can cause significant amounts of water to move from the roots into the tubers and decrease specific gravity. For potatoes going directly into processing, it may be preferable to dig the vines green without vine killing to prevent excessive tuber hydration.

Even if the crop is still actively growing, harvest date can influence specific gravity as the result of changes in the relative amounts of water and dry matter accumulation in tubers.

In areas with relatively long growing seasons, specific gravity can actually decrease during the latter stages of growth if tuber water uptake exceeds dry matter accumulation. Although this late season growth period can increase total yields, the decrease in specific gravity can significantly decrease processing quality if it drops too low.

Tuber Sugar Content

Allowable sugar levels in potatoes—standards set by processors—continue to decline. This is directly related to consumer preference for light-colored, attractive fried products. Given this trend, it is critical that potato growers and storage managers understand the principles and techniques for controlling the sugar content of potato tubers.

The predominant sugars found in potato tubers are sucrose, glucose, and fructose. Sucrose formed during photosynthesis is translocated to the tubers during bulking. In very general terms, non-stressed healthy tubers convert the sucrose into starch during development. Stresses (e.g., heat, disease, water) can interfere with this process, resulting in breakdown of sucrose to its component sugars—fructose and glucose. The involvement of sugar as a substrate in the synthesis of starch means that anything affecting one of these compounds will also have some influence on the other.

Measuring Tuber Sugars

Tuber sugars can be measured directly using one of many methods. The invention of the Yellow Springs Instrument (YSI) by the Yellow Springs Instrument Company in Yellow Springs, OH, has simplified sugar measurements to the point where YSI is now the method of choice (Fig. 15.6). Most processing companies now use the YSI to test for sugars in raw product.

A sample of tubers is ground in a buffer solution, filtered, and injected into the YSI, and a sugar value is obtained that can be expressed as either percent or mg/g. Values for both sucrose and glucose (the main reducing sugar in potatoes) can be obtained.

An indirect measure of sugar content can be obtained by frying tuber slices at 350–375 °F for 3–4 min and comparing the cooked color with publicly available USDA color charts or reflectance readings made with a Photovolt reflectance meter that are calibrated with reducing sugar concentrations (Fig. 15.7). Although this will only give an estimated sugar content value, it will provide good information about potential processing quality.

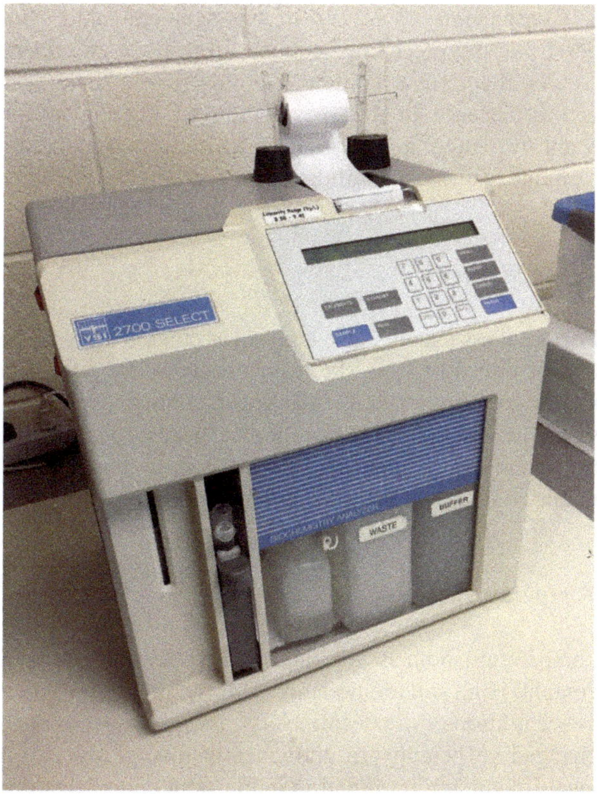

Fig. 15.6 Yellow Springs Instruments glucose analyzer. (Photo credit: Jeffrey Stark)

Fig. 15.7 Photovolt reflectance meter used to determine fry color (Photo credit: NR Knowles, Washington State University)

Factors That Affect Tuber Sugars

Because of the close relationship between sugars and starch, any environmental factor that affects tuber specific gravity also creates changes in sugar concentration. As a general rule, any environmental or management factor that increases specific gravity reduces sugars, and vice versa.

Consequently, a review of management factors for optimizing tuber specific gravity in the preceding section will provide an understanding of how these same factors affect sugar content. For this reason there is no need to revisit all of the factors covered in the specific gravity section as they relate to sugar content. Maturity, temperature, variety, storage stress, and handling influence tuber sugars in a unique fashion and will be discussed here.

Tuber Maturity

Potato tubers usually have high sugar content early in their development because the rate of transport from the leaves exceeds the rate of conversion to starch. As the tubers grow and mature, the sugar content decreases, reaching the lowest point when the vines are nearing complete senescence. For the tubers, this point is known as physiological maturity. Tubers left in the field after reaching physiological maturity generally begin to increase in sugars.

Temperature

Temperature both in the field and in storage has a large impact on tuber sugar content. High soil temperatures shortly after tuber set can lead to higher tuber sugar concentrations. In extreme cases, the physiology of the stem end of the tuber is changed sufficiently to permanently disrupt starch synthesis. This leads to high levels of sugar accumulation in the stem end of the tuber and the development of a condition known as sugar ends (also known as translucent ends and when severe enough to cause tissue death, jelly ends). Sugar ends are considered a severe defect in the processing industry and can lead to rejection of the crop.

The other temperature response that is important to understand is cold-induced conversion of starch back to reducing sugars. This can occur in the field or in storage. At temperatures from 50–55 °F, the balance between starch and sugars remains relatively static. As temperatures drop below this range, starch conversion to sugar becomes evident in most varieties. As temperatures decrease, the conversion of starch to sugars increases resulting in a higher final concentration.

The intended market and end use of the potatoes dictates the appropriate storage temperature. See Chap. 17 for detailed information on appropriate storage temperatures.

Potatoes that have accumulated excess sugars after exposure to low temperatures in the field or storage experience a partial recovery toward lower sugar concentrations when exposed to temperatures above 55 °F (60–65 °F is considered optimal). This process is termed reconditioning and is occasionally used to restore acceptable sugar levels in a lot of potatoes that has become unacceptable for processing during extended storage.

Reconditioning has limitations and cannot restore tubers to their pre-harvest low sugar levels. It is also more effective within the first 6 months of storage when the tubers are young and physiologically responsive.

After 6–10 months of storage, sugar accumulation may occur that is not temperature induced; it is due simply to age. This is called senescent sweetening and is a result of the tuber's inability to control its sugar metabolism. Reconditioning will not reverse senescent sweetening and, in fact, may make it worse.

Variety

Potato varieties exhibit large differences in sugar content, especially after storage. For that reason, it is critical to match varieties with intended use. In general, potatoes bred for the chipping industry are lowest in sugars. Potatoes bred for french fry processing typically have intermediate sugar contents, while those bred for the fresh market usually have the highest.

Potato breeders are currently making a concerted effort to develop varieties for chip or french fry processing that can be stored at temperatures as low as 42 °F and still maintain sufficiently low sugar levels. Several such varieties are now available. As they become recognized and accepted, temperature requirements for stored potatoes will need to be adjusted accordingly.

Storage Stress

In addition to cold-induced sweetening, a few other conditions in storage can produce an increase in tuber sugars. The most important of these is insufficient air movement.

Tubers require oxygen for respiration and low-level physiological activity. If a pile of potatoes becomes oxygen starved because of infrequent operation of the storage air system or because of excess dirt or other air blockage, the normal physiology of the tubers can be disrupted and sugar levels increase. Other problems that increase sugars include sprouting due to inadequate inhibition and the development of "hot spots" due to the presence of rot.

Handling

Normal tuber handling as part of moving potatoes into or out of storage has been shown to cause a slight increase in tuber sugars. This increase is usually short-lived, and the tuber sugars will decline to pre-handling levels after a week or so if they are not subjected to other stress factors.

Using Sugar Measurements to Predict and Manage Tuber Quality

Because sugar content is directly related to fried product quality, monitoring sugars is a valuable tool for predicting and maintaining quality. Growers and storage managers can use sugar measurements to assess current quality status and to predict possible changes in quality during storage. Sugar measurements can also be used to optimize harvest timing and make correct decisions on storage temperature protocols.

Assessing Current Quality

Using the YSI, a measure of reducing sugars (dextrose) can be obtained for any lot of potatoes. This can give a measure of current fry potential. Potatoes intended for chip production should have a reducing sugar level below 0.35 mg/g (or 0.035%) of fresh tuber weight. Potatoes intended for processing as french fries should have less than 1.0 mg/g (or 0.10%) of tuber fresh weight. Potatoes with higher values than these will usually show color problems after cooking.

Chemical Maturity and Storage Monitoring

Probably the most valuable use of tuber sugar measurements is monitoring of sucrose for the purpose of evaluating the potential for color problems, which can be used to make proper harvest and storage management decisions.

During tuber growth, the enzyme that converts sucrose to reducing sugars, acid invertase, is inhibited. Therefore, even if tuber sucrose concentrations are high during this time, there are usually no color problems evident in cooked products.

In storage, however, acid invertase becomes active and, if there is a sufficiently large pool of sucrose available, sucrose conversion results in a high level of reducing sugars. In simpler terms, high levels of sucrose at harvest can potentially result in high levels of reducing sugars in storage in varieties with high acid invertase activities, which results in poor frying quality.

Any stress on tubers in storage, such as low temperatures or insufficient air supply, can also cause an increase in the sucrose pool with associated reducing sugar and color problems. By knowing the relationship between sucrose concentrations and the future potential for quality problems, management tactics can be designed that will minimize the problems before they become detrimental to quality.

Maturity Monitoring

Growers and storage managers can measure tuber sugars to assess maturity and optimize the timing of harvest. If weather allows, harvest should be delayed until sugar levels cease to decline. If the potatoes have not been stressed during the growing season, sugar levels should fall below the levels indicated in Table 15.2.

Again, the most critical factor is the sucrose level. If it is below the indicated levels, harvest can occur, and the tubers can be stored in a normal fashion with the final holding temperature dependent on the variety and intended market.

Determination of Early Storage Condition

Tubers that come out of the field with sugar levels that are above the target values shown in Table 15.2 are preconditioned to having color problems. However, storage managers can consider one of the following strategies that are based on manipulation of early storage temperatures. These economically critical decisions are based on levels of sucrose and glucose at harvest.

Scenario 1 Sucrose levels are acceptable (<0.15%), but glucose levels are too high (chips >0.035%, fries >0.1%). The immediate fry color may be too dark, but the potential for long-term storage can still be good.

Action: During the wound-healing period at the beginning of storage, the temperature should be held at 60 °F for 2 weeks or until the glucose concentrations drop to acceptable levels. The temperature can then be ramped slowly downward to 45–48 °F for frying potatoes or 50–52 °F for chipping potatoes. Glucose levels should subsequently be determined at regular intervals to ensure they remain within the acceptable range.

Scenario 2 Sucrose levels are too high (>0.15%), but glucose levels are acceptable (chips <0.035%, fries <0.1%). The immediate fry color may be good, but long-term storage may be negatively impacted as sucrose is converted to reducing sugars.

Table 15.2 Target maximum sucrose and glucose concentrations at harvest and in storage for potatoes intended for chip and french fry processing based on fresh weight values

Intended market	Sucrose (mg/g fresh weight)	Glucose (mg/g fresh weight)
At harvest		
Chips	1.5 (0.15%)	0.35 (0.035%)
French fries	1.5 (0.15%)	1.0 (0.10%)
During storage		
Chips	1.0 (0.10%)	0.35 (0.035%)
French fries	1.5 (0.15%)	1.0 (0.10%)

Adapted from Sowokinos and Preston (1988)

Action: The same wound-healing conditions should be used as given for *Scenario 1*. The sucrose levels should be determined at the end of the wound-healing period. If the sucrose levels are still too high, a higher than normal holding temperature (possibly 55 °F for chipping potatoes and 50 °F for frying potatoes) may be required. It may be necessary to sell these potatoes before others that have better sugar indicators.

Scenario 3 Both the sucrose and glucose levels are too high. Both the immediate fry color and long-term frying potential may be poor.

Action: The recommendations described for *Scenario 2* should be followed. A wound healing temperature of 60 °F should be maintained until both the sucrose and glucose levels are acceptable. A more intensive monitoring program will be required, with sugars being measured at least every 5 days. The storage manager should consider moving these potatoes to market as early as feasible.

Storage Maintenance

As mentioned earlier, storage conditions can cause potatoes to accumulate unacceptable quantities of sugars, even when the levels are acceptable at harvest. Sugar analysis can be used to indicate when conditions need adjustment.

Sugar accumulation in storage can generally be attributed to low temperatures, inadequate supply of air to the pile, or senescent sweetening. These conditions can be detected by sugar monitoring, usually before any obvious decline in quality.

Low-Temperature Stress

When storage temperatures are too low, both sucrose and glucose levels will climb simultaneously into the unacceptable range. This can occur within a few days if the temperature is several degrees below optimum, or it can occur slowly when the temperature is only a few degrees too low.

Problems with low temperature sweetening can usually be solved with a 2- to 4-week period of reconditioning at 60 °F, followed by a slow return to the desired holding temperature.

Inadequate Air

Oxygen deprivation caused by inadequate air movement in the storage causes sucrose levels to slowly increase. Later, glucose levels follow the same pattern, and the fry color goes off-grade. Another typical symptom of ventilation stress is that individual tubers may fry darker in the middle than around the outside. Early detection of the rise in sucrose levels can help resolve this problem. Solving the problem may be as simple as increasing the frequency or length of ventilation to the pile.

If the problem is one of inability to move air through the pile due to obstructions or dirt, more drastic measures may be required, such as early marketing of the potatoes or movement to a different storage building. If ventilation stress is the culprit, an increase in air supply will result in an immediate response to corrective action, but the return to acceptable sugar levels may be slow.

Senescent Sweetening

The maintenance of acceptable sugar concentrations during the first 5–8 months of storage, followed by a slow increase in sucrose levels over the next several months, may be an indication of senescent sweetening. If no temperature stress or ventilation problems can be identified, and a sample of potatoes removed from storage does not respond to reconditioning, then the potatoes should be marketed as quickly as possible.

Senescent sweetening is a permanent condition and only gets worse with time. If an entire pile of potatoes is suspected of being affected by age-related sweetening, it is critical that no attempt be made to recondition the potatoes. Warm temperatures will only speed up the aging process and make the problem worse.

References

Hassel RL, Kelly DM, Wittmeyer EC, Wallace C, Grassbaugh EM, Elliott JY, Wenneker GL (1997) Ohio potato cultivar trials. Ohio State University Horticulture Series No. 666

Kleinkopf GE, Westermann DT, Wille MJ, Kleinschmidt GD (1987) Specific gravity of Russet Burbank potatoes. Am Potato J 64:579–587

Sowokinos JR, Preston DA (1988) Maintenance of potato processing quality by chemical maturity monitoring (CMM), Minnesota Agricultural Experiment Station Bulletin No. 586, University of Minnesota

Stark JC (1985) Management factors affecting potato quality. In: Proceedings of the University of Idaho Winter Commodity Schools, Vol 17, pp 139–141

Chapter 16
Harvest Management

Mike Thornton, Pamela J. S. Hutchinson, and Nora Olsen

Contents

Introduction

The focus of harvest management should be optimizing crop quality and maximizing value. The tools available to accomplish this include selection of a vine kill method and timing, modification of factors that determine tuber susceptibility to bruise damage, adjustments to equipment operation, and prevention/removal of

M. Thornton (✉)
Department of Plant Sciences, Parma Research and Extension Center, University of Idaho, Parma, ID, USA
e-mail: miket@uidaho.edu

P. J. S. Hutchinson
Department of Plant Sciences, Aberdeen R&E Center, University of Idaho, Aberdeen, ID, USA
e-mail: phutch@uidaho.edu

N. Olsen
Department of Plant Sciences, Kimberly R&E Center, University of Idaho, Kimberly, ID, USA
e-mail: norao@uidaho.edu

J. C. Stark et al. (eds.), *Potato Production Systems*,
https://doi.org/10.1007/978-3-030-39157-7_16

foreign material. This chapter focuses on management practices that can be used to optimize crop quality prior to, during, and after harvest.

Importance of Harvest Management

Virtually every market in which potatoes are sold into (seed, fresh, chips, fries, and dehydration) has incentives for quality characteristics that impact grower returns. These incentives may include direct adjustments to the base price due to qualities such as tuber size, bruise level, and specific gravity above or below the target levels. They may also take the form of an option to reject a crop if it does not meet minimum specifications. Therefore, the focus of harvest management should be optimizing these crop quality characteristics, as well as minimizing the presence of foreign material in order to maximize value.

Vine Kill Method and Timing

Potato plants can be allowed to senesce and die naturally or be killed by frost, followed by harvest. However, most of the potato industry relies on chemical, mechanical, or a combination of those two practices for vine kill. The most common mechanical methods include rolling with heavy tires or other implements to break the stems and flailing or chopping to remove part or all of the stems and leaves. Mechanical vine kill has several advantages, including relatively low cost, slow acting to allow tuber size and specific gravity to continue increasing, and the ability to break up vines into small pieces that are easier to remove during harvest. The disadvantages of mechanical vine kill include the relatively long time it takes to treat a field, the slow rate of plant death (for vine rolling), and the tendency of vigorous plants to develop new growth after treatment. Because of these factors, chemical and combinations of chemical and mechanical vine kill methods are most common.

Chemical Vine Kill Options (Also Referred to as Desiccation)

Carfentrazone-ethyl (Aim® EC40% WDG)

Rate: 2–3.6 oz/A 3.2–5.8 fl oz/A (0.05–0.09 lb ai/A). Do not exceed 11.6 oz/A 7.2 oz/A (0.018 lb ai/A) per crop season as a desiccant.

Timing: For best results, apply when potato crop is in early stages of natural senescence. Vine kill generally is adequate for harvest within 14 days after initial treatment. If potato crop vegetation is actively growing when application begins, two sequential applications may be required to desiccate leaf and stem tissue. If a second application is necessary, wait 7–14 days after the first application. The pre-harvest interval is 7 days.

Application Methods: Ground or aerial application is recommended. Do not apply this product through any type of irrigation system. Use enough water to thoroughly cover potato leaves and vines. Aim® EC should be applied with crop oil concentrate (COC) at 1% v/v or a methylated seed oil (MSO) at a minimum of 1 quart/A or 1% v/v when applied in volumes of more than 20 gal/A. Ground: Apply Aim® EC in at least 20 gal/A of water using 80 or 110 flat-fan nozzles. Vary spray volume and spray pressure according to density of potato canopy to assure thorough spray coverage. To enhance performance, increase spray volume and pressure if potato canopy is dense or if weather is cool, cloudy, or dry. Aerial: Apply Aim® EC with aerial equipment in 5–10 gal/A of water using higher volumes on dense potato canopies and vines. Apply 10 ft. or less above potato canopy and use low drift nozzles.

Remarks: Dense potato canopy, large plant size, and environmental conditions not conducive to product absorption or activity will reduce initial application efficacy and increase the need for a second application. Use a higher rate during cool or cloudy weather or when vine growth is heavy. Tuber stem ends may discolor if a high rate is used when soil moisture is low or temperatures are high. When Aim® EC is not tank-mixed with another vine kill product, grazing and hay operations may proceed with no restrictions.

Caution: Do not exceed 11.6 fl oz/A Aim® EC (0.181 lb ai) total product per crop season. This total allowable usage applies to all applications made to the field per calendar year, including preemergence burndown treatments at the beginning of the season and vine kill. Do not exceed 11.6 fl oz/A of Aim® EC per crop season for vine kill. Personal protective equipment must be worn for handling and application. See label for specific instructions.

Tank-Mixtures: Aim® EC may be applied in a tank-mix or as a sequential application with other potato desiccants. Refer to the other product's label for restrictions on tank-mixtures, and observe all label precautions, instructions, and rotational cropping restrictions.

Rotational cropping restrictions: A crop also registered for Aim® EC application may be planted at any time in the treated field. All other crops may be planted after 12 months.

Site of Action: Group 14: protoporphyrinogen oxidase (Protox) inhibitor.

Chemical Family: Triazolinone.

Diquat (Reglone® Desiccant; or Others, 2 lb Diquat Cation/gal)

Rate: 1–2 pints/A (0.25 lb ai/A–0.5 lb ai/A) in a single application. May be applied in two applications, but do not exceed a total of 4 pints/A (1 lb ai/A) per season before harvest.

Timing: Preharvest interval is 7 days. Do not exceed 2 pints/A (0.5 lb ai) in a single application. A 5-day interval between sequential applications is recommended.

Application Methods: Ground or aerial application is recommended. Do not apply through any type of irrigation system. Apply with a nonionic surfactant (NIS) at 0.5–4 pints/100 gal spray mix.

Remarks: Do not apply to drought-stressed potatoes. Diquat is rainfast 30 minutes after application. Cool, cloudy weather slows diquat activity but will not affect performance.
Caution: Diquat is considered a "moderately toxic" product that requires protective gear for handling and application. Follow all use restrictions and precautions on the label. Make the last application at least 7 days before harvest. Do not feed forage from treated potatoes to livestock. Avoid applications in extremely dusty conditions because dust coating the plant surface can reduce desiccant activity.
Tank-Mixtures: Diquat can be applied in a tank-mixture at time of potato desiccation with fungicides listed on the label and a nonionic surfactant to facilitate harvest. Otherwise, no tank-mixture restrictions are listed on the label.
Rotational Cropping Restrictions: None listed on label.
Site of Action: Group 22: photosystem I electron diversion.
Chemical Family: Bipyridilium.

Glufosinate-Ammonium (Rely® 280, 2.34 lb ai/gal)

Rate: 21 fl oz/A Rely® 280 (0.38 lb ai/A) in a single application only; do not split application.
Timing: Beginning of natural senescence of potato vines. Preharvest interval is 9 days.
Application Methods: Apply in 20–100 gal/A of water by ground or 5–10 gal/A by air. Use enough water to thoroughly cover potato vines. Spray coverage on a dense crop canopy will be better with higher water volumes.
Remarks: Potato varieties with heavy or dense vines may require an application of another desiccation product to complete desiccation.
Caution: Do not apply to potatoes grown for seed.
Tank-Mixtures: No tank-mix restrictions on the label.
Rotational Cropping Restrictions: Wait at least 30 days after application to plant wheat, barley, buckwheat, millet, oats, rye, sorghum, or triticale. Wait at least 120 days after application to plant rotation crops other than those listed. Corn and soybeans may be planted any time after application.
Site of Action: Group 10: glutamine synthase inhibitor.
Chemical Family: Phosphinic acid.

Paraquat (Firestorm®; or Others, 3 lb Paraquat Cation/gal)

Rate: 0.7–1.3 pints/A (0.26–0.49 lb ai/A). Use two applications of 0.6 pint/A if vine growth is dense. Use 1.3 pints/A where quick vine kill is desired. Note: Gramoxone® SL or Inteon are not labeled for use as a potato desiccant.
Timing: Begin application when foliage is in early stages of natural senescence. Immature potato foliage may tolerate paraquat. Make split applications at least 5 days apart. Potatoes must be harvested promptly after desiccation and processed immediately or consumed without storage.
Application Methods: Ground application only. Use a nonionic surfactant (NIS) with 75% or more surface-active agent at 0.125% v/v (1 pint/100 gal spray mix) or crop oil concentrate (COC) at 1% v/v (1 gal/100 gal spray mix).

Remarks: For fresh-market potatoes only, including those sent directly from the field to a processor. Do not use in potatoes that will be stored or used for seed.

Caution: Paraquat is a restricted-use pesticide. Personal protective equipment must be worn by applicators, handlers, mixers, and loaders. See label for specific instructions. Do not apply to drought-stressed potato vines. Do not pasture livestock in treated potato fields. Do not exceed 2.7 pints/A per season. Do not use on muck or peat soils.

Tank-Mixtures: No tank-mix restrictions are listed on the label.

Rotational Cropping Restrictions: All rotational crops may be planted immediately after the last paraquat application.

Site of Action: Group 22: photosystem I electron diversion.

Chemical Family: Bipyridilium.

Pyraflufen-Ethyl (Vida®, 0.208 lb ai/gal)

Rate: 2.0–5.5 fl oz/A (0.00325–0.00894 lb ai/A) in tank-mix with another desiccant; 5.5 fl oz/A alone. May be applied sequentially (split application), but do not make more than two applications or exceed a total of 11 fl oz/A (0.0179 lb ai/A) per year for potato desiccation. Note: The annual maximum is 11 fl oz/A for all applications combined (preplant burndown + after planting before potato emergence + desiccation).

Timing: For best results, apply when potato crop is in early stages of natural senescence. A second application of Vida® at a minimum 7-day interval or another desiccation product may be needed under certain climatic conditions to ensure complete desiccation. The preharvest interval is 7 days.

Application Methods: Ground or aerial application is recommended. Do not apply Vida® through any type of irrigation system. Apply in at least 5 gal/A spray mix by air or 20–50 gal/A using ground equipment. Higher water volumes must be used in dense canopy conditions. Note: Use an approved agricultural buffering agent buffering to less than pH 5 or less if using Vida® in a water source of pH 5 or more. Always buffer the water source BEFORE adding Vida®. Addition of an adjuvant is recommended—nonionic surfactant (NIS) at 0.25% v/v (1 quart/100 gal spray mix) or crop oil concentrate (COC) at 1–2% v/v (1–2 gal/100 gal spray mix).

Remarks: Vida® is rainfast within 1 h after application. High temperatures and sunlight following application generally will enhance performance and speed desiccation. Do not apply within 7 days of harvest.

Caution: Vida® is a corrosive product that can cause irreversible eye damage; requires goggles or a face shield and other protective gear for handling and application. Follow all use restrictions and precautions on the label.

Tank-Mixtures: Vida® may be applied in tank-mixtures or sequentially with other desiccant/harvest aides, such as diquat or glufosinate-ammonium, for improved desiccation. Weather, crop conditions, or the presence of certain weeds, crop-damaging insects, or diseases will indicate the inclusion of other products in desiccation tank-mixtures. Read and follow label directions and restrictions for each tank-mix product.

Rotational Cropping Restrictions: Do not plant any rotational crops except corn, cotton, grapes, olives, pome fruit, pomegranates, potatoes, soybean, stone fruit, tree nuts, wheat, or triticale for 30 days after the last Vida® application.

Site of Action: Group 14: protoporphyrinogen oxidase (Protox) inhibitor.
Chemical Family: Phenylpyrazole.

Sulfuric Acid (93%)

Rate: 17–28 gal/A.
Timing: Repeat application after 5 days if vines are not completely desiccated. Preharvest interval is 5 days.
Application Methods: For retail sale and use only by certified applicators or appropriately licensed persons directly under their direct supervision. Apply undiluted sulfuric acid product.
Remarks: This material is not as dependent on temperature or other environmental conditions as are most other desiccants.
Caution: A restricted-use herbicide. Sulfuric acid is very caustic. Requires protective clothing, including dust/mist filtering respirator, chemical-resistant headgear, protective eyewear, chemical-resistant boots, and waterproof gloves. Applicators must use a closed system/enclosed cab meeting Worker Protection Standards for agricultural pesticides. An adequate supply of water should be immediately available for drenching.
Tank-Mixtures: None listed on label.
Rotational Cropping Restrictions: None listed on label.

In general, vine-kill speed with chemical products ranges from relatively fast to slow, in the following order: sulfuric acid > diquat, paraquat > glufosinate ammonium ≥ carfentrazone, pyraflufen-ethyl.

Combinations of mechanical and chemical methods can greatly speed the rate of vine kill and increase efficacy. This is especially true when the crop canopy is dense, making it difficult to get good coverage of leaves and stems with ground or aerial applications. In those situations, some growers prefer to flail or chop the potato foliage before applying chemical vine-kill products. Likewise, chopping or splitting the vines with a colter ahead of the harvester can be used to aid separation of vine material and reduce carryover of large tubers on the deviner chain.

There are several reasons why growers use mechanical and/or chemical methods to kill vines prior to natural senescence. Those reasons include: (1) regulating skin set and physiological maturity, (2) controlling tuber size, (3) coordinating harvest activity, and (4) reducing the amount of vine material and weeds flowing through the harvester.

The transition to the crop maturity stage (defined as growth Stage V in Chap. 2) signals the beginning of several changes in physical as well as chemical characteristics of potato tubers. The periderm starts to thicken and adhere more tightly to the underlying tissues, resulting in resistance to skinning damage during handling. This process is referred to as skin set or "physical maturity," a progression that is hastened by vine kill. Rapid vine kill methods (mechanical removal, sulfuric acid) result in more rapid development of skin set compared to slower methods (vine rolling). Skin set occurs more slowly when the vines are growing vigorously and under cool or damp soil conditions. See Sidebar 16.1.

Sidebar 16.1: Skin Development and Set

Well-developed, mature, attractive skin is not only important for marketing purposes, but also for disease prevention and minimizing weight loss in storage. Several environmental and cultural factors can affect tuber maturity and skin set.

Skin Development

For russet-skinned varieties, moderate soil temperature and moisture conditions favor development of a uniform russet skin. Extremely high (>90 °F) or low (<45 °F) soil temperatures slow the rate of skin development, while temperatures in the 55–75 °F range are most conducive to russeting. Soils that remain wet during tuber maturation reduce the supply of oxygen necessary for proper tuber skin development. Dry soils also reduce russeting. Soil moisture should be maintained between 65 and80% for optimal russet skin development. This will require appropriate reductions in irrigation late in the growing season consistent with declining evapotranspiration rates. Very light or heavy textured soils are more conducive to developing hot and dry, or cool and wet conditions, respectively. As a result, these soils generally produce tubers that have poorer russeting than medium-textured, well-aerated soils.

Fertility management is also important for proper skin development. Studies conducted in southeastern Idaho show that Russet Burbank fields with petiole NO3-N concentrations appreciably above 15,000 ppm during the second week of August generally had poorer russeting than those with lower petiole nitrate levels. Proper phosphorus fertilization hastens maturity and enhances russeting, but cannot completely counteract the negative effects of excess N fertilization.

Red-skinned varieties are also influenced by soil and growing conditions. Attractive red skin is judged by intensity of color and lack of brown, scurfy layers on the skin that give a russeted appearance. Hot soil temperatures tend to diminish red skin color and contribute to russeting. For this reason, very sandy soils are a poor choice for growing red potatoes. Soils that remain relatively cool and have high organic matter content tend to be conducive to more intense red coloration. Deleterious russeting on red potatoes becomes progressively worse the longer they remain in the field after vine death. It is important to complete harvest as soon as skinning is no longer an issue.

Skin Set

Potatoes cannot be harvested and handled without damaging the skin unless they are properly matured. Excessive amounts of N and K fertilizer can slow tuber maturation and reduce skin set. This is particularly true for applications made after the second week of August, under southern Idaho conditions.

Proper vine condition and vine kill management are also important factors in skin set. In most cases, vines need to be killed at least 2–3 weeks before harvest to allow sufficient time for tubers to mature and skin to set. This usually allows adequate time for tuber skins to set, which makes them less susceptible to shatter bruise and skinning damage. However, the rate of skin set is very dependent on variety, vine kill method, vine condition, and environmental factors. Any condition that keeps vines excessively green late in the growing season delays tuber maturation, which will lengthen the time needed for skin to set.

Likewise, tubers reach the peak of specific gravity and yield during the crop maturity phase, while the concentration of sucrose and reducing sugars reach their lowest levels. This process has been called "physiological or chemical maturity." Allowing the crop to become overmature can result in an increase in reducing sugars and changes in cell composition that increase susceptibility to blackspot bruise. The vine senescence level, or date at which overmaturity starts to cause changes in tuber quality, is dependent on variety, cultural practices (such as fertilizer rates), and environment. See Chap. 15, for a more detailed discussion on physiological maturity.

The stage at which tubers become more susceptible to blackspot bruise has been determined for two common varieties; Russet Burbank and Ranger Russet. Russet Burbank vines can reach about 60% green (i.e., 40% dead or dying) before vine kill without significantly increasing blackspot bruise potential; whereas vines of Ranger Russet should be mostly green (about 95%) at vine kill time to minimize blackspot bruise (Fig. 16.1). Ranger Russet potatoes harvested under relatively green vines also tend to have better fry color and quality.

Management of tuber size profile is another consideration in vine kill method and timing decisions. This is especially critical for seed potato crops where there may be tolerances for oversize tubers, as well as for red-skinned and other specialty potatoes where there is a considerable price differential between small- and large-sized tubers. An example of how quickly a tuber size profile can change during growth of the variety Red Lasoda, is given in Fig. 16.2. Rapid vine kill methods that

Fig. 16.1 Vine maturity influences bruise susceptibility

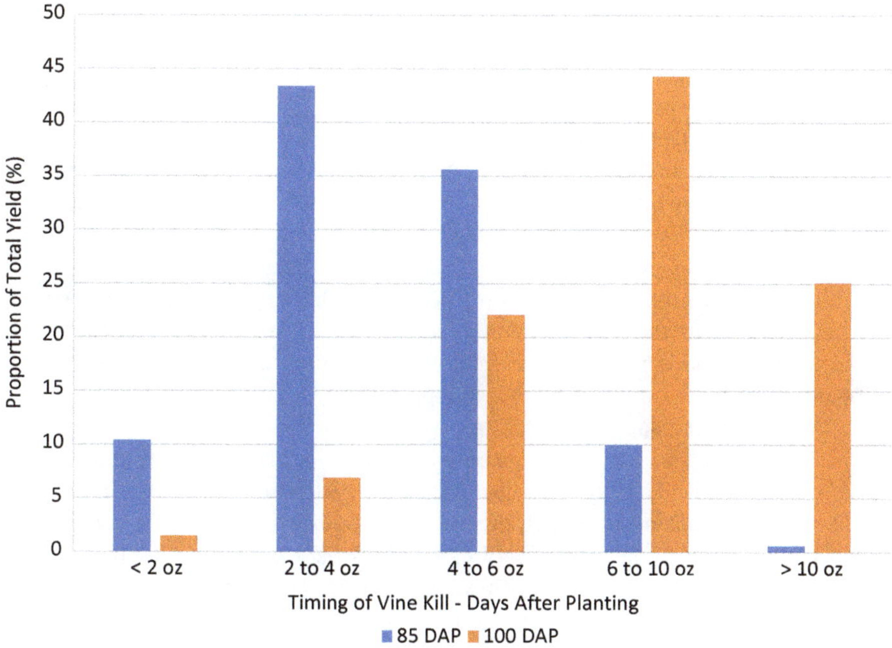

Fig. 16.2 Effect of vine kill timing on tuber size profile of Red Lasoda

remove most of the foliage will result in the quickest cessation of tuber growth, while slow methods will allow some further size increase to occur.

As the acreage of individual potato farms has increased, the need to plan and coordinate harvest operations within a fairly narrow window of time has also increased. In most northern production regions, there is usually a relatively short period of time after tuber bulking slows in late summer to the first frost that might damage tubers to the extent that storage is impacted. As a result, scheduling vine kill of each field to optimize equipment use and allow the entire crop to be harvested within that period is a critical step in potato production.

Some potato varieties can produce so much vine material that it becomes difficult to separate from tubers on the harvester. Weeds growing in the crop at the end of the season can also make harvesting operations difficult. Application of vine kill chemicals, mechanical vine removal, or both, are the only way to reduce the volume of plant material to improve harvester operations.

Factors in Tuber Susceptibility to Bruising

Potatoes may be damaged any time they are handled from harvest through storage, and during movement to the end market. There are four types of bruises that result from handling injury, which can reduce potato quality and economic value;

skinning, shatter bruise, blackspot, and pressure bruise. See Sidebar 16.2 for a detailed discussion about each type of bruise damage.

Sidebar 16.2: Types of Bruises in Potatoes

Tubers can be affected by four types of bruises; three of the types result from harvesting and handling operations, while the fourth occurs in storage. See Chap. 14, for more information on appearance of these physiological disorders.

Skinning

Rough handling, or handling tubers that are immature and have poor skin set, results in the outer skin becoming scuffed, which exposes the tuber flesh. The exposed flesh turns dark when subjected to wind, sun, and air. This makes the tuber less acceptable for the fresh market, increases moisture loss, and opens the tuber to infection by rot pathogens.

Blackspot Bruise

This type of bruise results from an impact that does not break the skin, but damages several layers of cells just beneath it. After an impact, a series of chemical reactions occur, which involve a substrate (tyrosine) that mixes and reacts with an enzyme (polyphenol oxidase). A dark, gray to black pigment called melanin forms. This reaction is complete in 24–48 h, and the damage is not evident unless the skin is removed.

Shatter Bruise

An impact causes the skin and several layers of cells beneath it to break or crack. These broken areas are good entryways for diseases, such as Fusarium dry rot, early blight, and bacterial soft rot. The cracks become more visible after drying, and are sometimes referred to as thumbnail cracks.

Pressure Bruise

Potatoes that were harvested when dehydrated (limp) or become dehydrated in storage as a result of low humidity or poor air ventilation will, after several months in storage, form a flattened area on the tuber called pressure flattening. Discoloration under this flattened area may or may not develop. Pressure bruise results when a flattened area becomes discolored under the tuber skin as a result of cell damage, and these tubers will be graded as defective.

Bruising has been estimated to cost the potato industry well over $300 million annually. This economic impact is due to a combination of lost revenue from processing contract incentives, rejections at market, increases in shrink and rot during storage, and costs associated with labor and machinery to remove bruised tubers from the finished product.

Susceptibility to bruising is affected by variety, soil conditions, and tuber conditions. Minimizing tuber bruising requires awareness of each of these factors, and a season-long management plan beginning before potatoes are planted and extending through harvest.

Variety

Potato varieties must be managed differently to ensure minimum bruising. Some varieties tend to be very susceptible to shatter bruise, but show almost no blackspot bruise. Conversely, some varieties tend to develop almost all blackspot and no shatter bruise, and some are susceptible to both types of damage. Table 16.1 provides specific management practices to minimize bruising for several common varieties based on their relative susceptibility to bruise damage.

Soil Condition

Field Selection and Tillage

Not all fields suited for growing potatoes are equal with respect to obtaining a minimum amount of bruise damage. Fields with numerous sharp rocks or soil that easily forms clods will increase the risk of bruising at harvest. Fields destined for potato production should be in a rotation that increases soil organic matter and alleviates compaction, which, in turn, will help minimize formation of clods.

Tillage practices can greatly influence potential tuber damage at harvest. Heavy soils easily form clods and should not be tilled in the spring when the soil is wet. Clods formed from spring tillage will usually not break apart before harvest and have the potential to damage tubers in much the same way as rocks. Where wind erosion is not a serious consideration, tillage in the fall is preferred because clods that form while working the soil will usually break up from freezing and thawing cycles during winter. Rocks should be removed before planting rather than during harvest to avoid tuber damage.

Soil Moisture at Harvest

Soil moisture at harvest is an important factor with regard to bruising. A light irrigation should be applied before harvest to condition the soil so clods will break apart. The soil should be just moist enough to carry to the secondary conveyor on the harvester (this is typically between 60–75% available soil moisture), where it should separate completely from the tubers.

Soil that is too wet may not separate from the tubers; whereas soil that is too dry will sift out too quickly, thus reducing the total soil plus tuber load on the primary and secondary conveyors, which increases bruising.

Timing of the pre-harvest irrigation depends on soil type. A conditioning irrigation is typically applied to sandy soils within 1 or 2 days of harvest, while soils containing more silt or clay are irrigated 3–7 days before harvest.

Table 16.1 Management practices to minimize tuber bruise damage of several commonly grown potato varieties

Variety	Susceptibility to shatter bruise[a]	Susceptibility to blackspot bruise[a]	Management practices to minimize bruise damage[b]
Bannock Russet	Very susceptible	Not susceptible	Vine maturity: Little importance; blackspot bruise is not an important consideration. Tuber hydration: Very important; apply an irrigation to condition soil for harvest only; harvest tubers slightly dehydrated to minimize shatter bruise. Wound healing: Critical; provide optimum wound-healing conditions in storage to rapidly heal shatter bruises that may lead to tuber decay
Ranger Russet	Moderately susceptible	Very susceptible	Vine maturity: Critical; kill vines while most (about 95%) are still green Tuber hydration: Critical; must apply an irrigation 8 days before harvest to hydrate tubers if soil is less than 60% ASM to minimize blackspot bruise Wound healing: Important; provide optimum wound-healing conditions in storage to rapidly heal shatter bruises that may lead to tuber decay
Russet Burbank	Moderately susceptible	Susceptible	Vine maturity: Very important; kill vines when about 40% are matured (dead or dying) Tuber hydration: Very important; apply an irrigation 8 days before harvest to hydrate tubers if soil is less than 60% ASM to minimize blackspot bruise Wound healing: Important; provide optimum wound-healing conditions in storage to rapidly heal shatter bruises that may lead to tuber decay
Russet Norkotah	Moderately susceptible	Moderately susceptible	Vine maturity: Somewhat important as blackspot bruise is not a major concern, but shatter bruise and skin set can still be an issue for immature vines Tuber hydration: Somewhat important; apply an irrigation 8 days before harvest if soil is less than 60% ASM to help minimize blackspot bruise Wound healing: Important; provide optimum wound-healing conditions in storage to rapidly heal shatter bruises that may lead to tuber decay
Shepody	Susceptible	Not susceptible	Vine maturity: Little importance; blackspot bruise is not an important consideration Tuber hydration: Important; apply an irrigation to condition soil for harvest only; harvest tubers slightly dehydrated to minimize shatter bruise Wound healing: Very important; provide optimum wound-healing conditions in storage to rapidly heal shatter bruises that may lead to tuber decay

(continued)

Table 16.1 (continued)

Variety	Susceptibility to shatter bruise[a]	Susceptibility to blackspot bruise[a]	Management practices to minimize bruise damage[b]
Umatilla Russet	Susceptible	Moderately susceptible	Vine maturity: Somewhat important; blackspot bruise is not a major concern, so time of vine kill has only minor influence on blackspot bruise Tuber hydration: Important; apply an irrigation 8 days before harvest if soil is less than 60% ASM to help minimize blackspot bruise, but handle carefully to minimize shatter bruise Wound healing: Very important; provide optimum wound-healing conditions in storage to rapidly heal shatter bruises that may lead to tuber decay

[a]Susceptibility is a subjective ranking using the following order from least to most susceptible: not susceptible < moderately susceptible < susceptible < very susceptible

[b]Management practices are rated by relative importance using the following order from least to most important: little importance < somewhat important < important < very important < critical; *ASM* available soil moisture

Tuber Condition

Fertilizer Management

Maintaining a balanced fertility program that results in the proper level of plant nutrition during the growing season will help lessen tuber susceptibility to bruise damage. Mineral nutrition may directly influence the susceptibility of tubers to bruising, or have an indirect effect by affecting tuber size, dry matter content, or plant maturity.

While there are numerous reports on the impact of both macro and micronutrients on quality, most of the consistent results are associated with just four main nutrients: nitrogen, phosphorus, potassium, and calcium. Nitrogen fertilizer timing and amount can have significant effects on bruise susceptibility due to their relatively large impact on maturity. Inadequate nitrogen can result in early crop senescence and an increase in susceptibility to blackspot bruise if the tubers sit under dying or dead vines for a long period prior to harvest. In contrast, excessive nitrogen (especially late in the season) can delay crop maturity, resulting in increased susceptibility to skinning and shatter bruise.

Phosphorous tends to have an opposite effect on maturity and skin development compared to nitrogen. Research in Idaho has shown that higher nitrogen rates, and later applications of nitrogen, require higher soil P concentrations to maximize skin development.

One of the earliest reports on blackspot bruise noted a direct relationship between potassium fertilizer amounts and susceptibility. When potassium is deficient in the plant, the tubers produce more tyrosine. This is the compound that is oxidized into the black pigment seen when bruised tubers are peeled. Many growers have increased their potassium fertilizer programs in an attempt to optimize yield, but also to reduce bruise susceptibility. The problem is that research has shown that when soil K concentrations are adequate for yield, additional fertilizer applications do not further reduce bruise susceptibility. However, in some studies excessive K fertilizer has been shown to also reduce skin development and specific gravity, so over-fertilizing can have negative impacts on tuber quality.

Calcium is thought to have an influence on bruise susceptibility through its effect on the strength of cell walls. Cells with higher calcium content have been shown to have higher resistance to deformation and fracturing from impacts that occur during handling. Most of the data on the relationship between calcium and bruise susceptibility comes from central Wisconsin, where potatoes are grown in very sandy soils with low Ca content. It is not known if low tuber calcium level plays much of a role in bruise susceptibility of tubers grown in the heavier soils that are commonly found in many U.S. potato production regions.

Hydration Level

Figure 16.3 shows the relationship between tuber hydration level and the type of bruise that occurs when harvested at 42 °F. When tubers are dehydrated, blackspot bruise is more prevalent; whereas hydrated tubers have a tendency to have more shatter bruise. An intermediate level of hydration results in the least amount of tuber bruising. This level is dependent on tuber pulp temperature (Fig. 16.4).

If it becomes necessary to increase the tuber hydration level, about 8 days are required to rehydrate tubers in the soil after the tuber skins have set. Irrigation practices used to condition soil, typified by a light application of water 1–4 days before harvest, may not impact tuber hydration. For this reason, soil moisture should be monitored during vine kill and maturation.

Pulp Temperature

Ideally, potatoes should be harvested when pulp temperatures are between 45 and 65 °F. Cold tuber pulp temperatures increase both blackspot and shatter bruise (Fig. 16.4), but the type of bruise damage also depends on tuber hydration level (Fig. 16.3). Cold, hydrated tubers tend to shatter bruise more readily; whereas warm, dehydrated tubers develop blackspot bruise more easily.

Because temperature has such a big influence on bruise susceptibility, it is important to have a thermometer and use it regularly to monitor pulp temperatures during harvest. A good practice is to record the temperature of every load going into storage. This information can be used to determine when to stop harvesting if temperatures become too warm. Tubers warmer than 65 °F may actually have less bruise than cold tubers, but removal of field heat in storage requires cooling air or refrigeration, and tubers with warm pulp temperatures are more susceptible to decay problems. See Chap. 17, for more information on the implications of high pulp temperatures on storage management.

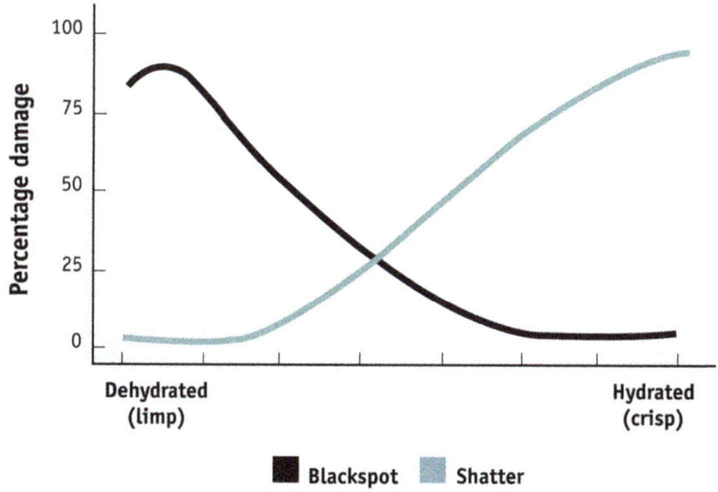

Fig. 16.3 Effect of tuber hydration level on blackspot bruise and shatter bruise of Russet Burbank potatoes at 42 °F. (Adapted from Thornton et al. 1973)

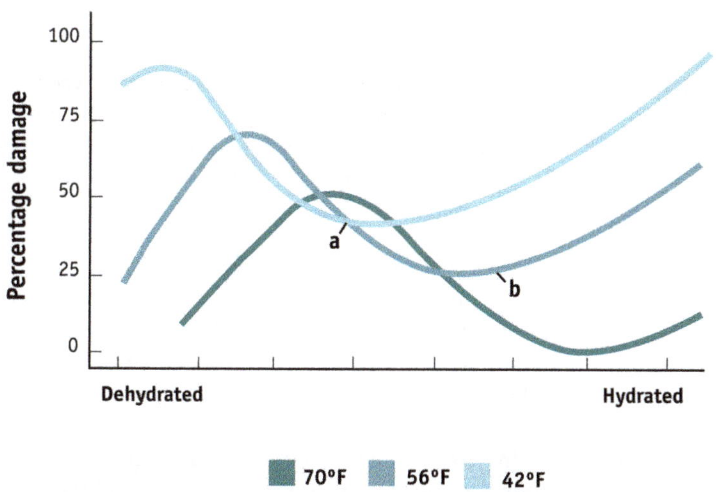

Fig. 16.4 Effect of tuber hydration level on total bruise damage in Russet Burbank potatoes. (Adapted from Thornton et al. 1973)

Equipment Operation

Surveys indicate that a large percentage of bruise damage in most handling operations typically occurs on the harvester. This is not surprising, because harvesters are used to accomplish three main tasks: eliminate soil, separate vines, and convey tubers from the ground up into a truck. A harvester with the blade set at 8 in depth moves about 480 tons of soil per acre, or approximately 10–20 times as much soil as tubers. Once the soil is removed, separation of vines from tubers and conveying tubers becomes a much easier task to accomplish.

Regardless of the type or brand of equipment, the key to reducing bruise is minimizing large impacts. The following is a list of common areas that should be evaluated in harvesting operations each year to minimize bruise damage.

Tuber Volume

The volume of tubers moving through any piece of equipment should match the capacity of that equipment. This is because tubers falling onto other tubers are less likely to be damaged than tubers falling onto a conveyor. For that reason, the speed of all conveyors on the harvester in relation to the harvester ground speed must be adjusted to keep conveyors full of potatoes. A properly adjusted harvester will not only minimize tuber damage but will eliminate nearly all soil before it reaches the truck.

Impact Points

The harvester should be checked for impact points where tubers experience large drops or strike bare steel. These potential damaging sites should be covered with padding. A thin layer of rubber material is not sufficient padding for most potential damage points. It is critical to use padding material that will adequately protect tubers. See Sidebar 16.3 for more information.

Sidebar 16.3: Conveyor Padding Material

Dropping potato tubers onto hard surfaces may result in a large proportion of the crop exhibiting either shatter bruise or blackspot bruise damage. More damage occurs when the surface on which the tubers land is hard, such as when tubers land on poorly padded areas on harvesters and other handling equipment. Cushioning materials protect tubers from damage by slowing the rate at which a tuber comes to a stop and spreading the point of impact over a larger area of the tuber, both of which reduce the impact force. However, not all cushioning materials are of equal benefit in minimizing bruising. While each type of material or design has its proper place, using the right material and design can go a long way to reducing tuber bruising.

Keep these points in mind when selecting conveyor cushioning:

1. The cushioning material should not flex to the point where it bottoms out when a tuber strikes the material.

2. The material and design of the padding should conform to the tuber shape when a tuber lands on it.
3. A wear-resistant surface will increase the useful life of the cushioning material.

Materials and designs that provide more cushioning will cost more, but will also result in less tuber damage.

The drop from the boom into the truck is one area where padding is not an option. The harvester operator has full control of boom height and needs to be trained to constantly adjust the boom to minimize the drop from the end of the boom into the truck.

Blade

The harvester digger blade should be positioned such that the potatoes flow evenly up the blade onto the primary conveyor and do not bump into the primary conveyor links (Fig. 16.5). The backside of the blade must be even with the top of the primary conveyor. If adjusting the backside of the blade causes the angle to be too steep, the best solution is to elevate the front side of the digger blade and lower the front of the harvester.

Fig. 16.5 Potatoes should flow evenly from the blade onto the primary conveyor

Conveyor Chain Padding and Flights

Padding on all conveyors should be inspected to ensure it is in good condition. Tubers dropping onto a bare steel chain will be damaged the most, and having only a rubber coating over steel offers little to no bruise reduction. Cushioning materials and designs with the most protection generally cost more, but the price received for a higher-quality crop will quickly offset the added expense.

All inclined conveyors, particularly the side elevator and boom, have the potential to cause bruising because of tuber rollback. On harvesters using flights to minimize rollback, the flights should be checked regularly and worn ones replaced immediately.

Roller Table

Rollers on the clod eliminator table should be inspected for wear and damaged rollers replaced. The off-load end of the table should be set lower than the front so the tubers will flow smoothly across. The speed of the rollers should be adjusted so tubers move across the rollers without bouncing.

Windrower Operation

Windrowers are used to increase harvesting efficiency by increasing the volume of potatoes flowing into a harvester. This increased flow of potatoes helps keep the conveyors full, which helps to minimize tuber bruising (Fig. 16.6). However, the windrower needs to be inspected and adjusted in the same manner as a harvester.

Tuber pulp temperatures are also influenced by windrower operation. The windrower should not be allowed to get too far ahead of the harvester on warm days because tuber pulp temperature can increase if tubers are left in the sun for more than a few minutes.

Bruising After the Harvesting Operation

Not all tuber bruising occurs during the harvesting operation. Bruising can occur after loading potatoes into a truck and while potatoes are moving through equipment into and out of storage. The following is a list of additional areas to focus on when seeking to minimize bruise damage.

Fig. 16.6 Windrowers increase the volume of tubers per unit area, which keeps conveyors full and helps minimize bruising

Tarping Trucks

A good practice is to tarp trucks used to deliver potatoes; some companies require tarping. Tubers can be damaged if a worker steps on them when tarping a truck, especially if the person is wearing hard-soled shoes. The best way to avoid walking on tubers is to have an automatic tarping device. However, if this is not possible, workers should be advised to not walk on the tubers during tarping.

Unloading Trucks

The stinger should be kept as close to the truck as possible; this may require remodeling some equipment. It should be remembered that potatoes hitting potatoes are less likely to result in bruised tubers, so conveyors should be kept filled to capacity at all times (Fig. 16.7).

Even-Flow Bins

Even-flow bins will help maintain a smooth flow of potatoes going into the storage building and allow for faster unloading of trucks. They help to keep conveyors full, but this is an advantage only if the drop from the conveyor into the even-flow bin is kept to a minimum (Fig. 16.8).

Fig. 16.7 Keep conveyor belts full as potatoes are off-loaded from trucks

Fig. 16.8 Using an even-flow bin means faster truck unloading and efficient flow of tubers into storage

Conveyors

Any point where tubers off-load from one conveyor to another should be checked for excessive drop height (Fig. 16.9). Slides can be installed at places with large drop distances so tubers roll from one conveyor to the next.

Dirt Eliminator Table

The slope of the dirt eliminator table should be checked. If the off-load end is higher than the inlet end, tubers will not move smoothly across the table and will be damaged. The eliminator table should slope downward so tubers quickly move from one end to the other without excessive rolling and tumbling.

Piler

The piler should be operated at full capacity to reduce the effective drop height at transfer points. The end of the piler should be kept as close to the pile as possible. Optimally, the pile should be built in a stair-step manner to reduce the number of tubers rolling down the face of the pile (Fig. 16.10). One person should be designated to operate the piler. That person should not be expected to perform other duties that would be a distraction from this operation.

Fig. 16.9 Adjust drop distances from one conveyor to another to minimize bruising

Fig. 16.10 Keep end of boom close to the pile and build the pile in a stair-step manner

Educating Harvest Personnel

Even a properly adjusted piece of equipment can bruise potatoes if not operated correctly. Therefore, an important part of harvesting bruise-free potatoes is making sure that all workers know and understand their part in accomplishing this goal. It is critical to educate and re-educate personnel who work during harvest about practices that keep bruising to a minimum—it will be time well spent.

Foreign Material Prevention/Removal

Whether potatoes are destined for the fresh market or processing, foreign material is a great concern to the industry. Foreign material is any material that is not the potato tuber. Common contaminants in potato loads include golf balls, glass, stones, metal, plastic, rubber, bones, aluminum cans, paper, and crop residues. These materials would cause a real safety concern if they were to pass undetected into the food system. As a result, potato processing and fresh market businesses employ an extensive system to detect and remove these contaminants. That system includes working with growers to remove foreign materials before, during, and after harvest. By minimizing foreign material in the field and in storage, growers help maintain a high-quality product, safe standards, and consumer confidence.

Field Practices

Visually inspect the field prior to harvest and remove any foreign material. Concentrate along areas that border public roads where there tends to be hotspots of trash. Also look for any evidence of old homesteads or dump sites, where the level of contaminants can be very high. It is best to flag these areas and avoid harvesting potatoes from that region.

Harvest Practices

Inspect all harvest equipment for loose pieces and leaky hydraulic line connections before taking it to the field. Make sure blowers, air heads, and roller tables are operating properly to maximize separation of clods, rocks, and crop residues. Use shatter-resistant light covers on tractors, harvesters, and trucks.

Storage Practices

Equipment should be inspected to ensure all loose parts and fluid leaks are repaired. Provide adequate facilities to handle foreign material removed by equipment and people, and provide containers for disposal of trash generated by workers who operate equipment. The storage building, pilers, and conveyors should be fitted with shatter-proof light covers. Loose or damaged objects should be removed from the storage structure, and the floor should be dragged with a magnet to pick up metal, and if needed, scraped prior to filling to remove the top layer of soil.

References

Thornton RE, Smittle DA, Peterson CL (1973) Reducing potato damage during harvest. Washington State University, Pullman, Bulletin EB646

Chapter 17
Storage Management

Nora Olsen and Gale Kleinkopf

Contents

Introduction

Potato tubers are living, respiring, biologically active organisms that require optimal storage conditions to maintain the quality entering the storage at harvest. Successful storage requires an understanding of the factors that affect tuber health and quality. Proper storage conditions depend on the crop's growing and storing environment, time in storage, variety, use of the potatoes, and knowledge of key characteristics of the potatoes placed in storage.

Storage Principles

The overall objective of storage is to help maintain quality and minimize further quality reductions. Storage basics for potatoes include: a dark facility; store only sound potatoes and at the proper maturity; maintain desired temperature and relative

N. Olsen (✉) · G. Kleinkopf
Department of Plant Sciences, University of Idaho, Kimberly, ID, USA
e-mail: norao@uidaho.edu

J. C. Stark et al. (eds.), *Potato Production Systems*,
https://doi.org/10.1007/978-3-030-39157-7_17

humidity (RH); and provide adequate, well-distributed ventilation. Potatoes are constantly respiring and, therefore, require oxygen. In turn, they produce carbon dioxide, water, and energy in the form of heat. Heat and carbon dioxide need to be exhausted from the storage. Many storage objectives revolve around the need to maintain quality, minimize loss, and address the nature of respiration.

Managers make many important decisions as the potatoes come into storage. Decisions should be made regarding storage volume needed; contract or end use(s); and facility disinfection, conditions, and management to maintain desired quality, method of sprout control, and length of storage before any potatoes are brought into the storage facility.

This chapter describes the basic principles associated with potato storage and provides information for application of these principles. Chap. 15 contains information about using sugar monitoring techniques in making harvest and storage decisions. The information in Chap. 15, in conjunction with this chapter, will provide the storage manager with an effective arsenal of storage management ideas and tools. If the intended crop is for organic certification, additional storage information can be found in Chap. 6.

Storage Structures

This chapter will not provide detailed information on building and equipping a potato storage. Several companies design and build modern potato storages, and any one of these businesses can provide the latest information on construction. These

Fig. 17.1 For decades, commercial potato production has used many kinds of storage structures. Older storages lacked all but the most basic climate controls

Fig. 17.2 All facilities need a strong foundation, insulation, protection from the elements, climate controls, ventilation system, and temperature controls to provide a suitable environment for potatoes. Modern storage facilities integrate air and humidification systems that allow year-round storage with minimal loss

specifications will vary depending upon the environment and location of the storage facility. Outside temperature, humidity, and weather extremes will impact the design and function of the storage facility. Growers also may successfully use many types of storages, of different ages and designs, to maintain the intended quality (Figs. 17.1 and 17.2). General types of structures include partially underground, straight-walled, Quonset/curvette, and slant or inclined walls.

To provide optimum storage conditions for potatoes, certain essential design and equipment characteristics must be present. These include:

- Sufficiently strong foundation and lateral wall support to hold the weight of the pile and roof support for weight of any snow load.
- Adequate insulation and moisture barrier.
- An air circulation system capable of providing a uniform supply of air to the entire storage.
- A method for raising or lowering air temperature, or maintaining it within a desired range. This is accomplished by bringing in outside cooling air and/or use of refrigeration.
- Equipment for supplying moisture (humidity) to the circulation air.
- Adequate sensors and controllers to allow maintenance of optimal conditions.
- Ability to easily clean and disinfect.
- Convenient access and handling of the potatoes.

When building a storage, it is important to keep in mind that its capacity should match the volume of the projected crop, and the air and humidity equipment should be adequate for maximum capacity. The storage will be designed and built based upon the intended capacity of the storage. Storing more or less than the intended amount will alter the environmental conditions provided to the stored crop. The following formula can be used to determine the capacity of a storage facility:

$$\text{Length in feet} \times \text{width} \times \text{pile height} / 2.5 = \text{hundredweight}\left(\#\text{of cwt;1 cwt} = 100\,\text{lbs}\right)$$

Example:

$$100\,\text{ft.} \times 50\,\text{ft.} \times 18\,\text{ft.} / 2.5 = 36{,}000\,\text{cwt}\left(\text{or } 3.6\,\text{million lbs}\right)$$

General layout or design of a storage facility will differ depending upon need and location. A simplified description of a typical North American-style storage is as follows. Walls and ceilings are built with insulation (*r*-value) and strength appropriate for the region. Outside air is brought in when louvers or doors are opened to the outside. The incoming outside air is mixed with return air (if required; return air is recirculated), humidified via evaporative cooling pads or supplemental humidification systems, and fans force the air down a centralized plenum. In order to distribute air to the potatoes, the plenum supplies air to lateral ducts (Fig. 17.3.) This can be accomplished with aboveground lateral ducts on top of soil or concrete floors with built-in lateral ventilation ducts. Each of those ducts provide the desired amount of air to the bulk potatoes above the ducts (Fig. 17.4). The air is pushed through the bulk pile of potatoes (bottom to top), and the air is either recirculated, mixed with in-coming fresh air, or exhausted. The ventilation system is controlled to a desired temperature and humidity set point utilizing sophisticated electronic control panels that takes into account the outside air temperature and humidity, rate of airflow, and temperature and humidity near the potatoes.

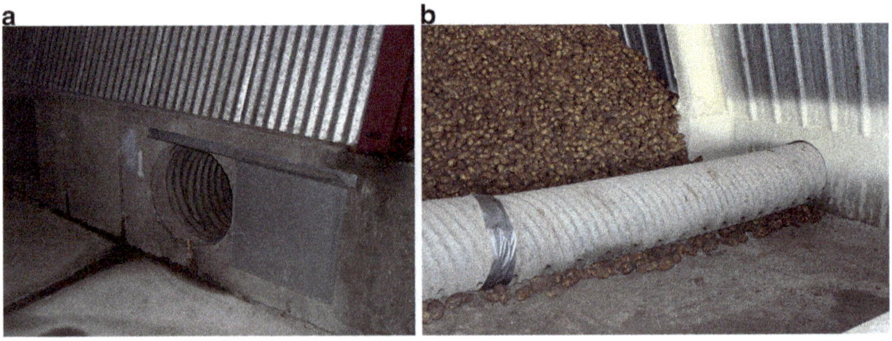

Fig. 17.3 Centralized plenum distributes air to lateral ducts (**a**) under bulk piled potatoes (**b**)

Fig. 17.4 Each of those ducts provide the desired amount of air to the bulk potatoes piled above the ducts

Fig. 17.5 Proper cleaning of a storage is important to minimize any disease or foreign material carryover into the in-coming crop

Storage Cleaning and Maintenance

Cleaning of the storage facility is a good practice for all storages and is essential for seed producers. The first step is to clear all debris associated with the previous year's operation. The second step is to clean facilities and equipment with a detergent and warm/hot water (Fig. 17.5). These procedures are particularly effective in eliminating bacterial problems that may have developed from a previous contamination. The third step is to rinse and apply a registered, labeled disinfectant. In most potato producing states, the Department of Agriculture maintains a current listing of available registered products that can be used for storage disinfection.

Off-season care of the potato storage facility is important to maintain its functionality and ensure long-term storage of high-quality potatoes. The period between storage crops is also a good opportunity for managers to repair equipment or add modifications that will improve efficiency and effectiveness of the storage. A short list of maintenance procedures is presented in the checklist at the end of this chapter.

Pre-harvest Decisions

Freshly harvested potatoes are extremely vulnerable to diseases (see Chap. 9) and physiological disorders (see Chap. 14), and evaluation must be made of best management practices for each incoming lot. Potatoes affected by temperature extremes, disease, nutrient excesses or deficiencies, water stress, physical damage, or other unfavorable growing conditions during the growing season may not respond to storage environments equally, and management can be altered to maintain quality as much as possible.

Harvest and handling operations also affect the storability of the crop (Fig. 17.6; Chap. 16). Potatoes that are bruised or damaged during any part of the harvesting, hauling, piling, or storing operations may require additional consideration for proper storage management (Fig. 17.7). Bruising creates entry points for diseases and increases tuber respiration and evaporation rates. Thus, specific storage conditions are needed to handle these situations. Be aware that tuber conditions vary with variety, growing location within or between fields, time of day harvested, and other factors that can add to the variability of the crop to be stored.

The first storage decision is whether to store potatoes from a particular field. Sampling potatoes from a field will provide the information needed to make an

Fig. 17.6 Proper storage begins with careful handling of potatoes from the field to the storage facility

Fig. 17.7 Bruising and damaging creates entry points for diseases, increases tuber respiration and evaporation rates

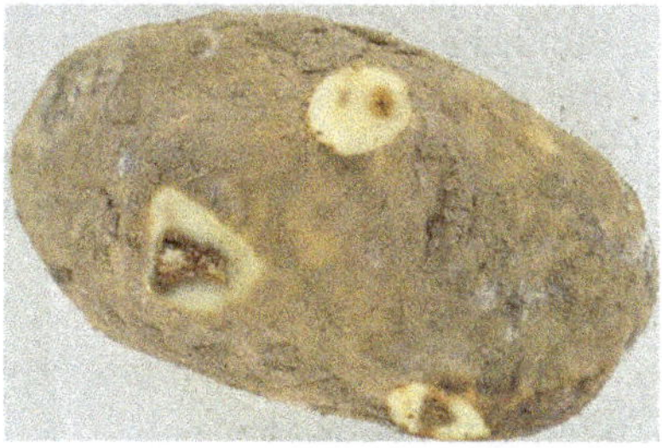

Fig. 17.8 Disorders of a tuber causing entry points, such as pink eye, can increase the potential for Fusarium dry rot infection

educated decision. Some tuber decay diseases, such as pink rot, late blight, soft rot, and bacterial ring rot, may be present in the crop before harvest. Most modern storages can provide conditions that will allow the presence of some decay at the beginning of storage. Unfortunately, some diseases, such as Fusarium dry rot and leak, will not commonly appear until after the potatoes are in storage. As little as 1–3% decay at harvest can make potatoes difficult to store. As a general rule, potatoes in a modern storage facility with up to 5% wet rot can be successfully stored if proper procedures are employed to minimize free and excess moisture and supply an aggressive ventilation program. The same is true for tubers damaged by frost. Other problems should be evaluated while potatoes are still in the field. This includes presence of potato virus Y, tobacco rattle virus, and potato mop-top virus that could cause tuber necrotic disorders, pink eye, and premature death. Symptoms associated with a virus or pink eye may or may not become more severe over time in storage, but if they cause any break in the skin they can become an entry point for other pathogens to enter (Fig. 17.8).

Potatoes with severe stress-related problems, such as sugar ends, jelly ends, or high overall sugar levels, should also be considered for immediate delivery. Potatoes with these problems can exhibit a rapid, unmanageable degradation of quality—especially processing quality—during the first few weeks of storage. Sugar monitoring can help with early detection of problems that may become serious later in the storage period. See Chap. 15, for details on sugar monitoring.

Once the decision is made to store potatoes, an understanding of the quality status of the crop will help with early management decisions. More information on this topic is available throughout this chapter and in Chap. 15.

Filling the Storage

Harvest potatoes with pulp temperatures between 45 and 65 °F. This range may change depending upon variety and harvest conditions. See Chap. 16. Attempt to minimize the amount of rocks, dirt, and debris entering the storage with the potatoes, while also minimizing bruise damage. This can be accomplished by the following practices:

- Use well-maintained unloading, even-flow bins, and/or sorting equipment for delivery of potatoes from trucks to the storage.
- Keep all drops to 6 in or less, and pad all sharp or hard surfaces on handling equipment.
- Keep all equipment running smoothly and full to capacity with potatoes.
- Use roll prevention belts on pilers and steep elevators.
- Eliminate dirt, rock, debris, and rotten potatoes.
- Pile the potatoes using a tier system. Start the piler low and to the front of the pile, then work up and back in a tiered fashion. Avoid rollback and bruising. Bulk pile no higher than the designed capacity of the building, typically 18–20 ft. (Fig. 17.9).
- Use only well-trained personnel to operate piling equipment.
- Ensure lateral ducts are properly aligned, connected, and sealed before piling potatoes over the ducts.
- Keep records of incoming tuber pulp temperatures and conditions, along with location of the crop within the storage facility.

Early Storage Management and Wound Healing Period

Freshly harvested potatoes can exhibit potential problems that are not apparent in the field. The first few days of storage are the time to recognize and address these problems. The first issue is elimination of field heat. Potatoes harvested at pulp temperatures above 65 °F can be much more prone to disease development and quality problems than cooler potatoes, unless the warm potatoes can be cooled immediately. The heat given off due to rapid respiration in freshly dug tubers can increase temperatures further. Every effort should be made to rapidly cool freshly harvested potatoes to 55–60 °F as soon as possible (within 2 or 3 days of harvest). This is accomplished by moving a maximum amount of cooling air through the pile. Use refrigeration or evaporative cooling to extend cooling time in these situations.

Unless there is a significant amount of wet rot present in the potato pile, high humidity (>95%) should be maintained during the early part of storage. If diseased or wet potatoes are present coming out of the field or a post-harvest problem with decay occurs (e.g., leak), it may be appropriate to dry the tubers during this early storage management period to prevent rapid development of disease problems. This is accomplished by moving a high volume of air through the pile with reduced

Fig. 17.9 Potatoes are bulked pile into a storage using a tier system and minimizing rollback and bruising

humidity. Reducing humidity can have consequences of increased weight loss (and pressure bruise), and this risk needs to be weighed when deciding on reducing humidification.

Potatoes coming from the field into storage usually go through a period of wound healing (curing period) before storage holding temperatures are attained. This is to promote healing of bruises, cuts, scrapes, and skinning damage. This wound-healing period may take from 2 to 3 weeks at 50–55 °F depending on the variety, disease presence, sugar content, and available cooling air. Wound healing occurs at a faster rate at warmer temperatures. Temperatures below 50 °F may reduce the rate of wound healing, thereby extending the length of storage periods required to provide adequate protection to the tubers. Temperatures above 60 °F may increase disease development before the wound healing process can be completed.

Wound healing is an extremely important component in reducing disease development and minimizing weight loss during storage (Fig. 17.10). However, it is also important to avoid keeping the tubers too warm for an extended period of time, because most disease organisms can multiply faster at warm temperatures. The storage manager must balance wound healing with slowing disease development and weight loss. Warmer temperatures promote greater evaporation and respirational losses. There are varietal differences in wound healing ability; therefore, wound healing temperature and duration may need to be modified during this early storage period. Usually, the best practice is to provide wound-healing conditions for an appropriate period of time, then reduce the temperature (ramp) to the desired holding temperature (0.1–0.5 °F/day). See the section on Storage Temperature for guidelines on ramping down the temperature.

Fig. 17.10 Proper wound healing conditions are necessary for potatoes to promote healing of bruises and cuts

The early storage and wound healing period is also an important time to assess and solve preexisting conditions that may lead to poor processing quality due to high sugar content. See Chap. 15. Sugar levels can be reduced by adjusting the temperature or length of the early storage period (preconditioning).

Holding Period and Removal from Storage

The holding period comprises the majority of the storage duration where potatoes are held at the desired temperature and humidity by modifying ventilation rates and air supply. Holding temperatures will vary with variety and desired end use. The holding period can last a few weeks to several months depending upon the time to market. Potatoes should be at least 45 °F prior to removal from storage to minimize the potential for damage when unloading, handling, and transporting. Below are additional specifics to the three basics of storage requirements: ventilation, temperature, and RH used during early storage and holding period management.

Ventilation

Storage managers are usually only able to control the supply air temperature, RH, and ventilation rate. However, by using these control features properly, adjustments can be made to the environmental conditions to provide optimum storage conditions. The ventilation system controls the temperature and distributes the humidity in the storage building and is a critical functioning system for maintaining quality in storage.

Proper storage management requires some specific knowledge of potato physiology and storage facility operation. Ventilation will provide conditions for the desired temperatures to the bulk pile, supply humidity to the potatoes, provide oxygen to the potatoes, and exhaust heat and carbon dioxide. Ventilation also provides conditions to dry out wet or decayed potatoes and a means for applying sprout control products. Some storages, particularly those holding process potatoes, are often ventilated to maintain carbon dioxide levels below the range of 1500–5000 ppm due to the potential for reducing sugar accumulation and/or changes in taste. Potatoes stored under refrigeration need periodic purging of carbon dioxide by bringing in outside air.

A consistent rate of forced-air supply is critical for maintaining tuber quality in potato storages. Most modern North American style storage air systems have been designed to supply 10–25 cubic ft. per minute (cfm) per ton of stored potatoes. In general, a storage design of 20+ cfm/ton air supply is required for handling potatoes that have been stressed or harvested in wet conditions. The ability of the storage facility to dry out "wet spots" or to remove free moisture is critical for preventing disease development in harvested potatoes. It is also important to have a sufficient air supply to remove water from decaying tubers. After the field heat and wound healing periods, ventilation rates are typically reduced.

Storage managers typically use variable speed fans, or control the number of operating fans, to alter ventilation rates to fine tune temperature differences between the top and bottom of the pile (Fig. 17.11). This differential, called delta-T or ΔT, should be maintained between 0.5 and 2 °F. Minimizing air flow when the ΔT is within the desired range is a good management tool for maintaining tuber quality, lowering risk for pressure bruise development, and reducing energy costs.

Basic principles of managing ventilation systems in a potato storage include the following: Ventilation fans should be operated to cool the pile, maintain the ΔT in the desired range, or provide fresh air. Use ventilation air at least 1–2 degrees cooler than the tubers at the bottom of the pile to avoid condensation within the pile. Operate humidifiers whenever bringing in outside air unless there is a need to remove excess moisture from the pile. Keep circulation air at or near 95 to 98 % RH.

Storage Temperatures

Optimal holding temperatures for potatoes in storage depend on the potato variety and its intended end use. Processing potatoes are generally stored between 45 and 50 °F to limit the concentration of reducing sugars in the tuber tissue. See Chap. 15, for more information. By comparison, potatoes intended for the fresh market are typically stored between 40 and 45 °F, while those intended for seed are usually stored at 38–40 °F. It is best to store potatoes at the lowest temperature specific to the variety, use, and desired quality. Storage temperatures are maintained with the use of outside cooling air mixed with recirculated air or refrigeration, if available or necessary.

Storage temperatures are also used to minimize weight losses caused by respiration and evaporation. Respiratory losses are often minimal near 45 °F (Fig. 17.12). Typical total weight loss over an 8–10-month storage season is 5–8%. Of that total, theoretically weight loss due to respiration alone can be 1.5%. The remaining loss is due to evaporation and decay. High rates of respiration can also reduce the amount

Fig. 17.11 Ventilation supplies oxygen, removes carbon dioxide and heat, and controls temperature of the potato pile

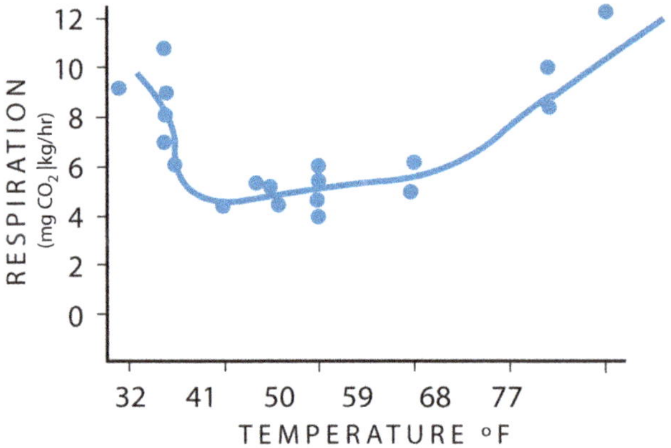

Fig. 17.12 Rate of respiration of potatoes at various storage temperatures. (Source: Burton 1978)

of carbohydrates in tubers, with a potential change in product texture and quality. Consequently, it is important for managers to consider minimum respiration rates for the variety when maintaining long-term storage.

An increase or decrease in storage temperatures can be used to minimize disease development. By reducing holding temperature, disease development and spread can be retarded, especially due to secondary infection from soft rot. However, managing storage temperature to control or prevent the spread of a disease may affect the tuber quality needed for processing or fresh market uses.

Sugar content is also an important consideration for potatoes stored for processing. Sugar monitoring can help establish appropriate early-storage conditions. See Chap. 15, for a discussion of this topic. Pre- and reconditioning refer to the use of elevated pile temperatures to help lower reducing sugar level in tubers. Higher temperatures increase the tuber respiration rate and allow for carbohydrate conversions, thereby decreasing reducing sugar concentrations so that the processed potatoes meet industry requirements. Preconditioning refers to using elevated storage temperatures at the beginning of the storage season compared to reconditioning, which is used before removal of the potatoes. Reconditioning, or increasing the temperature of potatoes in storage, is also used to stimulate sprouting of seed potatoes. Some varieties stored for seed at the normal temperatures of 38–40 °F may be very slow to sprout. Increasing the temperature of the seed potatoes in storage can help promote more consistent sprout development before planting, although there is the risk of too much sprout development, which alters the desired physiological age of the seed.

Standard guidelines for potato storage temperatures include many of the above-mentioned considerations. Temperatures outside of the optimal ranges may result in physical damage to stored potatoes. For example, potato tissue may freeze at temperatures below 30 °F. Also, a non-pathological breakdown of the tissue (chilling injury) may occur as the potato tissue approaches the freezing point. Most of the physical damage to potatoes at high temperatures is a result of increased disease activity. Blackheart, a physiological condition as a result of oxygen deprivation, may also increase as pulp temperatures rise above normal handling temperatures. See Chap. 14, for additional information.

Temperature changes in storage should be gradual and not exceed recommendations for various product uses. The rate of downward ramping of storage temperature for potatoes intended for processing should follow guidelines established by the processing industry. The typical rate of ramping is 0.1–0.5 °F per day to specified holding temperatures and will depend upon available outside cooling air. This gradual temperature reduction helps eliminate changes in the sugar content of tubers that can affect processed product quality. Guidelines for proper holding temperatures in storage may vary with the variety. However, storage managers should try to maintain a minimum temperature differential of less than 2 °F between the bottom and top of the pile.

For processing potatoes, it is critical that minimal sugar accumulation occurs. Cold sweetening is not the only concern with respect to sugar accumulation. Because potato tubers are alive, they age with time; a process that can be accelerated due to storage conditions. Higher temperatures result in more rapid aging.

One of the symptoms of aging is an uncontrollable increase in tuber sugar content. This is known as senescent sweetening. Once senescent sweetening begins to occur, the potatoes can no longer be warmed (reconditioned) to reduce sugar levels. See Chap. 15, for more discussion about senescent sweetening. Determination of optimum holding temperature consists of finding a balance between the temperatures at which cold sweetening becomes an issue and minimizing aging.

Relative Humidity (RH)

The impact of RH on storage weight losses can be substantial. Most of the tuber weight loss that occurs during the first month of storage results from water lost and tuber respiration. This early weight loss has the greatest impact on the overall total weight loss for the storage season. Maintaining high RH in storage lowers the vapor pressure deficit between the tuber and the surrounding air and helps control the total water loss during the storage season (Fig. 17.13).

Weight loss in storage is directly proportional to the length of the storage season and increases with time in storage. The rate of weight loss is inversely proportional to the RH conditions maintained within that storage, with lower RH promoting greater weight loss. The current recommendation is to maintain 95% RH or above (typical range is 90–98% RH). Equipment to maintain high RH is a standard part of the infrastructure of modern potato storages and can be constructed to maintain nearly any range of RH (Fig. 17.14). The necessary level to humidify the air will be dependent upon the humidity of the outside air. Arid conditions in the Pacific Northwest (PNW) mean that there are fewer grains of moisture in the air compared

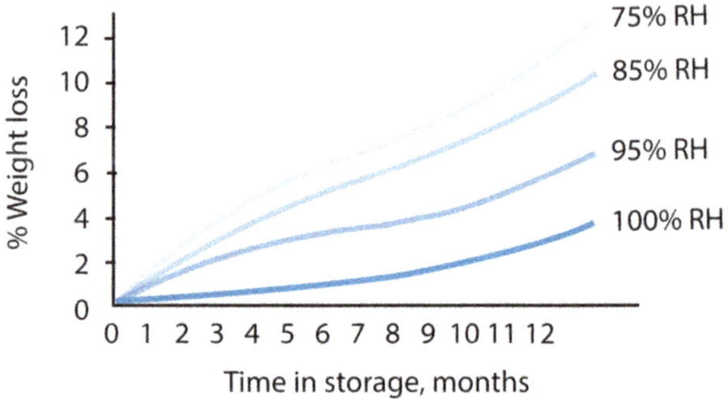

Fig. 17.13 Percent weight loss (shrinkage) of potatoes stored at various supply air humidity levels. (Source: Waelti 1989)

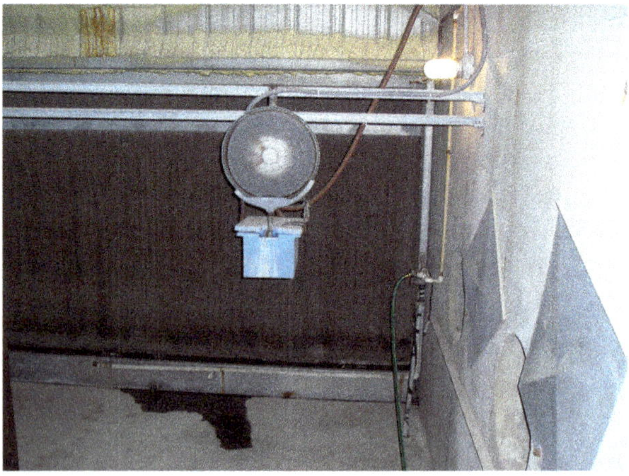

Fig. 17.14 Examples of humidification equipment in storages: evaporative cooling pad and centrifugal humidifier

Fig. 17.15 Avoid condensation from ceiling beams dripping free water onto potatoes in storage

to the more humid conditions of the Canadian Maritimes; therefore, storage facilities in the PNW require greater water input via humidification to reach 95% RH.

Another aspect of managing RH is to prevent condensation and free moisture on the potatoes. Free moisture on the surface of the potatoes can significantly contribute to disease breakdown in storage. Condensation can become a problem when it occurs directly on the tubers or on any inside surface of the storage. Condensation occurs when temperature differentials occur in the air circulating within the storage facility. If any of the air drops below the dew point, condensation will be the inevitable result. The "dew point"is defined as the temperature at which water vapor condenses at the same rate that it evaporates, which is just a more complicated way of saying the air is saturated with moisture or is at 100% RH. This means that the air simply cannot hold any more water vapor, and any further cooling of the air means that condensation will be the result. Localized areas within a storage facility where some of the air becomes cooled below the dew point can occur. Condensation on the tubers can also result when supplying moist circulation air that is warmer than the tubers at the bottom of the pile. As this air comes into contact with the tubers, the air rapidly cools, the RH reaches 100%, and moisture condenses on the tubers. Condensation on building surfaces usually results from inadequate insulation. If building surfaces are cooler than the air inside the storage, moisture condenses on the surface then drips onto the potatoes. This can be especially damaging if the moisture is on the ceiling of the storage and, therefore, becomes free moisture on the potatoes (Fig. 17.15).

Maintaining circulation air slightly cooler than the bottom of the pile will help prevent condensation directly onto the tubers. Likewise, condensation on building surfaces can be minimized by providing adequate insulation and making sure there is enough air movement to keep surfaces warm and to evaporate the moisture that collects before it drips onto the potatoes. By placing extra fans on top of the potato pile in strategic locations, condensation problems can be minimized in storages with historical problems or during extremely cold weather.

In any situation where there is a significant difference in potato or surface temperature (i.e., wall, ceiling, structural element, etc.) and air temperature within a humid storage, there is the potential for condensation to occur. This potential makes it very important to monitor and integrate the temperature of the crop into your ventilation system operation and storage management plans. Condensation that occurs on the interior of the storage, especially ceilings, beams, or walls, can drip free water onto the stored potatoes below. Warmer air holds more water vapor than cooler air, so if the warm air around the surface is quickly cooled, water may condense out if the temperature of the surface is below the dew point of the surrounding air.

Sprout Inhibition

Modern North American style potato storages are designed to store from 50,000 to more than 500,000 hundredweight (cwt) of potatoes in bulk piles. Successful long-term storage of fresh and process potatoes requires using a sprout inhibitor in combination with proper storage management to ensure control of sprouting. Cooler storage temperatures will slow down sprout development. Utilize temperature along with a sprout control program to successfully suppress sprout development and store potatoes to the preferred length of time. Decisions on what product to use will depend upon variety, the market use of the potatoes, if seed will also be stored in the same facility, and desired outcome. Each variety may react differently to sprout inhibitors. Knowledge of varietal differences in dormancy length is important for successful long-term storage (Table 17.1). This information will allow the appropriate timing of a sprout inhibitor.

Chlorpropham, or CIPC, is the most effective and commonly used post-harvest sprout inhibitor registered for use in potato storages in the U.S. This product has been used successfully as a sprout inhibitor for more than 60 years. CIPC inhibits sprout development by interfering with cell division. Cell division is not only important for sprout growth, but it is necessary to form the wound periderm during wound healing. Consequently, CIPC must be applied after the wound-healing period is over, but before dormancy break or initiation of sprout growth. Commercial applicators apply the CIPC aerosol formulation to bulk potatoes in storage. Other sprout inhibitor products are available in the U.S. and include maleic hydrazide, essential oils, naphthalenes, 3-decen-2-one, 1-Octanol. Others are currently under development.

Maleic hydrazide is applied as a foliar spray treatment during the growing season. Timing of application is important since the product alters cell division and can impact tuber development. Potatoes treated with maleic hydrazide have an approximate 30-day delay in dormancy break depending upon variety as well as rate and timing of application. Maleic hydrazide applications can be used alone for short-term sprout control or in combination with other sprout control products.

Several essential oils extracted from plant materials are effective potato sprout suppressants. They suppress sprouting by physically damaging the developing sprouts or buds before they can elongate. Repeated or continuous application of

Table 17.1 Approximate dormancy length in days after harvest of multiple varieties stored at various temperatures

	Storage temperature		
	42 °F	45 °F	48 °F
	Approximate dormancy length in days		
Ranger Russet	100	85	75
Alturas	100	90	75
Clearwater Russet	110	90	85
Blazer Russet	135	110	95
Russet Norkotah (standard)	130	115	100
Sage Russet	135	115	95
Teton Russet	135	115	100
Bannock Russet	175	125	110
Classic Russet	155	130	100
Umatilla Russet	145	130	100
Summit Russet	200	145	110
Russet Burbank	175	155	130
Alpine Russet	185	165	140

Note the longer dormancy at lower storage temperatures and the inherent differences between varieties

many of these essential oil-based sprout suppressants are necessary to achieve the desired sprout control. Carvone, the oil from caraway seed, is marketed in several countries as a potato sprout inhibitor. It is efficacious and can be applied to potato storages using both cold aerosol application and conventional thermal aerosol fogging. Clove, spearmint, and peppermint oils have been used successfully to suppress sprouting in potatoes for extended storage periods. These oils, which are derived and extracted from plants, are effective in suppressing sprouting in stored potatoes as long as the material is present in sufficient quantities in the head space of the potato storage. Essential oils can be applied in several ways. Conventional thermal fogging may not be quite as effective as a cold aerosol application or forced evaporation for spearmint and peppermint oils, but is the most effective way to apply clove oil (Fig. 17.16). A disadvantage to using essential oils is their high volatility. Normal air circulation can vent the material from the storage, and repeated or continuous applications are necessary to achieve the desired result. See Chap. 6, for additional information on the use of essential oils in organic potato production.

Substituted naphthalenes have been successfully used to suppress sprout development. Two products have been registered for application to potatoes; 1,4-dimethyl naphthalene and diisopropylnaphthalene. The naphthalenes probably assist in sprout suppression by hormonal action and act quite differently from the cell division inhibitor, CIPC. 1,4-dimethyl naphthalene is also registered and labeled for use on seed potatoes as a mild sprout suppressant. An unsaturated ketone, 3-decen-2-one, can be applied to potatoes in storage as a thermal aerosol. The product physically damages the emerging sprout, and repeated applications are necessary depending upon variety, storage temperature, and storage duration.

Occasionally, inadequate sprout inhibition occurs after treatment, and several factors may have caused the sprouting problems in storage: (1) Improperly designed air systems can cause the potato pile to have a temperature differential from top to bottom, with the top being 3 or more degrees warmer. This temperature differential causes increased respiration of the tubers, which may induce earlier sprouting; (2) Improper sizing, spacing, or placement of air ducts may result in non-uniform pile temperatures and uneven air circulation, which can produce poor sprout inhibitor distribution within the storage; (3) Hot spots in the pile caused by disease, excess dirt restricting air flow, and/or plugged air vents can cause elevated pile temperatures that may result in premature sprouting (Fig. 17.17). This sprouting may occur in a small spot or, if undetected, may spread to larger areas of the pile; (4) Field-stressed potatoes may respond differently to sprout inhibitor application in storage than potatoes grown under non-stressed conditions. Field stress conditions (e.g., disease, water balance, nutrition, temperature) may also reduce sprout inhibitor effectiveness; (5) Potatoes stored under fluctuating temperatures and humidity may physiologically age faster than those stored under more uniform conditions; (6) Late-season application (usually after dormancy break) of CIPC produces mixed results, ranging from adequate sprout inhibition to complete failure. In comparison, some sprout control products are effective at damaging an emerging sprout and can be used at this later date.

Internal sprouting is a disorder in which a lateral sprout grows inward into the tuber or outward into an adjacent tuber. This tuber defect occurs mainly in long-term storage, and then only occasionally. The causes of this disorder are not well understood, but appear to be related to a lack of sprout inhibitor on or around tightly

Fig. 17.16 Commercial clove oil application in a storage

Fig. 17.17 Sprouting problems may result from air ducts that are sealed, dirt in the pile, and potato rot

packed tubers. See Chap. 14. Environmental factors, such as storage temperature and pile pressure on tubers, can have significant effects on internal sprouting. Insufficient sprout inhibitor concentrations caused by pile settling or excess soil and debris in the pile, or late application, can accentuate the problem.

Disease Management

One of the most challenging storage management problems is controlling tuber diseases (Fig. 17.18). Some post-harvest disease decay problems are associated with field locations where disease pressure was high or soil saturation occurs. These areas need to be identified before harvest so that the resulting tubers are stored only if the facility is capable of handling problem lots, and is in a location where they can be easily removed from storage. Initially it is very important to properly identify the disease causing issues in storage to direct management to that specific disease. See Chap. 9, for information on properly identifying the casual disease.

Storage diseases that cause a wet decay are difficult to control unless the storage facility is equipped to supply high volumes of air. Most storage diseases that cause tuber decay or rot will release large volumes of moisture that must be removed before it contributes to the spread of soft rot. Generally, a 1% tuber loss in a 100,000 cwt storage will release 10,000 gallons of water over a rather short period of time. Engineers must design modern storages that can deal with this level of decay by providing high ventilation rates, along with proper temperature and humidity control. Soft rot, leak, pink rot, late blight, and dry rot, and are the most

Fig. 17.18 Diseased potatoes in storage can cause problems for the entire crop

common and problematic diseases in storages. Additional diseases, such as silver scurf, can impact quality of the crop, but will not cause loss due to decay. See Chap. 9, for additional information on causal organisms and management recommendations. In general, disease management in storage utilizes temperature, ventilation, humidity, and, if appropriate, post-harvest product application. Post-harvest products include spray applications going into storage, such as phosphite-based products or disinfectants, or applied via the humidification system or thermally as some disinfectants are currently. Regardless, ensure the product you are using targets the disease of concern. Not all products are effective against all storage diseases.

Potato soft rot, caused by several species of both *Pectobacteria* and *Dickeya*, is a serious storage disease due to the opportunistic nature of these bacterial pathogens. This disease will spread rapidly from tuber to tuber if the conditions are appropriate. These bacteria are present in tuber lenticels and can also infect the tuber skin at harvest when bruises, scrapes, and cracks occur. However, the most common way for infection to occur is where fungal diseases, such as dry rot, pink rot, leak, or late blight, are already present. Soft rot enters as the secondary invader after the initial infection and is often associated with "hot spots" or "sink holes" in the pile due to the rapid breakdown of the tubers. Storage management includes use of cooler temperatures and high airflow to those infected areas to prevent spread. For more information, see Chap. 9.

Wet spots that appear in the pile at the beginning of storage season are often associated with leak (caused by *Pythium ultimum*), pink rot (caused by *Phytophthora erythroseptica*), and late blight (caused by *Phytophthora infestans*). The initial diseases may result in greater breakdown, depending upon the level of the disease and if secondary soft rot infection occurs. Late blight-infected tubers will decay slower in storage compared to leak and pink rot, but can become infected with soft rot that will accentuate tuber decay and allow the soft rot to spread rapidly in storage. Control measures include constant fan operation to dry out the infected tubers before they can become a problem and the use of cooler storage temperatures to keep the tubers in marketable condition.

Dry rot, caused by *Fusarium sambucinum,* can be a serious storage disease of potatoes, especially when susceptible varieties are grown, excessive bruising and wounding occurs at harvest, and if they are stored under sub-optimal storage conditions for wound healing. *Fusarium sambucinum* need an entry point to infect, such as a wound, which occurs mainly during harvest or handling. Disease progression is relatively slow and may not be evident for several months into storage. Some varieties, such as Clearwater Russet and Umatilla Russet, may be especially susceptible to dry-rot infection. Having knowledge of varietal susceptibility can help identify management conditions to minimize wounding at harvest and provide proper wound healing conditions specific to the variety.

Silver scurf, caused by the fungus *Helminthosporium solani*, is a disease that causes silvery blotches on the surface of the tuber. Although this disease is mainly considered a cosmetic problem with fresh market potatoes, it can cause problems during processing because of the thicker, corky periderm that results from the surface infection sites. This disease can also spread in storage if conditions are right for spore germination. Research has shown reduced infection from *Helminthosporium solani* when storage humidity was lowered to below 85% RH. Lowering humidity is not generally a recommended practice for managing silver scurf in storage. Crop shrinkage will be greater at lower RH conditions. Making these decisions requires knowledge of the intended market, the history and severity of potential problems, and the economic balance between determining factors.

Checklist for Storage Management

1–3 Months Before Harvest

- Repair all insulation materials to minimize the potential for condensation.
- Clean plenum and duct ports thoroughly.
- Replace worn humidity equipment and high-pressure nozzles.
- Check for corrosion on all surfaces that may limit the life of the storage facility.
- Service the air system and check all fans for proper balance. Check the air-delivery system by adjusting all ports or ducts for optimum, consistent air flow.
- Repair or replace worn components on air louvers, both fresh air and exhaust.
- Calibrate all computerized sensors that are used for control functions.
- Service the RH supply cell decks. Check for mineral deposits and eliminate clogged flow paths.

1 Week Before Storage

- Inspect the crop and look for preexisting conditions that may impact storability. Know the quality of incoming potatoes and the potential problems that might arise in storage. If potatoes are destined for processing, consider sampling fields and determining sugar levels as an indicator for physiological maturity.

- Condition the storage by operating the air circulation, humidity, and temperature equipment before delivery begins.
- If needed, wet the soil floor surfaces of the storage to help minimize dehydration of the tubers at the bottom of the pile.

At Potato Delivery

- Harvesting and handling operations should deliver the least amount of bruise and wounds as possible.
- Check and record pulp temperatures of potatoes going into storage. A minimum temperature of 45 °F to a maximum of 65 °F should be maintained. When possible, suspend harvest operations until pulp temperatures are within this temperature range.
- Limit potato pile height to 18–20 ft., or to the specifications of the storage design, to minimize pressure bruise (remember that pressure bruise can be variety dependent) and deliver the proper volume of air per cwt.
- Tape all duct seams to improve system performance. Open seams will reduce air delivery consistency.
- Operate fan and humidity systems as soon as the first few ducts are covered. Early fan operation helps to remove field heat and standardize pulp temperature differences that are inherent between fields, truckloads, and time of day.
- Remove clods, loose dirt, and debris from the incoming loads. This is critical to achieve optimum air circulation performance from the mechanical system.
- Fill each storage structure with potatoes destined for similar end uses. Close storages as soon as filled to achieve rapid temperature equilibration of the pile.

During the First 2 Months

- Immediately remove field heat and stabilize potatoes to 55 °F as soon as possible.
- Maintain pulp temperatures between 50 and 55 °F for 2–3 weeks for proper wound healing. Relative humidity of 95% is recommended for the wound-healing period and for continued storage.
- Reduce pile temperatures slowly (also known as ramping) approximately 2–3 degrees per week, to a general recommended holding temperature of 38–40 °F for seed, 40–45 °F for fresh pack, and 45–50 °F for processing (all variety dependent).
- If appropriate, sample and test for sugars and adjust conditions to minimize future problems.
- Shorter dormancy varieties may require a sprout inhibitor application.
- Observe pile for any hot spots, odors, or areas of wetness that may indicate breakdown due to disease.

During the Holding Stage

- Continue to monitor the storage daily for operational continuity and any problem that might occur.
- Ventilation should be set to maintain pile temperature variation of 0.5 to 2 °F from bottom to top. Continuous fan operation at reduced airflow or speed is capable of maintaining the desired temperature control of the pile.
- When appropriate, use sugar monitoring for early detection of deterioration in processing quality.
- Watch for condensation on walls and ceilings dripping onto potatoes. Increase air circulation in these areas, if necessary.
- Note any odors, sunken areas, or hot spots that may indicate breakdown due to disease.
- Make a timely decision on the length of storage period. If necessary, have a certified applicator make applications of sprout inhibitor. The type of inhibitor or time of application may change with different varieties, storage temperatures, and expected length of time in sprout control.

Before and During Removal

- In a situation where potatoes are destined for processing and sugar levels are too high, recondition the pile by raising the temperature to 55–60 °F for 3–4 weeks. During the reconditioning period, test samples to confirm an improvement in color and ensure that senescent sweetening is not contributing to the sugar problems. Sprouting and disease development are more likely to occur at these warmer temperatures; therefore, ensure adequate sprout inhibition and watch for any breakdown. Only sound potatoes can withstand reconditioning.
- Maintain storage air supply during unloading to minimize quality losses. Remember that good storage management during the unloading operation includes adjustment of duct airflow to maintain consistent supply to all areas of the remaining pile.
- Make sure pulp temperatures are at least 45 °F before handling to minimize damage.
- Operate the storage under optimum conditions until the last potatoes are removed.

References

Burton WG (1978) The physics and physiology of storage. In: Harris P (ed) The potato crop—the scientific basis for improvement. Chapman and Hall, London

Waelti H (1989) Potato storage and ventilation in the Pacific northwest. In: Potato storage technology and practice. American Society of Agricultural Engineers, St. Joseph, pp 1–89

Chapter 18
Principles of Economics and Marketing

Joseph F. Guenthner

Contents

J. F. Guenthner (✉)
Department of Agricultural Economics, University of Idaho, Moscow, ID, USA
e-mail: jguenthner@uidaho.edu

J. C. Stark et al. (eds.), *Potato Production Systems*,
https://doi.org/10.1007/978-3-030-39157-7_18

Introduction

The potato market is a complex, dynamic economic system. It can be viewed as a chain linking producers on one end to consumers at the other. Fresh packers, processors, wholesalers, transporters, retailers, and restaurants make up some of the other links in the chain. The information presented here is mostly from the growers' perspective, but it should also be useful for potato science students.

This chapter provides a general overview of the potato market. Although it begins with a discussion of grower marketing decisions, it is not a "how to" guide. Instead, it is a reference for potato marketing principles and issues.

Grower Marketing Decisions

Potato growers make many decisions during the production and marketing of a crop of potatoes. Management decisions for producing a crop of storage potatoes may be spread over nearly 2 years. The decision to grow potatoes in a particular field may need to be made during fieldwork the previous fall. The decision-making process continues through variety selection, planting, irrigating, fertility management, pest control, vine kill, harvest, storage management, and the sale, which could be as late as August the year after harvest. Some decisions are based entirely on marketing factors. Others may seem to be production decisions, but if the results impact product quantity and quality, the choices have implications on marketing.

What to Grow

The first marketing decision is whether to plant potatoes. If the answer to the first decision is "yes," the next decisions are where and how much to plant. Potato production is an expensive, risky enterprise that requires the right soil, climate, equipment, finances, labor, management, and marketing opportunities to be profitable. Potato producers who are thinking of entering, continuing, or expanding potato production should understand two important things:

1. Most potatoes are a commodity.
2. Historically, commodity prices trend downward.

Figure 18.1 shows more than a half-century of U.S. potato prices. After adjusting for inflation, potato prices trended downward until flattening in recent years.

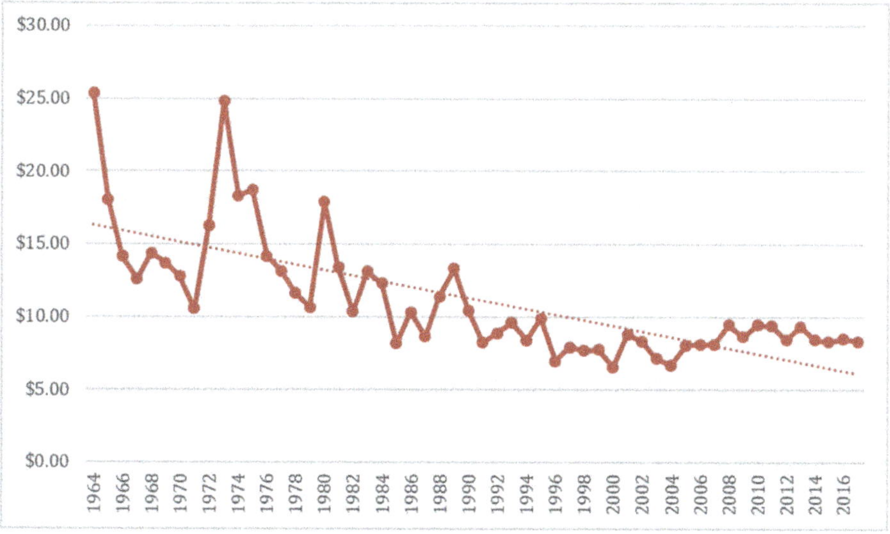

Fig. 18.1 U.S. potato price trend, 1964–2017 (1983 dollars)

Growers who intend to be in potato production for the long run should be prepared to reduce costs per unit in order to survive on downward-trending prices.

In the future, one way for potato growers to get out of the commodity business is to grow private varieties. Public varieties, such as Russet Burbank, can be grown by anyone without paying royalties. As a result, public varieties become commodities with, of course, downward trending prices.

Recent changes in property rights law and biotechnology have opened doors to privately owned varieties that indeed can be value-added, branded products if consumers view them as a superior good. With the supply of such varieties controlled by the owner, the excess supply problem that forces commodity prices down is a less powerful force. Growers can own varieties themselves or join alliances with others who hold varietal property rights.

Potato growers have many choices of public varieties. Market opportunities should be factors in grower variety selection decisions. The Russet Burbank has long been popular because it is a dual-purpose variety—well suited for both fresh and processed markets. Growing varieties that are suitable for only one mainstream market may be risky because of restricted marketing options. A single-purpose variety should have characteristics that make it well suited for market and production conditions.

Where and How to Sell

Producers who have not grown potatoes, and those who are considering expansion, should first think about where and how they can sell the potatoes they will bring to market. The main choices of market outlets are fresh, processed, and seed. More discussion of these choices follows in the Market Channels section later in this chapter.

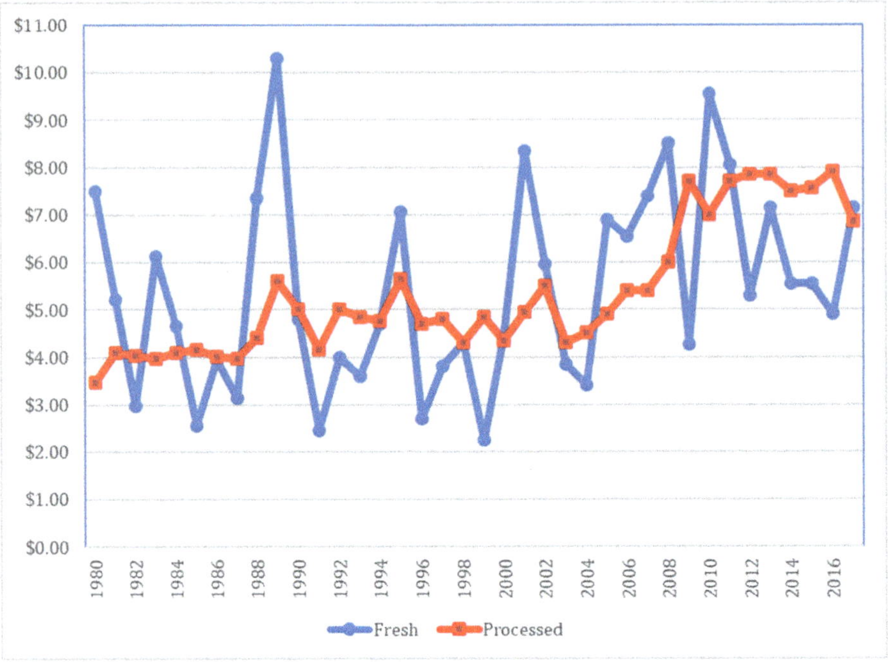

Fig. 18.2 Idaho fresh and processed potato prices, 1980–2017

Transport economics may be an important factor for some growers in the where-should-I-sell decision. Trucking potatoes long distances to a fresh packer or processor may put growers at a competitive disadvantage. Custom haulers in the main potato-producing regions can provide estimates of the costs for hauling to market.

The how-should-I-sell choices are generally limited to the open market or contracts. The open market offers the opportunity to sometimes sell at high, profitable prices, but the tradeoff is the risk of selling at low, unprofitable prices. The fresh potato channel has traditionally operated with open-market pricing. The processing industry contracts with growers for a large share of their raw product needs, but they also rely on the open market for some purchases.

Contract prices are more stable than open-market prices, but still vary from year to year. Figure 18.2 shows Idaho fresh and processed average annual prices. One example of price variability is in 1989 when the fresh market average price was $10.30/cwt and the processed price was $5.60. Two years later the fresh price had dropped to $2.45, and the processed price fell by a lesser amount to $4.15. For the entire 1980–2017 period, the fresh price average was $5.43, and the processed price average was not much lower at $5.34.

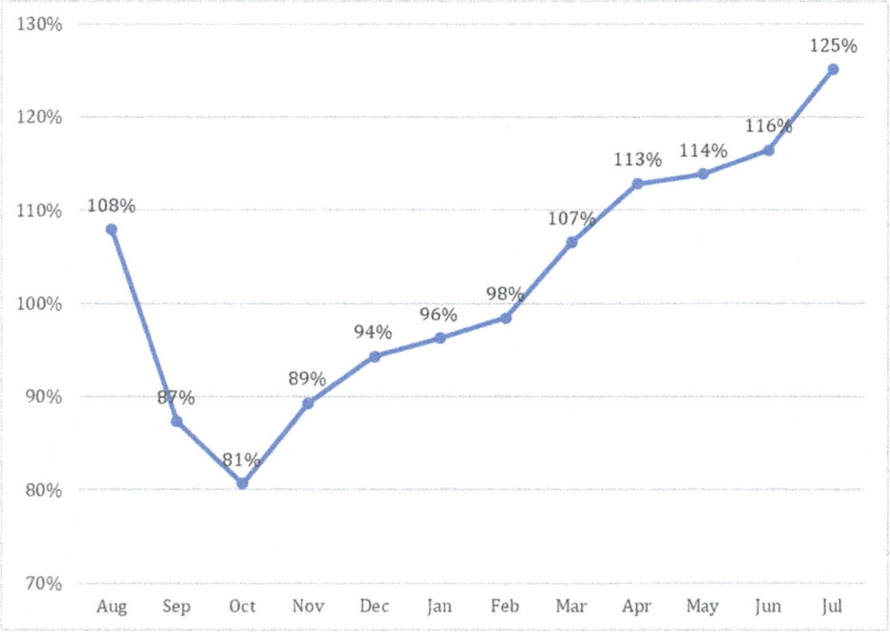

Fig. 18.3 Seasonal price patterns for U.S. fresh potatoes

When to Sell

Potato growers who contract with processors can sign contracts that call for delivery at harvest or during the storage season. Growers with their own storage facilities can estimate monthly storage costs and determine the type of contract they prefer. See Chap. 19.

Open-market growers should also estimate their storage costs, but they face the additional challenge of market uncertainty. For many fresh potato growers, the decision of when to sell can have huge financial implications, especially during years when open-market prices are quite volatile.

Figure 18.3 shows how U.S. fresh potato prices moved, on average, within the calendar years from 1980 to 2017. The percentages for each month measure that month's price in relation to the average price for the entire marketing season. For example, the October price is the lowest at 81% of the season average. That makes sense because that is when most potatoes are harvested, and the market supply is at its peak. Prices then rise steadily to a high in July of 125% of the season average.

Figure 18.3 shows averages only. In general, prices rise during the storage season or growers would not store for the open market. During some years the lowest prices of the season, not even accounting for storage costs, are at the end of the season. During other years, growers who sold at the end of the season reaped very profitable prices.

The challenge for growers is to estimate storage costs and try to sell when the storage enterprise shows a profit. They must do that without futures markets—a tool that producers of other commodities can use to lock in prices before delivery.

Principles of Economics and Marketing

Demand

In a free-market economy, the consumer is queen or king. Decisions made every day by consumers in grocery stores and restaurants drive the entire food production and marketing systems.

Price rations supply in a free-market system. When growers produce a small crop of potatoes, consumers bid up the price for the scarce good. Those who are willing to pay the higher price get the potatoes. When growers produce a large crop, potatoes are plentiful and can clear the market only at lower prices.

How consumers respond to changing prices is an important economic concept. For potatoes, consumers tend not to be responsive to price fluctuations and will buy about the same amount regardless of price, causing small supply reductions to push prices to high levels. Of course, it works the other way as well. A small increase in potato production can cause a large decrease in price. Researchers have found that a 1% increase in fresh potato supply can cause prices to change in the opposite direction by 7%.

In addition to price, other factors influence consumer demand. These can be grouped into four major categories, known as demand shifters: (1) population, (2) income, (3) prices of other goods, and (4) consumer tastes and preferences.

When a demand shifter causes an increase in potato demand, potato producer revenue increases. Consumers will purchase more potatoes at the same price or at a higher price.

Population

The U.S. population grew about 1% each year during the 1990s. The U.S. Census Bureau predicts that annual population growth is about 0.8%. This means that if nothing else changes, potato demand will slowly increase because of U.S. population growth. Although this is important, the characteristics of people making up the population also influence potato demand.

Demographics are the characteristics of population groups. People with northern European ancestry who live in the northern U.S. eat more potatoes than other ethnic groups. Age is another demographic. Young people eat many frozen potato products, mostly in the form of fries eaten away from home. Older people eat more fresh potatoes. As the baby boomer generation ages, their changing food preferences could influence the demand for some potato products.

Consumer Income

Price affects potato demand, but the impact varies across products. As income increases, consumers are likely to purchase more meals away from home. This increases demand for fries in fast-food restaurants, also known as quick-service restaurants (QSRs), but can decrease demand for fresh potatoes consumed at home.

Prices of Other Goods

This can have either a positive or a negative impact on potato demand. If the price of a substitute, such as rice, increases, the demand for potatoes increases because some consumers will switch from rice to potatoes. Researchers have found that the closest substitutes for potatoes are other potato products.

If the price of a complement, such as hamburgers, decreases, it too will cause potato demand to increase. When the worldwide fast-food restaurant chain McDonald's reduces its hamburger prices, people eat more fries with the less costly burgers.

Consumer Tastes and Preferences

These also influence potato demand. Families with both parents working outside the home tend to choose convenient food products. They purchase more services with their food, such as fries prepared at a fast-food restaurant. Another powerful trend is a preference for food that is healthy, such as fresh fruits and vegetables.

Advertising and Promotion

Firms and industries spend a lot of money attempting to change consumer tastes and preferences. If successful, increased sales revenue from higher prices and/or quantities more than pays for advertising and promotion.

Newspaper ads that feature a supermarket's potatoes, television ads that show people happily eating McDonald's fries, and magazine ads that feature a restaurant's Idaho baked potatoes are some examples of advertising by consumer-oriented businesses (Fig. 18.4). This is called "brand advertising." The businesses want consumers to come to their store or buy their particular potato product.

Potato growers are involved in another type of advertising. They use generic advertising to try to increase demand for all potatoes rather than one particular brand. The U.S. Potato Board collects money from potato sales to fund programs

Fig. 18.4 McDonald's television advertising featured some of its potato growers, including Frank Martinez in Washington State

Fig. 18.5 Potatoes USA develops generic potato promotion programs at sports events, including Ironman triathlons

designed to increase demand for all potatoes (Fig. 18.5). The Idaho Potato Commission and other state organizations do the same to attempt to increase demand for potatoes grown in their state (Figs. 18.6 and 18.7).

During tough economic times, some business firms reduce or even eliminate advertising expenditures. These occurrences may be a matter of trading long-run increases in demand for short-run survival. When potato prices are low, some

Fig. 18.6 A displays calls attention to the "Grown in Idaho™" brands of potatoes and entice shoppers to purchase fresh potatoes

Fig. 18.7 The Idaho Potato Commission conducts national advertising campaigns, including a multi-year program featuring a giant Idaho potato hauled by an 18-wheeler

growers talk about cutting the advertising funding mechanism for the same reason—to keep more money on the farm in the short run. Some are willing to accept lower prices in the future for short-run savings.

Advertising and promotion should be evaluated in terms of return on investment. For potato industry groups, it is a matter of putting money into programs that give

Fig. 18.8 Potatoes USA international marketing programs increase the global demand for US potatoes

the most return for the investment. For example, if running a promotional program on fry management for Asian fast-food restaurant managers provides a larger return than the same money spent on television time in a U.S. city, that is where the advertising money should go (Fig. 18.8).

Potato snack food processors use brand advertising to increase demand. Frito-Lay®, the top potato chip firm in the world, has a long history of effective advertising that helps build the value of the brand. Advertising among frozen potato processors is less common, but some feature their sustainability programs (Fig. 18.9).

Supply

Potato prices are sensitive to changes in supply within a given production year. Potato prices influence the amount of potatoes produced the next year. When potato prices are high throughout the marketing year, growers tend to plant more potatoes the following spring. They may not plant a lot more, but it takes only a small increase in supply to make a big change in price. High prices also entice other farmers, or even people who have never farmed, to plant potatoes. Economists call this behavior the "naïve price expectations model." See Sidebar 18.1.

Fig. 18.9 Potato processors include environmental messages in their marketing programs

Sidebar 18.1: Naïve Price Expectations

Since small decreases in supply cause large price increases, all the potato industry has to do is keep the supply down and prices will be profitable. Sounds simple doesn't it? It's not.

Naïve price expectation is one reason this is not so simple. Prices serve valuable purposes. First, they are needed to complete transactions between sellers and buyers. Second, they are signals. Our lives are full of signals. Traffic light signals tell us when to stop, when to go, and when to use caution. Alarm clocks signal people when to wake up. The referee's whistle signals when to stop the sporting event.

What do prices signal? They signal when to increase or decrease production. People in a free market read a high-price signal as an opportunity to make money. So they produce more. A low-price signal sends the opposite message; to grow less.

Each year the potato market sends new signals. Some growers don't trust the signals and stick to their rotations, but others do respond. The problem comes when a high proportion of growers responds to high-price signals. They seem naïve—which *Webster's Dictionary* defines as, "lacking critical ability or analytical insight."

Don't potato growers know that larger supplies drive down prices? Sure they know it, but they can't resist. The signal light has turned green.

Although individual grower behavior is unpredictable, the overall market impact is that higher prices bring higher plantings. Of course input costs, other crop prices, seed availability, processor contracts, government policy, and weather also influence plantings. Growers may see conflicting signals, but the price they received for their last crop is usually the strongest signal.

Some business people operate under a contrary principle. They try to make decisions that are the opposite of others in their industry. Contrary potato growers plant opposite the price signals. Realizing that naïve price expectations will influence the price of the next crop, contrary potato growers decrease plantings when prices are high and increase plantings when prices are low.

The Costermonger's Lesson

Long ago in England a new type of apple was developed—the costard. The new variety became so popular that the people who sold apples from their carts became known as costermongers.

The costermongers discovered an important economic principle that applies to modern produce markets—branding by variety increases demand. The apple vendors of old England realized the economic advantage of showing the good things about each variety to consumers.

The costermongers of today are supermarkets. They understand the lesson from the costermongers of old. You can see it in their fresh produce departments, where they display apples by variety.

Variety branding creates economic opportunities for growers to get out of the commodity business, where prices trend downward. You can see those opportunities in apple prices paid to growers.

The lowest apple price is usually for the Red Delicious, a variety that has a lot in common with the Russet Burbank potato variety. Both varieties date back to the 1870s when they were discovered as mutants. After 130 years, Red Delicious and Russet Burbank are still the most popular varieties among growers, but they are also public varieties that are in excess supply.

Apple growers are ripping out old Red Delicious orchards to plant varieties that bring higher prices. They have the opportunity to produce higher-valued apples because the industry brands by variety all the way to the consumer.

One apple variety that sells for high prices is Cameo, a private variety. An apple grower discovered Cameo as a chance seedling in the 1980s, protected it, and trademarked the name. Owners of this variety can control the supply and prevent it from becoming a commodity.

Meanwhile fresh potato varieties continue to be sold as commodities. They are red, white, russet, or yellow, but they are still commodities. The lesson from the costermonger says there is an economic advantage to branding by variety at the consumer level.

Researchers have found that the previous year's potato price does indeed influence supply the next year. It may not be that growers are truly naïve, but they get optimistic when prices are profitable and pessimistic when they are not. Non-price factors that influence supply include seven supply shifters: (1) input costs, (2) alternative crop prices, (3) technology, (4) joint product prices, (5) risk, (6) government programs, and (7) weather and pests.

Input Costs

Variable production costs, such as for seed, fertilizer, chemicals, and energy, can influence the supply of potatoes. The high fixed costs of entering the potato growing business are also a consideration. Large investment in specialized potato equipment (e.g., planters, harvesters, and storage buildings) create "asset fixity." Since these investments are "fixed" into potato production and are not easily used for other crops, growers who own them are likely to plant potatoes year after year.

Alternative Crop Prices

Commodity prices for other crops influence potato supply. High grain prices tend to increase grain plantings and reduce potato plantings. Different crops influence potato plantings in different parts of the country. Prices of wheat, barley, corn, alfalfa, sugar beets, onions, and other vegetable crops all affect potato plantings in some parts of the U.S.

Technology

Acreage planted is only half of the potato supply puzzle. Yields are the other half. Technology is an important force in the potato yield situation. Advances in irrigation and nutrient and pest management have helped growers increase yields so that they can grow more potatoes on fewer acres. Precision agriculture and biotechnology can also help yields increase.

U.S. potato yields have long been on an upward trend (Fig. 18.10). For the 1980–2018 period, average annual increases were 4.7 cwt/ac. A shift in potato plantings from low-yielding, non-irrigated areas to high-yielding irrigated areas has contributed to yield increases. Also, field selection decisions may be a factor, as growers choose to plant potatoes only in high-yielding fields well suited for potatoes.

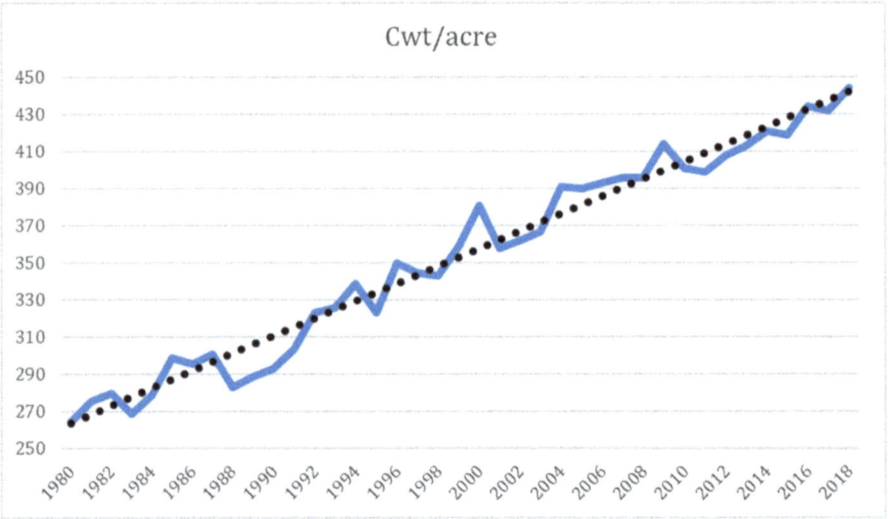

Fig. 18.10 Trends in U.S. potato yields, 1980–2018

Joint Product Prices

The sheep industry, which sells lambs and wool, is often cited as an example of the fourth supply shifter. In the Pacific Northwest (PNW) potato industry, fresh and dehydration markets have similar relationships. When Idaho growers sell open-market potatoes to fresh shippers, some go fresh, but the off-grade potatoes go to dehydrators. The growers may be paid a scoop-up price, but a joint product relationship is built into it. A high price paid by dehydrators for off-grade potatoes can increase the supply (planted acreage) of fresh potatoes the following year.

Risk

Potato growers face two types of risk: (1) production risk and (2) price risk. Growers who can reduce some of the production risk may choose to expand potato production. Increased risk can have the opposite effect. For example, when late blight first became a pest in the 1990s in Idaho and Colorado, some growers reduced plantings.

Price risk can be reduced with processor contracts. Researchers found that increases in potato contract prices cause increases in potato plantings. Even though acreage contracted may not change, the higher contract price reduces the risk of financial losses for individual growers, encouraging them to plant some open-market potatoes.

Open-market growers and contract growers seem to have different risk attitudes. Growers who prefer the volatility of the open, fresh market are willing to trade off the risk of low prices for the opportunity to sell at high prices.

Government Programs

Government involvement in the potato market has been relatively minor in recent decades, but programs designed for other crops do influence the supply of potatoes. For example, when the government removes price supports for wheat, some growers will switch wheat acreage to potatoes. When potato prices are disastrously low, the federal government sometimes provides support in the form of potato purchases or direct payments. This may help financially strapped growers survive, but it could increase potato supplies in following years.

Weather and Pests

Since potato prices are sensitive to changes in supply, market analysts closely watch conditions in potato-producing regions. Some forecasting models use weather variables to help predict potato supplies. Growers joke about disasters somewhere else helping the price situation, but they sometimes overestimate the impact of weather problems on potato supply. Potato plants are tough survivors that can produce higher-than-expected yields despite problems, such as frost early in the growing season.

Market Structure

Markets for agricultural products may be classified in three categories: (1) competitive, (2) oligopsony, and (3) monopsony. Competitive markets have many buyers and sellers, oligopsony markets have few buyers, and monopsony markets have one buyer.

Competitive markets also have homogenous products, no government intervention, no collusion, no barriers to entry, and accurate market information. Under competitive conditions, both buyers and sellers are price takers. No one individual is large enough to impact price. They all simply accept the prices determined by aggregate market forces.

Fresh potato markets have historically been competitive. Recent buyer consolidations have been a move toward oligopsony, which has long existed for processors. When there are only a few buyers, each one may have the power to influence price.

When a large number of growers sell to a few, powerful buyers they may receive lower prices than they would in a more competitive market.

Growers can gain market power by consolidating their sales. The federal government provides agricultural producers an exemption from antitrust laws. This allows farmers and ranchers to form cooperatives and bargaining associations that create a few large sellers out of many small sellers.

One challenge is getting growers to give up their independence to the organization. Another challenge is to manage the "free rider" incentive for those who want to enjoy the higher prices without paying organization membership costs.

Market Information

One of the requirements for competitive markets is accurate market information that is available to both buyers and sellers. Since large buyers can afford to assemble market information that individual growers cannot, the federal government provides unbiased market information. This is one example of official U.S. government policy to provide market power to growers.

The U.S. Department of Agriculture (USDA) and its network of state agricultural statistics services produce public information about potato acreage, yields, production, storage stocks, and other relevant market variables. One branch of USDA, the Agricultural Marketing Service, provides daily information on potato prices. The North American potato industry has a long history of private market information and analysis available for a subscription fee.

Pricing Mechanisms

Farm commodities are sold under five general types of pricing arrangements: (1) individual negotiations, (2) organized auctions or exchanges, (3) formula pricing, (4) collective bargaining, and (5) administrative decisions.

Individual Negotiations/Collective Bargaining

The most common pricing mechanisms in the potato industry are individual negotiations and collective bargaining. Individual fresh growers negotiate prices with individual fresh shippers. Although little opportunity exists to change competitive price levels, growers decide when and with whom they negotiate terms of sale.

Growers in Idaho, Washington, and a few other states have a history of collective bargaining with processors. When a large majority of growers are united in the

Table 18.1 Formula pricing example for fresh potatoes

	Share (%)	Price ($/cwt)
10# Film bags	28	$12.50
40-count cartons	1	$22.00
50-count cartons	2	$23.50
60-count cartons	3	$25.75
70-count cartons	5	$27.50
80-count cartons	4	$28.50
90-count cartons	5	$25.25
100-count cartons	5	$23.00
US# 2	7	$14.50
Total/weighted average	60	$18.18
Packing fee		$6.50
Return for fresh		$11.68
Processing grade	40	$4.00
Total grower return		$8.61

association, they can influence commodity prices. If the association represents an inadequate share of production, it has little influence on contract prices.

Organized Exchange Pricing

This mechanism is not available to U.S. potato growers. Potato futures contracts fit into this category, but U.S. futures exchanges no longer offer potato contracts.

Formula Pricing

The pack-out pricing method used by some fresh shippers is a type of formula pricing. With this method, the net price to the grower is a function of shipper prices for each type of pack, percent of the grower's potatoes going into each pack, and a packing cost. An example for a typical Idaho Russet Burbank packout of 60% fresh and 40% processing grade is presented in Table 18.1.

Administrative Pricing

Most potato marketers are "price takers" and do not have the power to set potato product prices. Those who acquire that market power are "price makers." In the public sector, the federal government has the constitutional power to set price floors

Fig. 18.11 Domesticated potatoes grown in South America have many colors and shapes

and ceilings, but has rarely done that in the potato industry. In the private sector, a firm can set its own prices if it has exclusive property rights, such as a patent, or the market power that comes with large size.

Market Channels

Fresh Potatoes

For many years fresh potatoes have been a staple in American diets. A traditional evening meal consisted of "meat and potatoes" eaten at home. With the increasing popularity of fast-food restaurants, the traditional meal seems to have shifted toward "burger and fries" eaten away from home. This shift in preferences partially resulted in a decline in per-capita fresh potato consumption in recent decades.

Several market forces could cause fresh potato consumption to increase. Potato packers are providing more convenient products, some of which are easy to prepare in microwave ovens. Successful promotions by Potatoes USA and state grower associations are changing public opinion to an understanding that potatoes are healthy. Still another factor is the large number of restaurants that include baked potatoes and fresh-cut fries among their menu choices.

Fresh market potatoes in North America are classified as russets, reds, and whites, based on skin color. Potatoes sold in North America are traditionally white fleshed. Yellow-fleshed potatoes, however, have been popular in Europe for many years. South American consumers have eaten blue, purple, and even black potatoes for centuries (Fig. 18.11). Although U.S. markets for these "exotic" potato types are small, they are growing, especially for yellow varieties.

Fresh potato sales are made in a variety of containers and grade specifications. Russet potatoes are usually packed in three general size categories: (1) consumer

Fig. 18.12 Processed potatoes now come in literally dozens of forms and use a significant portion of the commercially grown product

packs, (2) count cartons, and (3) institutional packs. Consumer packs consist mainly of 4–8-oz (non-size A) potatoes packed in plastic, paper, or mesh bags.

The most valuable potatoes are the 8–14-oz tubers that are packed in 50-lb cardboard boxes. These are called "count cartons." Each carton has a number that tells how many tubers are in a box (e.g., 60, 70, 80, 90, 100, 110, 120). Retail stores and restaurants buy count carton potatoes, which are typically used for baking. Institutional buyers, such as military bases, buy 100-lb bags of large potatoes to minimize packaging, handling, and peeling costs.

Although some reds and whites are sold in the same three size categories as russets, it is more common to pack them in a wider range of sizes in one container. Only the very largest (jumbos) and smallest (B-size) are typically sold separately. The russet consumer packs are usually the closest competitors for the reds and whites.

Some fresh potato production areas have mandatory inspection for fresh shipments. The rules, usually administered through marketing orders, require all fresh shipments of potatoes leaving the state to meet grade standards. Strict quality control has been an important component in some state advertising and promotion campaigns.

Processed Potatoes

Growth in potato processing in the later 1900s was rapid. By 1970 U.S. processed utilization was nearly equal to that of fresh. Major processing uses of potatoes now are chips, dehydration, and frozen-fried, which together account for about 98% of U.S. potato processing (Fig. 18.12).

Location of processing facilities varies according to product form. Freezing and dehydration plants are near growers in the Northeast, upper Midwest, and PNW. In Canada, processing is near producers from east to west along a band close to the U.S. border.

In contrast, potato chip plants are near consumers. Since chips are fragile and their low density makes them expensive to transport, chipping plants are in heavily populated areas. Chip growers are also widely dispersed. Before drought and flood problems, North Dakota was the largest chipstock producer in the U.S. Florida has replaced North Dakota as number one. Other major producing areas include Arizona, central California, Maine, and Michigan, but pockets of chip production are scattered across the continent.

Frozen processor contracts with growers emphasize factors that influence finished product quality. Specific gravity is of particular concern; processors pay premiums for high specific gravity and discounts for low specific gravity. See Chap. 15. Other characteristics, such as level of bruising, as well as tuber size and grade, are subject to incentives and may allow the grower to earn premiums. Contract provisions change in response to changes in technology and production practices.

Contracts allow growers to concentrate on production practices that improve yields and quality. Over time, contracts stabilize grower prices and, to a lesser extent, profits, which are subject to the vagaries of weather.

Frozen processors make fries, hash browns, and other products from the usable potatoes that growers deliver to the plant. The dehydration industry operates differently, purchasing much of its raw product as off-grade potatoes from fresh packers. Growers produce potatoes specifically for frozen and fresh markets, but returns are usually too low to attract growers to specialize in the dehydration market. Future development of varieties designed specifically for high yields of dehydration-quality potatoes could change this situation.

Seed Potatoes

A small, but important, component of the potato marketing picture is seed. As discussed in Chap. 4, use of high-quality seed potatoes is one of the most important practices in potato production. Most of the seed produced in the U.S. and Canada is used in North America, but some is exported outside the Western Hemisphere.

Marketing Issues

Regional and Seasonal Production Patterns

North American potato production has been shifting from East to West and from South to North. The East to West shift is the result of several economic forces. First, modern transportation makes it possible for growers in remote regions in the West to compete with growers closer to big cities.

Second, potatoes are less expensive to grow in the PNW because of favorable climate, plentiful irrigation water, few pests, lower power costs, and lower taxes. Another factor is consumer willingness to pay a higher price for what is perceived to be a premium product, such as the Idaho Grown Potato™.

The South to North shift results from enhanced storage technology that lengthened the potato storage season. This allowed frozen potato processing, the most rapidly growing market segment, to operate year-round on northern-grown potatoes, which increased efficiency.

Recent potato expansion in several regions of North America suggest that the trends may have ended. In the future, global markets and mobile resources could mean that potato production shifts could be more rapid than in the past.

Global Markets

Most potato processing firms began a few decades ago as small operations that sold product in regional markets. Some have evolved into large corporations with global operations. World-wide growth in the popularity of quick-service restaurants (QSRs) has fueled the growth of frozen fry processing. Consumers all over the globe have developed a taste for traditional burger-and-fries meals from North America.

Japan has long been the top customer for U.S. frozen fry exports. Greater China (Mainland China, Hong Kong, and Taiwan) has become another important market and is viewed by some as a major growth market of the future. The rapid growth of the Japanese fast-food industry and U.S. frozen fry exports to Japan may be repeated on a larger scale in China.

On the import side, growth in Canadian frozen fry imports has also been rapid. The unusually high value of the U.S. dollar in relation to the Canadian dollar provided a powerful economic incentive for processors to expand in Canada. And they did, but changes in currency exchange rates continually shift competitive positions.

Expansion of global frozen fry trade is causing a dispersion of processing plants. As more capacity is needed, processors spread their production facilities around the globe. This geographic diversity gives them flexibility to move production according to changes in currency exchange rates, consumer demand, transportation costs, potato yields, and quality.

Global markets provide opportunities not only for QSRs and processors, but also for growers. Many North American growers have expanded their operations along with the growth in the global market. Some have become international growers. Since they have the expertise to grow top-quality potatoes at home, they have ventured out to do it in other parts of the world.

Market Power

Growers are concerned about the increasing consolidation of buyers. It is true that buyers are getting larger and fewer, but the same thing is happening with growers. As grower numbers decrease, it becomes easier for them to organize. Some growers see the citrus cooperative, Sunkist, as the ideal they would like to work toward. Sunkist is an innovative global processing and marketing firm that is entirely owned by its grower members.

A group of Idaho growers formed United Potato Growers of Idaho in 2004 to gain market power in the fresh potato industry. Potato growers in other states formed similar regional co-ops and helped develop United Potato Growers of America. Development of the United Potato Growers of Canada soon followed. United succeeded in bringing higher, more stable prices to the fresh potato industry. An expensive lawsuit brought by a potato buyer led to a shift of United programs away from supply control and toward market information.

Futures Markets

Some commodity producers and buyers rely on futures markets to reduce price risk. This opportunity was available to potato growers in the past. The New York Mercantile Exchange, the Chicago Mercantile Exchange, and the New York Cotton Exchange provided potato futures trading, but dropped them due to low trading volume.

New Products

Thousands of new food products are introduced to U.S. consumers each year. They go through what is known as the product life cycle, which consists of four stages:

1. Introduction: The firm spends money to get people to try the product.
2. Growth: This can be quite rapid and profitable if the product is successful.
3. Maturation: When growth slows and product sales peak.
4. Decline: Sales decline, but innovative firms replace old products with new ones.

Fresh-Cut

Some fresh-cut vegetable products are in stage 1 while others are in stage 2 of the product life cycle. Convenience has driven peeled carrots and bagged salads into stage 2, while fresh-cut potatoes lag behind in stage 1. There could be unmet demand for potatoes peeled and cut for consumers who want to eat fresh potatoes with little

Fig. 18.13 Potandon Produce's branded, proprietary Klondike Rose® potato variety

preparation effort. The category could get a boost from the introduction of non-browning Innate® potatoes that can be packaged without preservatives.

Foodservice demand for fresh-cut potatoes also shows signs of growth. Some expanding restaurant chains, including Five Guys and In-N-Out Burger, feature generous portions of fresh-cut fries. While some foodservice buyers choose to buy raw product, others may develop a preference for peeled and cut refrigerated potatoes delivered to their doors.

Branded Potato Varieties

The U.S. potato market includes both private and public varieties, such as the Russet Burbank. Most retail stores put fresh potatoes into four categories: russets, reds, whites, and other. Unlike apples, potatoes have not been marketed by variety. That is one reason that fresh potato consumption is declining, and fresh potatoes, in general, are in stage 4 of the product life cycle.

Developers of new varieties can apply for patent-like rights, called plant variety protection (PVP) that allows 20 years of exclusive property rights. During this period the developer can operate like a monopoly with that variety. After 20 years it becomes a public variety.

Since PVP is relatively new for U.S. potatoes, there is not a long history of marketing proprietary varieties. Some entrepreneurs, including Discovery Gardens with Sierra Gold® and Potandon Produce with Klondike Rose® (Fig. 18.13) have marketed their products at price premiums. Innovations like these could lead to changes in retail potato marketing that rewards branded varieties with higher prices. This would allow some fresh potato products to get out of stage 4 of the product life cycle.

Sizing

Potato markets express size preferences through prices. In the fresh market, the highest prices are for count-carton size potatoes, followed by consumer-pack sizes. Potatoes too large or too small to fit into these categories are usually sold at lower prices to dehydrators or as low-priced fresh packs.

Size preferences have changed, especially for small potatoes, which facilitates moving some potatoes out of stage 4 and into stages 1 and 2 of the product life cycle. Fresh potatoes that formerly sold as low-priced Bs (USDA B-size), now often sell for prices higher than cartons. Even smaller C-sized potatoes, also known as Creamers, can bring the highest prices. In the frozen processing market, Simplot's Baby Bakers product line also offers price premiums for small potatoes.

Some potato packers have taken on small potato product lines to supplement their conventional packs. Others, such as The Little Potato Company, are specializing in small potatoes sold in attractive, small packages at high prices.

Size preferences reflected through prices can provide economic opportunities to growers. Those who can produce a smaller potato size profile through variety selection and cultural practices may find that producing small potatoes can provide big profits.

On the large end of the size profile, increasing demand for fresh-cut potatoes and fresh fries in QSRs helps support prices. Reduced peel loss is one attribute that can contribute to price strength for large tubers.

Organic Potatoes

Since consumer demand for organic fruits and vegetables has been increasing, that market offers another way to put some potatoes into stage 2 of the product life cycle. Prices for organic potatoes throughout the marketing chain are higher than for conventional potatoes. Along with opportunities for higher prices, organic potato producers also face higher costs and higher risks.

The USDA allows use of the Certified Organic label for potatoes produced in fields where no prohibited materials (synthetic fertilizers and pesticides) have been applied for 36 months prior to harvest. This means that organic potato fields must go through 3 years of transition from conventional to organic. While those three crops cannot be sold as certified organic, growers must follow organic production practices. The transition period adds to the costs and risks of entering the organic potato market.

University of Idaho researchers conducted a survey of organic potato growers in Idaho and other states. Low yields, insect control, weed control, storage losses, labor shortages, and market risk were among the growers' concerns. Growers also mentioned a need for longer rotations. Some cited strong demand for organic rotation crops, especially alfalfa hay. Respondents also pointed out the need for developing marketing plans.

Fig. 18.14 Organic potato packaging, which includes three brands: USDA Certified Organic, Grown in Idaho, and Wada Farms

A University of Idaho survey of 23 organic potato buyers revealed some market insights. All of the respondents said that they had experienced a shortage of organic potatoes. Tuber size preferences were for medium-sized potatoes. Preferences were for russet-type potatoes, followed by reds, then yellows, but some said there was also demand for fingerlings, purples, and other specialty organic potatoes. Packaging preferences were for 50-lb cartons, followed by 5-lb and 3-lb poly bags (Fig. 18.14).

Biotechnology

Potatoes were near the front of the development of food products made with genetically modified (GM) organisms. Calgene was first with the Flavr Savr tomato in the mid-1990s, but Monsanto followed shortly after with Newleaf® potatoes. Producers readily accepted the concept of potatoes that were genetically modified to protect themselves against pests. The problem was not with producer acceptance, but consumer acceptance.

As Monsanto was withdrawing from the biotech potato business, the J. R. Simplot Company began development of its biotech potatoes. Lessons learned from Monsanto's efforts led to Simplot doing three things differently.

First, Simplot focused on consumer traits rather than producer traits. Second, they used cisgenic rather than transgenic technology. Since many anti-biotech activists define GM as transferring genetic material across species lines, Simplot's Innate® potatoes, which only use potato genes, might be more readily accepted than Monsanto's Newleaf® potatoes. Third, Simplot implemented an identity preservation program in which only licensed growers and marketers can handle the product. This reduces the risk of Innate® potatoes getting into unwanted market channels.

Fig. 18.15 Simplot Plant Sciences point of sale (POS) material for its White Russet® potatoes

Three federal government agencies—USDA, EPA, and FDA—approved Simplot's Innate® Generation One potatoes for commercialization in 2015. One trait is a low level of acrylamide, a substance linked to cancer and birth defects in rats and common in foods cooked at high temperatures. Other traits include bruise resistance and non-browning when cut or peeled. In 2015 Simplot entered the fresh market with Innate® potatoes branded as White Russets™ (Fig. 18.15).

Innate® Generation Two has traits for late blight resistance and cold storage. Both generations will start in stage 1 of the product life cycle but could move rapidly into stage 2. These products could open the door for more new potato varieties that are better for producers, consumers, and the environment.

Chapter 19
Cost of Production

Paul E. Patterson and Ben Eborn

Contents

Introduction

The purpose of this chapter is to provide an example of the methods and procedures used by university specialists in developing potato production cost estimates for potato growers. The methods and procedures described are those used by the University of Idaho, illustrating a typical approach used in developing cost of production estimates. Also discussed are procedures that a grower might use to develop cost-of-production estimates for individual farms.

Growers who use published cost-of-production estimates need to understand their intended use and limitations. It is equally important that producers follow appropriate procedures when constructing cost-of-production estimates, which provide important information that can help growers manage their potato operations.

P. E. Patterson
Department of Agricultural and Rural Sociology, University of Idaho (Emeritus), Moscow, ID, USA
e-mail: pattersn@gold.uidaho.edu

B. Eborn (✉)
Department of Agricultural Economics and Rural Sociology, University of Idaho, Montpelier, ID, USA
e-mail: beborn@uidaho.edu

J. C. Stark et al. (eds.), *Potato Production Systems*,
https://doi.org/10.1007/978-3-030-39157-7_19

The terms “cost of production,” “costs and returns estimates,” and “budgets” will be used interchangeably throughout this chapter.

Costs and Returns Estimates

Many individuals and groups, including producers, processors, politicians, and consumers, have interest in the costs associated with producing agricultural commodities and whether profitable returns can be achieved. Commodity costs and returns estimates (CARs) are used to characterize the economic performance of a single commodity for an individual, a region, or even a nation. However, the intended use of a CAR estimate will influence how the costs and revenues are calculated and organized.

Availability and accessibility of data can also influence the process. Even when CAR estimates are prepared for the same intended use, many differences of opinion exist as to which costs to include, how the costs should be calculated, and how the costs should be organized. A direct comparison of CAR estimates is appropriate only when they are prepared using similar procedures. To minimize the chance of misinterpretation, the procedures and assumptions used to develop a CAR estimate should be clearly stated, along with the intended use.

Publication of the *Commodity Costs and Returns Estimation Handbook* in 1998 by the American Agricultural Economics Association (AAEA 1998) has helped to standardize procedures used in the development of cost-of-production estimates among land-grant universities and the USDA. The handbook summarizes issues associated with construction, use, and interpretation of CAR estimates. The handbook also discusses alternative methods for estimating cost of production and identifies conceptual and practical issues faced when evaluating alternative estimating methods. The *Commodity Costs and Returns Estimation Handbook* discusses the relative merits of alternatives and suggests guidelines to apply when preparing cost estimates for alternative uses.

The University of Idaho CAR estimates conform to the AAEA recommendations, except in use of nominal, rather than real, interest rates. Since not all the issues related to cost of production could be discussed in this chapter, individuals interested in more information should refer to the AAEA Handbook (AAEA 1998).

CAR estimates can be constructed using either historic or projected cost data. The scope of the CAR estimate can be narrow and represent an individual grower, for example, or it can be a composite that represents the costs for a region, state, or nation. The cost data can be from actual farm records or can be synthesized or “generated” for a model farm using a standard set of assumptions and procedures.

Growers with an interest in calculating cost-of-production estimates need to keep this use in mind as they develop their record-keeping system. Even with a detailed enterprise accounting system, however, certain costs will be tracked only on a whole-farm basis. These whole-farm costs will need to be allocated to different enterprises, an issue that will be discussed later.

Enterprise Budgets

Budgeting is a systematic approach to organizing revenue and cost data used in comparing and analyzing alternatives and in making management decisions. Budgets provide revenue and cost estimates or projections and should be an integral part of any planning process. An enterprise budget format is generally used for cost-of-production estimates. An enterprise is any coherent portion of a farm business that can be separated and analyzed as a distinct entity.

Traditionally, each crop is treated as a separate enterprise. Different enterprise designations can be made, however. Each field or pivot, for example, could be treated as a separate enterprise. The record system for the farm would have to be organized with this in mind so that the account structure would support the enterprise structure.

The enterprise budget tracks one production cycle—usually a 12-month period—and lists all expected revenue and costs. The enterprise budget can also include the quantity, time of use, and cost of each input used, along with expected yield and price.

Idaho's Costs and Returns Estimates

Understanding the budgeting procedures used by University of Idaho specialists will help commercial potato growers worldwide understand the potential uses and limitations of these cost estimates. It should also help if growers choose to modify these costs to fit individual farm situations.

The University of Idaho's crop CAR estimates are revised and published biennially in odd-numbered years. UI specialists use a computer program, Budget Planner, to generate individual crop CAR estimates. Crop CAR estimates are developed for four distinct geographic regions of the state. Three of these are located in the potato-producing areas of southern Idaho—Southwestern, Southcentral, and Eastern Idaho. Climate and soil conditions not only influence which crops are produced in each region, but also influence the specific production practices for those regions.

Even within a region where production practices are similar, costs can and do vary from farm to farm. Each farm has a unique set of resources with different levels of productivity, specific pest problems, and grower management skills. While the University of Idaho CAR estimates serve as useful benchmarks, they represent only single-point estimates that can't possibly capture the inherent variability that exists in production costs. These potato production cost estimates are representative or typical for a region. They are not, however, the average cost of producing potatoes.

Basic Assumptions

The University of Idaho cost-of-production estimates are affected by the assumptions made in depicting a representative farm for a region. Each region has a model farm (or farms), with assumptions about farm size, crop rotation, typical production

Table 19.1 2018 Idaho Russet Burbank potato costs and returns estimates, farm size, and potato acres by production region and use of fumigation

Region	Farm size (acres)	Potato acreage	Storage	Fumigation
Commercial potatoes				
Southwestern	1600	500	Yes	Yes
Southcentral	2200	550	Yes	No
Southcentral	2200	550	Yes	Yes
Eastern				
Southern region	2400	800	Yes	No
Southern region	2400	800	Yes	Yes
Northern region	2400	800	Yes	No
Seed potatoes				
Eastern				
G3 Russet Burbank[a]	1600	400	Yes	No

[a]*G3* third generation seed potatoes

practices, equipment used, and irrigation system. Budget Planner calculates machinery costs and labor requirements using standard engineering equations developed by the American Society of Agricultural Engineers.

The potato production costs published by the University of Idaho are based on survey data collected from Idaho farmers, farm supply businesses, and Extension faculty, as well as private consultants and industry representatives. Information on tillage, planting, fertilization, pest control, irrigation, and harvesting is collected from growers. In addition to the type of machinery and the number of workers used to perform field or custom operations, the type and quantity of inputs used is also collected. Survey information is then used to construct a model farm and develop typical production practices that are replicated by the computer program to generate costs on a per-acre basis.

The University of Idaho currently publishes seven potato budgets (Table 19.1). A sample cost-of-production estimate for Eastern Idaho's southern region is shown in Table 19.2 (operating costs) and Table 19.3 (ownership costs). Some potato budgets include the cost of on-farm storage and/or fumigation, while others do not. The cost of potato storage for the Eastern Idaho sample budget is shown in Table 19.4. The cost per hundredweight (cwt) of potatoes produced are shown both for field-run and paid yield.

Budget Procedures and Assumptions

Historical input prices are used to generate the University of Idaho's costs and returns estimates. Input prices come from surveys of farm supply businesses collected in the year when the CAR estimates are revised. The potato price used to calculate revenue in the budgets with on-farm storage is a 3-year average based on the most recent Idaho Agricultural Statistics Service (IASS) All Potato Prices seasonal average.

Table 19.2 2018 Cost of production operating costs estimates for commercial Russet Burbank potatoes in the southern production region of Eastern Idaho

Item	Quantity per ac	Unit	Price or cost	Value or cost/ac
Operating inputs				
Seed				$289.80
G-3 russet Burbank seed	21.00	cwt	11.95	250.95
Seed cutting	21.00	cwt	1.85	38.85
Fertilizer				$317.10
Dry nitrogen—preplant	135.00	lb	0.41	55.35
Dry P2O5	160.00	lb	0.43	68.80
K2O	195.00	lb	0.32	62.40
Sulfur	85.00	lb	0.23	19.55
Liquid nitrogen	100.00	lb	0.49	49.00
Liquid P2O5	60.00	lb	0.50	30.00
Micronutrients/humic acid—CP	1.00	acre	32.00	32.00
Pesticides and Chemicals				$232.62
Seed treatment	21.00	cwt	0.70	14.70
Admire pro®	8.00	fl oz	1.44	11.52
Moncut 70DF®	1.00	lb	29.00	29.00
Metribuzin 75DF®	0.67	lb	13.25	8.88
Eptam 7E®	3.50	pt	6.30	22.05
Prowl H2O®	2.00	pt	5.70	11.40
Quadris Flowable®	8.00	fl oz	1.25	10.00
Omega 500DF®	5.50	fl oz	3.00	16.50
Endura®	5.50	oz	4.25	23.38
Bravo Weatherstik®	1.00	pt	5.55	5.55
Dithane F45 Rainshield® (2×)	3.20	qt	8.75	28.00
Gavel 75DF®	2.00	lb	8.95	17.90
Agri-Mek .75SC®	3.50	fl oz	2.50	8.75
Brigadier®	6.00	fl oz	1.35	8.10
Reglone®	2.00	pt	8.45	16.90
Custom and consultants				$68.00
Custom fertilize: 400–800 lbs	1.00	acre	8.00	8.00
Custom fertilize: 0–400 lbs	1.00	acre	7.00	7.00
Custom air spray: 5.0 gal	3.00	acre	9.00	27.00
Consultant and soil/pet. Test	1.00	acre	26.00	26.00
Irrigation				$96.32
Water assessment	1.00	acre	38.00	38.00
Irrigation repairs—center pivot	24.00	acre-inch	0.54	12.96
Irrigation power—center pivot	24.00	acre-inch	1.89	45.36
Machinery				$144.05

(continued)

Table 19.2 (continued)

Item	Quantity per ac	Unit	Price or cost	Value or cost/ac
Fuel—gas	4.52	gal	3.05	13.79
Fuel—farm diesel	20.47	gal	2.80	57.32
Fuel—road diesel	1.92	gal	3.30	6.34
Lube	1.00	$	10.56	10.56
Machinery repairs	1.00	$	56.05	56.05
Labor				$176.82
Equipment operator labor	3.88	hrs	21.10	81.87
Truck driver labor	1.86	hrs	16.45	30.60
Irrigation labor—center pivot	0.96	hrs	21.10	20.26
Irrigation labor—Chem-Fert	0.80	hrs	21.10	16.88
General farm labor	2.24	hrs	12.15	27.22
Sorting				$62.37
Sorting labor	385.00	cwt	0.126	48.51
Sorting equipment repairs and power	385.00	cwt	0.036	13.86
Other				$142.46
Crop insurance	1.00	acre	80.00	80.00
Fees and assessments	347.00	cwt	0.18	62.46
Interest on operating capital at 6.75%				$51.65
Total operating costs				$1,581.19
Operating costs per unit (based on 385 cwt field-run yield)				$4.11

Includes cost to grow, harvest, and sort potatoes
ac acres, *ac in* acre inches, *cwt* hundredweight, *gal* gallons, *hrs* hours, *lb* pounds, *qt* quarts, *CP* center-pivot, *5G* 5 gallon rate

Table 19.3 2018 Cost of production ownership cost estimates for commercial Russet Burbank potatoes in the southern production region of Eastern Idaho

Ownership costs	Value or cost/ac
Cash ownership costs	
General overhead	40.00
Management fee	126.00
Land rent	535.00
Tractors and equipment insurance	5.50
Total cash ownership costs	706.50
Non-cash ownership costs	
Tractors and equipment depreciation ands interest	182.00
Potato handling equipment depreciation and interest	60.00
Total non-cash ownership costs	242.00
Total costs per acre	$2,529.69
Total cost per unit (based on 385 scwt)	$6.57

Table 19.4 2018 Storage costs per cwt for commercial Russet Burbank potatoes in the southern production region of Eastern Idaho

	Storage costs	Field-run cost per cwt	Paid-yield cost per cwt
Field-run yield		385.00	
Paid yield %	90%		346.5
Base cost to grow, harvest, and sort		\$6.57	\$7.30
Storage system annual ownership costs	\$0.380	\$0.380	\$0.422
Base cost + storage ownership costs		\$6.95	\$7.72
Storage system annual repairs	\$0.043	\$0.043	\$0.048
Base + storage system ownership and repairs		\$6.99	\$7.77
	Cumulative storage op. costs	*Cumulative base + all storage costs*	*Cumulative base + all storage costs*
October	\$0.237	\$7.23	\$8.03
November[a]	\$0.432	\$7.43	\$8.25
December	\$0.531	\$7.52	\$8.36
January	\$0.629	\$7.62	\$8.47
February	\$0.729	\$7.72	\$8.58
March	\$0.827	\$7.82	\$8.69
April	\$1.041	\$8.03	\$8.93
May	\$1.162	\$8.16	\$9.06
June	\$1.303	\$8.30	\$9.22

Data include costs to grow, harvest, sort, and store potatoes based on both field run and paid yield
Base cost of production includes cost to grow, harvest, and sort potatoes, both operating and ownership. Ownership costs for potato handling equipment are included in the base cost of production
Storage system includes: Storage facility, air system, and the equipment used to place potatoes.
Storage operating costs include: Repairs (shown separately), plus monthly operating costs: Labor, power, chemicals, interest, shrink, and insurance.
Storage costs do not include the cost of removing potatoes from storage.
Cumulative storage operating expenses are calculated to the end of the month.
Note: Copies of potato and other crop costs and returns estimates are available on the UI's Agricultural Economics and Rural Sociology Department Idaho AgBiz website: https://www.uidaho.edu/cals/idaho-agbiz (click on Crop Budgets). Additional information can be found under the Potato Bulletins and Reports, which is located in the Publications section
[a]Month when sprout inhibitor applied

A background and assumptions page for each budget describes the key assumptions used in developing Idaho's potato costs and returns estimates. This section describes the model farm's size, irrigation system, water source, and crop rotation, as well as its tillage, fertilization, pest management, and irrigation practices. If the CAR estimate includes on-farm storage, the length of storage and the type of storage facility are described. The machinery, labor, land, and capital resources used in production of the crop are also described. This information is critical to understanding how the costs are generated and the uses and limitations of these cost estimates.

The yield in a CAR estimate is used to calculate gross revenue. It can also be used to calculate break-even prices needed to cover various costs. Yield is also the basis for some costs, such as promotion and inspection fees paid by growers, as well as storage costs. The yields used in most crop budgets are 5-year rolling averages based on historical IASS data. Yields used in the potato budgets are based on a 3-year rolling average.

A microcomputer program, called Budget Planner, from the University of California at Davis is used to calculate the cost estimates. The computer program replicates each field operation using tractors and equipment typically used by producers. The cost to own and operate machinery is computed by the program and summarized for the model farm.

Model Farm

The cost-of-production estimate presented in Tables 19.2, 19.3, and 19.4 is typical for growing, harvesting, sorting, and storing irrigated Russet Burbank commercial potatoes for a 3-county area of Eastern Idaho. A 2400-ac model farm grows 800 ac of potatoes, using a typical 3-year rotation of potatoes followed by 2 years of grain. The farm uses a center-pivot irrigation system, pumping surface water from a canal so that irrigation power includes only the cost to pressurize the water. The farm is located in an irrigation district where a flat fee per acre is paid for water.

After the stubble from the preceding grain crop is chopped, the potato ground is irrigated, disked, and ripped in the fall, and subsequently chisel plowed and marked out in the spring before planting. Potatoes are planted in early May using two 6-row planters with a 36-in row spacing. The seeding rate is 21 cwt per acre that includes an additional 5% (1 cwt) to account for waste during cutting and planting.

Potatoes are cultivated and hilled in late May with a basin tillage tool. In September, vines are rolled and sprayed with a desiccant. Potato harvest begins 3 weeks later using a 4-row harvester, 4-row windrower, and six 10-wheeler trucks. Potatoes are hauled from the field to a location where they are sorted and cleaned. From there they can go directly into on-farm storage or be transferred to a semi and transported to a processor or fresh pack shed.

Most fertilizer is custom applied in a split preplant application in fall and spring. A starter fertilizer is applied at row mark out, and additional nitrogen is applied, as needed, through the irrigation system.

The weed control program uses cultural, mechanical (tillage and cultivation), and herbicidal control methods. For insect control, a systemic insecticide is banded at planting, and four foliar insecticides are applied by air in July and August. Six foliar fungicide applications are made for blight, white mold, and other diseases, starting in July. Some fungicide applications are made by chemigation, while others are applied by air in a tank mix with an insecticide.

Potatoes in this model farm receive 21 in of water during the growing season and 1.0 in pre-harvest in September. Water applied to the grain stubble the previous fall before tillage is also credited to potatoes, bringing the total to 24 in.

Enterprise Budget Structure

The CAR estimates produced by the University of Idaho are based on economic costs, not accounting costs. Accounting costs typically include only out-of-pocket costs and ignore opportunity costs. Economic costs place a market value on all inputs, regardless of whether they are purchased (an out-of-pocket expense) or provided by the producer (a foregone opportunity). For resources supplied by the farmer, such as land or labor, there is foregone income, or an "opportunity cost." For example, owned land could be leased to someone else, and the farmer could be working for wages.

Crop costs and returns estimates are developed on a per-acre basis, providing a common production unit for making comparisons between different crops. Gross returns or revenue is the first category in an enterprise budget. While it seems obvious, units for price and yield should correspond. Potato yield is generally measured in cwt or tons, so the price should also be per cwt or per ton. If the yield is field run, the price should be for field-run potatoes. If storage costs are not included, then a harvest-time price should be used. The price should correspond to the actual or assumed time of sale.

Operating and Ownership Costs

Costs in an enterprise budget are classified as either operating (variable) or ownership (fixed). Operating costs are those incurred only when production takes place, and they are typically used up or transformed during the production cycle. Seed, fertilizer, fuel, pesticides, hired labor, and water are all operating costs.

With the exception of labor and machinery costs, it is relatively easy to assign operating costs to a particular crop enterprise. It is also fairly easy for a grower to modify the operating costs in a published CAR estimate to match those on an individual farm.

In contrast to operating costs, ownership costs are associated with assets used in the production process that last for more than one production cycle. Many of these costs will continue even when production doesn't take place, hence the term "fixed costs." Ownership costs include the DIRTI-five: Depreciation, Interest, Repairs (that are a function of time and not of use), Taxes, and Insurance.

The assets generating fixed costs include machinery, buildings, and land. In addition to lasting more than one production cycle, these assets are typically used for more than one enterprise. Several different procedures can be used to allocate these costs over time and between different enterprises (crops) on the farm.

Custom Operators

Many growers find it more cost effective to use a custom operator than to own all the equipment or to supply all the needed labor. Money paid to a custom operator is classified as an operating cost. Where the cost shows up on a CAR estimate is different when a grower performs the service than when a custom operator is used.

The custom charge includes machinery costs that would be classified as ownership costs if the grower owned the equipment and provided the service. This can make a significant difference when comparing only operating costs or only ownership costs, especially when one CAR estimate uses owner-operator costs and another CAR estimate uses custom-based costs.

Estimated Operating Costs

The CAR estimates published by the University of Idaho list all inputs used in the production process. This makes it easier for users to modify these cost estimates to fit their situation, and it also makes it easier to update and revise the cost estimates. The individual inputs are listed along with the quantity applied and the cost per unit of input and per acre. The computer program used to calculate production costs, however, does place certain constraints on how inputs are classified or the sequence in which they appear on the printed copies.

Similar inputs are grouped together under a common heading. These headings include fertilizers, pesticides/chemicals, seed, irrigation costs, custom and consultant fees, field labor, and machinery costs. The quantity applied per acre, the unit of measure, and the cost per unit of the input are also listed. Multiplying the quantity applied by the price per unit gives the cost per acre. This is a fairly straightforward process for most operating inputs, especially purchased inputs.

Seed Costs

The sample budget operating costs (Table 19.2) shows a total seed expense of \$289.80 for 21 cwt of G3 (third generation) seed valued at \$11.95 per cwt, plus the cost of cutting seed, estimated at \$1.85 per cwt. The cost of seed should include transportation cost as well as the cost of the seed itself. The cost of seed treatment is listed in the pesticide/chemicals category and will vary considerably based on the type of product used.

While many commercial potato growers cut and treat their own seed, accurately calculating seed cutting costs can be difficult. The type and quantity of fertilizer applied is listed, except for the micronutrients. Micronutrient application varies substantially by grower and even by field. The \$26 per ac charge covers the cost of micronutrients and other soil amendments typically applied.

Fertilizer and Pesticide Costs

Fertilizer and pesticides are typically the two biggest operating cost categories in a potato budget. In the example potato budget, fertilizer ($317/ac) and pesticides ($233/ac) together account for 35% of operating costs (Table 19.2). Nitrogen fertilizer comes from several different products that have different costs. The price per pound of pre-plant nitrogen ($0.41) is based on 46-0-0-0, while the price per pound for liquid nitrogen ($0.49) applied post planting is based on 32-0-0-0. Pesticides include all products applied to control weeds, insects, nematodes, and diseases. The total quantity of product listed may represent one application, as with the 3.5 pints of Eptam® 7E, or multiple applications, as with Dithane®, where two applications of 1.6 quarts each are being applied.

Irrigation Costs

Irrigation costs ($96/ac) include the cost of water, power, and irrigation system repairs. Irrigation labor is included in the field labor category. Irrigation water for the model farm is delivered through a canal with a fixed water assessment fee charged per acre. The water assessment is the average charge made by irrigation districts/canal companies in Southeastern Idaho that are surveyed each time the crop budgets are revised.

Since the model farm uses surface water, the $1.89 per ac-inch power charge is only for pressurization. For a field using groundwater, the cost will vary based on the lift as well as service provider.

Irrigation costs are calculated using information from University of Idaho irrigation cost publications. Irrigation power costs are calculated using Idaho Power rates and the 160-ac center pivot with a corner system described in *Bulletin 787* (Patterson et al. 1996). The effective energy charge per kilowatt hour used in 2018 was $0.6052, the demand charge per kilowatt was $6.97, and the monthly service charge was $22.00.

Season-long irrigation power costs and repairs are calculated for the entire field and then converted to an acre-inch basis. The 24 in of water includes 21 in applied during the growing season, 1 in applied before harvest, and 2 in applied the previous fall prior to tillage. The center-pivot irrigation system application efficiency is assumed to be 80%. The pumping plant efficiency (electric motor and pump)—used to calculate kilowatt-hours—is 62%.

Custom and Consulting Costs

In addition to hiring custom operators, many growers also hire a consultant to provide water and fertility management. This charge is grouped with custom charges. While it isn't possible to tell from Table 19.2 which inputs are being applied by which custom operation, the Background and Assumptions page in each sample budget typically provides this information.

Some of the abbreviations can also be confusing. The 5 g designation on the custom aerial application, for example, is for the 5 g application rate charged by an aerial applicator. The 5 g rate is often used for fungicide applications, while a 3 g rate is charged for applying most insecticides.

Machinery Costs

Machinery operating costs ($144/ac) include fuel (gasoline and diesel), lube, and machinery repairs. All these values are calculated by the computer program using equations derived by the American Society of Agricultural Engineers. Basically, the computer program farms the field with the selected implements and tractors.

Most producers accumulate fuel and repair costs for the entire farm. The allocation of these whole-farm expenses to specific crops can be made using several allocation schemes. Growers should use or develop a scheme that is both simple and reasonably accurate. An equal distribution per acre, regardless of crop, may be simple, but not that accurate. Weighting the distribution based on expected gross revenue may improve accuracy and still meet the criteria of being simple.

Field and Sorting Labor Costs

Unlike growers who typically don't track labor to individual crops, the simulation approach used by the computer program accomplishes this by basing labor hours on the machinery hours calculated by the program. Based on speed, width, and field efficiency, the program calculates and accumulates machinery hours associated with each field operation.

Machine labor is calculated by multiplying the machine hours by 1.2. This accounts for time spent getting equipment to and from the field as well as time spent servicing it. Machine labor is calculated for all tractors, trucks, and self-propelled equipment. A market value is attached to all labor. No distinction is made between hired labor and unpaid family labor.

Sorting labor is based on the time needed to harvest all the potato acres, multiplied by the number of workers and their pay rate, divided by the number of potato acres. To base the labor sorting cost on per cwt as shown in Table 19.2, simply divide by the field-run yield.

General farm labor is the category name given to less skilled workers who do not operate machinery during planting and harvesting. The hourly labor charges include a base wage plus a percentage for various payroll taxes, workers' compensation, transportation, and other benefits. The overhead charge applied to the base wage used in the University of Idaho program calculates amounts of 15% for general farm labor, 25–30% for irrigation labor, and 25% for equipment operator labor.

Other Production Costs

The "Other" cost category contains two items: crop insurance and promotion fees and dues. Crop insurance is self-explanatory. Promotion fees and dues includes the grower's share (60%) of the advertising tax assessed by the Idaho Potato Commission, the promotion tax paid to the National Potato Promotion Board, 50% of the dock-side inspection fee, and membership dues in grower organizations. This works out to approximately $0.18 per cwt.

Interest

The last operating cost item listed is interest on operating capital. Producers use a combination of their own money and borrowed money and only pay interest on what they borrow. Since the University of Idaho's cost estimates are based on economic costs, no distinction is made as to the source of the capital. A market rate of interest is charged against all expenditures from the month the input is used until the harvest month.

Not all the interest on operating capital is listed in this last category. In the potato budgets with storage costs, an opportunity cost is calculated on the value of the potatoes during storage.

Calculating Individual Operating Costs

The type of accounting system used by a grower will determine how easy or difficult it is to derive enterprise-specific costs. Many growers have accounting systems that are designed to merely collect the cost information required to fill out Internal Revenue Service (IRS) Schedule F (Form 1040). Most growers do not use enterprise accounting, and it is not worth the effort to use enterprise accounting if the additional information available is not used for management decisions.

The question to ask is: How much does it cost to keep enterprise accounts compared to the value of the information? A sophisticated enterprise accounting system will have only limited value if the invoices from vendors don't provide the necessary detail needed to allocate the costs. Even without an enterprise accounting system, it is possible to develop reasonable, easy-to-use allocations for the different costs.

Costs, such as fuel or labor, are always going to present a problem unless growers log each machine operation and worker by field, which is an unlikely scenario. Until producers develop something specific to their operations, they might use the values in published enterprise budgets as proxy values or to calculate a percentage for allocation.

Using the University of Idaho Southeastern Idaho budget (Table 19.2), for example, fuel use per acre in potato production is roughly three times the amount used to grow an acre of wheat. If the total fuel bill for an individual's 2400-ac farm was

$125,000, and 800 ac of potatoes and 1600 ac of grain were grown, 60% of the fuel should be allocated to the potatoes and 40% to wheat, or roughly $75,000 and $50,000, respectively. On a per-acre basis for potatoes this totals $77.44. A potato grower may allocate general farm labor using the same method, or even the same percentages.

A grower may have to allocate costs for fertilizer, irrigation power, machine repair, interest on operating capital, and many other inputs using an arbitrary allocation system until an enterprise accounting system is developed. While a percentage allocation may not be as precise as an enterprise accounting system, it is better than making no attempt to allocate expenses to specific crops, and it may prove to be the best alternative.

Ownership Costs

Ownership costs, as stated earlier, are associated with assets that can be used for more than one production cycle. These include machinery, buildings, and land. Ownership costs are further divided into cash and non-cash costs. This section explains how ownership costs are calculated in the University of Idaho CAR estimates.

Ownership costs are based on the assumed values of the machinery and equipment. While not shown in this section, the published CAR estimates contain a table listing all the equipment used in producing that crop. The current replacement cost (or value) of all equipment is also included. The standard practice in the University of Idaho CAR estimates is to calculate ownership costs based on 75% of the replacement cost of new machinery and equipment.

Tax Life Vs. Useful Life

A distinction should be made between tax depreciation and management depreciation when discussing ownership costs. Depreciation is a measure of the reduction in value of an asset over time. For tax purposes, depreciation is spread over the tax life of an asset as defined by the IRS. Management depreciation, in contrast, spreads depreciation over the expected useful life.

The tax life of most farm equipment is currently defined as 7 years. The useful life could be 10–20 years. Management depreciation should be used in constructing enterprise budgets. For growers, this means keeping two sets of depreciation records.

Interest Cost

An interest charge based on the value of the equipment should also be calculated. It makes no difference whether the money is borrowed or supplied by the grower. In the first instance, the interest charge would be an actual cash expense. In the second, the interest calculation is a non-cash opportunity cost. The money could have been invested elsewhere, so the cost to the grower is the foregone income from this alternative investment.

The Budget Planner software used by the University of Idaho uses the capital recovery method to calculate the depreciation and interest on machinery. The total for all equipment used in potato production is listed as Equipment under the Non-Cash Ownership Costs (Depreciation and Interest). Depreciation and interest for the potato storage facility and potato storage equipment are listed separately, but in the same category. Since the Budget Planner software calculates costs on a field-operation basis, storage costs are calculated outside the program.

Taxes and Insurance

Taxes and insurance are the other two ownership costs. In the University of Idaho costs and returns estimates, these are based on the average level of investment, which is calculated by dividing the sum of the purchase price and the salvage value by two. Idaho eliminated property taxes on farm equipment in 2001, so there is no property tax shown in the CAR estimate. The annual insurance cost for each piece of equipment is calculated and then allocated to the appropriate crop(s) based on the percentage of use.

For equipment that is used 100% on potatoes, all the ownership costs are assigned to potatoes. But certain equipment, such as tractors and trucks, are used in producing other crops as well. The ownership costs for this equipment need to be allocated to the different enterprises in proportion to their use. This means that the ownership costs will not be simply divided by the total farm acres. For example, while the farm may have twice as many acres of grain as potatoes, the potato crop may account for 50% of the ownership costs for trucks and tractors based on use.

Value of Land

Unlike other capital assets, land is not a depreciable asset according to the IRS. And unless the land is being farmed in such a way as to degrade its productivity, excessive erosion for example, the land should last forever. But the money invested in land could be invested elsewhere. To avoid the issue of whether land is owned or leased and to be consistent with calculating economic costs, the land cost in University of Idaho crop budgets is a 1-year cash rent that includes an irrigation system. Repair costs for the irrigation system are classified as an operating cost under the Irrigation heading.

Other Ownership Costs

Two costs not related to land or equipment also appear as ownership costs. The first is general overhead. This is calculated at 2.5% of cash expenses and serves as a proxy for general farm expenses that are not typically assigned to a specific enterprise. It includes such things as legal fees, accounting and tax preparation fees, office expenses, and general farm utilities.

The second non-land, non-equipment expense is the management fee. This is an opportunity cost, and it is a residual in many costs and returns estimates. Because the University of Idaho budget specialists choose to include a management fee as an economic expense, all costs are accounted for except a return to risk. The management fee is calculated as 5% of total expenses (without management). The charge for the land and irrigation system is typically the largest ownership cost.

Calculating Individual Ownership Costs

While not as precise as the capital recovery method, calculating depreciation on a straight-line basis over the years of useful life is appropriate. This should be done for each piece of equipment. In a similar vein, interest can be calculated on the average level of investment.

Calculating annual ownership costs may be time consuming, but it is not difficult. The purchase price minus the expected salvage value gives total depreciation. Depreciation should be spread over the years of expected life to arrive at annual management depreciation.

Exclusive or Multi-Use Equipment

If the machine is used exclusively for one crop, the entire amount is allocated to that crop. The annual depreciation can then be allocated on a per-acre basis by dividing by the number of acres of that crop. If the machine is used on more than one crop, then part of the annual depreciation needs to be allocated to each crop. This value is then spread over the relevant acres.

For example, a \$170,000 potato harvester is expected to last 10 years and have a \$20,000 salvage value at the end of 10 years.

$$\text{Annual depreciation} = (\text{Purchase price} - \text{Salvage value}) \div \text{Useful life}$$

$$\text{Annual depreciation} = (\$170{,}000 - \$20{,}000) \div 10 \text{ or } \$15{,}000$$

If the harvester is used on 800 ac, the annual per-acre management depreciation is \$18.75.

Calculating annual depreciation for a tractor on this farm could follow the same procedure. The annual depreciation, however, would need to be allocated to different crops based on the hours the tractor is used on each crop.

Since most farms don't track machine time to specific crops, an approximation (informed guess) will suffice. The crop-specific depreciation can be allocated per acre in the same manner as the harvester.

While the interest on investment calculation is slightly different, the allocation procedure to the different crops on which the machine is used is the same. Interest should be calculated on the average level of investment, or the purchase price plus the salvage value divided by two. Using the potato harvester example:

$$\text{Average investment} = (\$170{,}000 + \$20{,}000) \div 2, \text{or } \$95{,}000$$

The interest rate can either be what is charged on a machinery loan or what a grower could earn on that money if invested in an alternative investment. Using a 6.0% interest rate, the annual interest charge would be:

$$\text{Annual interest} = \text{Interest rate} \times \text{Average investment}$$

$$\text{Annual interest} = 0.06 \times \$95{,}000, \text{or } \$5700$$

Again, this can be allocated on a per-acre basis.

Calculating Costs for Taxes and Insurance

The remaining ownership costs, property taxes and insurance, can be the actual costs taken from records and allocated to the appropriate equipment, or they can be calculated costs using an insurance rate and tax rate applied to the average investment, as calculated previously. The simplest method is to allocate these costs on an equal per-acre basis across the farm.

To deal with the obvious bias in this method, a crop such as potatoes might be assigned a higher percent of the costs. Using our example farm, potatoes might receive 50% of the cost even though they account for only 33% of the acres. The tradeoff in choosing between different allocation or calculation methods is often between time and precision. Producers should try to find a method that minimizes time yet provides a reasonably accurate estimate.

Storage Costs

Like most farm commodities, potatoes are placed in storage after harvest. The additional costs incurred storing potatoes should be added to the field production costs to get a total cost of production that the grower needs to recoup. This would include both operating and ownership costs. Operating costs would typically include labor, power, insurance, interest, sprout inhibitors, sanitation, and shrink. In the example, labor costs are included in the base cost of production (sorting labor) and not in the storage operating costs. Like field machinery, discussed earlier, the storage facility and the equipment used to place the potatoes in storage (even-flow bin, sorters/sizers, conveyers, and pilers) have an ownership cost component and would be calculated in a similar fashion. While the basic production unit found in Tables 19.2,

19.3, and 19.4 was an acre, storage costs are much easier to deal with on a per-cwt basis. Once calculated, storage costs can simply be added to the cost per cwt to grow, harvest, and sort potatoes based on a field-run yield basis from Table 19.4. Because potato growers are not paid for all potatoes that they deliver to a fresh-pack shed or processor, it is important to calculate costs on a paid-yield basis as well as the original field-run value. The paid yield shown in Table 19.4 is 90% of the field-run yield.

The storage system ownership and repair costs per cwt are added to the base cost of production. These costs are basically fixed and don't vary based on the length of the storage period. Table 19.4 also shows a cumulative storage operating cost by month from October to June. These values are added to the base cost of production and storage ownership and repair costs. As is apparent from an examination of Table 19.4, storage costs are a very significant component in the overall cost of potato production.

If the storage facility has a separate meter, calculating power cost per cwt is simply dividing the monthly cost by the size of the storage facility. There are two types of insurance that should be accounted for in storage costs. The cost of insuring the storage facility itself should be included as an ownership cost. The cost of insuring the stored potato crop should be included in the monthly operating costs. A sprout inhibitor is applied once or twice based on the length of storage. Application costs for many products are based on a cwt of stored potatoes, which makes it an easy value to include in the monthly storage costs. Don't forget to include the cost of chemicals used to sanitize the storage facility and storage equipment. Shrink and interest on the value of the crop are two of the biggest cost components. The base value of the cost of producing the crop is used in calculating interest in Table 19.4. This is a cash cost if the operating line of credit has not been paid. It is an opportunity cost of capital if the grower has not borrowed money to raise the potato crop. Potatoes respire and, therefore, lose moisture while in storage. The amount of loss will vary by variety, the condition of the crop going into storage, the type of air system in the storage facility, and the length of storage. The monthly value of shrink is based on the initial cost of production times a percentage shrink and other deterioration loss. The initial 3–4 weeks of storage has a larger shrink value than subsequent months (2%). November through March have a monthly shrink loss of 0.5%, while April through May have a shrink value of 0.75%. By June, this has increased to 1%. These are typical for Russet Burbank in a modern above-ground storage without a refrigeration unit and two applications of a sprout inhibitor.

Summary

The cost of potato production is influenced by all factors that determine the productivity of land, the type of resources committed to the production process, and the alternative uses of these resources. There is no single cost of potato production that fits all Idaho growers or even growers in one region.

Table 19.5 provides a summary of 2018 operating and ownership costs per acre by major cost category, as well as cost per cwt for field run and paid yield for

Table 19.5 2018 cost of potato production per ac comparison by state for major cost categories and per cwt for field run and paid yield

	Non-fumigated Colorado	Fumigated Wisconsin	Idaho fumigated			Idaho non-fumigated		
			Southwest	Southcentral	Eastern-S	Southcentral	Eastern-S	Eastern-N
Variety	R. Norkotah	R. Norkotah	R. Burbank	R. Burbank	R. Burbank	R. Burbank	R. Burbank	R. Burbank
Field run	395	400	505	470	420	430	385	365
% paid yield	85%	92%	90%	90%	90%	90%	90%	90%
Paid yield	336	368	455	423	378	387	347	329
Operating costs								
Seed	$454	$291	$353	$329	$290	$329	$290	$280
Fertilizer	$370	$377	$417	$383	$341	$353	$317	$306
Pesticides/chemicals	$343	$567	$568	$490	$442	$282	$233	$228
Custom/consultants	$45	$98	$133	$128	$111	$85	$68	$59
Irrigation	$267	$88	$129	$118	$100	$113	$96	$69
Other	$120	$136	$132	$141	$148	$135	$142	$134
Field labor	$181	$197	$254	$201	$180	$195	$177	$179
Machinery: FOLR	$143	$141	$175	$144	$145	$144	$144	$144
Sorting	$68	$72	$82	$76	$68	$70	$62	$59
Interest	$71	$73	$89	$81	$74	$57	$52	$49
Total operating costs per acre	$2,063	$2,038	$2,332	$2,091	$1,898	$1,762	$1,581	$1,507
Operating/cwt: F-R	$5.22	$5.10	$4.62	$4.45	$4.52	$4.10	$4.11	$4.13
Operating/cwt: P-Y	$6.14	$5.54	$5.13	$4.94	$5.02	$4.55	$4.56	$4.59
Ownership costs								
General overhead	$51	$51	$58	$52	$47	$44	$40	$38
Management fee	$138	$145	$177	$161	$144	$143	$126	$117
Land	$250	$395	$700	$650	$535	$650	$535	$440
Equip. tax and insurance	$11	$12	$6	$6	$6	$6	$6	$6

(continued)

Table 19.5 (continued)

	Non-fumigated	Fumigated	Idaho fumigated			Idaho non-fumigated		
	Colorado	Wisconsin	Southwest	Southcentral	Eastern-S	Southcentral	Eastern-S	Eastern-N
Sorting equip. D & I	$70	$72	$81	$74	$66	$66	$60	$56
Field equip. capital recovery	$192	$200	$215	$200	$184	$198	$182	$182
Total ownership costs per acre	$712	$874	$1,237	$1,143	$982	$1,107	$949	$839
Ownership/cwt: F-R	$1.80	$2.19	$2.45	$2.43	$2.34	$2.57	$2.46	$2.30
Ownership/cwt: P-Y	$2.12	$2.38	$2.72	$2.70	$2.60	$2.86	$2.74	$2.55
Total costss								
Total cost per acre	$2,775	$2,913	$3,569	$3,234	$2,880	$2,869	$2,530	$2,345
Total cost/cwt: F-R	$7.02	$7.28	$7.07	$6.88	$6.86	$6.67	$6.57	$6.43
Total cost/cwt: P-Y	$8.26	$7.92	$7.85	$7.65	$7.62	$7.41	$7.30	$7.14

Includes Colorado, Wisconsin, and Idaho

F-R field-run yield, *P-Y* paid yield, *Machinery FOLR* fuel, oil, lube, and repairs, *D & I* depreciation and interest, or capital recovery for machinery and equipment, *EI-N* eastern Idaho northern counties (north of Blackfoot), no fumigation, *EI-S* eastern Idaho southern counties (south of Blackfoot), no fumigation

Note: Cost to grow, harvest, and sort potatoes only. Storage costs are not included in this table. Cost of production studies for Idaho were funded by the Idaho potato commission. Cost of production studies for Colorado and Wisconsin were funded by united potato growers of America. Cost of production studies were conducted by ben Eborn, extension agricultural economist, University of Idaho

Colorado, Idaho, Washington and Wisconsin. While the values in any given cost category vary considerably between states, the overall cost per cwt on a paid-yield basis shows less variability. The high cost of potato production should encourage every grower to calculate their individual production costs so that their management and marketing decisions are based on reality.

Acknowledgement Unless otherwise noted, data were adapted from collections of University of Idaho Extension educators, scientists, and researchers, who wrote the chapters of the first edition of this textbook.

References

AAEA (1998) American agricultural economics task force on commodity costs and returns commodity, costs and returns estimation handbook. USDA, Ames

Patterson PE, King BA, Smathers RL (1996) Economics of low-pressure sprinkler irrigation systems: center-pivot and linear move. Bulletin 787. University of Idaho Extension, Moscow

Chapter 20
Potato Nutrition

Martha A. Raidl

Contents

Introduction

Potatoes are the most popular vegetable consumed by Americans. They contain carbohydrates, protein, fiber, numerous vitamins and minerals, and are a heart healthy (no saturated fat, trans fat, or cholesterol) food. This chapter covers per-capita consumption of potatoes in the U.S. and the nutrition profile of potato products.

Potato Consumption

Potatoes are the most popular vegetable consumed by Americans. Figure 20.1 shows that over the last 40+ years, potato consumption has varied, averaging between 110 and 130 lbs/year. From 1970 to 1990, potato consumption was approximately 120 lbs/person; it increased to approximately 130 lbs/person from 1991 to 2004 (peaking in 1996 at 145 lbs/person). It then dropped back to 120 lbs from 2005 to 2008, and further decreased to approximately 110 lbs from 2009 to 2014.

M. A. Raidl (✉)
University of Idaho, Moscow, ID, USA
e-mail: mraidl@uidaho.edu

J. C. Stark et al. (eds.), *Potato Production Systems*,
https://doi.org/10.1007/978-3-030-39157-7_20

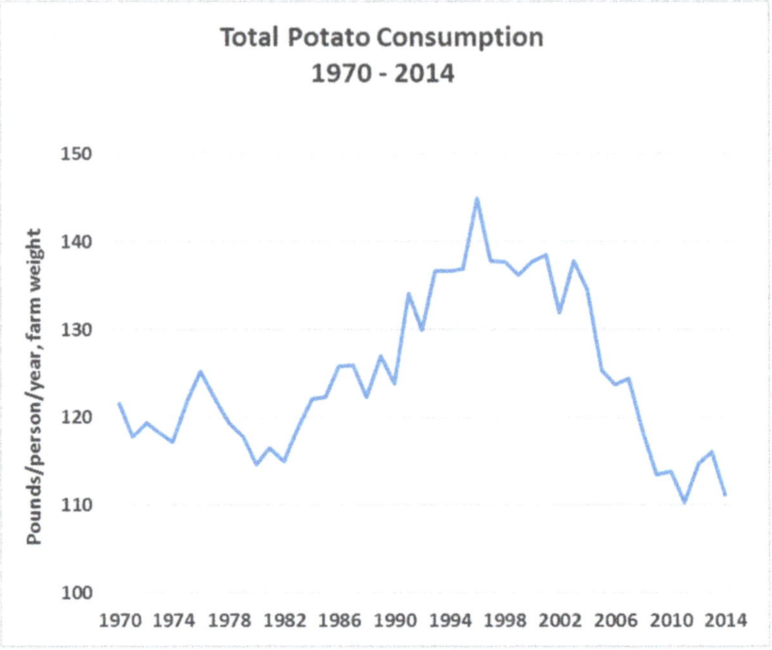

Fig. 20.1 Total U.S. potato consumption 1970–2014. Source: USDA, ERS 2015. http://www.ers.usda.gov/datafiles/Vegetable_and_Pulses_Yearbook_Tables/General/YRBK2015_Section%201_General.pdf

Decrease in Potato Consumption

Potatoes are composed mainly of carbohydrate, and the decrease in potato consumption may be related to the low carbohydrate diet weight loss programs that became extremely popular in the late 1990s and early 2000s. Studies show that a low carbohydrate diet results in a short-term weight loss, but individuals on these diets eventually regain the weight.

Nutrient Profile

The nutrient profile of potatoes is based on one serving and includes their calories, as well as macronutrient and micronutrient content.

There are numerous varieties of potatoes in the U.S., which can vary in their nutritional content. However, nutrition labeling is not provided based on variety, but is based on an average established by the U.S. Food and Drug Administration (FDA). Therefore, in this chapter, the macro- and micronutrient content is based on FDA data that is used in food labeling for one serving of potatoes which is 5.3 oz or 148 g.

Serving Size

Two ways to determine the serving size of potatoes is by weight and where they fit on MyPlate (Fig. 20.2). MyPlate, an icon chosen by the U.S. Department of Agriculture to help individuals eat healthy, illustrates the five food groups: fruits, vegetables, grains, protein, and dairy, using a familiar place setting icon. On MyPlate, potatoes fit into the vegetable portion of a healthy plate.

Potatoes are categorized as starchy vegetables. Two popular potato products, baked and fried, vary in their serving size. One serving of baked potato weighs 148 g or 5.3 oz, which is approximately 1 cup of cooked potato. Serving size for french fries, by weight, is 70 g prepared and 85 g frozen.

It is recommended that adults consume between 2 and 3 cups of vegetables daily and 14–21 cups weekly. On a weekly basis, adults should consume between 5 and 6 cups of starchy vegetables. Depending on how the potato is prepared, 1 cup of potatoes could be:

- 1 cup diced or mashed potatoes.
- 1 medium boiled or baked potato that is between 2 1/2–3″ in diameter.
- Approximately 20 medium fries that would be 2 1/2–4″ long.

Calories: The 40, 100, and 400 Rule

When checking a food's calorie content, keep the 40, 100, and 400 rule in mind: 40 calories per serving is low, 100 calories per serving is moderate, and 400 calories per serving is high.

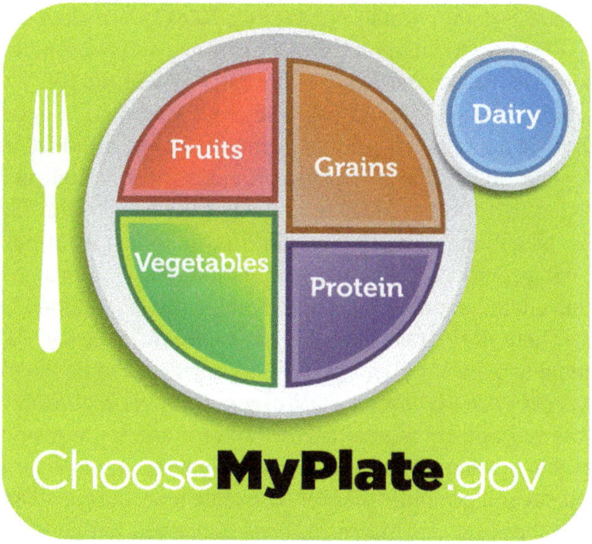

Fig. 20.2 MyPlate. (Source: US Department of Agriculture 2015. http://www.choosemyplate.gov/vegetables)

Table 20.1 Calorie comparison between small, medium, and large servings of baked and fried potatoes

Method of preparation	Calories per 100 g	Small serving calories	Medium serving calories	Large serving calories
Baked	97	134	168	290
French fries	323	229	378	495

Source: USDA, ARS, 2015 National Nutrient Database https://ndb.nal.usda.gov/

Table 20.1 shows how the calorie content of potatoes varies by size and method of preparation. Depending on the size of a plain baked potato, the calories vary, from 134 calories (potato small, 1 3/4–2 1/2″ diameter) to 168 calories (potato medium, 2 1/4–3 1/4″ diameter), to 290 calories (potato large, 3–4 1/4″ diameter). Frying potatoes significantly increases their calorie content. When comparing the calories per 100 g, fried potatoes are 2 and 1/3 times higher in calories than baked potatoes. A small serving of french fries contains 229 calories, a medium serving has 378 calories, and a large serving has almost 500 calories.

Based on the 40, 100, and 400 rule, a large serving of baked potato, at 290 calories, would not be considered high in calories, where as a large serving of french fries, at 495 calories, would be considered high in calories.

Macronutrients, Micronutrients, and Percent Daily Value

Macronutrients are defined as substances that are needed by the human body in relatively large quantities. Carbohydrates, protein, fat, and fiber are macronutrients found in potatoes. Micronutrients are substances that are required by the human body in smaller quantities and include vitamins, minerals, and phytochemicals. Each nutrient has a recommended daily value (RDV) or amount that individuals should consume on a daily basis. For example, the RDV for fiber is 25 g. The RDV is based on individuals consuming 2000 calories per day.

One way to interpret the macronutrient and micronutrient content of a food product is to use the Percent Daily Value (%DV). Use the %DV to choose foods that are high in nutrients you should increase (e.g., vitamins, minerals, fiber), and limit or avoid those in nutrients you should decrease (fat, saturated fat, trans fat, cholesterol, sodium).

The %DV is a guide that provides information on how much of each nutrient found in a food product is contained in one serving. This information is located on the Nutrition Facts label, which lists nutrition information and is required on most packaged foods. Food companies use %DV to promote a product as a "good" or "excellent" source of a specific nutrient; e.g., a "good" source of vitamin C.

There is a "low," "good," and "excellent" rule for %DV. If a food has less than or equal to 5%DV of a nutrient, it is considered to be "low" in that nutrient. If the food has between 10 and 19%DV of a nutrient, it is considered to be a "good" source of that nutrient. If it contains 20%DV or more, it is considered to an "excellent" source of that nutrient. The FDA has not proposed a %DV for three macronutrients—trans fat, sugars, protein—and for micronutrients classified as phytochemicals. Table 20.2 lists the %DV for macronutrients contained in one serving of potato.

Table 20.2 Macronutrient content and %DV in one serving of potatoes

Macronutrient	Amount per serving	%DV
Carbohydrate	26 g	9
Dietary fiber	2 g	8
Sugars	1 g	No DV established
Protein	3 g	No DV established
Total fat	0 g	0
Saturated fat	0 g	0
Trans fat	0 g	No DV established
Cholesterol	0 mg	0

Source: United States Department of Agriculture, Agricultural Research Service (USDA, ARS). National Nutrient Database for Standard Reference Release 28, Basic Report: 11352, Potatoes, flesh and skin, raw Available at: https://ndb.nal.usda.gov/ndb/foods/show/3080?manu=&fgcd=

Macronutrients

Potatoes contain 26 g of carbohydrate, 3 g of protein, and 0 g of fat per serving. Thus, most of the potato is composed of carbohydrate. But this does not mean that potatoes are too high in carbohydrate, since one serving of potato contains just 9%DV for carbohydrate. Individuals following a 2000 calorie diet should consume approximately 300 grams of carbohydrate daily.

Carbohydrate: Starches, Sugar, and Fiber

Potatoes contain three types of carbohydrate—starches, fiber, and sugar. One serving of potatoes contains 26 g of carbohydrate or starch, 2 g of dietary fiber, and 1 g of sugar.

Starch: Amylose and Amylopectin

As stated earlier, potatoes are composed mainly of carbohydrate or starch. There are two types of molecules found in potato starch, amylose and amylopectin. On average, potatoes contain 20% amylose and 80% amylopectin, but this ratio can vary significantly by variety. As shown in Fig. 20.3, amylose and amylopectin molecules are composed of repeating D-glucose units and differ in their linkages, which results in either a linear or branched molecule. Amylose is linear and its D-glucose units are linked by alpha 1,4 bonds. Amylopectin is formed from the amylose base polymers, and its D-glucose units randomly branch through its alpha 1,6 bonds.

Amylose and amylopectin differ in size and surface area. The linear amylose molecule contains between 500 and 2000 glucose units. The larger amylopectin molecule contains from 1000 to 10,000 glucose units, which assemble as short,

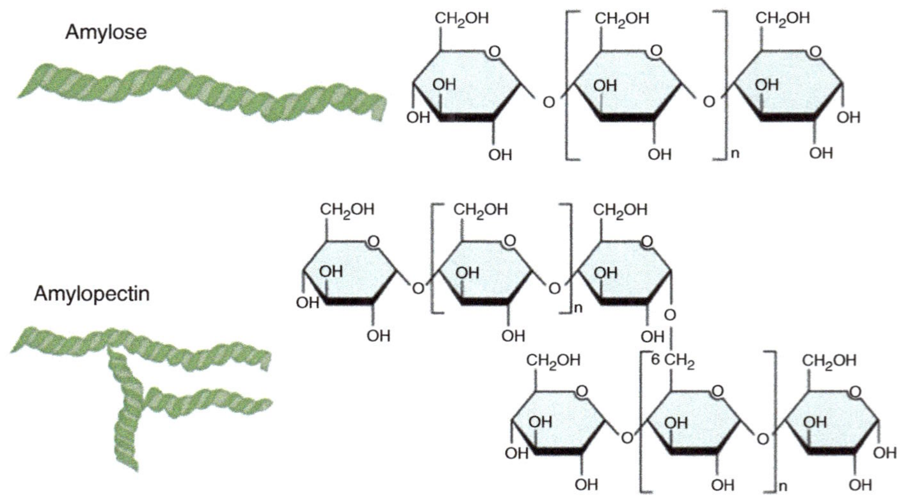

Fig. 20.3 Amylose and amylopectin molecules. (Source: https://online.science.psu.edu/chem005_wd/node/7882)

highly branched chains. Amylose has a smaller surface area because its glucose units are tightly packed together; this makes it difficult for the enzymes in the small intestine to reach and breakdown all of the glucose units. As a result, some of the undigested amylose molecules reach the large intestine and are called ‘resistant starch.’ Amylopectin is more easily digested because the alpha 1,6 linkages create a larger surface area making it more accessible by enzymes.

Resistant Starch

Resistant starch (RS) is undigested amylose that reaches the large intestine and is fermented by bacteria (that reside in the large intestine) to produce short chain fatty acids (SCFA). These SCFA exert potential effects on gut microbiota, which may play a role in prevention and control of colon cancer, diabetes, and obesity.

The amount of resistant starch found in potato products is small; 100 g of hot potato (baked, boiled, mashed, instant, fried) contains 0.59 g of resistant starch. This increases to 1.2 g if the products are cooled and then reheated.

Dietary Fiber

Dietary fiber, also called roughage or bulk, is found in plants such as fruits, vegetables, whole grains, and legumes. It is the part of the plant that the body cannot digest, traveling through the intestinal tract and exiting the body relatively intact.

One serving of potatoes contains approximately 2 g of fiber and meets 8%DV for fiber. Based on a 2000 calorie diet, an individual should consume 25 g of fiber daily. There are two types of fiber, soluble and insoluble. In potatoes, 74% of the fiber is insoluble and found in the peel; 26% is soluble and found in the flesh.

Both types of fiber have important roles in the body. Insoluble fiber binds to water and helps move food through the digestive tract easier. It makes stools softer, larger, and easier to pass. Soluble fiber dissolves in water and forms a gel-like substance, which decreases absorption of cholesterol and keeps blood glucose levels stable.

Health benefits of consuming soluble and insoluble fiber include lowering the risk of certain cancers, cardiovascular disease, and diabetes. It also helps in weight management since fiber helps fill people up, so they consume fewer calories.

Sugars

Sugars are classified as simple carbohydrates. This means they are composed of 1 or 2 sugar molecules, break down easily, and serve as a quick source of energy. A 5.3 oz serving of potatoes contains approximately 1 g of naturally occurring sugar.

Protein

One serving of potatoes contains 3 g of protein. The recommended daily intake (RDI) of protein varies by age and gender. Adults following a 2000 calorie diet should consume 50 g of protein daily. Even though the amount of protein in a potato is minimal, it is considered high quality, based on its essential amino acid content and digestibility. Potatoes contain four essential amino acids (lysine, methionine, threonine, and tryptophan), out of the nine essential amino acids, which the body cannot manufacture. In addition, the protein contained in potatoes is easily digested.

Fat: Total Fat, Saturated Fat, Trans Fat, and Cholesterol

The most recent Dietary Guidelines for Americans (DGA) recommends that individuals limit their intake of saturated fat to <10% of their calorie intake, and keep their intake of trans fat and cholesterol as low as possible. As Table 20.2 shows, potatoes are fat free and, therefore, have 0 g of saturated fat and trans fat, both of which are related to development of cardiovascular disease. Since potatoes contain no saturated fat, trans fat, or cholesterol, they are considered heart healthy. In fact, Idaho potatoes have been certified by the American Heart Association as being low in saturated fat and cholesterol (Fig. 20.4).

Fig. 20.4 Idaho potatoes are certified by the American Heart Association. (Reproduced from Idahopotato.com, 2011). https://idahopotato.com/pressroom/american-heart-association-certifies-fresh-idaho-potatoes-as-a-heart-healthy-food

Micronutrients

There are three categories of micronutrients contained in potatoes—vitamins, minerals, and phytochemicals. Micronutrients are nutrients that are required by the human body in small quantities. Potatoes contain 13 micronutrients—six vitamins and seven minerals. The six vitamins are thiamin, niacin, riboflavin, ascorbic acid, pyridoxine, and folate. The seven minerals are calcium, iron, potassium, zinc, phosphorous, magnesium, and copper. Table 20.3 shows the %DV of the vitamins and minerals contained in 1 serving (5.3 oz) of potato.

Using this rule, 5.3 oz of potatoes contains ≤5% DV and, therefore, low in three minerals (calcium, zinc, copper) and one vitamin (riboflavin). It is a good source of vitamin B6 (10%DV) and potassium (18%DV), and is an excellent source of vitamin C (45%DV). There is no category for %DV for nutrients that are between 6 and 9%DV, which includes thiamin and niacin at 8%DV and iron, folate, phosphorous, and magnesium, at 6%DV.

Even though the %DV of some of the nutrients in potatoes is low, it is important to remember that all of the nutrients found in potatoes help keep the body healthy and have specific functions, as listed in Table 20.4. Examples of how your body uses these vitamins and minerals include: (1) thiamin, niacin, and riboflavin to get energy from food; (2) potassium and magnesium to help your heart contract and relax; (3) zinc and vitamin C to help with wound healing; (4) iron, copper, folate, and vitamin B6 to maintain red blood cells; and (5) calcium and phosphorous to keep your bones strong.

Table 20.3 Micronutrient content, %DV, and rating in one serving of potatoes

Micronutrient	%DV	Rating
Vitamin		
Vitamin C	45	Excellent
Vitamin B6	10	Good
Thiamin	8	No rating established
Niacin	8	No rating established
Folate	6	No rating established
Riboflavin	2	Low
Mineral		
Potassium	18	Good
Phosphorous	6	No rating established
Magnesium	6	No rating established
Iron	6	No rating established
Copper	4	Low
Zinc	2	Low
Calcium	2	Low

Source: USFDA, 2017. http://www.fda.gov/Food/IngredientsPackagingLabeling/LabelingNutrition/ucm114222.htm http://www.fda.gov/downloads/Food/GuidanceRegulation/ucm063477.pdf

Table 20.4 Functions of vitamins and minerals contained in potatoes

Nutrients	Functions
Vitamin	
Thiamin (B1)	Assists the release of energy from carbohydrates and protein
Niacin (B3)	Assists the release of energy from fat, carbohydrates, and protein
Riboflavin (B2)	Assists the release of energy from fat, carbohydrates, and protein. Assists several antioxidant enzymes
C (ascorbic acid)	Antioxidant in blood and cells Augments functional activity of immune cells Assists collagen, carnitine, serotonin, and adrenaline production
Pyridoxine (B6)	Supports a wide variety of metabolic reactions Assists neurotransmitters, hemoglobin, and DNA production Influences steroid hormone action
Folate	Required for DNA synthesis Assists red blood cell production Prevents neural tube defects
Minerals	
Calcium	Structural component of bones and teeth Required for proper nerve transmission and muscle contraction Influences blood vessel construction and dilation; may reduce blood pressure
Iron	Component of hundreds of enzymes Needed for synthesis of hemoglobin Assists antioxidant enzymes Required for synthesis of DNA, amino acids, collagen, neurotransmitters, and certain hormones Critical for normal immune function
Potassium	Maintains fluid and electrolyte balance Required for proper nerve conduction and muscle contraction Lowers blood pressure
Zinc	Assists in hundreds of enzyme reactions Assists in hemoglobin production Assists antioxidant enzymes Supports immune function
Phosphorous	Structural component of bones and teeth Structural component of DNA Structural component of cell membranes Assists in energy production and storage
Magnesium	Structural component of bones Assists in hundreds of enzyme reactions involved in the synthesis of DNA and proteins Required for proper nerve conduction and muscle contraction
Copper	Assists in energy production and iron utilization Assists in neurotransmitter synthesis Maintains integrity of connective tissue Assists antioxidant enzymes

Source: Oregon State University. http://lpi.oregonstate.edu/mic

Phytochemicals

Phytochemicals are compounds that come from plants and are thought to decrease likelihood of developing heart disease and cancer. Thousands of phytochemicals have been identified in fruits, vegetables, beans, cereals, and plant-based beverages. Many consumers associate phytochemicals with brightly colored fruits and vegetables or green ones, like broccoli—but not potatoes. In 2007, researchers identified over 60 different phytochemicals in potatoes.

Three of the phytochemicals found in potatoes that have been studied include phenolics, flavonoids, and carotenoids. Phenolics are found in high concentrations in the peel and flesh of potatoes, with more in the peel. Potatoes have a better phenolic content than spinach, Brussels sprouts, or broccoli. In fact, they are the most important source of phenols in the diet, after apples and oranges. Flavonoid content depends on the color of the potato, with more being present in red skinned and purple potatoes. An example of a flavonoid in red and purple potatoes is anthocyanin. Researchers found that subjects who ate purple potatoes for 1 month lowered their systolic blood pressure by 3.5% and systolic by 4.3%. Carotenoids are highest in yellow potatoes. They are fat soluble phytochemicals that have strong antioxidant properties.

In summary, potatoes are an important component of a healthy diet. They contain nutrients that keep the body functioning well.

Further Reading

Chapter 1: A Short History of Potato Production Systems

Erickson CL (1988) Raised field agriculture in the Lake Titicaca Basin: putting ancient agriculture back to work. Expedition 30:8–16

FAO (2002) Ingenious: southern Peru's *waru-waru* farming system. Agricultural heritage systems. Spotlight/2002. Available at: http://www.fao.org/ag/magazine/0211sp1.htm

Fournet A, Rojas de Arias A, Charles B, Bruneton J (1996) Chemical constituents of essential oils of Muña, Bolivian plants traditionally used as pesticides, and their insecticidal properties against Chagas disease vectors. J Ethnopharmacol 52(3):145–149

Graves C (ed) (2001) The potato: treasure of the Andes. International Potato Center (CIP), Lima, Peru

Guardia SB (2013) Origins and history: the potato in the ancestral memory of Perú. Universidad de San Martín de Porres, Peru. Available at: http://www.kennesaw.edu/chs/CUALLI%20Journal/NGuardia103.pdf

Guenthner JF (ed) (1993) Past, present, and future uses of potatoes. In: Proceedings of the symposium, July 1992. Potato Association of America 76th annual meeting, New Brunswick, Canada. University of Idaho misc. bulletin 164, Moscow, ID

Hawkes JG (1966) The history of the potato. J R Horticul Soc 92:207–302

Hawkes JG (1990) The potato: evolution, biodiversity, and genetic resources. Smithsonian Institution Press, Washington, DC

Hawkes JG, Francisco-Ortega J (1993) The early history of the potato in Europe. Euphytica 70:1–7

Manrique K, Egusquiza R (2009) Muña (s. *Minthostachis mollis*) as a natural alternative to control potato sprouting tested under different storage conditions. INCOPA project/papa andina initiative. International Potato Center. 15th Triennial Symposium of the International Society for Tropical Root Crops (ISTRC). Lima, Peru, 2–6 November 2009

Messer E (2000) Potatoes (white). In: Kiple KF, Coneè K (eds) The Cambridge world history of food. Cambridge University Press, Cambridge, UK

Yamamoto N (1982) A food production system in the southern central Andes. Senri Ethnological Studies 10:39–62

J. C. Stark et al. (eds.), *Potato Production Systems*,
https://doi.org/10.1007/978-3-030-39157-7

Chapter 2: Potato Growth and Development

Dean BB (1994) Managing the potato production system. Haworth, Binghamton, NY
Struik PC (2007) Above-ground and below-ground plant development. In: Potato biology and biotechnology: advances and perspectives. Elsevier, Oxford, UK, pp 501–524
Wohleb CN, Knowles NR, Pavek MJ (2014) Plant growth and development. In: Navarre R, Pavek MJ (eds) The potato: botany, production and uses. CAB International, Boston, MA, pp 64–82

Chapter 3: Variety Selection and Management

Johansen RH, Sandar N, Hoyman WG, Lana EP (1959) Norland a new red-skinned potato variety with early maturity and moderate resistance to common scab. Am Potato J 36:12–15
Johansen RH, Farnsworth B, Nelson DC, Secor GA, Gudmestad N, Orr PH (1988) Russet Norkotah: a new russet-skinned potato cultivar with wide adaptation. Am Potato J 65:597–604
Johansen RH, Farnsworth B, Secor GA, Gudmestad N, Thompson-Johns A, Holm ET (1994) Goldrush: a new high quality russet-skinned potato cultivar. Am Potato J 71:809–815
Johnston GR, Rowberry RG (1981) Yukon Gold: a new yellow-fleshed, medium-early, high quality table and french-fry cultivar. Am Potato J 58:241–244
Mosley AR, James SR, Hane DC, Rykbost KA, Shock CC, Pavek JJ, Corsini DL, Love SL, Thornton RE (2000) Umatilla Russet: a full season long russet for processing and fresh market use. Am J Potato Res 77:83–87
Novy RG, Corsini DL, Love SL, Pavek JJ, Mosley AR, James SR, Hane DC, Shock CC, Rykbost KA, Brown CR, Thornton RE (2003) Alturas: a multi-purpose russet potato cultivar with high yield and tuber specific gravity. Am J Potato Res 80:295–301
Novy RG, Whitworth JL, Stark JC, Love SL, Corsini DL, Pavek JJ, Vales MI, James SR, Hane DC, Shock CC, Charlton BA, Brown CR, Knowles NR, Pavek MJ, Brandt TL, Gupta S, Olsen N (2010) Clearwater Russet: a dual-purpose potato cultivar with cold sweetening resistance, high protein content, and low incidence of external defects and sugar ends. Am J Potato Res 87:458–471
Novy RG, Whitworth JL, Stark JC, Charlton BA, Yilma S, Knowles NR, Pavek MJ, Spear RR, Brandt TL, Olsen N, Thornton M, Brown CR, James SR, Hane DC (2014) Teton Russet: an early-maturing, dual-purpose potato cultivar having higher protein and vitamin C content, low asparagine, and resistances to common scab and fusarium dry rot. Am J Potato Res 91:380–393
Pavek JJ, Corsini DL, Love SL, Hane DC, Holm DG, Iritan WM, James SR, Martin MW, Mosley AR, Ojala JC, Stanger CE, Thornton RE (1992) Ranger Russet: a long russet potato variety for processing and fresh market with improved quality, disease resistance, and yield. Am Potato J 69:483–488
Stark JC, Novy RG, Love SL, Whitworth JL, Corsini DL, Pavek JJ, Mosley AR, Pavek MJ, Knowles NR, Thornton RE, James SR, Hane DC, Olsen N, Vales I, Brown C (2007) Blazer Russet: an early to mid-season potato cultivar with high U.S. No. 1 yields and good processing and culinary qualities. Am J Potato Res 84:447–457
Thompson AL, Novy RG, Farnsworth BL, Secor GA, Gudmestad NC, Sowokinos JR, Holm ET, Lorenzen JH, Preston D (2005) Dakota Pearl: an attractive, bright white-skinned, cold-chipping cultivar with tablestock potential. Am J Potato Res 82:481–488
Webb RE, Miller JC (1954) Red LaSoda: a mutation of LaSoda. Am Potato J 31:40–43
Webb RE, Wilson DR, Shumaker JR, Graves B, Henninger MR, Watts J, Frank JA, Murphy HJ (1978) Atlantic: a new potato variety with high solids, good processing quality and resistance to pests. Am Potato J 55:141–145
Weigle JL, Kehr AE, Akeley RV, Horton JC (1968) Chieftain: a red-skinned potato with attractive appearance and broad adaptability. Am Potato J 45:293–296

Whitworth JL, Novy RG, Stark JC, Pavek JJ, Corsini DL, Vales MI, Mosley AR, James SR, Hane DC, Shock CC, Charlton BA, Knowles NR, Pavek MJ (2010) Yukon Gem: a yellow-fleshed potato cultivar suitable for fresh-pack and processing with resistances to PVYO and late blight. Am J Potato Res 87:327–336

Young DA, Tarn TR, Davies HT (1983) Shepody: a long, smooth, white-skinned potato of medium maturity with excellent French fry quality. Am Potato J 60:109–114

Chapter 4: Nonapplicable

None

Chapter 5: Field Selection, Crop Rotations, and Soil Management

Kotuby-Amacher J, Koenig R, Kitchen B (1997) Salinity and plant tolerance. Utah State University bull AG–SO–03

Richards LA (ed) (1969) Diagnosis and improvement of saline and alkali soils. USDA handbook 60

Thornton M, Stark J, Hopkins B, Thornton RE (2008) Selecting and preparing the planting site. In: Johnson DA (ed) Potato health management, 2nd edn. APS, St. Paul, MN, pp 23–30

Van Schilfgaarde J (ed) (1974) Drainage for agriculture. Agronomy monograph, series 17. American society of agronomy, Madison, WI

Chapter 6: Organic Potato Production

Andrews N, Sullivan D, Jensen J (2010) OSU organic fertilizer and cover crop calculator. Oregon State University Extension Service. https://extension.oregonstate.edu/organic-fertilizer-cover-crop-calculators. Accessed 20 Apr 2019

Collins D, Miles C, Cogger C, Koenig R (2013) Soil fertility in organic systems: a guide for gardeners and small acreage farmers PNW646. Washington State University, Pullman, WA

eOrganic website. eroganic.org. Hosted through Oregon State University, Corvallis, OR. Accessed 20 Apr 2019

Frazier MJ, Olsen N, Kleinkopf F (2004) Organic and alternative methods for potato sprout control in storage CIS 1120. University of Idaho, Moscow, ID

Greenway GA, Guenthner JF, Makus LD, Pavek MJ (2011) An analysis of organic potato demand in the U.S. Am J Potato Res 88:184–189

Hart J, Sullivan D, Meyers J, Peachey R (2010) Sweet corn (Western Oregon) EM 9010–E. Oregon State University, Corvallis, OR

Hunter L, Falen C, Moore A (2014) Cover crops for high-desert farming systems in Idaho, BUL 889. University of Idaho, Moscow, ID

Johnson, SB, Sideman E (2006) Producing potatoes organically in Maine, BUL 2419. University of Maine Cooperative Extension, Orono, ME

Moore A, Hunter L, Falen C, Shewmaker G (2013) University of Idaho cover crop calculator. University of Idaho Extension, Moscow, ID. http://www.extension.uidaho.edu/nutrient/CC_Calculator/Cover_Crop_Main_page.htm. Accessed 20 Apr 2019

Moore A, Olsen N (2010) Organic potato production and storage. In: Proceedings of the 2010 University of Idaho Winter Commodity Schools, Moscow, ID

Moore A, Olsen N, Frazier M, Carey A (2011) Organic potato production: nitrogen management and variety trials. In: Proceedings of the 2011 University of Idaho Winter Commodity Schools, Moscow, ID, pp 67–71

Moore A, Olsen N, Satterwhite M, Frazier M (2013) Organic potato production in Idaho: nutrient management and variety selection, BUL 885. University of Idaho, Moscow, ID

Murray T, Miles C, Daniels C (2013) Natural insecticides, PNW 649. Washington State University, Pullman, WA

Pscheidt JW, Ocamb CM (sr eds) (2015) Pacific Northwest plant disease management handbook. Oregon State University, Corvallis, OR. http://pnwhandbooks.org/plantdisease. Accessed 20 Apr 2019

Rosen C, Allen D (2007) Exploring the benefits of organic nutrient sources for crop production and soil quality. HortTech 17:422–430

Stone A (2014) Organic management of late blight of potato and tomato (*Phytopthora infestans*). eXtension website article. http://www.extension.org/pages/18361/organic-management-of-late-blight-of-potato-and-tomato-phytophthora-infestans#.VTVgipNBIFs. Accessed 20 Apr 2019

Sullivan D, Andrews N, Sullivan C, Brewer L (2019) OSU Organic fertilizer and cover crop calculator: predicting plant-available nitrogen, EM-9235. Oregon State University, Corvallis, OR

Sullivan D, McQueen J, Horneck D (2008) Estimating nitrogen mineralization in organic potato production, EM 8949-E. Oregon State University, Corvallis, OR

Thurston HD (2004) Ten thousand years of experience with sustainable plant disease control. Plant Dis 88:550–551

Weinert TL, Pan WL, Moneymaker MR et al (2002) Nitrogen recycling by nonleguminous winter cover crops to reduce leaching in potato rotations. Agron J 94:365–372

Chapter 7: Seed and Planting Management

Bohl WH, Nolte P, Thornton MK (1992) Potato seed management: seed certification and selection. CIS 974, University of Idaho, Moscow, ID

Bohl WH, Nolte P, Kleinkopf GE et al (1995) Potato seed management: seed size and age. CIS 1031, University of Idaho, Moscow, ID

Bohl WH, Love SL, Thompson AL (2001) Effect of seed piece removal on yield and agronomic characteristics of Russet Burbank potatoes. Am J Potato Res 78:397–402

Commercial Potato Production in North America (2010) In: Bohl WH, Johnson SB (eds) The Potato Association of America Handbook, pp 38–52. http://potatoassociation.org/documents/A_ProductionHandbook_Final-1.2012.pdf

Rykbost KA, Locke KA (1999) Effect of seed piece size on performance of three potato cultivars in the Klamath Basin of Oregon. Am J Potato Res 76:75–82

Schotzko TR, Iritani WM, Thornton RE (1984) The economics of Russet Burbank seed size and spacing. Am J Potato Res 61:57–66

Strand LL, Rude PA (2006) Integrated pest management for potatoes in the western United States, second edition. Agriculture and Natural Resources Publication 3316, University of California, Oakland, CA

Thornton RE, Hyde GM, Zhoa K (1986) Addressing the potato stand problems in the West. In: Proceedings of the North American seed seminar, National Potato Council, Denver, CO, pp 104–123

Chapter 8: Nutrient Management

Hopkins BG, Hirnyck RE (2007) Organic potato production. In: Johnson DA (ed) Potato health management. American Phytopathological Society, Minneapolis, MN, pp 101–108

Hopkins BG, Stark JC (2010) Potato nutrition. In: Bohl WH, Johnson SB (eds) Commercial potato production in North America, second rev of American Potato J Supplement, volume 57, and USDA Handbook 267. Potato Association of America, Orono, ME, pp 57–62

Hopkins BG, Horneck DA, Pavek MJ, Geary BD, Olsen NL, Ellsworth JW, Newberry GD, Miller JS, Thornton RE, Harding GW (2007) Evaluation of potato production best management practices. Am J Potato Res 84:19–27

Hopkins BG, Horneck DA, Stevens RG, Ellsworth JW, Sullivan DM (2007) Managing irrigation water quality for crop production in the Pacific Northwest, PNW 597-E. Oregon State University, Corvallis, OR

Horneck D, Rosen C (2008) Measuring nutrient accumulation rates of potatoes—tools for better management. Better Crops 92(1):4–6

Horneck DA, Ellsworth JW, Hopkins BG, Sullivan DM, Stevens RG (2007) Managing salt affected soils for crop production, PNW 601-E. Oregon State University, Corvallis, OR

Miller JS, Hopkins BG (2007) Checklist for a holistic potato health management plan. In: Johnson DA (ed) Potato health management. American Phytopathological Society, Minneapolis, MN, pp 7–10

Walworth JL, Muniz JE (1993) A compendium of tissue nutrient concentrations for field-grown potatoes. Am J Potato Res 70(8):579–597

Zebarth BJ, Rosen CJ (2007) Research perspective on nitrogen BMP development for potato. Am J Potato Res 84(1):3–18

Chapter 9: Disease Management

Agrios GA (2004) Plant pathology, 5th edn. Elsevier Academic Press, Burlington, MA

Banks E (ed) (2004) Potato field guide: insects, diseases and defects. Publication 823, Ontario Ministry of Agriculture and Food, Guelph, ON

Banttari EE, Ellis PJ, Khurana SMP (1993) Management of diseases caused by viruses and virus-like pathogens. In: Rowe RC (ed) Potato health management. American Phytopathological Society, St. Paul, MN, pp 127–134

Fry WE (1982) Principles of plant disease management. Academic Press, New York, NY

Gudmestad NC, Secor GA (1993) Management of soft rot and ring rot. In: Rowe RC (ed) Potato health management. American Phytopathological Society, St. Paul, MN

Maloy OC (1993) Plant disease control: principles and practice. Wiley, New York, NY

Powelson ML, Johnson KB, Rowe RC (1993) Management of diseases caused by soilborne pathogens. In: Rowe RC (ed) Potato health management. American Phytopathological Society, St. Paul, MN

Stevenson WR (1993) Management of early blight and late blight. In: Rowe RC (ed) Potato health management. American Phytopathological Society, St. Paul, MN

Stevenson WR, Loria R, Franc GD et al (2001) Compendium of potato diseases, 2nd edn. American Phytopathological Society, St. Paul, MN

Wale S, Platt HW, Cattlin N (2008) Diseases, pests, and disorders of potatoes. Academic Press, Boston, MA

Zehnder GW, Powelson ML, Jansson RK et al (eds) (1994) Advances in potato pest biology and management. American Phytopathology Society, St. Paul, MN

Chapter 10: Nematode Management

Faulkner LR, Darling HM (1961) Pathological histology, hosts and culture of the potato rot nematode. Phytopathology 51:778–786

Fourie H, Ahuja P, Lammers J, Daneel M (2016) Brassicacea-based management strategies as an alternative to combat nematode pests: a synopsis. Crop Prot 80:21–41

Hafez SL, Sundararaj P (2001) Chemical nematicides for the suppression of *Meloidogyne chitwoodi* and *M. hapla* on potato. Int J Nematol 11:192–194

Hafez SL, Thornton MK (1992) Potato nematodes and their control. CIS 925. University of Idaho, Moscow, ID

Hafez SL, Ojala JC, Mohan SK (1990) The potato-rot nematode. CIS 868. University of Idaho, Moscow, ID

Hafez SL, Davis JR, Love SL, Ojala JC (1992) Corky ringspot of potatoes. CIS 914. University of Idaho, Moscow, ID

King BA, Teberna JP Jr (2013) Site specific management of *Meloidogyne chitwoodi* in Idaho potatoes using 1,3–dicholoropropene: approach, experiences and economics. J Nematol 45: 202–213

MacGuidwin AE, Slack SA (1991) Suitability of alfalfa, corn, oat, red clover, and snapbean as hosts for the potato rot nematode, *Ditylenchus destructor*. Plant Dis 75:37–39

MacGuidwin AE, Wixted DJ, Hudelson BD (1992) Aboveground infection of snap bean by *Ditylenchus destructor*, the potato rot nematode. Plant Dis 76:1097–1102

Mwaura P, Niere B, Vidal S (2014) Resistance and tolerance of potato varieties to potato rot nematode (*Ditylenchus destructor*) and stem nematode (*Ditylenchus dipsaci*). Ann Appl Biol 166:1–14

Riga E (2011) The effects of Brassica green manures on plant parasitic and free living nematodes used in combination with reduced rates of synthetic nematicides. J Nematol 43:118–120

Smart GC Jr (1959) *Ditylenchus destructor* from grass, dahlia and gladiolus infecting potato tubers. Plant Dis Rep 43:1212

Thorne G (1945) *Ditylenchus destructor*, n. sp., the potato rot nematode, and *Ditylenchus dipsaci* (Kühn, 1857) Filipjev, 1936, the teasel nematode (Nematoda: *Tylenchidae*). Proc Helminthol Soc Wash 12:27–34

Chapter 11: Insect Pests and Their Management

Alvarez JM, Srinivasan R (2005) Evaluation of hairy nightshade as an inoculum source for the aphid-mediated transmission of potato leafroll virus. J Econ Entomol 98:1101–1108

Alyokhin A (2008) Colorado potato beetle biology and management. http://www.potatobeetle.org/

Blodgett S, Denke PM, Knerr V (2010) Blister beetles of Montana. MontGuide, Montana State University Extension. MT200209AG

Blua MJ, Rondon SI, Jensen A, Bell N (2018) Irish potato pests. In: 2018 Pacific Northwest insect management handbook. Oregon State University, Corvallis, OR

Dicklow MB, McKeag L (eds) (2015) 2016–2017 New England vegetable management guide. University of Massachusetts-Amherst, Amherst, MA

Egel DS (2016) Potato insect control. In: Midwest vegetable production guide for commercial growers 2016. Purdue University, West Lafayette, IN

Giordanengo P, Vincent C, Alyokhin A (eds) (2013) Insect pests of potato: global perspectives on biology and management. Academic Press, Oxford, pp 11–29

Hoy CW, Boiteau G, Alyokhin A, Dively G, Alvarez JM (2008) Managing insect and mite pests. In: Johnson DA (ed) Potato health management. The American Phytopathological Society Press, St. Paul, MN, pp 133–147

IRAC International (2018) Insecticide resistance action committee. http://www.irac-online.org/

Kuhar T, Alvarez JM (2008) Timing of injury and efficacy of soil-applied insecticides against wireworms on potato in Virginia. Crop Prot 27:792–798

Mac Gillivray ME (1979) Aphids infesting potatoes in Canada: a field guide. Minister of Supply and Services, Canada, p 23

Mondal S, Wenninger EJ, Hutchinson PJS, Weibe MA, Eigenbrode SD, Bosque-Pérez NA (2016) Contribution of noncolonizing aphids to *Potato virus Y* prevalence in potato in Idaho. Environ Entomol 45:1445–1462

Nolte P, Alvarez JM, Whitworth JL (2009) *Potato virus Y* management for the seed potato producer. University of Idaho, Current Information Series 1165, p 8

Radcliffe EB, Ragsdale DW, Flanders KL (1993) Management of aphids and leafhoppers. In: Rowe RC (ed) Potato health management. APS, St. Paul, MN, pp 117–126

Rondon SI (2012) Pest management strategies for potato insect pests in the Pacific Northwest of the United States. In: Perveen F (ed) Insecticides - pest engineering. In-Tech, pp 309–333. https://doi.org/10.5772/31023

Rondon SI, Gao Y (2018) The journey of the potato tuberworm around the world. In: Khan Perveen F (ed) Pests of potato, maize, and sugar beet. https://doi.org/10.5772/intechopen.81934

Rondon S, Schreiber A, Hamm P, Olsen N, Wenninger E, Wohleb C, Waters T, Cooper R (2017) Potato psyllid vector of zebra chip disease in the Pacific Northwest: biology, ecology, and management. PNW 633, Pacific Northwest Extension Publication. Oregon State University, University of Idaho, Washington State University

Schreiber A, Jensen A, Rondon S, Wenninger EJ, Reitz S (2018) Integrated pest management guidelines for insects and mites in Idaho, Oregon and Washington potatoes. http://nebula.wsimg.com/4396df733a0416f105d18451 1b94d2ef?AccessKeyId=5CF4B076522A4DFDD6B1&disposition=0&alloworigin=1

Srinivasan R, Alvarez JM (2008) Hairy nightshade as a potential *potato leafroll virus* (Luteoviridae: Polerovirus) inoculum source in Pacific Northwest potato ecosystems. Phytopathology 98: 985–991

Srinivasan R, Cervantes F, Alvarez JM (2013) Aphid-borne virus dynamics in the potato-weed pathosystem. In: Giordanengo P, Vincent C, Alyokhin A (eds) Insect Pests of Potato. Global perspectives on biology and management. Elsevier, Oxford, pp 311–339

Stoltz RL, Baird CR, Alvarez JM (2002) Beneficial organisms associated with Pacific Northwest crops. PNW 343, University of Idaho Cooperative Extension System

Strand LL (ed) (2006) Integrated pest management for potatoes in the western United States. University of California Agriculture and Natural Resources Publication 3316, p 167

Wyenandt A, Kuhar TP, Hamilton GC, VanGessel MJ, Sanchez E, Dugan D (2016) Insect management. In: 2016 Mid-Atlantic commercial vegetable production recommendations. Rutgers NJAES Cooperative Extension, New Brunswick, NJ

Chapter 12: Weed Management

Alvarez JM, Hutchinson PJS (2005) Managing nightshade plants to reduce potato viruses and insect vectors. Outlooks Pest Mgmt 16(6):249–252

Bechinski EJ (1998) Integrated pest management and the Idaho potato industry—results of grower surveys. Available at: http://www.uidaho.edu/ag/environment/ipm/reports/potato.html

Berlin, K. (2003) Phenology and shadow response of *Solanum physalifolium* var *nitidibaccatum* and *Solanum nigrum* ssp. *Nigrum*. Examensarbeten/Seminarieuppsatser. 58

Buhler DD (1999) Weed population responses to weed control practices. I. Seed bank, weed populations, and crop yields. Weed Sci 47:416–422

Callihan RH, Bellinder RR (1993) Management of weeds. In: Rowe RC (ed) Potato health management. APS, St. Paul, MN

Eberlein CV, Patterson PE, Guttieri MJ, Stark JC (1997) Efficiency and economics of cultivation for weed control in potato (*Solanum tuberosum*). Weed Technol 11:257–264

Edmonds JM, Chweya JA (1997) Black nightshades, *Solanum nigrum* L. and related species. Promoting the conservation and use of underutilized and neglected crops. 15. Institute of Plant Genetics and Crop Plant Research, Gatersleben/International Plant Genetic Resources Institute, Rome, Italy

Fenner M (1985) Seed bank dynamics. In: Seed Ecology. Chapman Hall, New York, NY, pp 82–96

Hillhorst HWM, Toorop PE (1997) Review on dormancy, germinability, and germination in crop and weed seeds. Adv Agron 61:111–165

Hutchinson PJS (2005) Outlook herbicide for weed control in potatoes. CIS 1126. University of Idaho Ed Comm, Moscow, ID, p 4

Hutchinson PJS (2007) A comparison of flumioxazin and rimsulfuron tank mixtures for weed control in potato. Weed Technol 21:1023–1028

Hutchinson PJS (2007) Chateau® herbicide for use in potatoes. CIS 1136. University of Idaho Ed Comm, Moscow, ID, p 8

Hutchinson PJS (2011) Hairy nightshade (*Solanum sarrachoides*) competition with two potato varieties. Weed Sci 59:37–42

Hutchinson PJS (2012) Common lambsquarters control in potato with dimethenamid-p and a comparison of hairy nightshade control with dimethenamid-p, s-metolachlor, or metolachlor. Weed Technol 26:279–283

Hutchinson PJS (2014) Hairy nightshade critical interference period in potato. Weed Technol 28:543–551

Hutchinson PJS, Tonks DJ, Beutler BR (2003) Efficacy and economics of weed control programs in glyphosate–resistant potato (*Solanum tuberosum*). Weed Technol 17:854–865

Hutchinson PJS, Eberlein CV, Tonks DJ (2004) Broadleaf weed control and potato (*Solanum tuberosum*) crop safety with postemergence rimsulfuron, metribuzin, and adjuvant combinations. Weed Technol 18:750–756

Hutchinson PJS, Ransom CV, Tonks DJ, Eberlein CV (2004) Russet Burbank potato (*Solanum tuberosum*) tolerance to dimethenamid-p. Weed Technol 18:850–852

Hutchinson PJS, Boydston RA, Ransom CV (2005) Weed management in potatoes with Spartan herbicide. PNW Ext Bull 577. University of Idaho Ed Comm, Moscow, ID, p 6

Hutchinson PJS, Boydston RA, Ransom CV, Tonks DJ, Beutler BR (2005) Potato (*Solanum tuberosum*) variety tolerance to flumioxazin and sulfentrazone. Weed Technol 19:704–717

Hutchinson PJS, Hancock DM, Beutler BR (2005) Efficacy of reduced sulfentrazone rates preemergence with metribuzin in potato (*Solanum tuberosum*). Weed Technol 19:954–958

Hutchinson PJS, Ransom CV, Boydston RA, Beutler BF (2005) Dimethenamid-p: weed control and potato (*Solanum tuberosum*) variety tolerance. Weed Technol 19:966–971

Hutchinson PJS, Eberlein CV, Kral CW, Guttieri MJ (2006) Using matrix in weed management systems for potatoes. CIS 1037. University of Idaho Ed Comm, Moscow, ID, p 8

Hutchinson PJS, Hancock DM, Beutler BR (2006) Weed control and potato crop safety with low rates of sulfentrazone and metribuzin applied postemergence. Weed Technol 20:1023–1029

Hutchinson PJS, Beutler BR, Farr J (2011) Hairy nightshade (*Solanum sarrachoides*) competition with two potato varieties. Weed Sci 59:37–42

Hutchinson PJS, Beutler B, Miera C (2014) Linuron for weed control in potatoes—welcome back to the West! Am J Potato Res 91:50. [abstract]

Miller TW, Parker R (2006) Nightshade biology and control in the Pacific Northwest, PNW0588. Oregon and Washington State University Extension Services, p 8

Ogg AG Jr, Dawson JH (1984) Time of emergence of eight weed species. Weed Sci 32:327–335

Ogg AG Jr, Rogers BS (1989) Taxonomy, Distribution, Biology, and Control of black nightshade (*Solanum nigrum*) and related species in the United States and Canada. Rev of Weed Sci 4:25–58

Ogg AG Jr, Rogers BS, Schilling EE (1981) Characterization of black nightshade (*Solanum nigrum*) and related species in the United States. Weed Sci 29:27–32

VanGessel MJ, Renner KA (1990) Effect of soil type, hilling time, and weed interference on potato (*Solanum tuberosum*) development and yield. Weed Technol 4:299–305

Zimdahl RL (1980) Weed-crop competition—a review. International Plant Protection Center. Oregon State University, Corvallis, OR

Zimdahl RL (1988) The concept and adaptation of the critical weed-free period. In: Altieri MA, Leibman M (eds) Weed management in agroecosystems: ecological approaches. CRC, Boca Raton, FL, pp 145–155

Chapter 13: Potato Irrigation Management

Ashley R, Neibling WH, King BA (1996) Irrigation scheduling using water-use tables. CIS 1039. University of Idaho, College of Agriculture

Clapp RB, Hornberger GM (1978) Empirical equations for some soil hydraulic properties. Water Resour Res 1464:601–604

Ley TW, Stevens RG, Topielec RR, Neibling WH (1994) Soil water monitoring and measurement, PNW 475. University of Idaho, College of Agriculture

Ojala JC, Stark JC, Kleinkopf GE (1990) Influence of irrigation and nitrogen management on potato yield and quality. Am Potato J 67:29–43

Chapter 14: Physiological Disorders

Eberlein CV, Westra P, Haderlie LC, Whitmore JC, Guttieri MJ (1997) Herbicide drift and carryover injury in potatoes. Pacific Northwest Ext Bull No 498

Hiller LK, Thornton RE (2008) Management of physiological disorders. In: Johnson DA (ed) Potato health management, 2nd edn. American Phytopathological Society, St. Paul, MN, pp 235–245

Kleinkopf GE, Lewis MD, Shock CC, Thornton MK, Thornton RE, Thornton RK, Westermann DT (1992) Management practices and sugar end development. Pacific Northwest Ext Bull No 427

Sowokinos JR (2007) Internal physiological disorders and nutritional and compositional factors that affect market quality. In: Potato biology and biotechnology: advances and perspectives. Elsevier Ltd, Oxford, pp 501–524

Chapter 15: Tuber Quality

Iritani WM, Weller LD (1980) Sugar development in potatoes. Washington State University Ext Bull 0717, Seattle, WA

Kleinschmidt GD, Kleinkopf GE, Westermann DT, Zalewski JC (1984) Specific gravity of potatoes. University of Idaho Current Information Series No. 60, Moscow, ID

Schreiber L, Nader-Nieto AC, Schönhals EM, Walkemeier B, Gebhardt C (2014) SNPs in genes functional in starch-sugar interconversion associate with natural variation of tuber starch and sugar content of potato (*Solanum tuberosum* L.), G3. Genes, Genomes. Genetics 4(10):1797–1811. https://doi.org/10.1534/g3.114.012377

Stark JC (1987) Effect of late season management on tuber quality. In: Proceedings of the University of Idaho Winter Commodity Schools, vol 19, pp 82–84

Werner BK, Love SL, Shafii B (1998) Comparison of seasonal development of tuber specific gravity among seven potato clones. Am J Potato Res 75:121–127

Chapter 16: Harvest Management

Agricultural Communications (2002) Continuing to manage foreign material for quality Idaho potatoes. DVD available from the University of Idaho

Brook RC (ed) (1996) Potato bruising: how and why emphasizing blackspot bruise. National Potato Anti-Bruise Committee. Potato Assn of America. Running Water Publ, Haslett, MI

Corsini D, Stark J, Thornton M (1999) Factors contributing to the blackspot bruise potential of Idaho potato fields. Am J Potato Res 76:221–226

Smittle DA, Thornton RE, Peterson CL, Dean BB (1974) Harvesting potatoes with minimum damage. Am Potato J 51:152–164

Thornton M, Bohl W (eds) (1998) Preventing potato bruise damage. University of Idaho. Bull 725 (rev)

Chapter 17: Storage Management

Cargill GF, Brook RC, Forbush TD (eds) (1989) Potato storage technology and practice. ASAE 01–89, Michigan State University, East Lansing, MI

Olsen N, Nolte P (2011) Cleaning and disinfecting potato equipment and storage facilities. University of Idaho Extension CIS 1180, Moscow, ID

Pringle B, C Bishop C, Clayton R (2009) Potatoes postharvest. CAB International, Cambridge, MA

Chapter 18: Principles of Economics and Marketing

Greenway G, Guenthner J, Makus L, Pavek M (2011) An analysis of organic potato demand in the US. Am J Potato Res 88(2):184–189

Guenthner J (2001) The international potato industry. Woodhead Publishing, Cambridge, UK

Guenthner J (2012) The development of United Potato Growers cooperatives. J Coop 26:1–16

Halterman D, Guenthner J, Collinge S, Butler N, Douches D (2015) Biotech potatoes in the 21st century: 20 years since the first biotech potato. Am J Potato Res 92(5):551–618
Toevs E, Guenthner J, Johnson A, McIntosh C, Thornton M (2011) An industry perspective of "all native" and transgenic potatoes. AgBioforum 14(1):14–19
Toevs E, Guenthner J, Johnson A, McIntosh C, Thornton M (2011) Identity preservation systems for genetically modified potatoes. Am J Potato Res 88(4):303–308

Chapter 19: Cost of Production

Eborn B (2018) Eastern Idaho costs and returns estimates for northern region Russet Burbank commercial potatoes: production and storage costs. EBB4-Po2-18. Department of Agricultural Economics and Rural Sociology, University of Idaho, Moscow
Eborn B (2018) Eastern Idaho costs and returns estimates for G3 Russet Burbank seed: production and storage costs. EBB4-Po4-18. Department of Agricultural Economics and Rural Sociology, University of Idaho, Moscow
Eborn B (2018) Eastern Idaho costs and returns estimates for South District Russet Burbank commercial potatoes: production and storage costs. EBB4-Po5-18. Department of Agricultural Economics and Rural Sociology, University of Idaho, Moscow
Eborn B (2018) Eastern Idaho costs and returns estimates for South District Russet Burbank commercial potatoes with fumigation: production and storage costs. EBB4-Po6-18. Department of Agricultural Economics and Rural Sociology, University of Idaho, Moscow
Eborn B (2018) Southcentral Idaho costs and returns estimates for Russet Burbank commercial potatoes: production and storage costs. EBB3-Po2-18. Department of Agricultural Economics and Rural Sociology, University of Idaho, Moscow
Eborn B (2018) Southcentral Idaho costs and returns estimates for Russet Burbank commercial potatoes with fumigation: production and storage costs. EBB3-Po3-18. Department of Agricultural Economics and Rural Sociology, University of Idaho, Moscow
Eborn B (2018) Southwestern Idaho costs and returns estimates for Russet Burbank potatoes with fumigation: production and storage costs. EBB2-Po2-18. Department of Agricultural Economics and Rural Sociology, University of Idaho, Moscow
Painter K (2011) The costs of owning and operating farm machinery in the Pacific Northwest. PNW 346. University of Idaho, Oregon State University, Washington State University
Patterson PE, Painter K (2015) 2013-2014 Custom rates for Idaho agricultural operations. Bull 729. University of Idaho College of Ag, Moscow

Chapter 20: Potato Nutrition

Bentley J (2015) Potatoes and tomatoes account for over half of U.S. vegetable availability. Amber Waves, September. Available at: http://www.ers.usda.gov/amber-waves/2015-september/potatoes-and-tomatoes-account-for-over-half-of-us-vegetable-availability.aspx#.VmiXlWWFO73
Keenan MJ, Martin RJ (2013) Dietary resistant starch: investigating the mechanism for beneficial health effects. Available at: http://www.lsuagcenter.com/en/communications/publications/agmag/Archive/2013/Fall/DietaryResistantStarchInvestigatingtheMechanismforBeneficialHealthEffects.htm
Suszkiw J (2007) Agricultural Research Service Information Staff. Agricultural Research Service, USDA phytochemical profilers investigate potato benefits. Agricultural Research/September 2007. Available at: http://agresearchmag.ars.usda.gov/AR/archive/2007/Sep/potato0907.pdf

United States Food and Drug Administration (USFDA) How to understand and use the nutrition facts label. Percent daily value (%DV). Available at: http://www.fda.gov/Food/IngredientsPackagingLabeling/LabelingNutrition/ucm274593.htm

Vinson J, Demkosky S, Navarre D, Smyda M (2012) High-antioxidant potatoes: acute in vivo antioxidant source and hypotensive agent in humans after supplementation to hypertensive subjects. J Agric Food Chem 60(27):6749–6754. Available at: http://pubs.acs.org/stoken/presspac/presspac/abs/10.1021/jf2045262

Index

J. C. Stark et al. (eds.), *Potato Production Systems*,
https://doi.org/10.1007/978-3-030-39157-7

B

C

I

L

Q

R

W

Y

Z

GPSR Compliance
The European Union's (EU) General Product Safety Regulation (GPSR) is a set of rules that requires consumer products to be safe and our obligations to ensure this.

If you have any concerns about our products, you can contact us on

ProductSafety@springernature.com

In case Publisher is established outside the EU, the EU authorized representative is:

Springer Nature Customer Service Center GmbH
Europaplatz 3
69115 Heidelberg, Germany

www.ingramcontent.com/pod-product-compliance
Lightning Source LLC
Chambersburg PA
CBHW061651050126
37739CB00003B/98

9783030391560